*Kin Recognition
in Animals*

Kin Recognition in Animals

Edited by

David J C Fletcher
Department of Entomology
University of Georgia

and

Charles D Michener
Departments of Entomology and
and of Systematics & Ecology
University of Kansas

A Wiley–Interscience Publication

JOHN WILEY & SONS
Chichester · New York · Brisbane · Toronto · Singapore

Copyright © 1987 by John Wiley & Sons Ltd.

All rights reserved.

No part of this book may be reproduced by any means, or
transmitted, or translated into a machine language
without the written permission of the publisher

Library of Congress Cataloging-in-Publication Data:

Fletcher, David J. C.
  Kin recognition in animals.

  Includes index.
  1. Kin recognition in animals. I. Michener,
Charles D. II. Title.
QL761.5.F57   1987   591.5'1   86–15748
ISBN 0 471 91199 2

British Library Cataloguing in Publication Data:

Fletcher, David J. C.
  Kin recognition in animals.
  1. Recognition (Psychology)
  I. Title II. Michener, Charles D.
  591.5   QL775

  ISBN 0 471 91199 2

Printed and bound in Great Britain by Anchor Brendon Ltd., Colchester, Essex.

QL
761.5
.K5
1987

Contents

List of Contributors vii

Preface ix

1 Introductory Remarks 1
   David J. C. Fletcher and Charles D. Michener

2. Kin Recognition: An Introductory Synopsis 7
   Edward O. Wilson

3. The Behavioral Analysis of Kin Recognition:
   Perspectives on Methodology and Interpretation 19
   David J. C. Fletcher

4. Genetic Aspects of Kin Recognition: Concepts,
   Models, and Synthesis 55
   R. H. Crozier

5. Discrimination Among Prospective Mates in *Drosophila* 75
   Eliot B. Spiess

6. Kin Recognition in Subsocial Arthropods, in Particular
   in the Desert Isopod *Hemilepistus reaumuri* 121
   K. E. Linsenmair

7. Kin Recognition in Primitively Eusocial Insects 209
   Charles D. Michener and Brian H. Smith

8. Kin Recognition in Highly Eusocial Insects 243
   Michael D. Breed and Beth Bennett

v

9. **Kin Recognition in Vertebrates (Excluding Primates): Empirical Evidence** 287
   Andrew R. Blaustein, Marc Bekoff and Thomas J. Daniels

10. **Kin Recognition in Vertebrates (Excluding Primates): Mechanisms, Functions, and Future Research** 333
    Andrew R. Blaustein, Marc Bekoff and Thomas J. Daniels

11. **Kin Recognition in Non-human Primates** 359
    Jeffrey R. Walters

12. **Kin Recognition in Humans** 395
    P. A. Wells

13. **Discriminating Nepotism: Expectable, Common, Overlooked** 417
    W. D. Hamilton

Author Index 439

Index of Scientific and Common Names 449

Subject Index 453

List of Contributors

M. Bekoff *Department of Environmental, Population and Organismic Biology, University of Colorado, Boulder, Colorado 80309, USA.*

B. Bennett *Department of Environmental, Population and Organismic Biology, University of Colorado, Boulder, Colorado 80309, USA.*

A. R. Blaustein *Department of Zoology, Oregon State University, Corvallis, Oregon 97331, USA.*

M. D. Breed *Department of Environmental, Population and Organismic Biology, University of Colorado, Boulder, Colorado 80309, USA.*

R. H. Crozier *School of Zoology, University of New South Wales, Kensington, NSW 2033, Australia.*

T. J. Daniels *Department of Environmental, Population and Organismic Biology, University of Colorado, Boulder, Colorado 80309, USA.*

D. J. C. Fletcher *Department of Entomology, University of Georgia, Athens, Georgia 30602, USA.*

W. D. Hamilton *Department of Zoology, Oxford University, South Parks Road, Oxford, OX1 3PS, England.*

K. E. Linsenmair *Zoologisches Institut (III), der Universität Würzburg, Röntgenring 10, D-8700 Würzburg, West Germany.*

C. D. Michener *Departments of Entomology and of Systematics & Ecology, University of Kansas, Lawrence, Kansas 66045, USA.*

B. H. Smith *Department of Entomology, University of Kansas, Lawrence, Kansas 66045, USA.*

E. B. Spiess *Department of Biological Sciences, University of Illinois at Chicago, PO Box 4348, Chicago, Illinois 60680, USA.*

J. R. Walters *Department of Zoology, North Carolina State University, Campus Box 7617, Raleigh, NC 27695, USA.*

P. A. Wells *Department of Psychology, University of London Goldsmiths' College, Lewisham Way, London, SE14 6NW, England.*

E. O. Wilson *Museum of Comparative Zoology, Harvard University, Cambridge, Massachusetts 02138, USA.*

Preface

The wide variety of animal groups in which kin recognition has been found is noteworthy. After participating in a symposium on kin recognition in insects at a meeting of the Entomological Society of America, and considering not only its common features but also its diversity in different animals, we concluded that a broader presentation in book form, not limited to and not unduly emphasizing insect examples, would be a useful contribution. This book is the result of that decision. It is intended for graduate students, researchers and teachers interested in behavior, population biology and evolution. Persons with such interests should also be aware of the excellent treatment and bibliography provided by P. Colgan in his general account of recognition (see Chapter 10 for reference).

As editors we have been dependent upon and greatly appreciate the contributions of the other authors, whose chapters make up most of this book. A word about the selection of contributors may be appropriate. It would have been easy to invite contributions from numerous researchers, each writing on his or her narrow speciality. It seemed to us, however, that the book would be most useful for our prospective readers—shorter, better organized and more cohesive—if written by a moderate number of authors, and we developed a list of about a dozen chapters. We therefore could not include the principal researcher on every topic, interesting as such a series of papers would be. Indeed in some cases we bypassed eminent researchers who seemed to have 'had their say' in readily available publications in favor of persons whose fresh ideas or approach we think will be appreciated.

As we planned this book it was obvious that some chapters needed to be based on particular taxonomic groups. To understand kin recognition and its functions one must often know something of the natural history of the animal species concerned. The same background is often needed to understand the experimental approaches. The life histories of animals have been important in determining what studies of kin recognition have been practical for particular animal species. The studies by any one author have generally been based on one or a few related species amenable to the experimental or observational techniques favored by that author. The chapters on various taxonomic groups left some other groups without great emphasis. For example, there are no chapters on sessile colonial invertebrates. Chapters 2, 3, 4 and 13 provide some information on such taxa.

Of course, there are common features of kin recognition cutting across taxonomic groups, especially common theoretical material of genetic, evolutionary and sociobiological interest. We therefore decided to include various chapters concerning such topics rather than provide specific coverage of all taxonomic groups of animals.

Authors of several chapters place emphasis on methods and appropriate viewpoints, theories and terminology for studies of kin recognition and, indeed, Chapter 3 relates largely to such general matters. Most of the authors did not see the chapters by others before submitting their manuscripts. The result has been some duplication of general material, definitions of terms, and the like. In some cases we consider this desirable, to remind readers, to better serve those concerned only with a particular taxon and to indicate the authors' differences in terminology, approach and philosophy. For the same reasons most authors preferred to retain such material in their chapters. To guide the reader to similar material in different chapters, we have sometimes added parenthetical references such as '(Chapters 3 and 13).'

Of course most of the chapters on kin recognition in various taxonomic groups (i.e. Chapters 5 to 12) are reviews of what has been published concerning given groups. Chapter 6 by K. E. Linsenmair on an isopod, *Hemilepistus*, is an exception in that it contains much hitherto unpublished material and detailed documentation, as well as details of published material not previously available in English. It seemed appropriate to give *Hemilepistus* special treatment. Thanks to the work of Linsenmair and his associates, it is the animal whose kin recognition system has been most intensively studied. A detailed presentation is needed to explain their investigations, which have involved many thousands of isopods and analyses of over 50,000 interactions. Thus Chapter 6 records the results of a remarkable *tour de force* in behavioral biology. More important, it suggests methods that might be applicable to studies of kin recognition in diverse other small animals that can be obtained in large numbers.

In addition to authors, we thank those who initially contributed ideas about the content and organization of the book, especially Marc Bekoff, and those who read either various chapters or the whole manuscript and offered helpful suggestions, especially William Wcislo and John Wenzel. For secretarial and editorial help, such as checking bibliographic references, we much appreciate the help of Joetta Weaver and Tim Brooks.

DAVID J. C. FLETCHER
Athens, Georgia

CHARLES D. MICHENER
Lawrence, Kansas

Kin Recognition in Animals
Edited by D. J. C. Fletcher and C. D. Michener
© 1987 John Wiley & Sons Ltd

CHAPTER 1

Introductory Remarks

DAVID J. C. FLETCHER[1] and CHARLES D. MICHENER[2]
[1] *Department of Entomology, University of Georgia, Athens, Georgia 30602 USA.*
[2] *Departments of Entomology and of Systematics & Ecology, University of Kansas, Lawrence, Kansas 66045, USA.*

The introductory material for this book consists of Chapters 1 and 2 and to a considerable extent Chapter 3 as well. We have placed here in Chapter 1 a series of remarks that relate to the subject at hand, kin recognition, but that we regard as too important to relegate to a preface. In themselves these remarks by no means constitute an adequate introduction.

Discrimination among conspecific individuals is associated with social behavior. Mating in some species may occur whenever individuals of the opposite sex and in proper physiological condition meet; no discrimination other than of species, sex and condition is necessary. But for most other social interactions, and for mating also in many species, animals make use of the ability to recognize individuals (distinguished as such), or to discriminate among groups of individuals that are familiar versus unfamiliar, or among close kin, distant kin and unrelated conspecifics. Virtually all social interactions (territorial behavior, care of young, maintenance of social hierarchies, pair bonding, colony defense, etc.) involve such recognition, i.e. the ability to discriminate is a prerequisite to most kinds of social behavior.

As the bibliographic material for the chapters of this book suggests, the study of kin recognition has exploded in the last few years. Previously, the number of papers devoted primarily to this topic was small and a thorough review of what was known would have been brief and easy to assemble. The impetus provided by the theoretical work of W. D. Hamilton and others, and the strong interest in animal behavior in general and the evolution of social behavior in particular, are some of the factors that have led to the intense activity in the field in recent years.

The heart of the idea of kin recognition is differential treatment of conspecific individuals based on kinship relationships. If we do not see any difference in treatment of individuals across the range from non-kin to full-siblings or parents and offspring, then we say there is no evidence of kin recognition. Of course, such recognition may exist, even though the animals have chosen not to reveal it to us. The ingenuity of the observer or experimenter is often taxed as he or she tries to devise studies that will cause the animal to reveal its recognition of kin, if it has such ability. Thus it may be impossible to prove that any species lacks the ability to recognize kin, even though we may discover many circumstances under which we can find no evidence of such recognition.

The reverse side of this coin consists in making the twin assumptions that differential treatment of kin and non-kin does in fact constitute evidence of kin recognition, and that such recognition has biological meaning. The first assumption is entirely reasonable and may be supported by careful analysis of the sensory mechanisms involved, whereas the second assumption presents problems of a kind already familiar to evolutionary biologists.

Nearly all authors explain kin recognition as adaptive, and given the theoretical reasons for seeking evidence of kin recognition in the first place, it is not surprising that most explain it as a mechanism that increases inclusive fitness benefits or optimizes the level of outbreeding, or both. Such explanations are satisfying and may well be correct, but we must always remember that they often are not verified. Many of the data on kin recognition in some animal species have been obtained in the laboratory and may or may not indicate that kin recognition is prevalent and important under natural conditions. Even if it occurs in the field we do not know whether it is adaptive, or is just something that happens, for example, as a by-product of something else that is important. This caveat notwithstanding, it is of course entirely reasonable to postulate a function, or functions, for kin recognition in any species on the sensible grounds that if one does not seek a function, one will not find it, even if it exists. Moreover, it should be legitimate for anyone to do this without being labeled a panselectionist, provided of course that he or she does not insist that each case of kin recognition must of necessity have a function. We are convinced that kin recognition is prevalent and is often adaptive, but we do not think that the most obvious theoretical explanation is necessarily the correct one in every case; neither do we think that lack of a prediction of kin recognition in a given case means that it does not happen, is unimportant, and should not be looked for.

In the analysis of kin recognition, investigators have employed a method common to all areas of human inquiry; they have broken down concepts into constituent parts and sought relationships of various kinds between the parts. While such methodological reductionism is the essence of analysis, it yields only a limited view of reality and, more importantly, there is frequently more than one valid way of breaking down a concept. The different views of the same reality thus obtained may be complementary or in opposition, but few are heuristically useless. Nevertheless, discussion usually develops as to whether one view is 'better', i.e. more useful, than another, and all too often whether one is right and the other wrong.

Among the more obvious results of methodological reductionism are classification systems, and of course the same set of objects or ideas may be classified in different ways according to the purpose of the classification. In this context attempts to classify the many known examples of kin recognition as social behavior or mating preference on the one hand, and as adaptive or neutral on the other, may be readily understood. It seems unlikely that anyone would wish to contest the utility of the two classifications even though some biologists may disagree with the premises on which they are based. Neither, we presume, would anyone wish to deny that the two are complementary, since they may easily be combined into a single classification system. As we have indicated, we favor open-mindedness on the question of whether kin recognition is adaptive in every case, but more may be said about the logic of separating mate selection from the other presumed functions of kin recognition.

In animal groups in which there is no parental care, as for example in most insects, mate selection is seen to be the most primitive type of kin selection. The simplifying assumption of random mating is made in much of the theoretical and modeling work in population genetics. Yet random mating is probably not usual for most animals. There is preferential mating with close relatives, distant relatives, unrelated individuals or whatever, depending on the species and population concerned. Such behavior is only possible if there is some sort of kin recognition or at least a tendency to mate with individuals that go to or are found in certain places. It follows that kin recognition, at least in an incipient stage, is extremely widespread in animals.

The results of studies that have been carried out with *Drosophila* (see Chapter 5) give an idea of the sort of kin recognition that may be widespread in insects and other small-brained creatures that have no obvious social behavior except that associated with mating. We are fortunate that the genetics and behavior of *Drosophila* have been studied in enough detail that its discrimination ability and mate selection can be elucidated, because it can serve as an exemplar for so many other animals.

Drosophila has a rather elaborate courtship behavior that precedes mating; one might suppose that mate discrimination such as occurs in that insect might be possible only in forms with courtship. However, in the mating of sweat bees (*Lasioglossum*, Chapter 7), which probably has nothing to do with their colony life but is similar to that of their solitary relatives, mating is quick, with no suggestion of courtship. Yet there is discrimination and mate selection by the males, and females frequently refuse to mate. Again we emphasize the widespread occurrence among diverse animal groups of some sort of kin recognition.

Most of the principles and behavioral mechanisms of kin recognition are the same in small-brained arthropods and in large-brained vertebrates. Indeed brain size seems to have little relation to the potential for kin recognition. The animal whose kin recognition has been most fully studied, the isopod crustacean, *Hemilepistus reaumuri*, is said to have only about 10^4 neurons in its brain. (Chapter 6 gives a detailed account not only of this animal's rather elaborate social behavior and kin recognition but also of the methods used in a major scientific investigation.) For comparison with the *Hemilepistus* brain, the number of neurons in the brain of the honey bee worker (*Apis mellifera*) is about 10^6 and that of man about 10^{10} (see Chapters 8 and 12, respectively). But consider that some coelenterates recognize kin in spite of lacking a brain altogether (Chapter 4), and that many plants recognize self versus more or less unrelated pollens with no nervous systems at all.

Individual recognition, of course, is well known in higher vertebrates. One can argue about the definition, but something very like it has been found in arthropods (Chapters 6, 7). Human beings are experts at individual recognition—even at a distance the manner of walk or something about a voice often permits us to identify an individual. We use this ability so easily that it may largely have replaced kin recognition *sensu stricto*, i.e. recognition of family groups and degree of relationship.

It is important to recognize that there is a continuum from recognition of individual differences to recognition of kin, of members of different but conspecific populations and on to recognition of specific differences including reproductive isolating mechanisms. Thus principles applicable at one level are likely to be relevant at others.

Biologists have long since learned to expect interspecific isolating mechanisms and even some degree of isolation among populations considered to be conspecific. Until recent decades, however, most biologists did not anticipate that kin recognition in its usual sense, i.e. recognition of intrapopulational kin groups, or of individuals, would be common except among 'higher' animals. The finding that such recognition occurs even in small-brained animals like isopods and insects was surprising to many and doubtless is one reason for the intense current interest in the field of kin recognition.

Some authors exclude parent-offspring interactions from kin recognition. Possibly this is because it is commonplace to realize that parents of many animals recognize their own offspring and offspring recognize their parents, but in some cases it may reflect the types of data available to these authors from the particular animal group(s) they have studied. For example, animals that show no parental care are unlikely to provide data on parent-offspring interactions. Other authors regard parent-offspring recognition as a part of kin recognition, perhaps because parent-offspring interactions are particularly important in the animal species with which they are most familiar, as for example when the young form mixed groups with unrelated conspecifics while they are still in need of parental care. These divergent biases among authors are another example of different approaches to breaking down a concept into constituent parts and they suggest that in origin such differences may be philosophical, or practical, or both.

Another aspect of kin recognition is also potentially confusing to non-specialists as a consequence of the particular methodological reductionism employed by different authors. This problem concerns analyses of the sources of recognition. There is no general agreement concerning the number and types of sources. At the root of this difficulty is the old problem of whether it is valid to analyze behavior in terms of components that are of genetic origin on the one hand and environmental origin on the other (i.e. innate or inborn behavior and learned or acquired behavior), or whether one should adopt a developmental approach, i.e. break down behavior according to arbitrarily defined stages in the continuum of its ontogeny. The vigorous debate on this subject that took place over a period of two decades is well known and when pressed, most authors nowadays side with the developmentalists. Nevertheless, many of them, including a number writing on kin recognition, continue to make use of the innate/learned dichotomy in their interpretations of behavior, although granting that nearly all behavior contains elements from both innate and learned sources. The persistence of this model demonstrates that to some it remains genuinely useful.

We again wish to emphasize that when methodological reductionism yields two or more analytical models, we do not necessarily believe that one of them is right and the other(s) wrong, only that one of them may ultimately be found to be more useful. The diversity of opinion in the field of kin recognition, of which many examples will be found on comparing the following chapters, is characteristic of all areas of scientific inquiry in their most exciting stages. Such variation may be especially appreciated by evolutionary biologists who know better than most that for anything to evolve directionally there must be variation upon which selection can act.

Kin Recognition in Animals
Edited by D. J. C. Fletcher and C. D. Michener
© 1987 John Wiley & Sons Ltd

CHAPTER 2

Kin Recognition: An Introductory Synopsis

EDWARD O. WILSON
Museum of Comparative Zoology, Harvard University, Cambridge, Massachusetts 02138, USA.

> *As it was written, so it has been done.*
>
> The Koran

It is fair to use this Koranic prescription to characterize the subject at hand, for rarely in the history of biology has a domain of empirical knowledge followed so closely and fruitfully upon an abstract theoretical idea. Research on kin recognition, the patterns and mechanisms entailed in the discrimination of kin from non-kin, has been shaped to a remarkable degree by the concept of kin selection. The theory of kin selection was developed most originally and fully by W. D. Hamilton during the 1960s, and further developed by R. H. Crozier, G. F. Oster, R. L. Trivers and others over a period of 20 years. It has been augmented by new and exciting studies of incest avoidance, which is equally relevant to kin recognition.

Also, rarely has a subject accelerated so abruptly over so short a time as kin recognition. When I accepted the invitation of the editors to write a synopsis for this book, during a conference in the summer of 1982, the subject still seemed small and easily managed. However, during the late 1970s, when kin selection and other aspects of sociobiology became a much-discussed part of academic biology, many field and laboratory studies had been initiated, and these began to bear fruit during the early 1980s. The result today is a spate of important discoveries and new insights, which in turn are altering the basic theory. So now, as it is done, so shall it be written.

In order to keep this synopsis reasonably brief and accessible to those not in the field itself, I have organized the material in didactic fashion and eliminated references. The reader who starts here can then proceed to other chapters in the book, which have been prepared by some of the most active investigators in the field.

FUNCTIONS OF KIN RECOGNITION

Kin selection can be defined as the alteration of the frequencies of genes shared by relatives through actions that favor or disfavor the survival and reproduction of the relatives; it is inferentially a powerful force in evolution. To take an extreme imaginary case: if an allele appears in a population that causes its bearer to act so as to triple the reproduction of the bearer's siblings, the allele will spread rapidly through the population. This will occur even if the allele-bearer completely sacrifices itself in the process, because many of the siblings will also carry the altruistic allele.

If kin selection works as effectively as such arguments imply, in other words as a strong 'ultimate' factor, then kin recognition is the immediate causal ('proximate') factor that can be expected to evolve as the means for discriminating between kin and non-kin. In the studies of kin recognition to date, a wide range of functions have indeed been documented. They can be loosely classified under the following ten categories.

1. *Altruism toward siblings.* The most striking examples of altruism (or nepotism as some writers prefer to call it) occur in the advanced social insects, including ants and termites, where non-reproductive workers toil on behalf of their sisters and brothers. It is not surprising to find finely developed mechanisms of kin recognition in these organisms, as well as its inverse: alien workers are rejected from the nests, typically with lethal violence.

2. *Parent-offspring relations.* Throughout the animal kingdom, species that practice parental care also display special sensory skills by which they distinguish their own from alien young.

3. *Alloparental care.* 'Aunting,' the care of immature nephews, nieces, and other close relatives in addition to direct offspring, occurs in social insects, mole rats, some species of birds, canids and primates. The advantage of directing aid of this nature toward kin instead of non-kin seems clear enough. In addition, the practitioners also benefit from experience with infants as well as from the alliances such behavior helps to cement.

4. *Adoption.* Closely related to alloparental care is the adoption of orphaned or abandoned offspring, a phenomenon known in a few social insects and higher primates. In these species the adoptees are typically close kin of the adults who care for them.

5. *Optimal outbreeding.* Incest avoidance at the level of $r = 0.5$, in other words between siblings or between parents and offspring, is prevalent but not universal in the animal kingdom; and it is accompanied by devices of kin recognition that appear tailored to the social and population structures of each of the species in turn. Captive Japanese quail (*Coturnix coturnix*), for example, prefer to mate with first cousins. The discoverer of the phenomenon, P. P. G. Bateson, has proposed a condition of

'optimal outbreeding,' in which preferred mates are closely enough related to enhance cooperation and the preservation of coadapted gene complexes but not so close as to cause inbreeding depression.

6. *Formation of schools and other aggregations.* Aggregate formations of vertebrates appear at first glance to be unstructured, but recent studies have shown that they are affected to some extent by the tendency of individuals to associate with close relatives. Such preferential recognition has been demonstrated in salmon, tuna and the tadpoles of frogs and toads. The resulting segregation might serve to enhance the genetic advantage of alarm signals, aposematic signals, and cooperative feeding.

7. *Grooming.* In groups of rhesus macaques and other higher primates, grooming is often preferentially directed towards close kin. Since the practice results in the removal of ectoparasites and dead tissue, in addition to serving a purely communicative role, discrimination in favor of kin has a potentially adaptive value.

8. *Alarm signals.* Social birds, rodents, antelopes, cervids and a wide range of other group-forming animals emit special alerting calls when a predator is sighted. Some species employ stereotyped movements or pheromones to the same end. Dissenting authors have attributed alternative functions to this class of signals that benefit the signaler directly, such as informing the intruder that pursuit is now less likely to succeed or attracting competing predators to interfere with one another. But it is also true that alerted animals are less likely to be caught, and those saved are usually close to the signaler within organized groups and hence more likely to be relatives.

9. *Avoidance of cannibalism.* Organisms that are cannibalistic increase their inclusive genetic fitness when they feed preferentially on non-kin. A case in point is provided by wasps whose larvae are internal parasites of other kinds of insects: those species in which non-kin regularly occupy the same host body are more prone to aggression and cannibalism than those in which siblings are associated. Similarly, guppy females (*Poecilia reticulata*) eat unrelated fry of their own species in preference to related ones.

10. *Habitat selection.* Siblings of voles and spruce grouse tend to settle in the same places, resulting in associations more likely to reward forebearance and cooperation on the part of the members.

To summarize this section on function, kin recognition of one kind or another has been implicated in most kinds of social behavior. Where a discriminating capacity was expected from kin selection theory, it has always been found to exist in fact. Moreover, the linkage has proved heuristic: the mediating behavior is often complex and highly effective, entailing little-understood or wholly unknown physiological responses.

GLOSSARY

Words, as opposed to equations and data, are especially important in the early stages of exploration of a subject. The ideas being newly expressed are still weak and volatile. They depend heavily upon precise definition of freshly coined phrases as well as specialized meanings given to old terms. To produce a glossary is therefore in substantial degree to summarize the principal concepts. It also serves to expose different shades of meaning and even contradictions among different writers. In preparing the following list I have tried to express the consensus of the researchers most responsible for the terminology. Following each definition are the chapter numbers in the present volume that deal with the expressions and their patron data in greatest detail. I have placed the glossary near the front of this chapter instead of at the usual appendicial rear because some familiarity with it will ease the reading of the synopsis as well as the entire book.

Altruism. Self-denying or self-destructive behavior performed for the benefit of others. Some authors have begun to drop this admittedly value-laden word (the use of which also ignores the fact that the prescribing genes are selfish rather than altruistic) in favor of expressions such as nepotism and reciprocation. In my opinion we should nevertheless keep it, especially in human studies. (Chapters 3 and 13.)

Cue bearer (recognized individual). The individual that is classified by other members of the species according to kinship. (Chapters 3 and 7.)

Decision rule. The pattern of response to other individuals according to their possession of stimuli identifying them as kin or non-kin (Chapter 4).

Discriminating individual. See 'recognizing individual'.

Discriminators (recognition pheromones). The genetically determined chemical cues that permit individuals to be classified as kin or non-kin. (Chapters 6 and 7.)

Eusocial. The evolutionarily advanced condition, limited among higher animals almost entirely to termites, ants, bees and wasps, in which there is an overlap of generations, care of young, and existence of reproductive and non-reproductive castes. (Chapters 7 and 8.)

Extrinsic cue. A label acquired from the environment that serves to identify the bearer to a discriminating individual as kin on a probabilistic basis. (Chapter 3.)

Gestalt model. The hypothetical system in which a common odor is created by pooling the recognition pheromones of some or all of the individuals belonging to a group. Others are classified as kin or non-kin according to the degree to which they possess the odor. Compare with 'individualistic model' (q.v.). (Chapters 4 and 8.)

Green-beard effect. A metaphor introduced by Richard Dawkins which says, essentially, 'I have a green beard and will be altruistic to others who have green beards'. The phrase refers in particular to the use of 'recognition alleles' (q.v.) in kin recognition. (Chapters 4, 12 and 13.)

Inclusive fitness. The sum of an individual's own genetic fitness plus all of its influence on the genetic fitness of relatives other than direct descendents; hence the total effect of both ordinary individual ('Darwinian') selection and kin selection with reference to individual genetic fitness. (Chapters 3, 4 and 13.)

Individualistic model. The hypothetical system in which individuals judge others to be kin or non-kin according to whether they possess certain alleles that encode a particular recognition pheromone. The alleles may be 'recognition alleles' (q.v.) in the strict sense, in other words they control both the production of the pheromone and its perception, or they may prescribe a less direct phenotype matching system. Compare 'Gestalt model' (q.v.). (Chapters 4, 7 and 8.)

Intrinsic cue. A phenotypic expression of a cue bearer's genotype such that differences in a particular character, or characters, among conspecific animals reflect genotypic differences between them, and these differences are used by discriminating individuals in the assessment of relatedness. (Chapter 3.)

Kin recognition. The recognition of and discrimination toward various categories of kin. (All chapters of this volume.)

Kin selection. Differential survival or reproductivity that changes the proportion of genes through time due to the circumstance that individuals favor or disfavor relatives other than direct offspring. The relatives are assumed to possess at least some of the same genes through common descent. See also 'inclusive fitness' (q.v.). (Chapters 3, 4 and 13.)

Kingram. A graphical technique introduced by W. M. Getz (1981) for evaluating decision rules and corresponding maximum efficiencies in various recognition systems. (Chapter 4.)

Nepotism. The favoring of kin over non-kin, or else of certain categories of kin over other categories. (Chapters 3 and 13.)

Optimal outbreeding (or **optimal inbreeding**). Mating among kin sufficiently distantly related to enhance cooperation and the preservation of coadapted gene complexes but not close enough to produce significant degrees of inbreeding depression. (Chapters 9, 10 and 12.)

Phenotype matching. The process by which an individual learns cues (such as 'recognition pheromones') from either itself or its kin and then matches them with cues provided by other individuals in order to classify them as kin or non-kin. The set of learned cues used in phenotype matching is sometimes referred to as the 'template' (q.v.). Contrast with discrimination by 'recognition alleles' (q.v.). (Chapters 4, 7, 9, 10, 11 and 12.)

Pheromone. A chemical substance or mixture, usually a glandular secretion, that is used for communication within a species. One individual releases the material as a signal and others respond after tasting or smelling it. (Most chapters in this volume.)

Recognition alleles. Alleles hypothesized to encode the production of a recognition cue and simultaneously the ability to recognize the cue in others, leading to the discrimination of kin from non-kin. The system is sometimes described metaphorically as the 'green-beard effect' (q.v.). Recognition alleles are considered very difficult either to demonstrate or disprove in nature. Contrast with 'phenotype matching' (q.v.). (Chapters 4, 9, 10, 11, and 12.)

Recognition pheromones. See 'discriminators'.

Recognizing individual. An individual that responds to appropriate cues by distinguishing kin from non-kin. (Chapters 3 and 7.)

Referents. The organisms chosen by an individual as the 'standard' by which strangers are classified as kin or non-kin. The referents can be the self, or a subset of the social group such as the family, or even the entire group. The set of cues they offer is referred to as the 'template'. (Chapter 4.)

Template. The representation in memory of a set of cues, either strictly innate or learned, by which a responding individual recognizes others as kin or non-kin. In the case of pheromones, the template can consist of a set of individually recognized cues that vary among group members but still identify them as kin (this hypothesized system is called the 'individualistic model'), or it can be a blend of substances shared by the group as a whole (the 'Gestalt model'). (Chapters 4, 7 and 8.)

BIOASSAYS

No single behavioral response has emerged as the standard by which all kin recognition is measured, nor is one desirable, due to the great diversity of the organisms and varying roles of recognition. Instead, bioassays have been tailored by researchers to fit each species in turn. In the case of social insects, although mate choice has been used in some studies, investigators have favored aggression within or near the nests. They use the varying degrees of rejection, from the mildest of threat displays to all-out attack, as a scale to indicate the divergence of the intruder's odor from that of the resident colony. In honey bees positive responses are also employed. The treatment of larvae, including the propensity to rear them as new queens, is considered to reflect the similarity of the odor of the larvae to that of their attending worker nurses.

Students of fish and amphibian sociality have relied on virtually the only trait available to them, the degree of association in aggregations and schools. Primate sociobiologists, in contrast, have a much greater array of behavioral categories from which to choose. In one study or another they have utilized almost the entire set, including aggression, alliance formation, proximity, grooming, adoption, mate choice and the composition of play groups.

Finally, some anthropologists have begun to measure interactions to explore more deeply the effects of kinship on social structure and reproductive success. The Ye'kwana Amerindians have a verbal kin classification that suggests only relatively coarse distinctions. They lump persons ranging in degree of relatedness from full-sibling to first cousin into the categories of 'brother' (*udui*) or 'sister' (*yaya*). Yet they make finer, unspoken distinctions when choosing partners for such coordinated activities as conversation, the sharing of meals, and play. In short, behavior can outrun language in the most important aspects of social life.

MECHANISMS OF RECOGNITION

The cues employed by animals of different kinds have proved to be almost as diverse as the imagination allows. In fact, they closely parallel the signals employed in other categories of communication and reflect the ecology and sensory capabilities of each group in turn. Lower invertebrates, for example, are limited almost entirely to chemical signals (pheromones), while primates use sophisticated combinations of vision, sound, and smell.

Because the vast majority of species of organisms utilize pheromones, and many rely on them almost exclusively, these substances have received particular attention during the exploration of kin recognition. It is clear that among social insects at least, recognition pheromones form a different class from primer pheromones, which alter the physiology of the recipients and prime them for a new repertory. They also differ from releaser pheromones, which evoke responses in a direct manner. In contrast, recognition substances identify their bearers as a member of one class or another, and the responses that follow depend upon the prior learning experiences of the animals. Some progress has been made in characterizing these substances. Female sweat bees (*Lasioglossum zephyrum*) vary in the proportions of four macrocyclic lactones in a manner correlated with degree of kinship. A similar relation has been demonstrated in the mandibular gland keto- and hydroxy-acids of queen honey bees (*Apis mellifera*), which appear to evoke different degrees of aggression from workers. Fire ant species (*Solenopsis*) differ among themselves in the composition of their cuticular hydrocarbons. These substances are acquired by symphilic beetles of the genus *Myrmecophodius* after they penetrate the host fire ant colonies.

Which cues denote a relative? Which a non-relative? The animals themselves classify the signals by means that also vary enormously. In sorting these association phenomena out it is useful to employ a mnemonic device such as that used for partitioning isolating mechanisms in the study of species formation. In the latter case we ask, what is the origin of the mechanisms? The answer is 'anything that can go wrong during mating and reproduction'. Any genetic divergence between two populations that interferes with free interbreeding is *ipso facto* an intrinsic isolating mechanism. Thus any difference in habitat preference, time of mating, courtship ritual and so forth, can serve. In an analogous fashion, most means of identifying relatives postulated by theoreticians have in fact been documented in one species or another. Almost any device that can be imagined as working seems in fact to have evolved somewhere.

Useful classifications of the labeling phenomena have been prepared by P. W. Sherman and W. G. Holmes for animals generally, and by B. Hölldobler and C. D. Michener for social insects in particular. They are ably discussed and in some cases modified by the authors of the present volume. The categories that appear to me to have been most usefully distinguished are the following:

1. *Purely spatial distribution* can be used as a rough cue as to whether individuals are kin or not, especially when animals show high degrees of site fidelity. Adult bank swallows (*Riparia riparia*), as one well-studied example, learn the locations of the nest holes they excavate, and they feed any chicks found in their personal holes (including alien swallow chicks introduced by the investigator) up until the time their own offspring fledge at about two weeks of age. Conversely, they ignore their own chicks if these are transferred experimentally to nearby burrows. At about the time of fledging the young birds begin to emit individually distinctive 'signature' calls, which the adults learn and use afterward. In other words the swallows switch from a purely spatial cue to a behavioral signal.

2. *Pure allelic recognition*, that is, recognition without the intervention of learning, is an intriguing possibility, but most writers agree that it is less likely to evolve as the primary system. In any case it would be technically difficult either to demonstrate or to disprove. This hypothesized phenomenon, also known metaphorically as the green-beard effect, entails the possession of individually distinctive sets of genes that prescribe both the recognition cue and the sensory apparatus used to recognize it (see Chapters 3, 4, 13 etc.). The principal difficulty in documenting such an ultrasimple arrangement is that learning has been implicated in most cases of animal behavior, even though it is often of a very constrained, predictable nature. No one has invented a way to control for all conceivable learning possibilities, including especially the familiarization of an individual animal with its own cues.

The recognition allele controversy, perhaps better termed the recognition allele confusion, is thus a microcosm of that dreadnaught of all difficulties in behavioral biology, heredity versus environment. Even so, it is reasonable to look for cases in the simplest of organisms, where interactions take the form of growth and tissue rejection rather than conventional neuron-mediated learning. Such is the case of corals, sponges and other colonial invertebrates that fuse with genetically identical tissue and reject tissue differing even slightly.

3. *Phenotype matching* means the use of learned cues as a template by which other individuals are compared and judged. The cues may come from the animal's own body or from other individuals with which it interacts during some critical learning period. To take one of many excellent recent examples, spiny mice (*Acomys cahirinus*) prefer to associate with other individuals raised with them as littermates whether or not the littermates are siblings. Conversely they tend to avoid their own siblings that were raised in alien litters.

A great deal of thought and experimentation has gone into the concept of phenotype matching, particularly that based on learned cues of a chemical nature. Two competing models have been proposed that help to clarify the issue. One, the 'individualistic model', has the responding animal accept another as a relative if the two share some particular allele or set of alleles (Chapters 3 and 4). The alternative scheme, the 'Gestalt' model, has individuals deriving an odor as a mix of pheromones under multilocus control; the animal classifies another as kin if the two share at least some threshold amount of the mix. The Gestalt system would appear to be the more efficient of the two, in the sense that it requires fewer alleles to achieve some arbitrary level of accuracy (say, 95%) when distinguishing true kin from non-kin.

There is a way in which the whole evolution of kin recognition by pheromones can be short-circuited to make a relatively simple system. If all the members of the group acquire the same odor from a single individual, which in turn differs genetically from referent individuals in other groups, the system becomes much easier to manage. And indeed, recent research has revealed that the colony odor of monogynous ants in the genus *Camponotus* is a hierarchically organized complex of environmental cues and genetic labels derived from workers but dominated by discriminators produced by the queen. Workers of alien species of *Camponotus* adopted by the colony acquire the distinctive queen mix, while siblings exposed earlier to alien queens are rejected. Under such a regime, relatively few loci and alleles would be required to differentiate colonies. For example, with two alleles at each of n loci and hence 3^n combinations possible, only 15 loci are needed to generate 10 million colony odors. However, such a system can be expected to break down in species whose colonies have multiple queens. This has been shown to be the case in *Pseudomyrmex ferruginea*.

THE GENETICS OF KIN RECOGNITION

Although a great deal has been learned about kin recognition, a full understanding cannot be achieved without a genetic analysis of the signals employed during discrimination. A promising beginning has been made in several species of insects and mammals. The examination of hybrid *Teleogryllus* crickets, for example, has shown that both the production and recognition of some song characteristics are encoded on the X-chromosome. If individual songs are discriminated, they are likely to be differentiated by allelic differences at these loci. The incest-avoiding behavior of *Drosophila melanogaster* is mediated by pheromones, which are prescribed by genes on both the X- and autosomal chromosomes.

The best known and most intriguing genetic system is the major histocompatibility complex (MHC) of the house mouse (*Mus musculus*), which imparts an individual scent to most of the animals. The MHC, which is believed to occur in one form or other throughout the vertebrates, is a chromosomal region occupied by a string of genes with immunologic functions. The mouse MHC, known specifically as H-2 (the human system is HLA), is a genetic hot spot: its principal loci have over 50 alleles each in the population at large, and they are affected by mutation rates 25 or more times greater than those at non-MHC loci. Hence at any given time 2,500 or more allelic combinations can occur, and most probably do. Experiments with Y-maze olfactometers have demonstrated that the mice can distinguish other mice by differences in smell imparted by the MHC allele alone. When choosing mates, and also when tracking the odors of females in the olfactometer, males show a preference for females with a MHC genotype different from their own. MHC loci are known best to biologists for the specificity they impart to the glycoproteins inserted into the immune cells, or lymphocytes. But like many such allelic systems, they affect other phenotypes as well, including the size of the animal, the development of the thymus gland, and—most importantly for behavior—the products of steroid hormone metabolism that contribute to urinary odors. Not only do the distinctive smells serve to avoid incest, they also appear to act as primer pheromones. In the Bruce effect, pregnancy is blocked when recently impregnated females are exposed to males with an odor sufficiently different from that of the stud. Recent experiments have disclosed that the effect can be achieved with the odors of males or females that differ genetically at the MHC loci alone.

Mice can also distinguish the bearers of different genotypes at the T-locus, preferring to associate with members of the opposite sex that are homozygous wild type. The result is an innate avoidance of the debilitating t-alleles at the same locus.

As behavioral genetic analysis is extended more fully to kin recognition in these and other species, a much clearer picture of the origin and precision of the recognition systems will be obtained. In the case of pheromonal recognition, which is based upon discrete biochemical products, it should also be possible to trace the production of the cues all the way from the genes to the final phenotypes.

PRECISON AND ECONOMY

The past ten years of research have revealed kin recognition in animals to be more precise than was generally expected. In early studies the discrimination was perceived as generally dichotomous, meaning that distinctions were made between kin and non-kin or between offspring and alien young. More recently the ability to distinguish among kin at least to the level of full-siblings versus more distant relatives, such as half-siblings or first cousins, has been documented in the following animals: eusocial sweat bees (*Lasioglossum*), honey bees (*Apis mellifera*), frog tadpoles (*Rana cascadae*), quail (*Coturnix coturnix*) and ground squirrels (*Spermophilus beldingi*).

Human beings have also been found to rank high in the innate ability to recognize kin (Chapter 12). Infants can distinguish their own mother from other women by voice alone at 24 hours of age, the smell of their mother's breast pad within six days of age and a photograph of their mother when they are four to seven weeks old. For their part, mothers can recognize their infant's smell with only 30 minutes exposure as early as six hours post partum. These findings illustrate the phenomenon of 'prepared' learning, in which certain stimuli are perceived and remembered far more readily than others. The innate components, prescribed by the genes, are the distinctive qualities of sensory screening and cognition that affect the direction and timing of learning. The learned component is the information that the peculiarly skewed central nervous system acquires from the environment.

The preparedness of human learning in kin recognition is further illustrated by the pathological condition of prosopagnosia, caused by lesions at certain pinpoint locations on the ventral surface of the brain. Prosopagnosiacs cannot tell people apart by facial features alone, even though their vision is not otherwise impaired. But they can distinguish them by their voices, and, most remarkable of all, they can still identify other objects visually. It thus appears that the extraordinary concentration of human beings on the face during social interactions, entailing the ability to name literally thousands of other people from photographs alone, is abetted by a special circuitry in the visual cognitive portion of the brain.

Yet another example of prepared learning in kin recognition is the mechanism of sibling incest avoidance. Evidence from the Israeli kibbutzim and Taiwanese minor marriages shows that when children are associated in close domestic proximity during the first six years of life, they avoid full sexual liaisons at maturity. This aversion occurs even when the young people are unrelated and are encouraged by their families to marry. What is inherited in incest avoidance, then, is the propensity to use spatial proximity as the cue to identify kin. This learning rule applies indiscriminately, even when the persons affected are not, in reality, genetic kin. Because children are usually reared in close association with siblings and other close relatives but not with the children of other families, the rule works in almost all societies most of the time.

The human rule of sibling incest avoidance, which incidentally also occurs in chimpanzees, is an apparent example of the rule of economy in evolution. Genetic systems tend to evolve to the point of achieving a high level of adaptive performance, and then, true to natural selection theory, they stop. The reason is that complicated, energetically expensive mechanisms are likely to be replaced by simpler systems that produce the same result. Conversely, if the simpler systems appear first they are unlikely to be replaced. It is reasonable to suppose that early proximity works so well as a cue in human incest avoidance that it has not given way to more complicated forms of cognition.

This conception is worth bearing in mind in the future exploration of kin recognition systems. How well, we might ask, does each newly discovered system work? Theoretical comparisons are in order. Are alternative mechanisms conceivable that might work better at no greater energetic cost or risk to the individual? An ancillary problem is the degree to which the system has been hard-wired, so as to depend on automatic responses to allelic differences among individuals as opposed to open programs that make use of transient learned features of the environment. For example, *Lasioglossum* bees depend primarily on genetic differences among the workers, while *Polistes* wasps employ mostly environmental cues such as nest and brood odors. Ants of the genus *Camponotus*, as noted, depend primarily upon the transfer of the genetically determined odor of the mother queen to all of the workers. Virtually all animal species analyzed to date utilize both genetic and environmental variables to differentiate other members of the same species. (The only exception may be a purely allelic recognition system in the colonial invertebrates.) It is the relative magnitude of the two components and the manner in which they interact that varies so much from one species to the next. These relationships, and their evolutionary pattern within the social context, await definitive interpretation.

CHAPTER 3

The Behavioral Analysis of Kin Recognition: Perspectives on Methodology and Interpretation

DAVID J. C. FLETCHER
Department of Entomology, University of Georgia, Athens, Georgia 30602, USA.

THE RELEVANT QUESTIONS

A large part of the study of animal behavior is concerned with the identification of its evolutionary (ultimate) and immediate (proximate) causes, i.e. with analyses of functions and mechanisms. In order to test the ability of animals to discriminate between kin and non-kin and between kin of different degrees of genetic relatedness to themselves, knowledge of the probable functions of kin recognition is useful, because it suggests what test procedures are likely to be appropriate. However, it is also possible in some cases to test hypotheses about mechanisms while remaining agnostic on the question of function.

Two ultimate causes of kin recognition are frequently cited: kin selection (Hamilton, 1964) and mating preference (Bateson, 1983). Kin selection theory predicts that an animal will behave in such a way as to maximize its inclusive fitness, from which it follows that an ability to recognize kin would minimize the misdirection of parental care and of aid given to relatives other than offspring. Mating preference has usually been associated with the avoidance of inbreeding, which has been assumed to be deleterious in most animals. However, substantial genetic advantages may accrue from striking a balance between inbreeding and outbreeding (Bateson, 1983). Either way, selection favoring discrimination among potential mates may be expected in many animals. Strictly speaking, mating preference is an aspect of social behavior. The reason for separating it here is that mating preferences may be displayed by any sexually-reproducing animals, solitary as well as social, whereas other proposed functions of kin recognition (Chapter 2) are associated with more advanced levels of social behavior. Since the distinction is useful, it will be maintained throughout this chapter.

There are many questions of interest that may be asked about the proximate causes of kin recognition, but it is desirable that these be preceded by two preliminary questions.

1. Do interacting conspecifics in fact display the expected differential behavior toward each other according to their degree of relatedness, i.e. do they display kin-biased behavior?
2. Is it necessary for the performance of kin-biased behavior that kin should be recognized, or is there a more parsimonious explanation?

If it is concluded from preliminary investigations that kin recognition probably occurs, the more important questions concerning its proximate causes are the following, although there are also many additional questions subsidiary to these.

1. At what level does recognition occur—at the level of the group, subgroup, or individual?
2. Is kin recognition enough, or must an individual assess more than relatedness?
3. What sensory cues are employed in recognition, and are these part of an animal's phenotype or are they acquired from the environment, i.e. are they of intrinsic or extrinsic origin?
4. How do phenotypic cues and corresponding perceptual mechanisms develop in the individual?
5. How reliable are the recognition mechanisms, i.e. can they be subverted intra- and inter-specifically?

It is not necessary that any particular study should attempt to answer all of these questions, or even all aspects of any one of them. On the other hand, it may not always be possible, or convenient, to separate some of them. For example, the preliminary question concerning whether recognition is necessary for kin-biased behavior to occur is usually included in discussions of the mechanisms of recognition, e.g. by Holmes and Sherman (1982, 1983). Further, it may not be practical in some cases to determine whether suspected kin recognition occurs, without at the same time analyzing, at least in part, the nature of the cues involved, and it is obviously not possible to determine how a particular mechanism develops without first learning something about the nature of the mechanism itself. Clearly, the questions do not comprise rules of procedure for investigating kin recognition; rather they summarize the conceptual framework. In this chapter I shall discuss some of the theoretical and practical issues involved in obtaining answers to these questions. First, however, I will discuss the usage of several key terms frequently used in the kin recognition literature.

ALTRUISM, NEPOTISM AND INVESTMENT

Many words, such as searching and aggression, used in animal behavior studies are the same as those used to describe human behavior. In general, the adoption of familiar words into the scientific lexicon is preferable to the creation of jargon, although it has not always been without attendant problems. Thus anthropomorphism, beginning with the Darwinian search for animal intelligence, was virtually built into the science of animal behavior through adopted terminology. With time, however, even the most anthropomorphic terms have acquired a technical flavor through usage, i.e. they have undergone subtle changes of meaning. Some words undergo more substantial semantic changes than others as the concepts they designate evolve through continuing theoretical and experimental analysis. Altruism is one such word. In fact, altruism, nepotism and investment are all used to refer, in part, to the same phenomenon, and since they have a strong bearing on the literature concerning kin recognition, I propose to examine how they are currently used in order to avoid semantic confusion. Let us look at several definitions of altruism.

1. Devotion to the welfare of others, regard for others, as a principle of action; opposed to egoism or selfishness (*Oxford English Dictionary*).
2. Self-destructive behavior performed for the benefit of others (Wilson, 1975).
3. Dangerous behavior or behavior disadvantageous to the individual that benefits other individuals of the species. Often used with respect to reproduction, in which the altruistic individual reduces its reproductivity while enhancing that of another (Michener, 1974).

From these definitions it is evident that altruism involves aiding behavior in which the cost of helping others exceeds the benefit. Most often the cost can be calculated in terms of time and resources expended, but if the help provided involves physical danger, the cost may also include the risk to the altruist of suffering bodily injury or even death. Such costs are immediate and apparent, so that acts of altruism would seem to be easily identifiable, as indeed they often are in human societies, although hidden benefits may complicate matters.

The above definitions make no distinction between relatives and non-relatives among the recipients of altruism. However, it is clear that in human behavior the emphasis is on non-relatives from the meaning of altruism in ethics. This is given in Webster's *New World Dictionary* as: 'The doctrine that the general welfare of society is the proper goal of an individual's actions.' In animal behavior this doctrine is embodied in theories of group selection. For example, in his commentary on the work of Lack (1954, 1966, 1968) on population regulation in birds, Alexander (1979) wrote that there is no reason to suppose as many ecologists have (e.g. Wynne-Edwards, 1962) that when birds keep the size of their egg clutches small they do so 'in an altruistic effort to regulate their population at optimal levels.'

For an individual animal the cost of helping non-relatives translates into reduced reproductive success, since the time and resources spent in aiding others could have been invested in raising offspring of its own. On the other hand, if the animal aids its relatives, genetic benefits that exceed the costs can result via the inclusive fitness effect of Hamilton (1964). Thus altruism acquired a new and contradictory meaning in evolutionary biology, which contradiction can be made more obvious as follows. An animal may be said to increase its inclusive fitness by investing time and resources in the reproductivity of its relatives, and the dividend from this investment is a greater increase in the number of copies of its own genes in the next generation, i.e. the help it provides is not altruistic but capitalistic. Use of capitalist terminology in this context also helps to emphasize the close similarity between parental investment and investment in the reproductivity of kin.

Another way in which the contradiction may be demonstrated is by contrasting seeming altruism with gene selfishness. Alexander (1979) has pointed out that sexually reproducing organisms are altruists of a very special kind, because their altruism, whatever form it takes, is normally directed toward genetic relatives. 'Such altruism...,' he said, 'may be described as phenotypically (or self-) sacrificing, but genotypically selfish.' Dawkins (1976) made the contrast in the following way: 'The key point... is that a gene might be able to assist replicas of itself which are sitting in other bodies. If so, this would appear as individual altruism, but it would be brought about by gene selfishness.' Thus, in the context which most concerns us—the evolution of social behavior—the root cause of altruism is selfishness.

Perhaps the main reason why the word altruism is still used by many biologists is that the immediate costs of aiding behavior are so visible compared to its ultimate benefits. Nevertheless, certain changes in usage have been taking place. Dawkins (1976) continued to use altruism to denote all behavior that is phenotypically self-sacrificing regardless of whether it is directed toward relatives or non-relatives, but at the same time he drew a clear distinction between 'pure, disinterested altruism,' which he suggested is unique to humans, and 'kin-altruism,' in which the benefits are directed exclusively to relatives. Alexander (1974), having introduced the distinction between phenotypic and genotypic altruism, also distinguished three classes of genetically selfish behavior that 'appear to be altruistic because they cause the bearers to raise the fitnesses of other individuals either at the expense of the bearer's phenotype or at the expense of phenotypes and genotypes of some third parties or 'manipulated' (exploited) individuals.' The three classes of selfish behavior are 'reciprocity' (reciprocal altruism of Trivers, 1971), 'nepotism' and 'parental manipulation.' This useful classification offers a solution to our semantic problem.

Nepotism is another term adopted from the language of human behavior (see Chapter 13). Originally it meant: 'The practice, on the part of the Popes or other ecclesiastics (and hence of other persons), of showing special favour to nephews or other relatives in conferring offices; unfair preferment of nephews or other relatives to other qualified persons' (*Oxford English Dictionary*). As used in animal behavior it means favoritism toward or investment in kin. According to Alexander (1974) it includes not only giving aid to relatives on direct and indirect lines of descent with the nepotist, but also 'parental altruism' and the assistance of mates. If one excludes assistance of mates, therefore (except help given in raising offspring), nepotism is an exact synonym for kin-altruism and may be substituted for this term. Where it is convenient to distinguish offspring from other relatives as the beneficiaries of nepotism, the commonly used term 'parental investment' may be consistently employed instead of parental altruism. A few authors, e.g. Holmes and Sherman (1982, 1983), already use nepotism in preference to altruism where it is appropriate to do so. One may hope that the practice will spread, thereby restoring to altruism its original unambiguous meaning and validating the useful generalization that whereas nepotism can evolve, true altruism cannot. As previously suggested (Fletcher and Ross, 1985) altruism is best treated as a cultural concept that belongs to the world of human beings. When aid is given to non-relatives in the non-human world this usually implies either reciprocity, which is another form of investment (for discussions see Alexander, 1974, 1979), or social parasitism. Occasionally, however, instances may occur that have the character of 'mistakes' resulting from the malfunctioning of a recognition system, as for example when honey bee foragers take pollen or nectar into neighboring hives in crowded apiaries.

It will now be useful to examine some examples involving behavior that is actually or potentially suicidal even though it benefits relatives. A well-known example is the stinging of enemies by honey bees (*Apis mellifera*). When workers defend their nests and food stores by stinging to death conspecific robber bees, they rarely lose their stings and consequently do not die (Butler, 1967; personal observations). The selection pressure for the evolution of a barbed sting must, therefore, have been defense of the colony against larger (mainly mammalian) enemies. In Africa, where the honey bee is thought to have evolved, the most important of these enemies are humans and ratels, *Mellivora capensis* (Fletcher, 1978). They usually rob not only the stored honey and pollen, but also the brood (larvae and pupae), thereby destroying the combs and most of the bees, as well as the nest site itself. Under these circumstances a colony's chances of surviving are minimal, even if the queen and some of the workers escape, so that selection for the delivery of substantial doses of poison through a combination of mass attack and sting autotomy is intense. Since workers are essentially sterile and short-lived, the entire genetic investment of a current worker force would be lost if the colony did not survive long enough to swarm. Hence, the additional investment of their lives by some of the workers makes evolutionary sense and in this context their behavior is clearly nepotistic. Moreover, suicidal stinging makes even greater evolutionary sense when one recalls that the defenders of a colony are the older bees (foragers) that have already made a substantial investment in the colony as nurse bees, comb builders, etc., so that they are the most expendable class of members.

Another example is the performance of alarm behavior by a wide variety of animal species. Hamilton (1964) suggested that the function of such behavior is to warn relatives of impending danger. This hypothesis can only be tested if the genetic relatedness of associating individuals is known, if the test group contains unrelated individuals as controls, and if the observer himself can recognize the interacting individuals. These conditions were met in a study of alarm behavior in Belding's ground squirrels, *Spermophilus beldingi*, by Sherman (1977), who tested Hamilton's hypothesis in addition to five other hypotheses. His results supported the nepotistic hypothesis primarily, but also to some extent the alternative hypotheses that alarm calling serves to warn reciprocators and that it may help the group. However, Sherman also found that squirrels that gave alarm calls were stalked and chased more often than non-callers, but did not give data to show if this resulted in a significantly higher mortality among callers. If such a higher mortality did result, then alarm calling in this species would best be interpreted as nepotism mixed with reciprocity, since both relatives and non-relatives are among the beneficiaries. Moreover, it is unlikely that the life of more than one relative could be saved by the sacrifice, in which case the cost of calling would outweigh the benefit, since the maximum coefficient of relationship to a saved relative would be one half (most likely a mother or a sister, since males emigrate) and the average relationship would be much less. Thus pure nepotism could not be involved. On the other hand, if the mortality of callers is lower than that of non-callers in spite of their being hunted more often, this would tend to support the alternative hypothesis that alarm calling increases an individual's inclusive fitness by enhancing its own chances of survival as well as that of relatives. This could occur because a caller is likely to have a greater opportunity to escape. It occupies a position in which it can see the predator and it may be stalked more often simply because it is conspicuous, whereas many non-callers are cryptic. Under these circumstances, nepotism unmixed with reciprocity could be positively selected, but elements of both nepotism and reciprocity are probably present. Similar arguments may also apply to the evolution of alarm behavior in other animals, especially birds.

A final and much discussed example, first considered by Fisher (1930), is the aggregation of sibling groups of aposematic distasteful insects. A predator must taste an insect in order to learn to avoid nearby relatives, and such a tasted insect is said to be altruistic (e.g. Hamilton, 1964). However, being tasted is not a consequence of specific behavior in relation to the predator in the same sense as stinging or giving an alarm call. The approximate equivalent of such behavior is the possession of warning coloration, and each individual is, in effect, gambling that another member of the group rather than itself will be taken; it is playing a numbers game like a fish in a shoal (apart from the possibility that large groups may appear more formidable to a potential predator). Additionally, however, by aggregating with siblings, the gambler has bought insurance through kin selection against total genetic loss if it is eaten. Hence, it stretches the imagination to call a loser an altruist, although it may well be regarded as a nepotist that invests (inadvertently) in the reproductive potential of numerous kin.

We are often reminded that disputes about semantics are sterile (e.g. Popper, 1972; Hinde, 1968; Dawkins, 1976), to which view I fully subscribe. The purpose of this section, then, is not to dispute the meanings of words, nor yet to suggest that any particular usage of terms is either correct or incorrect (users will decide that); it is to show that the meanings of certain words that occupy important positions in sociobiology are evolving. Readers will certainly find diversity in the usage of many terms in this volume, but I hope that with the help of the glossary in Chapter 2 and of this section, semantic confusion will be minimized.

KIN-BIASED BEHAVIOR

In the behavioral analysis of kin recognition, the first question that logically requires an answer is whether members of the animal species concerned behave asymmetrically toward kin and non-kin. But what constitutes such kin-biased behavior? The answer to this subsidiary question depends, like so much else in the analysis of kin recognition, on the details of the life history and behavioral ecology of the species.

Kin selection and kin-biased behavior

For an assessment of kin-biased behavior under kin selection, two broad classes of social interaction need to be defined. In essence these are cooperative and uncooperative behaviors. The most important feature common to both these classes of behavior is that they involve interactions between conspecifics, even when no more than tolerance or avoidance are expressed. Such interactions may be one-on-one, or they may involve three or more individuals.

The task of the investigator is to determine whether specific cooperative and/or uncooperative behaviors occur that are consistently kin-biased, i.e. whether there are significant differences between quantitative measures of behaviors directed toward kin and non-kin or toward kin of varying degrees of relatedness. Where data are adequate, the analysis may include regression of the behavioral measures on coefficients of relatedness, as accomplished for example by Greenberg (1979) in his study of kin recognition in the primitively eusocial bee, *Lasioglossum zephyrum*. The choice of behaviors studied depends on such factors as the social context, ease of quantification (i.e. the frequency and/or variations in intensity or duration of performance of the behavior), ease of observation, etc. Some examples follow which show a progressive increase in the complexity of the kinds of behavior employed in the assays; for further details see Chapters 7 and 9. The reader should bear in mind that many experiments are designed to obtain data not only on kin-biased behavior, but on other aspects of kin recognition as well, so that the examples contain simplifications.

Unrelated adults of the American toad, *Bufo americanus*, breed together in the same ponds and the tadpoles form schools. It has been found, however (Waldman and Adler, 1979; Waldman, 1982), that schooling is kin-biased in that the tadpoles associate preferentially with siblings. This affinitive behavior was quantified by determining for many tadpoles whether their nearest neighbors were siblings or non-siblings.

The primitively eusocial bee, *L. zephyrum*, nests in soil and one bee guards the entrance to the nest by blocking it with its head. It will admit nestmates, but it will not permit alien conspecifics to enter (Michener, 1974). Since an established colony normally consists of a single queen and her daughters, this constitutes kin-biased behavior which is easily quantified by comparing the numbers of nestmates versus non-nestmates admitted by guard bees under controlled conditions, as was done by Greenberg (1979).

Belding's ground squirrels live in burrows. Males disperse permanently soon after weaning, but females remain in the home area surrounded by near and distant female kin with whom they interact throughout their lives. They generally mate with several males, with the result that they usually have multipaternate litters (Hanken and Sherman, 1981). Four kinds of interaction between females were investigated for kin bias by Sherman. They included two uncooperative behaviors (chasing or fighting between pairs of individuals) and two cooperative behaviors (cooperative chasing of a trespasser from a territory or assisting an individual being chased by chasing the pursuer). The numbers of encounters between pairs of females that resulted in chasing or fighting were recorded, as well as the numbers of occasions on which individuals involved in territorial interactions were assisted by various relatives and non-kin. Statistical comparisons demonstrated a strong kin bias in both cooperative and agonistic behaviors (Chapter 9).

In these examples, cooperative and uncooperative behaviors appear as reliable indicators of relatedness, but this is not always the case. Mild aggression often characterizes the weaning behavior of mammalian mothers (Dawkins, 1976) and in some social insects extreme aggression towards close relatives may be manifested. For example, in the fire ant, *Solenopsis invicta*, workers kill virgin queens under certain circumstances (Fletcher and Blum, 1981) even though they are their full sisters to whom they are maximally related ($r = 3/4$) by virtue of the mother queen being singly inseminated (Ross and Fletcher, 1985).

Obtaining genealogical data

Investigations of kin-biased behavior require precise knowledge of the genetic relatedness of the experimental animals, as do all other phases in the analysis of kin recognition. Also required are means of marking the animals so that either individuals or kin classes are recognizable to the observer. Several methods are employed to obtain accurate relatedness data and to mark the animals.

A common procedure is to rear animals from known parents under laboratory and/or field conditions, as was done by Waldman and Adler (1979) and Waldman (1982) in their studies of American toads. They captured amplectant pairs of toads from various breeding ponds and kept them in separate containers for 24 h. During this time most of the females laid eggs, which in *Bufo americanus* are externally fertilized, so that knowledge of parentage was certain. Tadpoles were reared in separate aquarium tanks and were marked by placing them in very dilute aqueous solutions of methylene blue or neutral red for 24 h. This procedure stained their translucent fins and did not influence their behavior in any detectable manner. It served to distinguish siblings as a class from non-siblings. Marked tadpoles used in field experiments were reared both in the laboratory and in separate compartments in a common outdoor tank (Waldman, 1982).

The honey bee, *Apis mellifera*, is of special interest since queens normally mate with many males (Taber and Wendel, 1958; Laidlaw and Page, 1984), so that the workers in a colony represent a number of patrilines. To obtain bees of known relatedness, Getz, Brückner and Parisian (1982), Page and Erickson (1984), and Visscher (1986) artificially inseminated queens with semen from single drones and introduced the queens into field colonies. Insects are easily marked with drops of enamel or acetate paints applied to the thorax.

A second method of obtaining information about the genetic relatedness of experimental animals is to observe naturally breeding populations either in captivity or in their native habitats, and to record genealogical data over a period of time. This method is used extensively in studies of non-human primates and, because these animals are long-lived, it often requires many years of careful observation to build up such genealogies. In his field study of Japanese macaques, *Macaca fuscata*, that began in 1972, Kurland (1977) had available to him birth records extending back for five years. Marking by means of facial tatoos permitted easy identification of individuals, but as young monkeys were typically not caught and tatooed until they were six to nine months old, all the genealogical data in fact consisted of 'behavioral matrilines', i.e. a newly marked animal and its mother were assumed to be the same pair first seen together a few days or weeks after the birth. Prior to 1968 close association and nipple contact between immature animals and adult females were used for identification of mothers and their offspring. This could involve circularity in studies of kin-bias in behavior, but was not considered by Kurland to be a serious source of error in view of the known rarity of adoption by Japanese macaques both in the laboratory and the field.

Genealogical data are rather more easily obtained for captive groups of primates, although of necessity the investment remains long-term. Such data are a valuable research resource that can be made available for short-term investigations. Thus in a study by Massey (1977) of aiding behavior associated with agonistic interactions in pigtail macaques, *Macaca nemestrina*, use was made of genealogical information that had been collected systematically over a period of more than twelve years through four generations in a group consisting of 44 monkeys at the time of the study.

To establish paternity of experimental animals when this is unknown, allozyme data may be used. Hanken and Sherman (1981) employed six polymorphic loci coding for blood proteins to determine electrophoretically that 78% of the litters of Belding's ground squirrels studied by them were multipaternate, and Small and Smith (1981) established the paternity of 46 members of a rhesus monkey *(Macaca mulatta)* study group using electrophoretic and serological techniques for 20 polymorphic loci.

Importance of variables other than relatedness

When an animal maximizes its inclusive fitness by rendering aid to conspecifics, it must assess (or behave as if it could assess) not only its genetic relatedness to potential recipients, but also what Dawkins (1976) has called their 'reproductive expectancy.' Many variables may contribute directly or indirectly to this expectancy, but a few are of greater importance than others. Thus an animal that is past the age of reproduction may be worth zero investment, even if it is a close relative, whereas an individual of the same degree of relatedness but on the threshold of its reproductive prime may be worth maximum investment. On the other hand, the individual that can no longer reproduce directly may nevertheless be worth investing in since it may contribute indirectly to the investor's reproductivity, e.g. by helping to care for young, as frequently occurs in human families. Other important variables are likely to be the sex of a potential recipient, its health and its 'ownership' of resources. No particular animal needs to be able to make optimal assessments of all, or even of any, important variables, but those that make the best average assessments will be the most successful in propagating their genes.

The significance of the above for the experimental analysis of kin recognition is readily apparent. Any variable, in addition to genetic relatedness, that may have sufficient influence on the behavior of test animals to affect the results significantly should be carefully controlled. Occasionally, however, circumstances may prevent the incorporation of proper controls. Such was the case in Kurland's (1977) field study of Japanese macaques, in which provisioning of the monkeys by park authorities, while facilitating observations, may also have caused behavioral bias. To quote the author: 'One can hope that only the frequency of social interactions changes with provisioning, and not the very structure of the social system.'

Mating preference and kin-biased behavior

In parallel with studies of kin recognition under kin selection, behavioral biases towards prospective mates usually provide the initial evidence that kin recognition may be involved in mate selection. To quantify such biases the same two classes of behavior (cooperative and uncooperative interactions) may be used, although non-behavioral criteria are also of value (see below). Assuming that a mating preference is found, an additional question is whether the male or female is responsible for making the choice or whether both sexes choose. However, few studies have been concerned solely with whether or not mating preferences occur. They are usually combined with investigations of the mechanisms involved and the nature of the sensory cues. Some of what follows therefore, is anticipatory to discussion of these topics.

Examples will serve to demonstrate the diversity of measurements that may be employed to assess mating preferences in different animal species.

In his study of Japanese quail, *Coturnix coturnix japonica*, Bateson (1982) made use of affinitive behavior, since he had previously found (Bateson, 1978) that the time spent near a particular female by a male bird was strongly linked to subsequent copulation preference. He tested the preferences of both sexes of varying degrees of relatedness, i.e. siblings, first cousins, third cousins and unrelated birds, none of which was familiar with any other (novel birds), and an additional category of familiar siblings. Tests were conducted in an apparatus in which individual test birds could view members of the five categories of stimulus birds through one-way screens. The stimulus birds were in separate compartments arranged in a circle, and the time spent by a test bird in front of each compartment during 30 minute test periods was recorded by automated means.

Male *Lasioglossum zephyrum* bees pounce on small dark objects in the presence of the female odor (Barrows, 1976). Greenberg (1982) used this behavior to determine the mating preferences of the males. He tested their responses to their own sisters and to unrelated females by concealing females behind black spots and comparing the numbers of pounces they elicited.

Methods that depend upon precopulatory behavior to indicate mating preferences do not show with certainty whether successful mating would have occurred, because some animals, e.g. *Drosophila* spp. (Chapter 5), make their final choice only after physical contact has been made. This drawback by no means invalidates the use of precopulatory criteria, but it emphasizes the need for caution in the interpretation of results. Postcopulatory criteria are preferable.

In the wasp, *Polistes fuscatus*, no differences were found by Larch and Gamboa (1981) in the frequencies of behavioral interactions, including mating, between related and unrelated pairs. However, Post and Jeanne (1983) reported that coupling of the male and female genitalia for as long as one minute or even more does not necessarily result in sperm transfer. They therefore used successful insemination, determined *post hoc* by dissection, in a re-examination of mating preference in this species. Their results showed no evidence of inbreeding avoidance, but as they placed equal numbers of males and females in the mating cages with half the females being siblings of the males and the other half unrelated to them, it would be interesting to learn whether a bias in mating preference would appear if the males were given a substantial, but equal, excess of related and unrelated females from among which to choose.

Dewsbury (1982) used several criteria, some of which were postcopulatory, to investigate the possible avoidance of inbreeding in two species of deermice, *Peromyscus maniculatus* and *P. eremicus*. These were the probability of breeding, latency to breed, litter size, number of litters, and success in rearing young, and the data he obtained suggest that deermice discriminate against siblings as mates.

With regard to obtaining genealogical data and the marking of experimental animals, the methods are similar to those employed in kin-selection studies. Special considerations may sometimes arise however. For example, *Drosophila* spp. make use of their wings in courtship behavior, for which reason the practice of marking flies by cutting notches in their wings in studies of the rare male mating advantage has been controversial. For a discussion of this see Chapter 5.

ANALYSIS OF RECOGNITION SYSTEMS

What constitutes recognition?

An act of recognition by an animal usually entails comparing a sensory cue (or cues) borne by a conspecific, with a neural template (or templates), following which comparison behavior is performed that is either cooperative or uncooperative, depending on the extent of the match or mismatch between the cue(s) and the template(s). The first animal may be referred to as the 'discriminating individual' and the second as the 'cue bearer', which terminology overlaps with, but is not identical to, that of Michener and Smith (Chapter 7). Interacting conspecifics may be both discriminating individuals and cue bearers simultaneously.

There are two aspects to the analysis of recognition systems: one is concerned with perceptual mechanisms, especially with how and when the neural templates are formed, and the other with the origin and nature of the cues involved. Because templates and cues comprise interacting systems, however, their separation for analytical purposes is artificial and is not always possible, as for example when considering the functioning of receptor organs.

Levels of recognition

From what is known about the structure of social groups and the nature of conspecific interactions, several levels of recognition may be hypothesized.

1. *Discrimination between members of one's own and an alien group.* Recognition of kin is not necessarily implied, because a group may be heterogeneous, i.e. it may consist of both relatives and non-relatives as in a colony of slave-making ants, e.g. *Harpagoxenus americanus*, and their slaves. On the other hand, it may be synonymous with kin recognition if the group consists exclusively of family members, as in fire ants and many other species of eusocial insects.

2. *Discrimination of classes within a group.* Within groups an individual may be capable of distinguishing various classes or subgroups, while not being able to recognize individuals. Recent data (Getz, Brückner and Parisian, 1982) suggest, for example, that in honey bees, different patrilines of half-siblings which result from multiple inseminations of the queen, tend to segregate when a colony swarms. Siblings can also be discriminated from both maternal and paternal half-siblings by tadpoles of the Cascades frog, *Rana cascadae* (Blaustein and O'Hara, 1982a), but from paternal half-siblings only by tadpoles of the American toad, *Bufo americanus* (Waldman, 1981). In the sweat-bee, *Lasioglossum zephyrum*, guard bees have the ability to discriminate among various classes of relatives over a range of coefficients of relatedness (Greenberg, 1979).

Barrows, Bell and Michener (1975) drew a distinction between homogeneous subgroups, the members of which are alike in a biological attribute by which they are grouped, and heterogeneous subgroups, which are not united by any common feature, but are grouped according to the experience of the individuals. The examples cited above are all of homogeneous subgroups. Two heterogeneous subgroups that are frequently distinguished are unfamiliar and familiar classes of individuals, but this can cause confusion. Unfamiliar individuals unquestionably form a heterogeneous subgroup, but if an individual responds differentially to members of a familiar subgroup, it must either recognize them as individuals (because they are not united by any common feature) or it must perceive different stimulus intensities associated with them, in which case any two members of the subgroup that exchange their stimulus intensities would be mistaken for each other. Consider the case of a linear dominance hierarchy in which all the members are familiar with each other. It might conceivably, though improbably, be sufficient for each member of the hierarchy to recognize only two classes of other members—dominants and subordinates. A signal, such as a pheromone, that could vary in intensity as a continuum between the most and least dominant members could provide a plausible mechanism, e.g. in a colony of the slave-making *Harpagoxenus* ants mentioned previously, in which the workers have been reported (Franks and Scovell, 1983) to form a linear hierarchy. However, an individual in the middle of the hierarchy that attempted to improve its position would need to be able to identify the particular ant in the position immediately superior to itself if it were to avoid repeated and unnecessary conflict with all other dominants. Hence, what is probably involved is a differential response to different stimulus intensities, and whether or not one regards this as individual recognition is a matter of taste, but it is clearly on a different level than individual recognition in, say, primates.

3. *Discrimination of individuals.* Recognition systems at the level of the individual were largely ignored until the beginning of the last decade, but interest in them was stimulated by the resurgence of the individual as the unit of selection (Myrberg and Riggio, 1985). Individual recognition in fact forms the very basis of the advanced organization found in vertebrate social groups (Wilson, 1975). It has most often been studied in the context of parent-offspring and/or offspring-parent recognition in birds (Beer, 1970; Miller and Emlen, 1975; Beecher, Beecher and Hahn, 1981; McArthur, 1982; Sieber, 1985; see also review by Falls, 1982) and mammals (Petrinovich, 1974; Poindron and Carrick, 1976; Kaplan, Winship-Ball and Sim, 1978; Cheney and Seyfarth, 1980) that possess acoustical means of communication. On the other hand, among adult vertebrates, individual recognition through chemosensory signaling has received most attention (see review by Halpin, 1980).

Among invertebrates individual recognition has rarely been reported except for compatibility in colonial forms like corals and sponges. This may be because chemosensory cues play a dominant role in their communication systems and, as discussed above, it is especially difficult to classify differential responses to varying stimulus intensities in this sensory mode. Examples include mate recognition as well as parent-offspring recognition in the desert woodlouse, *Hemilepistus reaumuri* (Linsenmair and Linsenmair, 1971; Linsenmair, 1972; Chapter 6), habituation to female odors by males of the sweat-bee *Lasioglossum zephyrum* (Barrows, 1975; Chapter 7), and discrimination among unfamiliar queens of differing fecundity levels in the fire ant, *Solenopsis invicta* (Fletcher and Blum, 1983). Caldwell (1985) showed that mantis shrimps, *Gonodactylus festae*, can distinguish between the odors of two conspecifics previously encountered in agonistic interactions.

This discussion of the various levels at which animals may distinguish among others of their species provides necessary background material for the formulation of the hypotheses concerning both the perceptual and the signaling aspects of kin recognition systems.

Analysis of perceptual mechanisms

Known and possible mechanisms of kin recognition have been discussed by Hamilton (1964), Alexander and Borgia (1978), Alexander (1979), Hölldobler and Michener (1980), Bekoff (1981), Beecher (1982), Dawkins (1982), Holmes and Sherman (1982, 1983), Lacy and Sherman (1983), and by various authors in this volume. The following classification has emerged from the theoretical and experimental work of these authors. It constitutes a set of working hypotheses.

1. *Location, or spatial distribution.* If the probability is high that an animal will encounter only close relatives at a particular place (e.g. a parent bird returning to its nest is likely to find its own offspring in the nest), it may not be necessary for it to distinguish between kin and non-kin in order that its investment behavior be appropriately directed. It could learn site-specific cues instead. This is an indirect, probabilistic mechanism that is vulnerable to subversion, as is illustrated by the example of the cuckoo.
2. *Familiarity.* If animals interact in social contexts that are unambiguously associated with relatedness, as for example interactions between parents and offspring or between siblings in a nest or burrow, they may learn to recognize one another through familiarity. This is referred to as 'association' by Holmes and Sherman, but to avoid any possibility of confusion with associative learning, only familiarity, which is an already commonly used term, will be used. This method of recognition is also probabilistic.
3. *Phenotype matching.* If phenotypic similarity is well correlated with genotypic similarity, recognition may result from comparing phenotypes (phenotype matching). One animal (the discriminating individual) assesses its relationship to an unfamiliar conspecific (the cue-bearer) by comparing phenotypic cues from the latter with those of a referent. The referent may be a familiar relative or the discriminating individual itself. In either case

the phenotype of the cue-bearer is compared with a learned 'template.' The essential feature that distinguishes this mechanism from recognition by familiarity is that no previous experience of the cue-bearer by the discriminating individual is required.

4. *Recognition alleles.* For this mechanism, alleles encode in bearers both phenotypic cues indicative of genotypic similarity and the ability to recognize these in others without having any previous experience of them, i.e. in the absence of a learned template against which to compare. As pointed out by Crozier (Chapter 4) this mechanism is sometimes confused with the green-beard effect (see below), the difference being that recognition alleles do not encode any behavior toward recognized individuals.

5. *The green-beard effect.* Hamilton (1964) speculated that 'altruism' could conceivably be effected by alleles if their expression resulted not only in the possession of a characteristic phenotypic cue, e.g. a green beard (Dawkins, 1976), and in the ability to recognize this, but also in the performance of 'altruistic' behavior toward recognized individuals. Hamilton considered it unlikely that such a mechanism could occur.

How does the analysis of perceptual mechanisms proceed? In practice it makes little difference which mechanism one tests for first, but the following arbitrary sequence will aid discussion. First one might determine whether post-natal (post-parturition or post-hatching) learning of any kind is involved. Where appropriate this would include tests for recognition by location, but more commonly one would wish to distinguish between recognition by familiarity and the kind of phenotype matching in which relatives act as referents. One would very likely find that post-natal learning does indeed play a role in a particular case, but this would not exclude the possibility that other mechanisms are simultaneously involved. Those remaining are: a pre-natal influence, phenotype matching with self as referent, recognition alleles, and a green-beard effect. Among these, one might test for pre-natal maternal and paternal effects, as will be discussed later, but no way has yet been devised of testing a green-beard hypothesis. Given the low probability that such a mechanism occurs anyway, only phenotype matching with self reference and recognition alleles would be left after testing for pre-natal effects. However, as pointed out by several authors (Kareem and Barnard, 1982; Blaustein, 1983; see also Chapters 4 and 10), it does not seem possible to distinguish between these two, nor is it important to do so. What is important is that testing for recognition alleles involves eliminating any possible role for experience, either pre-natal or post-natal. Thus, evidence favoring the existence of recognition alleles is ultimately negative in character, which although unavoidable is not very satisfactory (cf. discussions of the innate/learned dichotomy by Hinde, 1968, and Lehrman, 1970).

It is likely that in different animal species different levels of interaction between genetic potential and environmental input are required for the development of a fully functional recognition system. In view of the difficulties arising from attempts to divide recognition behavior into discrete mechanisms, such a developmental model seems worth elaborating.

A recognition system that requires, in addition to genetic potential, only material (mainly nutritional) input from the environment for completion of its development (if such exists) is categorized arbitrarily as an allelic mechanism. On both the perceptual and signaling sides, or on either one, such development may be completed either pre- or post-natally. In either case it is impossible to eliminate an animal's experience of itself; moreover, in the case of social species one has no way of knowing whether the development of test animals in isolation is 'normal.' Certainly the behavior of many species, e.g. primates, is likely to develop abnormally in isolation. To alleviate this problem Wu et al. (1980) permitted infant macaques, reared in isolation from their relatives, to spend time each day with a play group of their peers, but this experience could conceivably have influenced their subsequent behavior as test subjects.

On a slightly different level, the development of a recognition system might be influenced (hypothetically) by siblings in the same womb or by maternal influences such as hormones. Here the environmental input is experiential as well as material, but these variables are somewhat more accessible to control, e.g. by extra-uterine fertilization and transfer of ova to surrogate mothers. Aside from such practical issues, the dividing line between a mode of development that requires no experiential input and one in which there is some pre-natal influence is a fine one. An equally fine line divides this second mode of development from a third, in which early post-natal experience with parents and/or siblings is necessary for the complete development of a recognition template. In this case, the post-natal influence might be chemical (as pre-natal influences are likely to be) or it might be acoustical, visual, or even tactile.

This developmental sequence may extend to learning cues at a later post-natal stage, e.g. adult birds may learn to recognize their offspring individually, or a recognition template may be continuously modifiable. Correspondingly, the sources of the cues may become more varied. They may be phenotypic characters, extended phenotypes (sensu Dawkins, 1982), or environmental accidentals (defined in the section on the analysis of recognition cues).

In every case a recognition system develops as a consequence of continuous interaction between the genetic potential of the individual and its environment, a process that reveals a fundamental unity among the various recognition mechanisms that have been proposed. This unity may be appreciated more fully when one recalls that the learning capabilities of animals are as much an expression of their genetic potential as is the ability of some animals to recognize kin without having had any previous experience of them.

Testing for recognition by location

Possession of a fixed nest site is frequently an accurate predictor of genetic relatedness. When parent birds return to their nest, the probability that they will encounter and provide for their own offspring, before they become mobile, is very high (Hamilton, 1964; Burtt, 1977; Holmes and Sherman, 1983). A standard technique for testing whether parents are able to recognize their chicks is to exchange young between nests, taking care that the exchanged individuals are of the same age to control for possible developmental differences in recognition cues. As noted in Chapter 2, a particularly well studied example is that of the bank swallow, *Riparia riparia* (Beecher, Beecher and Hahn, 1981; Beecher, Beecher and Lumpkin, 1981). Parent birds readily accept and feed transferred chicks up to 15 days of age, but they either ignore or evict them if they are 17 or 18 days old. By that time the young birds have developed signature calls, so that when they fledge at the age of 18 or 19 days, their parents are able to continue to identify and feed them.

A similar technique was employed by Holmes and Sherman (1982) to study parent-offspring recognition in Belding's ground squirrels. When pups were transferred to the burrows of unrelated females, they were usually accepted if the pups were less than about 23 days old. This is about the time when they emerge from their parental burrows for the first time. Females attacked most pups transferred at the age of 25 days.

Orientation to and recognition of a fixed nest site containing immobile young does not necessarily render other means of kin recognition superfluous, so that tests for additional mechanisms are advisable. Honey bees, for example, locate their nest entrance using visual and chemical cues (Butler, Fletcher and Watler, 1970) but can also discriminate between kin (nestmates) and non-kin using both extrinsic and intrinsic cues (Ribbands, 1965; Breed, 1983). Similarly, many ants employ odor trails and visual cues as means of orientation to the nest in addition to possessing colony-identifying odors (Wilson, 1971; Chapter 8). Some species such as the African weaver ant, *Oecophylla longinoda*, also mark territories with a colony-specific pheromone (Hölldobler and Wilson, 1977).

Location may be an accurate predictor of genetic relatedness in a viscous population. Viscosity, as found in polistine wasps for example, would lead to a degree of inbreeding and, over a period of time, to generally increased levels of relatedness in local populations (Hamilton, 1964; but see Pollock, 1983). In *Polistes fuscatus*, inseminated females overwinter solitarily, and the following spring establish new nests, often cooperatively, near the site of their natal nest. Cofoundresses are siblings (Bornais *et al.*, 1983; Klahn and Gamboa, 1983; Chapter 7) and they recognize one another in part by means of odors that appear to be derived from their natal nest (Pfennig, Reeve and

Shellman, 1983). In this species, therefore, the level of uncertainty engendered by recognition of location is greatly reduced.

Testing for recognition by familiarity

Becoming familiar with kin by associating with them is probably the most common mechanism of kin recognition and is the usual means of identification between mothers and their offspring (Holmes and Sherman, 1983). As seen in the bank swallow and ground squirrel examples, simple transfer experiments are sufficient to indicate whether parents are able to recognize their own young, with results being assessed by the more or less immediate acceptance or rejection of alien young of the same age. The fact that the parents were able to make a discrimination only after their offspring had reached a certain age suggests that learning was involved. This may be tested by means of crossfostering experiments, in which young of equal age are reared by foster parents, leaving some individuals with their own parents to serve as controls. The transfers are usually, but not neccessarily, effected at birth (or before hatching). Then, after recognition has developed, the ability of parents to discriminate between their foster young and natural offspring is determined. If they favor their foster young, one may conclude that they learned to recognize them. Crossfostering is also used in other contexts (see below).

To determine whether other categories of kin, usually siblings, learn to detect relatedness through familiarity, separation experiments may be employed. Porter, Wyrick and Pankey (1978) and Porter and Wyrick (1979) tested the ability of weanling pups of the spiny mouse, *Acomys cahirinus*, to recognize their littermate siblings after being treated in various ways. They employed a simple affinitive behavior (huddling with siblings or non-siblings) to determine association preferences and inferred from these the recognition capabilities of the mice. The age and sex of the test subjects were standardized, while their social experiences were varied. In the initial experiment (Porter, Wyrick and Pankey, 1978) two pairs of siblings from different litters were caged together for five days, isolated for two days, then caged together again for a further three days. In nine replications, siblings huddled preferentially both before and after isolation. This was quantified by calculating the mean frequencies of huddling in pairs (dyadic pairing) with siblings or non-siblings, recorded at numerous observation times. In a follow-up experiment designed to show whether post-weaning association with non-siblings would alter their preferences, two pairs of siblings from different litters (nine replications) were separately caged, but one member of each pair was exchanged. After five days with non-littermates, they were isolated for two days and then all four mice were caged together for a further five days. In the controls, the procedure was identical, except that the mice were housed in littermate pairs instead of in non-sibling pairs during the five days prior to isolation. Various statistical comparisons between the huddling preferences in the experimental and control cages after isolation showed that post-weaning experience of non-siblings modified the discriminative response that otherwise favored siblings. For example, the mean frequencies (per cage) of huddling with siblings was significantly greater in the controls compared with the experimental cages, whereas there was no significant difference in huddling with non-siblings.

Porter and Wyrick (1979) examined the effects of isolation on sibling recognition in spiny mice, again using preferential huddling as the behavorial indicator. They found that weanlings could discriminate reliably between siblings and non-siblings after three to five days of isolation, but not after eight days. However, a single re-exposure period of four hours on the fifth day of isolation was sufficient to maintain sibling recognition on day eight. Porter and Wyrick believe that intermittent re-exposure occurs under natural conditions and that prolonged kin recognition is thereby maintained. These authors also investigated the development of recognition behavior in the mice and showed that while nine to eleven day old pups tended to huddle indiscriminately with littermates and non-littermates alike, a distinct preference for littermates developed by the time they were 14 to 15 days old. This suggested that sibling recognition in spiny mice probably results from learning during early development.

This hypothesis was tested by Porter, Tepper and White (1981) by means of a crossfostering experiment. They exchanged single pups of the same age between two litters, each consisting of two pups, either on the day of birth (day one) or on post-natal day ten or day 20 with ten replicates of each treatment. Tests for sibling recognition (begun on post-natal days 28–34) were conducted by placing the four animals of crossfostered pairs in a single cage and recording their huddling preferences during the ensuing five days. Recognition was defined as a greater relative frequency of dyadic pairing involving specific animals than would be expected by chance alone. Such pairing occurred significantly more often between unrelated littermates reared together than between siblings reared apart, regardless of the age at which they were crossfostered. This result supports the hypothesis that recognition develops in the mice as a result of familiarity, and is, in fact, readily modifiable by experience after the age at which it normally develops.

Testing for recognition by phenotype matching

Phenotype matching as used here refers only to a mechanism in which other individuals act as referents. The standard way of testing for such a mechanism is crossfostering, but with the difference that some of the experimental animals are denied post-natal experience of one another before their recognition capabilities are assessed. This is easily accomplished with some animal taxa such as birds, since eggs can be separated before hatching, but is more difficult with mammals. An elegant study that will serve to illustrate the procedure is that of Holmes and Sherman (1982) with the arctic ground squirrel, *Spermophilus parryii*, and Belding's ground squirrel.

Litters vary in size and composition with regard to the sexes, and littermates may be full-siblings or half-siblings as a result of polyandry in the case of *S. beldingi* (unknown for *S. parryii*), so that the term siblings means uterine siblings. Pups were toe-clipped for identification and crossfostered within three hours of birth. Four combinations of infant pairs were created:
1. Siblings reared together by their mother or a foster dam.
2. Siblings reared apart, one by its mother and the other by a foster dam.
3. Non-siblings reared together by the mother of one of them.
4. Non-siblings reared apart, one by its mother and the other by a foster dam.

Tests for recognition were conducted when *S. parryii* were about 51 days old and when *S. beldingi* were about eight months old.

Tests were conducted in an arena between pairs of conspecific animals, each test lasting five minutes. Since in the field females defend territories against conspecific intruders, agonistic (uncooperative) interactions were employed to assess recognition. These included 'threat' vocalization, withdraw, open mouth threat, paw swipe, lateral presentation of body, lunge strike, chase, squeak/squeal vocalization, bite, belly up, and fight, details of which are given by Holmes and Sherman. Comparisons were made in terms of the total number of interactions per five-minute test. They showed that siblings reared together were significantly less agonistic than those reared apart, and that there was no difference in agonism between siblings and non-siblings reared together. These results are consistent with recognition through familiarity, as was found in spiny mice. However, a further comparison showed that siblings reared apart were significantly less agonistic than non-siblings reared apart, and since most of the test animals in the former group had been reared with one or more siblings with whom they had become familiar, recognition by phenotype matching is also implied. Further analysis revealed that only sisters reared apart were significantly less agonistic than unrelated females reared apart, which accords with field observations that only females are territorial. The data on *S. beldingi* yielded similar results.

In their study of kin recognition in the sweat bee *Lasioglossum zephyrum*, Buckle and Greenberg (1981) also found good evidence of recognition by phenotype matching among unfamiliar sibling bees using a similar experimental design to that described above. It is of particular interest that in this case also the data show clear evidence of recognition by familiarity in that guard bees learned to recognize both sibling and non-sibling nestmates.

As might be expected the learning processes involved in recognition by familiarity and by phenotype matching are either the same or closely similar, but what is learned is somewhat different, namely, different kinds of phenotypic cues (see cue classification).

Testing for recognition without previous experience

The possibility that kin recognition may occur without the discriminating individual having had any prior experience of related conspecifics has been addressed experimentally by several investigators. The recurrent difficulty in all such studies is how to obtain naïve test subjects.

Blaustein and O'Hara (1981) collected egg clutches of the cascades frog, *Rana cascadae*, from the field. They reared a number of tadpoles in isolation in separate containers for comparison with others reared as groups in common containers. Rearing conditions (photoperiod, temperature and food supply) were uniform, but the sex of test individuals was unknown and was therefore random. The test procedure made use of an affinitive behavior to assess the recognition capabilities of the naïve tadpoles (for further details of which see Chapter 9). The results of their experiments were strongly suggestive of a recognition mechanism that develops without postnatal experience, but the possibility remained (later discounted, Blaustein and O'Hara, 1982a) that recognition might be influenced by a pre-natal maternal factor associated with the jelly surrounding the eggs.

The difficulty imposed by common pre-natal experience on tests for recognition alleles applies equally to other animal species, but it may be bypassed by using paternal half-siblings, since sperm may be considered as carriers of genetic information only. The well-known investigation of Wu *et al.* 1980) with pigtail macaques is an example.

Infant macaques of known genealogy were taken from their mothers within five minutes of birth and were reared without experiencing relatives, but were given daily access to peer play groups of non-relatives to ensure 'normal' social development. To test their preferences among relatives and non-relatives, they were matched by sex, age, weight, general activity level and appearance (subjectively assessed) with stimulus individuals reared identically to themselves. When given a simultaneous choice between entering a cage with a paternal half-sibling, an unrelated conspecific and an empty cage, 13 of 16 spent significantly more time with their unfamiliar half-sibling than with the unrelated individual, which result supported the hypothesis of recognition without post-natal experience or common pre-natal experience.

Blaustein and O'Hara (1982a; Chapter 9) also used half-siblings in further experiments with tadpoles of the cascades frog. To obtain both maternal (m) and paternal (p) half-siblings they stripped ripe ova from females and removed testes from males so that eggs could be fertilized with a sperm suspension in petri dishes. Their subsequent tests showed that tadpoles preferred to associate with full-siblings over either m or p half-siblings, but that they preferred both m and p half-siblings to non-siblings.

Kareem and Barnard (1982) made use of paternal half-siblings to determine whether they could detect in mice (*Mus musculus*) any kin recognition capability not attributable to the effects of familiarity. They felt, however, that one should employ an animal's total repertoire of social interactions as an index of kin recognition instead of only a generalized tendency to approach unfamiliar relatives, as did Wu *et al.* and Blaustein and O'Hara. Accordingly, they used both cooperative and uncooperative behaviors, the most useful of which were:

1. Sniff: olfactory exploration of the anogenital region.
2. Investigate: olfactory exploration of other body areas.
3. Aggression: a physical struggle between two animals, including wrestling, rolling, and biting.
4. Allo-grooming: one animal grooms the other by mandibulating or licking.
5. Touch: apparently undirected body contact that could not be classified as any of 1–4.

Pairs of mice were tested in a cage for five-minute periods and the frequencies of the various behavior patterns were recorded.

Among adult males, unfamiliar (naïve) paternal half-siblings showed significantly less investigate (2) and allo-groom (4) and significantly more touch (5) than unfamiliar non-siblings, which is of particular interest since investigate and allo-groom are preludes to fighting in mice, while touch is an indicator of spacing at times other than during specific interactions. Touch and allo-groom scores of the half-siblings were comparable to those of familiar full-siblings, but their frequencies of investigate and sniff were significantly higher. Aggression occurred only between adult non-siblings, but occurred too infrequently to be significant. This bears out the advantage of having recorded investigate and allo-groom, which are more sensitive measures of agonism. No aggression was observed in adult females, which was to be expected, as it is primarily males that defend the family territories.

These results, together with those of an additional experiment, led Kareem and Barnard to conclude that a kin recognition mechanism that does not depend on prior experience probably does occur in mice, but that 'prior familiarity completely overrides differences in the degree of relatedness in deciding patterns of interaction.' In addition to this significant result, however, the discovery that kin discrimination is not evident in all types of interactions is of particular interest; it shows that it is useful to employ several indices of recognition whenever possible.

Cue complexity

As we have seen in the previous section, analysis of the aspect of recognition mechanisms that concerns template formation may be accomplished in some cases with little or no information about the cues involved. The focus now shifts to the identification of cues and their sources.

Given the potentially high costs of misidentifying kin in terms of lowered reproductive success through the misdirection or usurpation of resources, or the suboptimal choice of a mate, it would not be surprising to find that many recognition cues are relatively (or extremely) complex.

Hypothetically, a simple recognition cue might consist of a signal in one dimension of a single sensory mode, e.g. a chemical signal (pheromone) consisting of only one compound, or an auditory signal consisting of a monotone. Even so, its potential complexity and corresponding informational content are appreciable—the concentration of a chemical could be varied, as could the amplitude of a tone, and both kinds of signals could be emitted in pulses in a regular or patterned manner.

A considerable increase in the complexity of a cue could theoretically result if even only one more dimension were added to it in the same sensory mode, e.g. a second compound to the pheromonal signal or another frequency to the monotone, yet in nature the signals are often much more complex. Thus, recognition pheromones may be multicomponent and sounds may contain several frequencies. The possible variations in composition of a chemical mixture are virtually infinite, so that the practical limits to the discriminatory power of a pheromonal signaling system are more likely to be set by the perceptual mechanism of the receiving animal than by the complexity of the signal that can be generated by the cue bearer. Similar considerations apply to an auditory signaling system, such as birdsong with its range of frequencies and potentially infinitely variable patterning, and they apply equally to visual systems.

While some kin recognition cues appear to function in one sensory mode only, e.g. nestmate recognition in many social insects appears to be mediated purely chemically (Hölldobler and Michener, 1980), others involve more than one of the senses (see Chapter 7). The different signals may come into play either approximately sequentially as in *Drosophila* mate selection (Chapter 5) or simultaneously, so that the amount of information they contain is indeed impressive. This order of complexity was vividly conveyed by Marler (1965).

> 'Perhaps the most striking generalization that can be advanced from [a] survey of the communication signals of monkeys and apes is the overwhelming importance of composite signals. In most situations it is not a single signal that passes from one animal to another but a whole complex of them, visual, auditory, tactile and sometimes olfactory. There can be little doubt that the structure of individual signals is very much affected by this incorporation in a whole matrix of other signals... these composite systems are a special feature of close-range communication, transmitting information between different members of the group.'

Although Marler was not discussing kin recognition *per se*, the potential for individual recognition of group members among non-human primates by means of composite signals is clearly evident (see also Chapter 11). It is also evident that the probability of a recognition signal being subverted is inversely related to its complexity, so that we might reasonably expect complex signaling systems to have evolved by natural selection among most animals.

Cue classification

Hölldobler and Michener (1980) and Michener (1982) distinguished two classes of chemical cues in their discussions of identification and discrimination of kin in the social Hymenoptera (see also Chapter 7). They said that odor signals that differ among individuals of a population, when not of extrinsic origin, can be called 'discriminators' or, citing Ehrman and Probber (1978), 'recognition pheromones', and that the expression 'discriminating substances' may be used when it is uncertain whether identification of an individual is based on discriminators or on environmental materials. The fundamental idea of this classification, the division of cues according to their sources, is retained here, but it is expanded in order to be applicable to cues in other sensory modes and to include cues that identify classes as well as individuals. Note that there is the potential for confusion between the terms discriminating individual and discriminator; the former is an animal that discriminates among cues and the latter is a particular kind of cue.

Broadly categorized, recognition cues are of either intrinsic or extrinsic origin with respect to the cue bearer.

An 'intrinsic cue' is a phenotypic expression of a cue bearer's genotype such that differences in a particular character, or characters, among conspecific animals reflect genotypic differences between them and these differences are used by discriminating individuals in the assessment of relatedness.

An 'extrinsic cue' is a label acquired from the environment that serves to identify the bearer to a discriminating individual as kin on a probabilistic basis.

Cues of intrinsic origin may be classified according to the manner in which they are utilized by discriminating individuals. There appear to be three possible kinds.
1. Cues of which no previous experience is required for a reliable assessement of relatedness, e.g. visual cues of pigtail macaques (Wu *et al.*, 1980).
2. Cues that provide a basis for individual recognition through familiarity, e.g. sounds made by young bank swallows in parent/offspring recognition (Beecher, Beecher and Hahn, 1981) and facial features among humans (Chapter 12).
3. Cues that provide a basis for class recognition via phenotype matching, e.g. recognition pheromones of *Lasioglossum zephyrum* (Greenberg, 1979).

Extrinsic cues may be classified according to their particular sources and appear to be of four possible kinds.

1. Maternal labels acquired pre-natally by recognized individuals, e.g. during embryonic development of American toad tadpoles (Waldman, 1981).
2. Labels acquired post-natally from mothers or other close relatives by recognized individuals, e.g. from the mother in goats (Gubernick, 1981) and from both parents and siblings in desert isopods (Chapter 6).
3. Environmental cues that serve indirect recognition by location, e.g. a nest or a territory-marking pheromone. Such cues are similar to those of the previous category, but are distinguished on the grounds that they do not label the recognized individual *per se*.
4. Cues acquired by recognized individuals from environmental sources other than conspecifics, e.g. colony specific odors dependent upon diet in honey bees (Ribbands, 1965). These may be called 'environmental accidentals'.

It is of interest that while maternal and other transferred cues are extrinsic with respect to the cue bearer, they are likely to be intrinsic to the donor, who is the discriminating individual, i.e. they are extended phenotypes of the discriminating individual. The probabilistic cues of category 3 are also extended phenotypes. The implications of this will be discussed later.

Identification of cues and their sources

The identification of recognition cues employs many well-known techniques developed during analysis of stimulus/response relationships and animal communication generally, so that it is not necessary to give much detail here, but a number of points may be made by discussion of examples. Of greatest theoretical interest are intrinsic cues that may possibly function without the discriminating individuals having had any previous experience of them.

In their study of pigtail macaques Wu *et al.* (1980) found that in 14 of 16 stimulus pairs the individuals that elicited the greater orientation time by test animals also elicited the greater amount of entry time into their compartments when these were opened. This suggested that visual cues probably played a role in establishing preferences among subjects before they were permitted to enter the compartments. However, this evidence is circumstantial only, as the experiment was not designed primarily to investigate the nature of cues and so lacked appropriate controls. It seems probable that under natural conditions non-human primates learn to discriminate among conspecifics; as Wu *et al.* point out, 'pigtails do prefer a familiar conspecific to an unfamiliar one.'

Blaustein and O'Hara (1982b) established that waterborne chemosensory cues only are used by tadpoles of the cascades frog (reared in sibling groups) to discriminate between siblings and non-siblings, from which it may be inferred that the same cues were probably used by naïve tadpoles to associate preferentially with kin in earlier experiments (Blaustein and O'Hara, 1981, 1982a). Since visual stimuli were known to be effective in causing aggregation among tadpoles, tests for visual cues were also conducted, but chemical cues alone were as effective as visual and chemical stimuli together.

It is well established that in mice odor cues mediate individual recognition under laboratory conditions. Bowers and Alexander (1967) used positive reinforcement of water-deprived mice in a Y-maze with a droplet of water as the reward to determine that both males and females can be trained to distinguish between two males of the same strain as themselves without having had prior experience of either (non-littermates). Using a similar Y-maze technique Yamazaki et al. (1979) showed that both males and females can discriminate between the odors of H-2 congenic mice (H-2 designates the major histocompatibility complex in mice), and subsequently Yamaguchi et al. (1981) demonstrated that the distinctive odors are contained in the urine. More recently, Yamazaki et al. (1983) found that inbred mice differing genetically only with regard to a single mutation (of the H-2K gene) produce urine of different odors, although the authors are careful to point out that mutant phenotypes do not necessarily imply that the odorants are structural derivatives of the H-2K molecule, or that the odorant molecules are structurally different, or that there is only a single difference rather than multiple differences in odor phenotype. Indeed, one might add that part of the odor difference may conceivably be the result of metabolic variations independent of genotype, so that it would be informative to test for such differences in homozygous mice of a single strain fed similar and different diets. Assuming that H-2 odor phenotypes dominate any other odor differences that may occur, another important question concerns how mice might make use of them.

Yamazaki et al. (1976) found that when male mice were offered a choice between oestrus females of the same H-2 type as themselves and congenic females of a different H-2 type, they preferred to mate with the latter. To test whether this sensory distinction had a genetic basis, an $F2$ linkage test was conducted (see review by Beauchamp, Yamazaki and Boyse, 1985), in which a pair of H-2 congenic strains were crossed to produce $F1$ heterozygotes. These were then bred to give an $F2$ containing homozygotes of each strain that were exposed both pre- and post-natally to an H-2 type different from their own. If these $F2$ homozygotes were unaffected by such experience, their mating preferences would be no different from those of homozygous mice of each strain that had never had experience of any H-2 type other than their own. When tested, their preferences were not, in fact, the same in all cases. From this it was tentatively concluded that the preferences were the result, at least in part, of familial chemosensory imprinting. Hence, there is no hard evidence that mice can use H-2 odor phenotypes as cues without prior experience of them, although in view of the evidence favoring recognition by familiarity in mice (see also Gilder and Slater, 1978) this would seem to be unnecessary under natural conditions. Nevertheless, it would be of interest to perform a complementary experiment to the above, incorporating crossfostering of the $F2$ homozygotes at birth with unrelated mice, in order at least to separate pre-natal from post-natal experience.

The other two kinds of intrinsic recognition cues and the several kinds of extrinsic cues listed earlier all require that discriminating individuals have prior experience of them before they can function in kin recognition. The life histories and social organization of most animal species offer only limited opportunities for young animals to learn, without ambiguity, the intrinsic cues of their genetic relatives. This suggests that a large majority of cues are likely to be learned early in life, often through an imprinting-like process, a deduction which appears to be supported by much of the available empirical data. Clearly, learned cues must not only provide information on the degrees of relatedness between the cue bearer and the discriminating individual, but in most cases they must also be stable through time, especially where they function in mate selection later in life and in recognition by phenotype matching. Only cues of the animal's own phenotype or that of associates are likely to have these properties. These cues include six of the seven categories listed earlier, leaving only environmental accidentals (cues acquired from environmental sources other than conspecifics) out on a limb. This invites comparisons of the ways in which recognition cues are used.

Environmental accidentals may plausibly function only in closely-knit animal groups, such as social insect colonies and primate groups, in which there are frequent opportunities for monitoring constantly changing cues. In that case they would serve to identify kin only in family groups.

It seems probable that environmental accidentals are mainly or exclusively chemosensory in nature; an example is the colony specific odor of honey bees. When Köhler (1955) fed three colonies on the same scented syrup, the bees entered each others' hives freely and removed food, but when he switched to feeding them differently scented syrups, guard bees began to recognize and attack the robber bees within a few hours. Presumably the colonies had acquired distinctive odors from the different scent sources. Butler (1967) also suggested that nest odors derived from floral sources may be adsorbed onto the bees' cuticle thereby contributing to individual colony odors. Environmental accidentals almost certainly act only in a supplementary capacity to intrinsic cues, with the result that two templates would exist simultaneously in discriminating individuals (Chapter 8).

Imprinting is not the only way in which intrinsic cues may be learned; at least some, like environmental accidentals, can be forgotten (or ignored) and replaced by different cues. This applies, for example, to queen recognition by workers of the fire ant, *Solenopsis invicta*. D. Fletcher and D. Cherix (unpublished) divided 60 colonies into queenright (with queen) and queenless halves, and after three days introduced an unrelated queen into each queenless half. Colonies were fed identically. After zero, two, four, eight and twelve days both queens of a pair (twelve pairs per day) were removed and immediately reintroduced to the half that had had the unrelated queen to determine which one the workers would reject. On day zero they unanimously killed the unrelated queens, but from days two to twelve an increasing proportion of half-colonies killed their own mother queen (Fig. 1). This suggests that while the workers can forget queen recognition cues, which are probably pheromones, and learn new ones, such cues are not as easily forgotten as environmental accidentals. In honey bees, in which daughter queens regularly replace their mother after swarming, the adaptiveness of this kind of learning is evident, but its advantage to fire ant colonies, if there is one, is obscure, since as far as is known, queens are seldom replaced (but see Tschinkel and Howard, 1978).

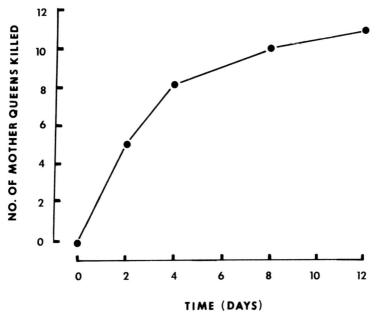

Fig. 1. *Solenopsis invicta*. Numbers of colonies in five samples of twelve each, in which the workers killed their mother queen after living with an alien conspecific queen for zero to twelve days

The general procedure for the identification of recognition cues has a number of similarities to analyses of kin-biased behavior, mating preferences, and recognition mechanisms. Assays depend upon the same kinds of cooperative and uncooperative behaviors and the same variables of age, sex, physiological state, and rearing conditions have to be taken into account. The most important additional consideration is that several cues may function simultaneously or sequentially or in combinations of these patterns. Analyses of cues involved in mate selection in *Drosophila* provide a good example (Chapter 5). Different cues may also be used at different times. For example, parent midas cichlid fishes and their fry recognize each other visually in clear water, but in turbid water and at night they employ chemosensory cues (Barnett, 1982). It is not known, however, whether both stimuli operate simultaneously under favorable conditions or whether additional stimuli are involved.

CONCLUSIONS

This review and discussion of the methods currently employed in the behavioral analysis of kin recognition has not included methods of testing the predictions of various genetic models such as those of Crozier and Dix (1979) and Getz (1982), as these are treated by other contributors to this volume. Moreover, the testing of hypotheses under field conditions has not been given particular prominence, because procedures depend to a large extent on the particulars of the behavioral ecology of the different animal species and these too are more appropriately dealt with in other chapters. Nevertheless, valid generalizations can only be made in the context of the natural circumstances under which kin recognition occurs.

In both laboratory and field experiments kin recognition is most often defined as the differential responses of individuals to conspecifics according to the degree of relatedness between them, i.e. recognition is inferred from the nature of behavioral interactions, rather than by obtaining physiological data. Diverse behavioral interactions are employed, but these are conveniently divisible into cooperative and uncooperative behaviors, with the most cooperative behavior usually being displayed by close relatives and the least cooperative between non-relatives. Exceptions include cooperative behavior between mates and, in some species, antagonistic interactions among the relatives in the context of parental manipulation, sibling rivalry and the formation and maintenance of dominance hierarchies. Test subjects are frequently given a simultaneous choice of two or more stimulus animals of varying relatedness with which to interact and data are usually recorded in terms of interactions between pairs of animals. These techniques are particularly useful in the analysis of kin-biased behavior, mating preferences, and the perceptual mechanisms of kin recognition. The analysis of cues, on the other hand, especially acoustical and chemosensory signals, is advancing mainly by analyzing the responses of test subjects to experimentally isolated signals. Tape recordings of animal sounds are particularly applicable to field situations, whereas chemosensory signals are best studied in the laboratory, where their release can be controlled and where stimuli in other sensory modes from living stimulus animals may be effectively blocked.

In the field, manipulations of the relatedness of experimental subjects is possible in some cases, e.g. through crossfostering and, in honey bees, through the introduction into colonies of instrumentally inseminated queens, but complete control over relatedness is generally much more easily achieved in the laboratory, especially through inbreeding and crossing according to a predetermined protocol. These procedures afford a sophisticated level of control over numerous variables, and valuable insights are gained, but the unavoidable gap that is created between laboratory and field conditions is difficult, if not impossible, to bridge. Consequently, the best hope of maintaining the current rapid rate of progress in the experimental analysis of kin recognition lies in building a large body of comparative data in both laboratory and field for diverse animal species. The comparative data already available encourage one to hazard certain generalizations.

The results of detailed investigations into *H-2* related odor differences in mice and the apparent functioning of these odors as recognition cues, as well as the discriminatory capabilities of naïve cascades frog tadpoles, are suggestive of the existence of recognition alleles, but one cannot fail to be impressed by the recurring evidence that evinces a dominant role for recognition by familiarity. It thus seems unlikely that a system dependent exclusively on recognition alleles will be found. But why, one may ask, would probabilistic mechanisms have evolved universally when they are vulnerable to subversion? Perhaps 'hard-wired' mechanisms are unnecessary, since for the most part probabilistic mechanisms work well. On the other hand, the classification of recognition cues and the available comparative data suggest another reason —the value of retaining a degree of flexibility.

Almost all cues employed in recognition systems appear to be of phenotypic origin, although many characters that permit individual recognition may not have evolved specifically under kin selection or mating preference. Often, phenotypic cues are part of the bearer's phenotype (intrinsic cues), but some originate as extended phenotypes of the discriminating individuals that are either transferred to the bearer or are associated with the bearer spatially. Thus in most recognition systems the cues are highly reliable, because they are 'created by' and are private to their users. Flexibility lies, for the most part in the perceptual mechanisms.

When relatives live together in groups, or encounter each other frequently over prolonged periods (e.g. spiny mice and primates), they can constantly renew their recognition templates or learn to recognize new offspring and other relatives. On the other hand, when relatives disperse but may encounter one another later in life as potential mates, as commonly occurs, inflexible recognition systems are advantageous and templates are likely to be imprinted early in life. Many animals may well have evolved both flexible and inflexible systems, serving kin selection on the one hand and mating preferences on the other. This hypothesis is capable of expansion and is easily testable.

ACKNOWLEDGMENTS

I thank C. D. Michener, R. W. Matthews, K. G. Ross and P. C. B. Fletcher for reading the manuscript and making many helpful suggestions.

LITERATURE CITED

Alexander, R. D. 1974. The evolution of social behavior. *Annual Review of Ecology and Systematics*, **5**, 325–383.
Alexander, R. D. 1979. *Darwinism and human affairs*. University of Washington Press, Seattle.
Alexander, R. D. and G. Borgia. 1978. Group selection, altruism, and the levels of organization of life. *Annual Review of Ecology and Systematics*, **9**, 449–474.
Barnett, C. 1982. The chemosensory responses of young cichlid fish to parents and predators. *Animal Behaviour*, **30**, 35–42.
Barrows, E. M. 1975. Individually distinctive odors in an invertebrate. *Behavioral Biology*, **15**, 57–64.
Barrows, E. M. 1976. Mating behavior in halictine bees (Hymenoptera: Halictidae): Patrolling and age-specific behavior in males. *Journal of the Kansas Entomological Society*, **49**, 105–119.
Barrows, E. M., W. J. Bell and C. D. Michener. 1975. Individual odor differences and their social functions in insects. *Proceedings of the National Academy of Sciences, USA*, **72**, 2824–2828.
Bateson, P. 1978. Sexual imprinting and optimal outbreeding. *Nature*, **273**, 659–660.
Bateson, P. 1982. Preferences for cousins in Japanese quail. *Nature*, **295**, 236–237.
Bateson, P. 1983. Optimal outbreeding, in Bateson, P. ed., *Mate Choice*. Cambridge University Press, Cambridge, England, and New York, pp. 257–277.
Beauchamp, G. K., K. Yamazaki and E. A. Boyse. 1985. The chemosensory recognition of genetic individuality. *Scientific American*, **253**, 86–92.
Beecher, M. D. 1982. Signature systems and kin recognition. *American Zoologist*, **22**, 477–490.
Beecher, M. D., I. M. Beecher and S. Hahn. 1981. Parent-offspring recognition in bank swallows (*Riparia riparia*): development and acoustic basis. *Animal Behaviour*, **29**, 95–101.
Beecher, M. D., I. M. Beecher and S. Lumpkin. 1981. Parent-offspring recognition in the bank swallow (*Riparia riparia*): I. Natural history. *Animal Behaviour*, **29**, 86–94.
Beer, C. G. 1970. Individual recognition of voice in the social behaviour of birds, in Lehrman, D. S., R. A. Hinde and E. Shaw, eds., *Advances in the Study of Behaviour*, Academic Press, New York, Vol. 3, pp. 27–74.
Bekoff, M. 1981. Mammalian sibling interactions: genes, facilitative environments, and the coefficient of familiarity, in Gubernick, D., and P. N. Klopfer, eds., *Parental care in mammals*. Plenum Press, New York, pp. 307–346.
Blaustein, A. R. 1983. Kin recognition mechanisms: phenotype matching or recognition alleles? *American Naturalist*. **121**, 749–754.
Blaustein, A. R. and R. K. O'Hara. 1981. Genetic control for sibling recognition? *Nature*, **290**, 246–248.
Blaustein, A. R. and R. K. O'Hara. 1982a. Kin recognition in *Rana cascadae* tadpoles: maternal and paternal effects. *Animal Behaviour*, **30**, 1151–1157.

Blaustein, A. R. and R. K. O'Hara. 1982b. Kin recognition in *Rana cascadae* tadpoles. *Behavioral and Neural Biology*, **36**, 77–87.

Bornais, K. M., C. M. Larch, G. J. Gamboa and R. B. Daily. 1983. Nestmate discrimination among laboratory overwintered foundresses of the paper wasp, *Polistes fuscatus* (Hymenoptera: Vespidae). *The Canadian Entomologist*, **115**, 655–658.

Bowers, J. M. and B. K. Alexander. 1967. Mice: individual recognition by olfactory cues. *Science*, **158**, 1208–1210.

Breed, M. D. 1983. Nestmate recognition in honey bees. *Animal Behaviour*, **31**, 86–91.

Buckle, G. R. and L. Greenberg. 1981. Nestmate recognition in sweat bees (*Lasioglossum zephyrum*): does an individual recognize its own odour or only odours of its nestmates? *Animal Behaviour*, **29**, 802–809.

Burtt, E. H. 1977. Some factors in the timing of parent-chick recognition in swallows. *Animal Behaviour*, **25**, 231–239.

Butler, C. G. 1967. *The world of the honey bee*. Collins, London.

Butler, C. G., D. J. C. Fletcher and D. Watler. 1970. Hive entrance finding by honey bee (*Apis mellifera*) foragers. *Animal Behaviour*, **18**, 78–91.

Caldwell, R. L. 1985. A test of individual recognition in the stomatopod *Gonodactylus festae*. *Animal Behaviour*, **33**, 101–106.

Cheney, D. L. and R. M. Seyfarth. 1980. Vocal recognition in free-ranging vervet monkeys. *Animal Behaviour*, **28**, 362–367.

Crozier, R. H. and M. W. Dix. 1979. Analysis of two genetic models for the innate components of colony odor in social insects. *Behavioral Ecology and Sociobiology*, **4**, 217–224.

Dawkins, R. 1976. *The selfish gene*. Oxford University Press, Oxford, England.

Dawkins, R. 1982. *The extended phenotype*. Oxford University Press, Oxford, England.

Dewsbury, D. A. 1982. Avoidance of incestuous breeding between siblings in two species of *Peromyscus* mice. *Biology of Behavior*, **7**, 157–169.

Ehrman, L. and J. Probber. 1978. Rare *Drosophila* males: the mysterious matter of choice. *American Scientist*, **66**, 216–222.

Falls, J. B. 1982. Individual recognition by sound in birds, in Kroodsma, D. E. and E. H. Miller, eds., *Acoustic communication in birds, Vol. II: song learning and its consequences*. Academic Press, New York, pp. 237–273.

Fisher, R. A. 1930. *The genetical theory of natural selection*. Clarendon Press, Oxford, England.

Fletcher, D. J. C. 1978. The African bee, *Apis mellifera adansonii*, in Africa. *Annual Review of Entomology*, **23**, 151–171.

Fletcher, D. J. C. and M. S. Blum. 1981. Pheromonal control of dealation and oogenesis in virgin queen fire ants. *Science*, **212**, 73–75.

Fletcher, D. J. C. and M. S. Blum. 1983. Regulation of queen number by workers in colonies of social insects. *Science*, **219**, 312–314.

Fletcher, D. J. C. and K. G. Ross. 1985. Regulation of reproduction in eusocial Hymenoptera. *Annual Review of Entomology*, **30**, 319–343.

Franks, N. R. and E. Scovell. 1983. Dominance and reproductive success among slave-making worker ants. *Nature*, **304**, 724–725.

Getz, W. M. 1982. An analysis of learned kin recognition in Hymenoptera. *Journal of Theoretical Biology*, **99**, 585–587.

Getz, W. M., D. Brückner and T. R. Parisian. 1982. Kin structure and the swarming behavior of the honey bee *Apis mellifera*. *Behavioral Ecology and Sociobiology*, **10**, 265–270.

Gilder, P. M. and P. J. B. Slater. 1978. Interest of mice in conspecific male odours is influenced by degree of kinship. *Nature*, **274**, 364–365.

Greenberg, L. 1979. Genetic component of bee odor in kin recognition. *Science*, **106**, 1095–1097.

Greenberg, L. 1982. Persistent habituation to female odor by male sweat bees (*Lasioglossum zephyrum*) (Hymenoptera: Halictidae). *Journal of the Kansas Entomological Society*, **55**, 525–531.

Gubernick, D. J. 1981. Mechanisms of maternal 'labelling' in goats. *Animal Behaviour*, **29**, 305–306.

Halpin, Z. T. 1980. Individual odors and individual recognition: review and commentary. *Biology of Behaviour*, **5**, 233–248.

Hamilton, W. D. 1964. The genetical evolution of social behavior. I, II. *Journal of Theoretical Biology*, **7**, 1–52.

Hanken, J. and P. W. Sherman. 1981. Multiple paternity in Belding's ground squirrel litters. *Science*, **212**, 351–353.

Hinde, R. A. 1968. *Animal behaviour: a synthesis of ethology and comparative psychology*. McGraw Hill, New York.

Hölldobler, B. and C. D. Michener. 1980. Mechanisms of identification and discrimination in social Hymenoptera, in Markl, H. ed., *Evolution of social behavior: hypotheses and empirical tests*. Dahlem Konferenzen, Verlag Chemie, Weinheim and Deerfield Beach, Florida, pp. 35–58.

Hölldobler, B. and E. O. Wilson. 1977. Colony-specific territorial pheromone in the African weaver ant *Oecophylla longinoda* (Latreille). *Proceedings of the National Academy of Sciences, USA*, **74**, 2072–2075.

Holmes, W. G. and P. W. Sherman. 1982. The ontogeny of kin recognition in two species of ground squirrels. *American Zoologist*, **22**, 491–517.

Holmes, W. G. and P. W. Sherman. 1983. Kin recognition in animals. *American Scientist*, **71**, 46–55.

Kaplan, J. N., A. Winship-Ball and L. Sim. 1978. Maternal discrimination of infant vocalizations in squirrel monkeys. *Primates*, **19**, 187–193.

Kareem, A. M. and C. J. Barnard. 1982. The importance of kinship and familiarity in social interactions between mice. *Animal Behaviour*, **30**, 594–601.

Klahn, J. E. and G. J. Gamboa. 1983. Social wasps: discrimination between kin and nonkin brood. *Science*, **221**, 482–484.

Köhler, F. 1955. Wache und Volksduft im Bienenstaat. *Zeitschrift für Bienenforschung*, **3**, 57–63.

Kurland, J. A. 1977. *Kin selection in the Japanese monkey*. Karger, Basel, New York.

Lack, D. 1954. *The natural regulation of animal numbers*. Oxford University Press, Oxford, England.

Lack, D. 1966. *Population studies of birds*. Clarendon Press, Oxford, England.

Lack, D. 1968. *Ecological adaptations for breeding in birds*. Methuen, London.

Lacy, R. C. and P. W. Sherman. 1983. Kin recognition by phenotype matching. *American Naturalist*, **121**, 489–512.

Laidlaw, H. H. and R. E. Page. 1984. Polyandry in honey bees (*Apis mellifera* L.): sperm utilization and intracolony genetic relationships. *Genetics*, **108**, 985–997.

Larch, C. M. and G. J. Gamboa. 1981. Investigation of mating preference for nestmates in the paper wasp *Polistes fuscatus* (Hymenoptera: Vespidae). *Journal of the Kansas Entomological Society*, **54**, 811–814.

Lehrman, D. S. 1970. Semantic and conceptual issues in the nature–nurture problem, in Aronson, L. R., E. Tobach, D. S. Lehrman and J. S. Rosenblatt, eds., *Development and evolution of behavior: essays in memory of T. C. Schneirla*. W. H. Freeman, San Francisco, pp. 17–52.

Linsenmair, K. E. 1972. Die Bedeutung familienspezifischer 'Abzeichen' für den Familienzusammenhalt bei der sozialen Wüstenassel *Hemilepistus reaumuri* Audouin u. Savigny (Crustacea, Isopoda, Oniscoidea). *Zeitschrift für Tierpsychologie*, 31, 131–162.

Linsenmair, K. E. and C. Linsenmair. 1971. Paarbildung und Paarzusammenhalt bei der monogamen Wüstenassel *Hemilepistus reaumuri* (Crustacea, Isopoda, Oniscoidea). *Zeitschrift für Tierpsychologie*, 29, 134–155.

Marler, P. 1965. Communication in monkeys and apes, in Devore, I. ed., *Primate behavior: field studies of monkeys and apes*. Holt, Rinehart, and Winston, New York, pp. 544–584.

Massey, A. 1977. Agonistic aids and kinship in a group of pigtail macaques. *Behavioral Ecology and Sociobiology*, 2, 31–40.

McArthur, P. D. 1982. Mechanisms and development of parent-young vocal recognition in the pinon jay (*Gymnorhinus cyanocephalus*). *Animal Behaviour*, 30, 62–74.

Michener, C. D. 1974. *The social behavior of the bees*. Harvard University Press, Cambridge, Massachusettes.

Michener, C. D. 1982. Early stages in insect social evolution: individual and family odor differences and their functions. *Bulletin of the Entomological Society of America*, 28, 7–11.

Miller, D. E. and J. T. Emlen. 1975. Individual chick recognition and family integrity in the ring-billed gull. *Behaviour*, 52, 124–144.

Myrberg, A. A. and R. J. Riggio. 1985. Acoustically mediated individual recognition by a coral reef fish (*Pomacentrus partitus*). *Animal Behaviour*, 33, 411–416.

Page, R. E. and E. H. Erickson. 1984. Selective rearing of queens by worker honey bees: kin or nestmate recognition? *Annals of the Entomological Society of America*, 77, 578–580.

Petrinovich, L. 1974. Individual recognition of pup vocalization by northern elephant seal mothers. *Zeitschrift für Tierpsychologie*, 34, 308–312.

Pfennig, D. W., H. K. Reeve and J. S. Shellman. 1983. Learned component of nestmate discrimination in workers of a social wasp, *Polistes fuscatus* (Hymenoptera: Vespidae). *Animal Behaviour*, 31, 412–416.

Poindron, P. and M. J. Carrick. 1976. Hearing recognition of the lamb by its mother. *Animal Behaviour*, 24, 600–602.

Pollock, G. B. 1983. Population viscosity and kin selection. *American Naturalist*, 122, 817–829.

Popper, K. R. 1972. *Objective Knowledge: an evolutionary approach*. Clarendon Press, Oxford, England.

Porter, R. H. and M. Wyrick. 1979. Sibling recognition in spiny mice (*Acomys cahirinus*): influence of age and isolation. *Animal Behaviour*, 27, 761–766.

Porter, R. H., V. J. Tepper and D. M. White. 1981. Experiential influences on the development of huddling preferences and 'sibling' recognition in spiny mice. *Developmental Psychobiology*, 14, 375–382.

Porter, R. H., M. Wyrick and J. Pankey. 1978. Sibling recognition in spiny mice (*Acomys cahirinus*). *Behavioral Ecology and Sociobiology*, 3, 61–68.

Post, D. C. and R. L. Jeanne. 1983. Relatedness and mate selection in *Polistes fuscatus* (Hymenoptera: Vespidae). *Animal Behaviour*, 31, 1260–1261.

Ribbands, C. R. 1965. The role of recognition of comrades in the defense of social insect communities. *Symposium of the Zoological Society of London*, 14, 159–168.

Ross, K. G. and D. J. C. Fletcher. 1985. Comparative study of genetic and social structure in two forms of the fire ant, *Solenopsis invicta* (Hymenoptera: Formicidae). *Behavioral Ecology and Sociobiology*, 17, 349–356.

Sherman, P. W. 1977. Nepotism and the evolution of alarm calls. *Science*, **197**, 1246–1253.
Sieber, O. J. 1985. Individual recognition of parental calls by bank swallow chicks (*Riparia riparia*). *Animal Behaviour*, **33**, 107–116.
Small, M. F. and D. G. Smith. 1981. Interactions with infants by full siblings, paternal half-siblings, and nonrelatives in a captive group of rhesus macaques (*Macaca mulatta*). *American Journal of Primatology*, **1**, 91–94.
Taber, S. and J. Wendel. 1958. Concerning the number of times queen bees mate. *Journal of Economic Entomology*, **51**, 786–789.
Trivers, R. L. 1971. The evolution of reciprocal altruism. *Quarterly Review of Biology*, **46**, 35–57.
Tschinkel, W. R. and D. F. Howard. 1978. Queen replacement in orphaned colonies of the fire ant, *Solenopsis invicta*. *Behavioral Ecology and Sociobiology*, **3**, 297–310.
Visscher, P. K. 1986. Kinship discrimination in queen rearing by honey bees (*Apis mellifera*). *Behavioral Ecology and Sociobiology*, **18**, 453–460.
Waldman, B. 1981. Sibling recognition in toad tadpoles: the role of experience. *Zeitschrift für Tierpsychologie*, **56**, 341–358.
Waldman, B. 1982. Sibling association among schooling toad tadpoles: field evidence and implications. *Animal Behaviour*, **30**, 700–713.
Waldman, B. and K. Adler. 1979. Toad tadpoles associate preferentially with siblings. *Nature*, **282**, 611–613.
Wilson, E. O. 1971. *The insect societies*. Harvard University Press, Cambridge, Massachusetts.
Wilson, E. O. 1975. *Sociobiology: the new synthesis*. Harvard University Press, Cambridge, Massachusetts.
Wu, H. M. H., W. G. Holmes, S. R. Medina and G. P. Sackett. 1980. Kin preference in infant *Macaca nemestrina*. *Nature*, **285**, 225–227.
Wynne-Edwards, V. C. 1962. *Animal dispersion in relation to social behaviour*. Oliver and Boyd, Edinburgh.
Yamaguchi, M., K. Yamazaki, G. K. Beauchamp, J. Bard, L. Thomas and E. A. Boyse. 1981. Distinctive urinary odors governed by the major histocompatibility locus of the mouse. *Proceedings of the National Academy of Sciences, USA*, **78**, 5817–5820.
Yamazaki, K., G. K. Beauchamp, I. K. Egorov, L. Thomas and E. A. Boyse. 1983. Sensory distinction between H-2^b and H-2^{bm1} mutant mice. *Proceedings of the National Academy of Sciences, USA*, **80**, 5685–5688.
Yamazaki, K., E. A. Boyse, V. Mike, H. T. Thaler, B. J. Mathieson, J. Abbott, J. Boyse, Z. A. Zayas and L. Thomas. 1976. Control of mating preferences in mice by genes in the major histocompatibility complex. *Journal of Experimental Medicine*, **144**, 1324–1335.
Yamazaki, K., M. Yamaguchi, L. Baranoski, J. Bard, E. A. Boyse and L. Thomas. 1979. Recognition among mice: evidence from the use of a Y-maze differently scented by congenic mice of different major histocompatibility types. *Journal of Experimental Medicine*, **150**, 755–760.

Kin Recognition in Animals
Edited by D. J. C. Fletcher and C. D. Michener
© 1987 John Wiley & Sons Ltd

CHAPTER 4

Genetic Aspects of Kin Recognition: Concepts, Models, and Synthesis

R. H. CROZIER
School of Zoology, University of New South Wales, Kensington, NSW 2033, Australia

MODELERS

As Dawkins (1982) notes, some animals have long been known to have strong powers of discriminating between individual humans and, perhaps, therefore between individuals of their own species. Yet it has not been until very recently that quantitative genetic models of recognition systems have been constructed. Why did it take so long? (See Chapter 13.)

One reason may have been a pre-eminence of environmental-determinist views on the mechanisms. Such views certainly prevailed for ant studies until the close of the 1970s.

Another reason may have been that reasons for kin recognition did not become apparent until Hamilton (1964a, 1964b) thoroughly aired the concept of kin selection, noting (Hamilton, 1964b) that 'discriminatory behavior' would greatly increase the effectiveness of such selection. In this section he further noted the likelihood of a genetic basis for recognition systems.

Despite Hamilton's clear signal of a subject of interest, no quantitative models appeared in the area until 1979. But these years did include some thought on the subject. Thus, Alexander and Borgia (1978) extended Hamilton's remarks in a thoughtful though purely verbal discussion.

Then quantitative modeling appeared in a rush (Crozier and Dix, 1979; Getz, 1981, 1982; Beecher, 1982; Lacy and Sherman, 1983), has replaced the previous intuitive approach and will form the basis for the following discussion.

WHY RECOGNIZE?

The kinds of recognition of others that an organism may be capable of can be divided roughly into kin recognition and individual recognition. Kin recognition is a class phenomenon (other individuals are recognized only by their membership in groups defined in terms of relatedness to the actor), whereas individual recognition involves the ability to distinguish among particular specimens.

The reasons why selection may have favored these abilities are in each case two-fold. As also noted in other chapters, kin recognition may evolve either to enable incest avoidance or in concert with kin selection (or both, of course).

Individual recognition may evolve in association with dominance hierarchies or in concert with social compensation ('reciprocal altruism'). Of course, individual recognition is not essential for the functioning of dominance hierarchies (Barnard and Burk, 1979), but it would enable an animal to avoid a few battles with stronger opponents.

Naturally, the forming and maintenance of relationships involves a number of skills in organisms with sophisticated behavior. Thus, in many species, mate choice involves first a form of class recognition (telling the boys from the girls), then choice of a mate from among the right class (on some basis of 'quality'), and then it is necessary to remember who it is (individual recognition). Probably relatively few bisexual species show all of these abilities, but all show some!

I will discuss kin recognition primarily in terms of its role in discrimination involving aiding behavior. Such behavior can occur without kin recognition, such as if parents treat individuals in their nest as their offspring, but recognition abilities will greatly improve the ability to discriminate accurately. For example, cuckoos would be ineffective social parasites if other birds didn't depend on the occupants of their nests actually being their own offspring.

'RECOGNITION ALLELES', LEARNING AND PSEUDOLEARNING

There are some confused distinctions drawn in the literature between 'green-beard' effects, 'recognition alleles' and 'phenotype matching'. The confusion is natural, because the distinctions are not absolute; to see this, let us redefine certain terms (see also Chapters 2 and 3) for purposes of our discussion.
1. *'Green-beard'* alleles (Dawkins, 1982) simultaneously code for a characteristic phenotype (such as a green beard), the ability to recognize this phenotype in others, and altruism toward other bearers of the same phenotypic label.

2. '*Recognition alleles*' do not encode any behavioral characteristics, but do confer the first two characteristics of the 'green-beard' alleles, namely the bearing of a phenotypic marker and the ability to recognize this in others.
3. '*Phenotype matching*' systems (Holmes and Sherman, 1982a) refer to the ability of an organism to learn aspects of its own characteristics or those of other models (termed below 'referents' after Lacy and Sherman, 1983), and to recognize such characteristics in the phenotypes of others. Dawkins (1982) referred to this as the 'armpit effect' (an individual smelling its own armpit to learn its own scent against which to compare that of others).

Part of the problem arises because the first two concepts have been referred to under both names (Alexander and Borgia, 1978; Dawkins, 1982), and part because of discussion of them under the confused 'outlaw' allele concept (better replaced by a levels of selection approach using genetic insights).

The real problem, however, is operational: in what ways can these three concepts be distinguished? Firstly, as Dawkins argues, green-beard alleles are not about kin recognition and, provided that the help given by altruists exceeds the cost to them, will soon be fixed in the population until a phenotypically identical allele arises that accepts help without giving it. The vulnerability of green-beards to such invasion may prevent them from ever becoming established in the long term.

Secondly, the recognition-allele and phenotype-matching systems overlap (Blaustein, 1983). Given the appropriate recognition rule, there is no way to distinguish between a recognition system involving phenotype matching in which each individual's model is itself, and one involving recognition alleles. For example, does an Aa individual 'recognize' AA and Aa individuals as siblings because it has become habituated to the A and a labels it carries itself, or because it is 'programmed' to recognize A and a labels? I find it hard to think of experiments, even in principle, that could solve such a question.

The difference between the genetic system required for phenotype matching and that for recognition alleles is only that the latter includes the intrinsic ability to recognize similar genotypes, whereas under phenotype matching the power of actual recognition is vested in a learning apparatus encoded separately. It seems worth stressing again that, without some quite fancy molecular biology, these distinctions are not useful when self matching remains a possibility. Some elementary considerations also blur the apparent behavioral distinction between the two concepts. Consider for example the mammalian blood-group antibody system: an impressive array of genotypes is possible, with the internal metabolism of each individual 'knowing' its own genotype through various mechanisms. Because an individual's antigen type also shapes its antibody production, such systems, were they to function in recognition, could be considered either as recognition allele or phenotype matching systems.

Immune system parallels with kin recognition show that the supposed greater complexity of recognition allele systems as against phenotype matching systems (e.g. Alexander and Borgia, 1978) is more apparent than real. A further case shows that the same locus may, possibly, encode both for transmission of a behavioral signal and its reception: behavioral analysis of hybrids in the cricket *Teleogryllus* showed that both the production and recognition of some song characteristics are encoded on the X-chromosome (Bentley and Hoy, 1974).

The models for phenotype matching systems discussed further below depend on genetic variation for phenotypic differences usable in kin recognition. While kin can be recognized individually but treated as a class because of internal programming (which would accord with traditional anthropological views of our own species, for example), the models of interest concern the class recognition of individuals as kin according to cues they present. Kin recognition systems thus work as long as kin have a greater similarity in terms of such cues than do non-kin.

The use of genetically based phenotypic cues in kin-recognition itself imposes selection on the alleles for these cues. The evolutionary dynamics of recognition alleles, and of their encoded cues, may therefore show similarities to those postulated for green-beard alleles. I will consider—briefly!—the problems imposed by such dynamics at the end of this chapter.

THE CROZIER–DIX MODEL REVISITED

I will use the ideas put forward by Mike Dix and me in 1979 to introduce the models for kin-recognition. This paper was, I believe, the first to consider the question quantitatively. The models considered were mostly simple precursors of those that followed, although one of them has never been so followed up.

The Crozier–Dix models concerned the possible genetic basis of colony distinction in social Hymenoptera. Hymenoptera have a male-haploid genetic system, and are the largest group among the 10–20% of metazoa which have this system. Under this system, normal males are haploid and arise from unfertilized eggs, whereas females are diploid and arise from fertilized eggs. In human terms, all hymenopteran genes are X-linked (but with several to many X chromosomes; the general genetics of male-haploidy are reviewed elsewhere, Crozier, 1985). The significance of this genetic system for pedigree analysis is that full sisters (remember that hymenopteran societies are essentially composed only of females) necessarily share one allele (the father's), plus one from the mother as well with a probability of 0.5. This constraint of similarity through the father's gene makes it more likely that a recognition-allele system will work well for Hymenoptera living in simple families than in creatures with both sexes diploid (such as termites).

In our models, kin recognition was the same as colony recognition, and concerned only mechanisms enabling workers to recognize others from foreign colonies as different. Our conceptual framework posited the cues involved to be genetically determined pheromonal ones, and the alleles at these colony-odor loci to be equal in frequency. As Getz (1981) notes, moderate deviation from this last point is not serious: a system with k alleles of which one is rare becomes similar to one with $k - 1$ alleles. Further loci can be added to the system, with nestmates being recognized by the sharing of at least one allele at each locus.

Under our 'Individualistic' model, individuals retain their pheromonal integrity and separate identities, scoring other individuals by comparison with themselves and accepting them as kin if they share at least one allele per locus with them. The system could be considered as either involving recognition alleles or, as noted above, as one involving phenotype matching and comparison with self.

Under our 'Gestalt' model, the pheromones produced by individuals are pooled to produce a common colony odor. The effect of this model is to introduce distinction in terms of relative pheromone concentration, whereas under the Individualistic model it was implicit that distinction is only by the kinds of molecule present, not their relative amounts. The result is that the Gestalt system is capable of much greater precision than the Individualistic one, for example consider the colonies resulting from the matings shown in Table 1.

Table 1. Colonies resulting from specific matings.

Colony	1	2	3	4
Matings	AB × C	DE × C	AD × E	AC × B
♀ Offspring	AC, BC	CD, CE	AE, DE	AB, BC

Under the Individualistic model, all workers of colonies 1 and 2 will accept each other, because they all have the C allele in common, but under the Gestalt model no worker of colony 1 will accept any from colony 2. Consideration of colony 3 brings out a further difference between the two models: under the Gestalt model all members of one colony will behave similarly to those of another, but under the Individualistic model differences may occur. Under the Individualistic model the AE individuals of colony 3 will accept the AC individuals of colony 1 and vice versa, but each genotype will be rejected by the other genotypes (BC and DE) in the other colony. Similarly, the Gestalt model would lead to the queens (mothers) being accepted or rejected at the same rate as the workers, whereas, in the colony line-up shown in Table 1, colonies 1 and 2 would each reject the queen of the other. (Colonies are regarded as large enough so that the queen's own contribution is minimal compared to that of the workers.) Finally, all members of colonies 1 and 4 will accept each other under the Individualistic model, but none would under the Gestalt model because the levels of the various pheromones differ between the colonies.

The next question is: how much genetic variation is needed for the two models to work? By 'work' is meant the probability of a colony being either completely or at least partially protected against mixing with one or more neighboring colonies (hostility of remaining genotypes will generally prevent actual fusion of partially similar colonies under the Individualistic model, but the colony could still suffer penalties such as from inefficient territorial defense). The Gestalt system is much more efficient than the Individualistic one, with a need for only three alleles at each of two loci to prevent fusion of a colony with any of ten randomly-assigned neighbors (with a probability of 95%). By contrast, the Individualistic scheme requires 17 alleles at each of two loci (or three at each of nine loci) to prevent fusion with any of ten neighbors, or 122 alleles at two loci (or eight alleles at 15 loci) to prevent genotype sharing with any of ten neighboring colonies.

We came down at that time on the side of the Gestalt model as the more likely to be a general explanation for two reasons. Firstly, the Gestalt model is more efficient (but nature is under no obligation to fit our notions of efficiency). Secondly, and more compellingly, the Individualistic model cannot work for more complex family systems, such as in organisms with diploid males, or in those many social Hymenoptera which have multiple queens or whose queens mate more than once, because in those cases many workers will not share any alleles.

A MATTER OF REFERENTS

The possible frameworks for phenotype matching systems are much more clearly understandable following the clear statement of Lacy and Sherman (1983) on the different possible classes of referents. Given that 'recognition alleles' have the same dynamics as one kind of cue in the phenotype-matching approach, the referents concept is thus of universal importance.

Lacy and Sherman point out that individuals may choose various classes of other individuals to form their reference points when assessing unknown individuals for relationship. The most important such possibilities seem to be:

1. Self.
2. Special subset of group, e.g. parent(s).
3. Group.

The Individualistic model of Crozier and Dix (1979) is a self-matching model. The Gestalt model of Crozier and Dix differs from other models involving group matching, because these other models involve learning each genotype individually rather than as a pooled result. However, in formal terms, for very large groups the Gestalt model would give the same results as matching both parents but with the father's sole allele (recall that he is haploid) given double the weight of each of the mother's two alleles. For small groups the Gestalt model differs from such a parent-matching model for two reasons, firstly stochastic variation in the composition of the group leads to deviations in the proportions of various genotypes present, and secondly, in social Hymenoptera, the presence of the queen, if she also emits the cue pheromones, will become increasingly important the smaller the colony.

This last consideration raises a point about phenotype-matching systems: different referents may be given different weights in forming a 'template' (Chapter 8). These weights might contribute to a single template, or to different templates for different purposes (such as incest avoidance as against kin-directed altruism) or differing degrees of kin-discrimination (recognition is necessary for discrimination, but discrimination may not follow from recognition).

AN ABSURDLY SIMPLE SYSTEM: CLONES

The simplest possible genetically-based recognition involves rejection of an individual unless it is of exactly the same genotype as one's own. Such a mechanism is not useful in social organisms whose groups are genetically heterogeneous, but is practical in organisms whose groups are composed of genetically-identical individuals (clones). In addition to many plants such as dandelions (Solbrig, 1971), clonal groups are common among animals, such as in anemones (Ayre, 1982; Lubbock, 1980), aphids (Blackman, 1979; Suomalainen et al., 1980), corals (Neigel and Avise, 1983a) and sponges (Neigel and Avise, 1983b). The anemone studies are of particular interest, because the animals actually make use of the information about identity: anemones of different clones attack each other with discharges from specialized structures (although habituation to repeatedly-encountered foreigners occurs, Sebens, 1984, and complicates interpretations, Stoddart et al., 1985). The parallels with the immunological mechanisms of mammals are striking, and indeed Lubbock (1980) refers to the mechanism as immunological.

The anemones, corals and sponges appear to have only transitive compatibilities, i.e. if individual A is compatible with both B and C, then B will also be compatible with C. The implication is that exact matching of recognition genotype is necessary before a non-agonistic interaction can occur. We can now briefly explore the characteristics of such a system.

Given k alleles, each at equal frequency, p, the probability of two randomly-chosen individuals being of the same genotype equals:

$$(2 - p)p^2 = (2 - 1/k)/k^2$$

which yields a requirement of seven alleles at this locus to distinguish an individual from one other with a probability of 95%. For comparison with the Crozier–Dix results given above, the numbers of alleles required for two-locus systems are three to distinguish a colony from one neighbor, five for five neighbors, and six for ten neighbors. The actual precision of the system seems often to be greater than this: Lubbock (1980) found that only one of 102 inter-clone interactions failed to elicit an agonistic response.

The ascidian *Botryllus schlosseri* has clonal colonies among which incompatibility reactions can be intransitive, i.e. if A is compatible with both B and C, then B and C may still be incompatible with each other (Scofield et al., 1982). The system indicated by these results is logically identical with the Individualistic model of Crozier and Dix (1979): animals show compatibility if they share one or both alleles at the incompatibility locus. Because every colony of these ascidians consists of only one genotype, as against the genetically-heterogeneous ones of social insects, the probability of fusion of different clones differs from that for social Hymenoptera, being for one locus (assuming again k alleles each of equal frequency, p):

$$p(4 - 6p + 3p^2) = (4 - 6/k + 3/k^2)/k$$

so that 79 alleles are required to prevent fusion between individuals from different clones with a probability of 95% if one locus is involved. For a system with two loci, the numbers of alleles per locus required to distinguish one colony from one, five and ten neighbors would be 17, 39 and 55, respectively. This system is thus significantly more efficient than that postulated for social insects by Crozier and Dix.

As remarked above, the recognition systems of marine colonial invertebrates has been explicitly compared to the histocompatibility systems of mammals. This is no idle comparison, because inbred mouse strains can distinguish each other according to genotype at the major histocompatibility complex (*H-2*) or at a closely-linked locus, Qa-Tla, and generally prefer to mate with mice of different genotypes at these loci (Boyse, Beauchamp and Yamazaki, 1983). W. D. Hamilton (pers. comm.) made the suggestion that this mating preference might be due to these loci leading to sufficient antigenic difference to result in different associated bacterial communities. The mating preference would then not be due to mice detecting the antigen genotypes themselves, but rather the different odorous metabolites produced by the various mixes of bacteria. Unfortunately, this reasonable suggestion is not supported by the fact that the preferences are demonstrable when urine samples are used instead of the mice themselves. Boyse, Beauchamp and Yamazaki (1983) list many effects of *H-2* genotype on phenotypic characteristics in mice, but clearly mice do not require these characteristics for recognition in view of the effectiveness of the urine samples in eliciting discriminating behavior. No specific pheromone has been isolated, but neither does it appear to have been looked for yet.

BASIC MODELS FOR GENETICALLY VARIABLE GROUPS

While the simple probability of the sharing of alleles or complete genotypes, as discussed so far, formed a useful approach, it is more likely that organisms have to use decision rules in estimating the degree of kinship to themselves of others. An animal's decision to treat another as being of one relationship to itself as against a different one will depend on the number of labels (values of the phenotypic traits) the two animals hold in common. The appropriate threshold value for treating a stranger as of one relationship instead of the other will depend on the frequency distribution of shared labels for the two relationships (Fig. 1). This joint distribution graph has been called a kingram by Getz (1981).

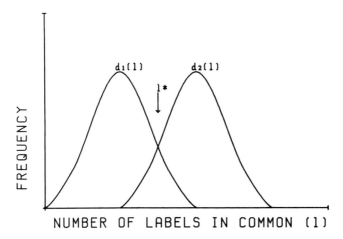

Fig. 1. Kingram showing the distributions of labels held in common with a referent by animals having one or the other of two relationships (such as full- and half-sisters) with the referent. For self-matching, for example, curve $d_1(l)$ is the frequency distribution of animals of relationship 1 to the referent with respect to the number of labels held in common with it, and curve $d_2(l)$ is the frequency distribution for relationship 2. Individuals with a number of labels in common with the referent above some critical value, l^*, will be considered as being of relationship 2, and if equal to or below this value, as being of relationship 1. In the diagram l^* is chosen to minimize the misassignment error, which will be equal to half the overlap of the two frequency distributions (Getz, 1981). As explained in the text, however, the determination of l^* will also depend on the relative consequences of misassigning an individual of relationship 1 as one of relationship 2, and vice versa

It is possible to choose a threshold value (of traits held in common between an individual being assessed and the referent) on a kingram so that as few individuals as possible are misassigned with respect to relationship.

This threshold value will depend on the distribution of the numbers of labels held in common with the referent for individuals of the relationships involved, and the frequency with which individuals of each relationship are encountered. But the actual threshold value finally employed by an animal will also depend on the relative importance of errors, such as of accepting a non-sibling as a sibling, or conversely, rejecting a sibling as a non-sibling (Getz, 1981). It seems unlikely that this critical value, 1* (Fig. 1), is actually set by selection so as to minimize the error, rather than minimizing the product of the error and its consequences.

The various genetic models that can be considered given the framework established so far are quite numerous, involving not only the referents but also the inheritance pattern (dominance or codominance) of the phenotypic cues. Getz's (1981) model is one of self-matching with co-dominance. Subsequently, Getz (1982) analysed three models of group-matching (involving siblings) under codominant inheritance:

1. *Genotype recognition.* An individual is accepted if it has a genotype present in the group (e.g. accept AB only if AB is already present in the group).
2. *Foreign-label rejection.* An individual is accepted as long as both labeling alleles are present in the group (e.g. AB is acceptable so long as group members have both of alleles A and B, even if as genotypes AC and BD).
3. *Habituated-label acceptance.* Accept an individual so long as at least one of its alleles is already held by others in the group (e.g. AB is acceptable if either or both A and B are already present in the group, even if the group consists of AC and CD).

Getz (1981) found that recognition systems distinguishing full- from half-sisters are intrinsically much more efficient for male haploid genetic systems than for those in which both sexes are diploid, and Lacy and Sherman (1983) point out that this results from the greater relatedness between full sisters under male-haploidy (0.75 as against 0.5).

Lacy and Sherman (1983) studied a family of models assignable to Getz's genotype-recognition category, examining both dominant and codominant inheritance modes. They also considered metric (continuous) traits as well as the previously modeled discrete traits, for a variety of referent types. They used the by now standard criterion of the number of alleles and loci required for recognition to exceed a certain level of reliability, and found that continuous traits confer more efficient recognition, and that codominance is generally more efficient than dominant inheritance, but that this is reversed for some relative-referent combinations. Notably, dominance increased the reliability of distinguishing between all categories examined, full-siblings, half-siblings, and non-relatives, under parent matching.

Lacy and Sherman (1983) also briefly considered the effect of environmentally induced variation on the efficiency of the system, remarking that the metric-trait system would be more easily degraded by such variation. Of course, such variation would probably enhance individual recognition, and this would also be kin recognition if individuals so learned were characterizable as kin according to the circumstances in which they were learned (such as being in the nest together while young).

Both Getz (1981, 1982) and Lacy and Sherman (1983) conclude that systems with moderate numbers (about five) of alleles per locus at a moderate number of loci are more efficient than systems with large numbers of alleles at a small number of loci or a small number of alleles at a large number of loci. Lacy and Sherman (1983) note that such loci are rare even in *Drosophila*, but given that only a few loci are required for the system to work, this observation does not provide any clue as to the likely genetic architecture of kin-recognition systems. In particular, humans and mice are less genetically variable overall than most *Drosophila* species, yet two of the loci of the major histocompatibility complex have over 50 alleles each (Steinmetz and Hood, 1983).

HOW MECHANISMS AFFECT MODELS

As Lacy and Sherman (1983) implied with their study of continuous-trait recognition systems, the form of the signal involved will have consequences for the efficiency of the underlying genetic mechanism. For example, a system relying on allelic variation in pheromonal molecular structure will intrinsically have more distinguishable genotypes than one relying on relative variation in pheromonal production levels. By this I mean that, if A and B encode high production, and a and b low levels of pheromones at different loci, then *AABB* and *aabb* individuals will have similar ratios of the pheromones involved, whereas a molecular-difference model would have these genotypes distinguishable.

A further effect in some signal systems but not others would be functional interference between signal strength, or other characteristics, and the array of genotypes. Variation in strength of call components will face the restriction mentioned above for pheromone level, and under both systems the phenotype involving no sound (or pheromone) for all components would presumably be ineffective in attracting any notice at all! Similarly, call variation, which appears to function as a kin recognition system in some birds at least (see Beecher, 1982, who also provides a genetic model similar to the Individualistic model of Crozier and Dix, 1979), may have to fit within 'sound windows' in the local sonic environment. Naturally, these are not serious difficulties with such systems—the number of possible genotypes is probably sufficiently large that subtraction of some would not disable the system. Might such non-functional genotypes be produced occasionally and form part of the genetic load?

Other influences of mechanism have already been alluded to. These include the referents involved and the transferability or otherwise of labels.

DIVERSITY IN UNITY

I suspect that, across large phylogenetic vistas, there is unity in the kinds of selection favoring the maintenance of kin recognition. The reasons for its maintenance are incest avoidance and kin discrimination in social settings. Kin discrimination itself is largely divisible into care of the young (such as, but not limited to, parental-offspring interactions) and cooperation in defense and maintenance of the group and its 'property'.

The mechanisms of kin recognition seem far more diverse, both in terms of the signal systems used and also the referents used.

Signal systems range from the immunological type seen in some marine animals and in mice, to calls, pheromones and appearance. Some signal systems involve non-transferable cues; others use cues that have to be acquired from other individuals (such as some pheromones or distinctive behaviors).

The referents involved range from self to various combinations of group members.

Some unity of mechanism occurs in that, apart from the self-matching (or recognition allele) possibility, similarities in ontogeny seem likely. New individuals have to learn the attributes of referents, and often the other colony members need to learn the attributes of newcomers. These aims can be fulfilled by each individual going through a 'protected' phase while young, during which it is both safe from attack and does not attack others, as reported for the isopod *Hemilepistus reaumuri* (Linsenmair, 1985; Chapter 6) and two ground squirrel (*Spermophilus*) species (Holmes and Sherman, 1982b).

TESTS

Studies on kin recognition mechanisms generally seek to determine the overall importance of innate (to the species) versus environmental influences, the signal system used, the referents involved, whether labels are endogenous or acquired, and, especially, the ontogenetic and genetic bases of the recognition system. A significant problem is that group composition is of course variable. Hence, matings involving exactly the same parental genotypes will yield offspring groups differing in composition; for small groups, these differences could mean that members of such groups can usually distinguish their group from others.

A case in point about the difficulties imposed by group size concerns the *Lasioglossum* work of the University of Kansas group, without doubt the best-characterized genetically-based system of discriminatory kin recognition known in terrestrial animals (Greenberg, 1979; Chapter 7). Buckle and Greenberg (1981) found that if a group of immature bees who are sisters have an alien immature bee added to them, this alien bee will admit to the nest further, unknown, sisters of its nestmates in preference to its own sisters reared separately from it. Buckle and Greenberg concluded that a bee learns the odors of its nestmates, but not its own odor. Getz (1982), however, points out that, of its own sister group, the lone bee has only itself to learn, whereas it has its nestmates as a larger sample of their sibling-group to learn. Getz (1982) concludes from the *Lasioglossum* data, using not only the admission rates for bees whose labels should be known, but also the admission rates for totally unknown bees, that the best candidate mechanism is group-referent and foreign-label rejection.

Getz's (1982) analysis is an example of fitting frequency data to the expectations based on various models. This approach should be followed wherever sample sizes permit (which may not be often in vertebrate studies but should be possible for most insect studies, for example).

Even if quantitative data-fitting is not possible, qualitative observations may be illuminating. One key observation would be whether or not relationships are invariably transitive or whether intransitivity can happen occasionally. Intransitivity, in terms of which strangers different members of a group accept, would rule out parent-matching and other common-label schemes (such as the Gestalt model of Crozier and Dix, 1979).

Mintzer (1982) studied colonies of an acacia-ant, *Pseudomyrmex ferruginea*, which were maintained on trees from the same *Acacia hindsii* clone, thus ensuring extreme environmental uniformity (the trees feed and house the ants). Mintzer's tests involved introducing small numbers of ants from one colony into another, and observing their reception. Although most introduction attempts showed effective recognition of strangers, there were some significant exceptions. Mintzer found some combinations in which ants from colony 1 were accepted by colony 2 but in which colony 2 ants were rejected by colony 1. In some other tests, some colony 3 (say) ants, but not others, would be admitted by colony 4. These intransitivities permit the rejection of the Gestalt model and a parent-referent model also invoking acquisition of labels by offspring from parents (see below), both of which demand uniformity of response to outsiders.

Another trap would be to assume that related species necessarily have similar kin-recognition systems. While *Drosophila melanogaster* has an incest-avoidance kin-recognition system involving both autosomal and X-linked pheromone loci (Averhoff and Richardson, 1976), *D. pseudoobscura* appears to lack any such system at all (Powell and Morton, 1979). Consequently, it is plausible that ants other than *Pseudomyrmex* may have a queen-referent system in which queen pheromones are transferred to and label the workers (Hölldobler and Michener, 1980). Carlin and Hölldobler (1983) were able to reject the self-matching model for various *Camponotus* ant species by showing that workers reared with queens and workers of a different species would reject their own, separately-reared, sisters in favor of non-conspecific nestmates. The small size of the groups used by Carlin and Hölldobler (1983) invalidated rejection of other models, due to expected chance effects as discussed above for the *Lasioglossum* work, but further experiments (Carlin and Hölldobler, 1986), using queenless and queenright groups with individuals of various origins, yielded evidence of both worker- and queen-derived cues. Carlin and Hölldobler found that queen effects dominated worker-derived ones in small groups of generally less than ten workers each, but the situation in full-sized colonies of several thousand workers is unknown; if all colony members contribute to a common odor pool, the relative contribution of the queen will fall as the colony size increases (Crozier and Dix, 1979; Carlin and Hölldobler, 1983, 1986).

BUT HOW CAN KIN RECOGNITION BE POSSIBLE ANYWAY?

As remarked above, a common feature of kin-recognition models is a reliance on frequency-dependence to maintain genetic variability at the loci postulated. This frequency-dependence does not seem possible for an important class of cases.

Kin recognition does seem likely to result in frequency-dependent selection favoring its own efficiency in incest avoidance and, less certainly, in parent-offspring recognition. In incest avoidance, unlike individuals are sought as mates, so that very common phenotypes will have to expend more time in searching than rare ones. In parent-offspring recognition, a common phenotype will have more variation (i.e. errors) in the assignment of young to parent after separations, reducing the efficiency of rearing, again favoring rare types for which such errors are rarer.

The problem emerges with respect to kin-discrimination behavior, such as in territorial defense or in general cooperative behavior. Kin selection for altruistic and kin-discriminative behavior may be frequency dependent (Templeton, 1979), but the alleles determining the cues are under no such selection. In fact, the system encounters the same problems seen by Alexander and Borgia (1978) as facing 'recognition alleles'.

Consider the genotype-recognition system suggested as operating for some sea anemones. An individual belonging to a rare recognition genotype will, on encountering members of other clones, nearly always be engaged in a fight. An individual belonging to a common recognition genotype will often be of the same such genotype as individuals of other clones, and hence be less subject to conflicts. Consequently, the rarer a genotype is, the lower will be its fitness. This intuitive argument is not of itself conclusive, however. Consider a locus with two alleles, and let each genotype have a fitness equal to its frequency, so that, letting 'W' stand for the fitness of a genotype, we have:

W_{AA}	W_{Aa}	W_{aa}
p^2	$2pq$	q^2

Over a range of frequencies ($1/3 < p < 2/3$) the heterozygote has a higher fitness than either homozygote, normally a sufficient condition for a balanced polymorphism. In this case, however, the equilibrium value towards which the population moves for any set of fitnesses is always further from 0.5 than the current frequency, as can be shown by replacing p by $(0.5 + d)$:

let $u = p/q = (0.5 + d)/(0.5 - d)$

u^* = the equilibrium value according to current fitnesses
  $= [2pq - q^2]/(2pq - p^2]$
  $= [q/p][(3p - 1)/(2 - 3p)]$

and $u^*/u = (q^2/p^2)[(0.5 + 3d)/(0.5 - 3d)]$
  which is > 1

Thus, despite overdominance, a two-allele system will move ever further from 0.5 until one allele or the other is fixed. It is unlikely that a multi-allele system will behave differently. It can be shown that the system proposed for the ascidian *Botryllus schlosseri*, which, as remarked above, is the same as the Crozier–Dix Individualistic model, is also unstable for two alleles.

Modeling recognition alleles under more conventional kin-selection models would almost certainly give similar results. Thus, assume a codominant locus determining the cues, with a sibling group-referent system involving foreign label rejection, and further assume that each animal donates aid to animals it 'recognizes' with some cost to itself. When A is exceedingly rare (occurring only in Aa and $Aa + aa$ sibships), it will be recognized by extremely few animals in the population (and therefore receive rather little aid) but will recognize most of the population (because nearly all are aa, sharing an allele with the Aa carriers of A). Consequently, A will have costs vastly in excess of aid. On the other hand, if A was close to fixation, then a would face these problems instead.

These considerations therefore make it unlikely that alleles selected solely for their determination of cues in social recognition for kin discrimination can be maintained as polymorphisms. How then can it be that genetically-based kin recognition has been so widely found?

One possibility is that unexpected behavior of the genetic system may emerge if the recognition-system alleles are considered in concert with alleles for kin-discriminative behavior (altruism). This possibility seems unlikely, but has not yet been investigated.

The other possibility is that kin discrimination 'piggy-backs' on other genetic kin-recognition systems, maintained by incest avoidance and parent-offspring recognition, and on genetic systems maintained by completely different kinds of selection. For example, the kin-recognition system of anemones may indeed be not just genetically similar to the histocompatibility systems of vertebrates, but actually function in the same way and give protection against pathogens. Frequency-dependent selection can easily act on immune-system variation if there is any matching of resistance genes (which I suggest are also 'kin recognition' ones) in the anemones with counter-resistance genes in the pathogens.

Might kin recognition in creatures other than marine invertebrates be based on systems maintained by selection for factors such as incest avoidance and disease resistance? More thought seems required, but I make two observations. Firstly, one of the best demonstrations of genetically-controlled mate-recognition pheromones also comes from *Lasioglossum zephyrum* (Smith, 1983), noted above as the species in which the genetic basis for kin-discrimination substances has been best elucidated. Secondly, and this will betray a sociohymenopterological bias, aculeate Hymenoptera frequently, probably generally, have heterozygosity-based sex-determination systems (see Crozier, 1979), which impose severe costs to inbreeding and provide the most plausible explanation (in combination with life-cycle variables) so far for multiple mating by females in various social species (Page and Metcalf, 1982; Crozier and Page, 1985).

There are two further possibilities for the maintenance of polymorphism for kin-discrimination cues. One is the 'piggy-backing' onto cues maintained by selection for individual recognition involved with dominance hierarchy maintenance. Alleles that are very common for such systems will result in numerous cases of mistaken identity in comparison with alleles that are rare. On average, such errors may cancel out, carriers being mistaken for high-ranking colleagues as often as for low-ranking ones, but even if so it seems likely that common alleles would then be selected against compared with rare ones. This result is expected because of a finding by Gillespie (1977): if two genotypes have the same average fitness, selection will usually favor that with the lesser variance for fitness (the case of honey bee polyandry provides an exception to this principle: Page and Metcalf, 1982; Crozier and Page, 1985).

To derive the closing possibility (and complete Table 2) I invoke the classic strategy of geneticists when faced with an intractable phenomenon—bring in another locus (or two). Instead of assuming a general and indiscriminate ability of organisms to learn cues, suppose that their abilities to learn certain kinds of cues is itself genetically influenced (and it is well-known that species vary in their relative abilities to learn from specific stimuli). Such a limitation may arise if there is a limit to the number of labels that an organism can learn, or to the number of genotypes that it can distinguish. In that case, although quantitative modeling is needed, it is plausible that there will be genetic variation for the cues used in recognition, and that genotypes relying on weakly polymorphic loci for kin recognition will do less well than those that rely on highly polymorphic loci. This recognition-switching phenomenon would not of itself maintain polymorphism, but rather would act to reduce selection stemming from kin recognition, enabling a locus to recover its polymorphism under selection under category 1 of Table 2.

Table 2. Possible modes of selection resulting in polymorphism of cues (labels) functioning in kin recognition systems leading to cooperative or competitive kin discrimination.

1. When the chief source of selection is not kin discrimination, but some other form of selection, such as:
    a. Other kin-recognition systems.
        (i) Incest avoidance.
        (ii) Recognition of related young.
    b. Immune-system loci.
    c. Individual recognition (dominance hierarchy maintenance).

2. Recognition-system switching (favors those genotypes using the more highly polymorphic genetic systems for recognition over those using less highly polymorphic ones). This mode is effective only in combination with modes listed under 1.

NOTE ADDED IN PROOF

Selection on the label alleles of marine invertebrates is considered further elsewhere (Crozier, 1986). I thank Rick Grosberg for pointing out that the equations on pp. 61–62 were also derived by Curtis et al. (*Transplantation*, **33**, 127–133).

ACKNOWLEDGMENTS

I thank D. T. Ayre, E. A. Boyse, M. W. J. Crosland, D. B. Croft, Y. C. Crozier, W. Getz, R. Lacy, P. Pamilo, C. Peeters and the editors for helpful comments on the manuscript. My work on evolutionary genetics is supported by the Australian Research Grants Scheme.

LITERATURE CITED

Alexander, R. D. and G. Borgia. 1978. Group selection, altruism, and the levels of organization of life. *Annual Review of Ecology and Sytematics*, **9**, 449–474.

Averhoff, W. W. and R. H. Richardson. 1976. Multiple pheromone system controlling mating in *Drosophila melanogaster*. *Proceedings of the National Academy of Sciences, USA*, **73**, 591–593.

Ayre, D. J. 1982. Inter-genotype aggression in the solitary sea anemone *Actinia tenebrosa*. *Marine Biology*, **68**, 199–205.

Barnard, C. J. and T. Burk. 1979. Dominance hierarchies and the evolution of 'individual recognition'. *Journal of Theoretical Biology*, **81**, 65–73.

Beecher, M. D. 1982. Signature systems and kin recognition. *American Zoologist*, **22**, 477–490.

Bentley, D. and R. R. Hoy. 1974. The neurobiology of cricket song. *Scientific American*, **231(2)**, 34–44.

Blackman, R. L. 1979. Stability and variation in aphid clone lineages. *Biological Journal of the Linnean Society*, **11**, 259–277.

Blaustein, A. R. 1983. Kin recognition mechanisms: phenotypic matching or recognition alleles? *American Naturalist*, **121**, 749–754.

Boyse, E. A., G. K. Beauchamp and K. Yamazaki. 1983. The sensory perception of genotypic polymorphism of the major histocompatibility complex and other genes: some physiological and phylogenetic implications. *Human Immunology*, **6**, 177–183.

Buckle, G. R. and L. Greenberg. 1981. Nestmate recognition in sweat bees (*Lasioglossum zephyrum*): does an individual recognize its own odor or only odors of its nestmates? *Animal Behavior*, **29**, 802–809.

Carlin, N. F. and B. Hölldobler. 1983. Nestmate and kin recognition in interspecific mixed colonies of ants. *Science*, **222**, 1027–1029.

Carlin, N. F. and B. Hölldobler. 1986. The kin recognition system of carpenter ants (*Camponotus* spp.) I: Hierarchical cues in small colonies. *Behavioral Ecology and Sociobiology*, in press.

Crozier, R. H. 1979. Genetics of sociality, in Hermann, H. R. ed. *Social insects*, Academic Press, New York, Vol. 1, pp. 223–286.

Crozier, R. H. 1985. Adaptive aspects of male haploidy, in Helle, W. and M. W. Sabelis, eds. *Spider mites. Their biology, natural enemies and control*, Elsevier, Amsterdam, Vol 1A, pp. 201–222.

Crozier, R. H. 1986. Genetic clonal recognition abilities in marine invertebrates must be maintained by something else. *Evolution*, in press.

Crozier, R. H. and M. W. Dix. 1979. Analysis of two genetic models for the innate components of colony odor in social Hymenoptera. *Behavioral Ecology and Sociobiology*, **4**, 217–224.

Crozier, R. H. and R. E. Page. 1985. On being the right size: male contributions and multiple mating in social Hymenoptera. *Behavioral Ecology and Sociobiology*, **18**, 105–115.

Dawkins, R. 1982. *The extended phenotype. The gene as the unit of selection*. Freeman, San Francisco.

Getz, W. 1981. Genetically based kin recognition systems. *Journal of Theoretical Biology*, **92**, 209–226.

Getz, W. M. 1982. An analysis of learned kin recognition in Hymenoptera. *Journal of Theoretical Biology*, **99**, 585–597.

Gillespie, J. H. 1977. Natural selection for variances in offspring numbers: a new evolutionary principle. *American Naturalist*, **111**, 1010–1014.

Greenberg, L. 1979. Genetic component of bee odor in kin recognition. *Science*, **206**, 1095–1097.

Hamilton, W. D. 1964a. The genetical evolution of social behaviour. I. *Journal of Theoretical Biology*, **7**, 1–16.

Hamilton, W. D. 1964b. The genetical evolution of social behaviour. II. *Journal of Theoretical Biology*, **7**, 17–52.

Harvey, P. H., 1980. Mechanisms of kin-correlated behavior, Group report, in Markl, H. ed. *Evolution of social behavior: hypotheses and empirical tests*, Dahlem Konferenzen, Verlag Chemie, Weinheim, pp. 183–202.

Hölldobler, B. and C. D. Michener. 1980. Mechanisms of identification and discrimination in social Hymenoptera, in Markl, H. ed., *Evolution of social behavior: hypotheses and empirical tests*, Dahlem Konferenzen, Verlag Chemie, Weinheim, pp. 35–57.

Holmes, W. G. and P. W. Sherman. 1982a. Kin recognition in animals. *American Scientist*, **71**, 46–55.

Holmes, W. G. and P. W. Sherman. 1982b. The ontogeny of kin recognition in two species of ground squirrels. *American Zoologist*, **22**, 491–517.

Lacy, R. C. and P. W. Sherman. 1983. Kin recognition by phenotype matching. *American Naturalist*, **121**, 489–512.

Linsenmair, K. E. 1985. Individual and family recognition in subsocial arthropods, in particular the desert isopod *Hemilepistus reaumuri*, in Hölldobler B. and M. Lindauer, eds., *Experimental behavioral ecology and sociobiology*, Gustav Fischer, New York, pp. 411–436.

Lubbock, R. 1980. Clone-specific cellular recognition in a sea anemone. *Proceedings of the National Academy of Sciences, USA*, **77**, 6667–6669.

Mintzer, A. 1982. Nestmate recognition and incompatibility between colonies of the acacia-ant *Pseudomyrmex ferruginea*. *Behavioral Ecology and Sociobiology*, **10**, 165–168.

Neigel, J. E. and J. C. Avise. 1983a. Clonal diversity and population structure in a reef-building coral, *Acropora cervicornis*: self-recognition analysis and demographic interpretation. *Evolution*, **37**, 437–453.

Neigel, J. E. and J. C. Avise. 1983b. Histocompatibility bioassays of population structure in marine sponges. Clonal structure in *Verongia longissima* and *Iotrochota birotulata*. *Journal of Heredity*, **74**, 134–140.

Page, R. E. and R. A. Metcalf. 1982. Multiple mating, sperm utilization and social evolution. *American Naturalist*, **119**, 263–281.

Powell, J. R. and L. Morton. 1979. Inbreeding and mating patterns in *Drosophila pseudoobscura*. *Behavior Genetics*, **9**, 425–429.

Scofield, V. L., J. M. Schlumpberger, L. A. West and I. L. Weisman. 1982. Protochordate allorecognition is controlled by a MHC-like gene system. *Nature*, **295**, 499–502.

Sebens, K. P. 1984. Agonistic behavior in the intertidal sea anemone *Anthopleura xanthogrammica*. *Biological Bulletin*, **166**, 457–472.

Smith, B. H. 1983. Recognition of female kin by male bees through olfactory signals. *Proceedings of the National Academy of Sciences, USA*, **80**, 4551–4553.

Solbrig, O. T. 1971. The population biology of dandelions. *American Scientist*, **59**, 686–694.

Steinmetz, M. and L. Hood. 1983. Genes of the major histocompatibility complex in mouse and man. *Science*, **222**, 727–733.

Stoddart, J. A., D. J. Ayre, B. Willis and A. J. Heyward. 1985. Self-recognition in sponges and corals? *Evolution*, (in press).

Suomalainen, E., A. Saura, J. Lokki and T. Teeri. 1980. Genetic polymorphism and evolution in parthenogenetic animals. Part 9: Absence of variation within parthenogenetic aphid clones. *Theoretical and Applied Genetics*, **57**, 129–132.

Templeton, A. R. 1979. A frequency dependent model of brood selection. *American Naturalist*, **114**, 515–524.

CHAPTER 5

Discrimination Among Prospective Mates in Drosophila

ELIOT B. SPIESS
Department of Biological Sciences, University of Illinois at Chicago, PO Box 4348, Chicago, Illinois 60680, USA.

INTRODUCTION

In many animal species, kin recognition may evolve as a behavioral advance leading to social organization; but in non-social insects like *Drosophila*, evidence is scanty that individuals can discriminate between a close relative and a more distantly related member of the species. Some data support arguments for predominant outcrossing trends in *Drosophila*; thus we might reason that such trends necessarily originate not only from an ability to discriminate but also from preference for unrelated prospective mates. Optimal outbreeding (Bateson, 1983) may be a consequence. (Other features of kin recognition that have evolved among advanced social species like group protection, altruism and sharing of resources seem completely unknown in *Drosophila*.) We are not yet in any position to say what signals flies may recognize as preferred in prospective mates, but enough information is available on genetic and behavioral details to explore the problem of discrimination and mate choice in these insects.

The aim of this chapter is to set forth *Drosophila* as a model experimental animal with behavioral features that can evolve through stages of discrimination among prospective mates to preferential choice and then to assortative systems including ethological isolation among species. Some evidence demonstrates glimmerings of these features. *Drosophila*'s mate recognition ability and its other behavioral traits that act as cohesive forces keeping the species as mating units ('ensuring effective syngamy within a population occupying its preferred habitat', Paterson, 1980) deserve study, particularly with respect to their genetic determination. Furthermore, the stages, or routes, leading to evolution of new mate recognition systems via discrimination, preferences and assortative mating need to be explored as potential sources of ethological (prezygotic) isolation. What mechanisms for nascent isolation exist or do not exist in the genetic-behavioral features of any species can be investigated using *Drosophila* with less technical difficulty and more informative conclusions than with most experimental animals, particularly because of its advantages in genetic analysis and culture. From ten to 15 years ago several reviews of the literature established the definitive features of mating behavior and its genetic basis (Spieth, 1968, 1974a, 1978; Spiess, 1970; Petit, 1972; Parsons, 1973; Burnet and Connolly, 1974). During the intervening years an extensive literature has grown, particularly in the genetic dissection of courtship and mating elements and the types of signals between prospective mates.

TYPES OF SIGNALS BETWEEN PROSPECTIVE MATES

To provide a background for an account of discrimination and choice, it is important to survey the types of signals that come into play during *Drosophila* courtship and mating. Males' courtship elements and females' responses toward them have been summarized by Spieth (1974a) in terms of behavioral activities observable to a careful investigator. Information exchange between prospective mates can be categorized in terms of four principal stimuli, which, though they must be integrated for normal fly behavior, have been analyzed separately in extensive detail over the past decade: acoustic, olfactory, visual, and tactile. The following sections constitute a brief overview of these sensory stimuli; for convenience, references from the past 15 years are listed in Table 1.

Acoustic signals

Courtship in most species of *Drosophila* includes a period of wing vibration by the male, producing an audible sound or sounds of specific frequency or pairs of frequencies. Owing to the predominance of wing vibration as a courtship element throughout the family Drosophilidae, it was investigated earlier and more extensively than the other courtship signals. Thus the literature is considerable (Table 1).

Recording of the sounds with high precision electronic devices has been carried out for several years since Shorey in 1962 first analyzed courtship sound (Bennet-Clark, 1975; von Schilcher, 1976a,b; Ewing, 1977a,b). Generally, male courtship sound ('love song') consists of a train of mono- or polycyclic pulses (e.g. pulse song of *D. melanogaster*), with a given species characterized by its particular sound that depends on such parameters as the length of a pulse train (in ms), number of pulses per train, length of pulse, number and frequency of cycles per pulse (ipf, or intrapulse frequency) and

Table 1. References to Literature on Sensory Stimuli Modes ('Signals') Mentioned in the Text and Not Cited in Spiess (1970) Review.

A. *Acoustic Signals*:
Bennet-Clark, 1971, 1975
Bennet-Clark, Dow, Ewing, Manning and von Schilcher, 1976
Bennet-Clark and Ewing, 1969
Bennet-Clark, Ewing and Manning, 1973
Bennet-Clark, Leroy and Tsacas, 1980
Burnet, Eastwood and Connolly, 1977
Chang and Miller, 1978
Cook, 1973a,b
Cowling and Burnet, 1981
Crossley and McDonald, 1979
Eastwood and Burnet, 1979 (see comment by Robertson, 1982b)
Ewing, 1969, 1977a,b, 1978, 1979
Ewing and Bennet-Clark, 1968
Hoikkala, Lakovaara and Romppainen, 1982
Hoikkala and Lumme, 1984
Ikeda, Idoji and Takabatake, 1981
Ikeda and Maruo, 1982
Ikeda, Takabatake and Sawada, 1980
McDonald, 1979
Miller, Goldstein and Patty, 1975
von Schilcher, 1976a,b, 1977
von Schilcher and Manning, 1975
Wood and Ringo, 1982

B. *Olfactory Signals*:
Antony and Jallon, 1982
Averhoff and Richardson, 1974, 1976
Ehrman, 1972
Gailey, Jackson and Siegel, 1984
Hall, Tompkins, Kyriacou, Siegel, von Schilcher and Greenspan, 1980b
Jallon, 1984
Jallon and Hotta, 1979
Leonard and Ehrman, 1976
Leonard, Ehrman and Pruzan, 1974
Leonard, Ehrman and Schorsch, 1974
Mane, Tompkins and Richmond, 1983
Shorey and Bartell, 1970
Siegel and Hall, 1979
Sloane and Spiess, 1971
Tompkins and Hall, 1981a,b
Tompkins, Hall and Hall, 1980
Tompkins, Siegel, Gailey and Hall, 1983
van den Berg, Thomas, Hendriks and van Delden, 1984
Venard, 1980
Venard and Jallon, 1980

Continued overleaf

TABLE 1 (continued)

C. *Visual Signals*:
   Burnet and Connolly, 1973
   Cook, 1979, 1980
   Grossfield, 1971
   Markow and Hanson, 1981
   Matsumoto, O'Tousa and Pak, 1982
   Pak, 1975
   Pinsker and Doschek, 1979
   Tompkins, Gross, Hall, Gailey and Siegel, 1982

D. *Summaries Including All Types of Signals*:
   Hall, Siegel, Tompkins and Kyriacou, 1980a
   Hall, Tompkins, Kyriacou, Siegel, von Schilcher and Greenspan, 1980b
   Siegel, Hall, Gailey and Kyriacou, 1984
   Spieth and Ringo, 1983
   Tompkins, 1984

interpulse interval (ipi, or time between major peaks of two successive pulses). Another type of song, the sine song of *D. melanogaster*, produces a single long polycyclic hum (with sine wave cycle) like the sound of flight but of lower frequency (von Schilcher, 1976a,b). Functionally equivalent sounds 'A' and 'B' were described in *D. mercatorum* (Ikeda, Takabatake and Sawada, 1980; Ikeda, Idoji and Takabatake, 1981; Ikeda and Maruo, 1982), characterized by a short train of rapid pulses and a longer train of more pulses with more cycles per pulse, respectively. Normally sound particle velocity levels are about 75 dB at 5 mm or 95 dB at 2.5 mm range, emanating from the vibrating wing of the male (Bennet-Clark, 1975).

Damaging males' wings or removal of portions of either or both wings up to 95% of wing area still allows sound production with pulse pattern largely unaffected but intensity of sound much diminished, a fact that indicates the sound is not entirely produced by air set in motion by the wing; evidence, on the contrary, points to sufficient sound being transmitted via the substrate for significant female response (Miller, Goldstein and Patty, 1975; Ikeda, Takabatake and Sawada, 1980). Mating speed for males with reduced wing area is slower than for normal males in 'no choice' mating tests (Bennet-Clark and Ewing, 1969; Robertson, 1982a,b); and in rivalry with normal males, those with wing area reduced below 50% are at a considerable disadvantage.

The antenna functions as an acoustic receptor (Burnet and Connolly, 1974; Ewing, 1978). Sound waves at an approximate frequency of 200 Hz (i.e. cycles per second) cause the branched arista of the funiculus (third antennal segment) to oscillate, sending resonance stimulation to the pedicel (second antennal segment) where it is received by sensilla of Johnston's organ

(Bryant, 1978). When receptive virgin females perceive a courting male, they slow down their wandering or agitated movements to a stop or slow advance by short determined movements in front of him (Ikeda, Takabatake and Sawada, 1980; Ikeda, Idoji and Takabatake, 1981; Ikeda and Maruo, 1982; Markow and Hanson, 1981; Robertson, 1982a; Tompkins et al., 1982). Males, in contrast, respond to other males' courtship by increasing locomotor activity, courting each other (*D. melanogaster*). While the females' response is probably the result of all four sensory stimuli, the auditory component has been demonstrated to be critical for efficient and normal mating through experiments where either artificial courtship sound was electronically simulated (Bennet-Clark, Ewing and Manning, 1973; Bennet-Clark, 1975; von Schilcher, 1976a,b) or natural courtship song was played for males with portions of wings removed (Robertson, 1982a). Simulated courtship sound enhanced receptivity of females, inducing them to reduce their locomotor activity, and it speeded up males' activity. When males' wings are amputated, their reduced mating success can be restored substantially by artificially providing sound of the appropriate pulse frequency at the natural level (von Schilcher, 1976a,b; Bennet-Clark et al., 1976). However, strains of *D. melanogaster* vary in their responses to auditory stimuli during courtship: in comparing females of Oregon K, Pacific and Novosibirsk strains, Eastwood and Burnet (1979) found females of the last strain to be less affected when deprived of their aristae and to mate nearly as often with wingless males as with intact males of their own strain. Similarly Robertson (1982b) corroborated the disadvantage for wingless males with Canton-S strain females. Conclusions by Eastwood and Burnet point out that while 'auditory cues are not essential for *D. melanogaster* to mate... females deprived of such stimulation are likely to be at a considerable disadvantage... in competition with normal females.' Thus the amount of auditory component necessary for efficient female response to a courting male varies considerably among strains.

Genetic variation in acoustic signals has not only been demonstrated from these experiments but also from quantitative polygenic changes in wing vibration selection experiments (McDonald, 1979; Wood and Ringo, 1982). McDonald's experiment was performed on the Or-R stock (*D. melanogaster*) over 44 generations of selection among males for highest and lowest percentage wing vibration tested with unselected females. Even though the base population strain was presumably relatively inbred and genetically uniform, the response to selection was significant, and after 35 generations of selection, crosses between high and low selected lines showed the genetic differences to be mostly additive with no maternal effects of interactions. Wood and Ringo (1982) found bouts of vibration and scissoring bout length to respond to selection in males of *D. simulans*. Isolation of a pulse-song mutant was made (von Schilcher, 1977) from an attached-X stock treated with the mutagen, EMS (ethyl methane sulfonate). This X-chromosome mutant *cacophony* (*cac*) increases the interpulse interval, pulse length, intensity and number of cycles per pulse in the pulse song while flight wing-beat frequency and sine song are unaffected. Mating success of *cac* males is about one-sixth that of normal males (Hall *et al.*, 1980a); however, wingless wildtype males are more successful at mating than wingless *cac* males, indicating either that *cac* males are defective in more properties than the courtship song or that indeed the transmission of song pulses does not depend entirely on wing vibration frequency in air.

Courtship songs of about 40 species have now been recorded. Species differences, particularly between sibling species, have been emphasized, first by Ewing and Bennet-Clark (1968) and by Ewing (1969) who compared *D. melanogaster* with *D. simulans* and *D. persimilis* with *D. pseudoobscura*. These pairs of species differ in several features of their wing displays, but their acoustical features are also distinct: for example in the pulse song, *D. melanogaster*'s interpulse interval (30 ms) is significantly shorter than that of *D. simulans* (55 ms) and with cycles of lower amplitude, while the sine song of the former is slower at about 170 Hz than the latter at nearly 200 Hz. *D. pseudoobscura* males produce two kinds of song:

1. A low repetition rate (*rep*) song with a train of pulses, each with 4–7 cycles at 525 Hz and repeat of about five per second; and
2. A high *rep* song with about 24 pulses per second, each pulse being about two cycles at 250 Hz and *ipi* about 38 ms, a value similar to *D. melanogaster*'s.

The low *rep* song begins courtship and is transformed into the high repeat song as wing vibration continues. In contrast, *D. persimilis* rarely peforms the low *rep* song, and its typical song has a high *rep* of about 15 pulses per second but the cycles within pulses (4–7) are the same as the low *rep* song and at 525 Hz. Genetic analysis of interspecific hybrids for both of these pairs of species has indicated largely X-chromosome determination for the song differences, though species hybrid analyses may be far more complex genetically than these few studies demonstrate (Ewing, 1969; von Schilcher and Manning, 1975).

These earlier studies testing courtship songs for other species, summarized by Bennet-Clark (1975) who included courtship song data for other members of the *D. melanogaster* subgroup, *D. obscura* subgroup, *D. melanica* subgroup, and the *D. paulistorum* species complex, indicated very strongly that ethological isolation of sibling species, particularly from those that are sympatric, can be attributed to detailed differences in courtship songs. Also see Spieth and Ringo (1983) for a summary of evidence on sexual isolation. Furthermore, later studies have tended to confirm this view; species and subgroups of the genus have now been explored considerably: *D. athabasca* and *D. affinis* by Miller and his colleagues, *D. mercatorum* by Ikeda and associates, the *D. funebris* subgroup by Ewing, the *D. melanogaster* subgroup by Cowling and Burnet and the *D. virilis* subgroup by Hoikkala and colleagues.

Olfactory signals

An observation by Sturtevant in 1915 that *D. melanogaster* couples mated more rapidly in a vial that had just been used for matings than in a clean vial was not confirmed by Ewing and Manning in 1963. Confirmation of *Drosophila* sex pheromones was thus delayed for several years. Shorey and Bartell in 1970 cited these earlier observations and described experiments, giving for the first time positive evidence that odors, predominantly from females and to a lesser degree from males, elicit male courtship activity in *D. melanogaster*. These investigators designed a device for olfactory assay, a glass observation cylinder into which either clean air or air from an odor source could be admitted. Their fundamental experiments can be summarized as follows:
1. With no odor source and five males either alone or with one female, results indicated that introduction of the female did not alter the males' time spent in locomotion but doubled their tapping and orientation time; however, these initial courtship elements were directed randomly toward any other fly irrespective of sex. Wing vibration and attempted copulation were directed mostly toward the female. Therefore tapping and orientation apparently were necessary for males to identify the sex of individuals they had begun to court.
2. With several females as an odor source and males without a female in the observation chamber, the males' orientation to each other increased from twice to three times that when only males were the odor source, a value six to eight times that when only water was the odor source. Locomotion and tapping were not changed by female odor compared with the male or water source odors.
3. Odor from females in the source chamber reduced the lag time to initial wing vibration directed by a male towards a female in the observation chamber.

In the following year, Sloane and Spiess (1971) obtained similar results using the Shorey and Bartell apparatus for two fast-mating strains of *D. pseudoobscura*: orientation and wing vibration among six males in the observation chamber increased significantly when females were in the source chamber over values obtained when males were the odor source. A slow-mating third strain did not show significantly different responses to odors of either sex. Interchanging slow and fast strains of females in the source had no single significant effect on opposite strain males, although fast-mating-strain females increased orientations among both types of males while wing vibration increased for the slow-mating males.

Ehrman (1972), who had consistently observed the rare male mating advantage (see section below) with strains of Arrowhead (AR) and Chiricahua (CH) chromosomal arrangements of *D. pseudoobscura*, found that carcasses of whole males or abdomens streaked on paper, ether or acetone extracts of males and in some cases saline or detergent aqueous extracts of males would reverse the rare male advantage when the strain of male that was rare (20 percent) was used to supplement the odor source. Also, when equal numbers of AR and CH males were present in the mating test, a rare male advantage could be induced by adding ether or acetone extract from one strain. Leonard, Ehrman and Schorsch (1974) then developed a bioassay for the extractable pheromones from CH males (ether soluble fraction) and from AR males (acetone soluble fraction) (Leonard, Ehrman and Pruzan, 1974).

Airborne odor signals as source material in olfactometer tests similar to those of Shorey and Bartell were carried out by Averhoff and Richardson (1974, 1976). These authors had been measuring mating among lines maintained by sibling-mating and had noticed a marked drop in mating frequency after eight generations. When individuals that were reluctant to mate were exposed to individuals of the opposite sex from lines different from their own, normal mating occurred. Odors of either live flies (of both sexes together, a technique that seems to me unwarranted because of confounding any differences between the sexes) or a cold condensate of odors (again of both sexes together) transferred to filter paper increased the numbers of matings with flies from the inbred lines. Males were apparently stimulated to normal courtship activity by odors from lines different from their own, whereas they had been ignoring females of their own line before. Thus the negative assortative tendency that increased with inbreeding was postulated to result from pheromonal diversity between lines but similarity or identity within lines. These authors' working model was stated as follows:

> 'Pheromones mediating male courtship and female acceptance are qualitatively polymorphic... At least some, if not all, of the pheromones produced by both males and females are identical... Probably due to saturation of chemoreceptors, an individual is less responsive, or non-responsive, to those pheromones which are constantly in highest concentration in that individual's environment. Therefore... similar males and females do not sexually excite each other... there is greater male courtship stimulation and female acceptance among lines than within lines.'

Criticism of this work has been deserved to some extent, though the evidence stands that odor sources did affect mating frequencies in the observation chamber of the olfactometer. On theoretical grounds Bryant (1979) predicted, as an alternative mechanism to that of Averhoff and Richardson, that when inbreeding brings down the average mating speed, the apparent level of heterogamic mating speed should increase. In computer simulations where the effective excitation between males and females was reduced within lines, Bryant achieved the Averhoff and Richardson effect entirely artificially without invoking pheromonal polymorphism in the model. Of course such simulation fails to explain the physiological basis for the lowering of effective excitation within lines during inbreeding. As far as experimental verification with actual flies is concerned, Averhoff and Richardson's results have not been confirmed when tested either with *D. pseudoobscura* strains (Powell and Morton, 1979) or with *D. melanogaster* (van den Berg et al., 1984). No evidence was obtained for an increase in negative assortative mating among inbred lines (a decrease in mating speed within the lines was not found). In the latter study, sex pheromone composition was analyzed by mass spectrometry in the manner of Antony and Jallon (see below). The sexes differ in a number of components of each pheromone, and there is little or no basis for the Averhoff and Richardson proposal that the sexes share identical pheromones. Polymorphism for the male pheromone was evident among the four strains of *D. melanogaster* tested, but the female pheromone was not variable among those strains.

Two avenues of research on olfactory stimuli that have established the chemical nature of the *D. melanogaster* sex pheromones have been developed since 1979, namely:
1. Perfection of bioassay techniques by Jallon, Venard and Antony that led to chemical analysis of both male and female aphrodisiac substances; and
2. Isolation and analysis of mutants that produce abnormal odors or that fail to produce or fail to respond to odors, studied by Tompkins, Hall and their associates. The following discussion summarizes the important details from these two groups.

Venard and Jallon (1980) and Venard (1980) improved on the Shorey and Bartell apparatus. Their technique was modified from that used by Jallon and Hotta (1979), who had studied 'sex appeal' (duration of wing vibration directed towards live females or gynandromorphs) in order to locate sources of female pheromones (principally in the abdomen). One hundred flies of either sex were allowed to run for an hour in a small vial which was then washed with hexane. The hexane-fly odor mixture poured into a small watchglass and evaporated rapidly became the fly concentrate, and the watchglass upside down served as a feminine or masculine 'sky' over a glass plate, thus enclosing a courtship chamber. Either one or two males four-days old were introduced into the chamber that contained either female or male extract, or no extract. In a blank chamber two males would in a few cases vibrate slightly, never more than a second or two. In a feminized chamber, a single male would also vibrate just slightly and for one second at most; but two males in such a chamber vibrated wings toward each other considerably, and the male toward whom the vibration was directed responded with rejection signals of wing flicking or kicking. If one of the males had been held in a feminine chamber and then was introduced with a naïve male into a blank chamber, the naïve male courted the other nearly as much as in the feminized chamber. Finally, masculinizing the chamber did not significantly affect wing vibration. Thus a male will vibrate toward another male in the presence of some odor emitted by females; however, the chemical stimulus alone was not sufficient to initiate the courtship, since a male alone did not vibrate to any degree. Thus sensory cues involving presence of a second fly in addition to the odor are necessary to evoke vibration. By diluting the hexane extract and testing males at various concentrations, Venard showed that an optimal concentration equivalent to the extract of a single female evoked the largest proportion of time spent in vibration and shortest latency time (1.5 min).

A second part of Venard's experiments measured the attraction of each sex towards a source of either sex's extract. By using an olfactometer Y-maze about 12 cm long with horizontal air flow, where 50 flies were placed in the starting position at the base of the Y and allowed to wander for 5 min either toward the odor source up one branch of the Y or toward the other clean branch, Venard showed that about 30 percent of the flies were not responsive to the odor flow. For the remainder, females do not attract either males or females over the distance of the Y-maze. Thus the female pheromone(s) is(are) attractive only over a short distance (probably less than 7 mm). However, significantly more females were attracted to the males than to the blank arm of the maze in 4–10 min time. Thus male odors appear to act over greater distances and are more volatile than female odors. This was the first good evidence of females' response to the presence of males before courtship.

Antony and Jallon (1982) settled many questions about *Drosophila* pheromones when they applied gas-liquid chromatography, sensitive to small amounts of airborne substances, for analysis of tissue compounds extracted with hexane from mature *D. melanogaster*. They found hydrocarbon peaks, 26 peaks in females and 15 in males, indicating compounds ranging from 20 to 30 carbons with larger molecules (27–29 carbons) predominating in females (heptacosadiene having the highest concentration among four dienes and some monoenes) and smaller molecules (23–25 carbons) in males (tricosene and pentacosene in highest concentration and lacking dienes entirely). As a bioassay for behavioral response to these substances, Antony and Jallon applied hydrocarbons extracted from each sex separately to the cuticle of a freshly killed male fly that had been rinsed in hexane, and measured wing vibration by naïve males that courted the decoy; also the separate compounds were applied individually to decoy flies and tested for induction of wing vibration. Female dienes and monoenes were most effective in inducing wing vibration by males while male extract, alkanes, and male monoenes did not have much effect on males' courtship. Thus the larger less volatile hydrocarbons were typical of females. Males responded to these only when close to the decoy fly. The more volatile compounds predominant in males could be sensed by females and males over greater distances. Jallon (1984) has summarized recent evidence on chemical cues and pointed out that evidence for male odors that influence female movement is less clear than evidence for female odors that influence male courtship positively.

Antony and Jallon's findings were largely confirmed by van den Berg *et al.* (1984), who analyzed four wild strains of *D. melanogaster* using ether extracts of each sex: male substances were predominantly single unsaturated alkenes (23–25 carbons) and female substances, saturated alkanes and double unsaturated alkenes (27–29 carbons). The strains varied in relative amounts of the male substances tricosene and pentacosene, but female substances were unvarying among the strains. The Gröningen strain males mated in greatest numbers with least courtship effort among the four strains and also produced the largest amount of pheromone, particularly *cis*-9-tricosene. Thus its lower courtship effort and greater mating frequency could be readily explicable. Jallon (1984) found females of three strains to show marked quantitative variation among the strains, particularly in the dienes; and males of one strain (Canton-S) do mate with each strain in proportion to aphrodisiac content of the females.

Genetic techniques involving isolation and analysis of mutants that either produce or fail to produce odors or fail to respond to odors have been employed to demonstrate olfactory signal mechanisms. Tompkins, Hall and Hall (1980) extracted flies of the mutant fruitless (*fru*), a recessive (chromosome 3, *D. melanogaster*) behaviorally characterized in males by active courting of wildtype males and the tendency to court other males rather than females. Immobilized *fru* males and isolated abdomens from *fru* males stimulate wildtype males' courtship. Gas–liquid chromatograms of extracts from *fru* males revealed that their hydrocarbon profile is remarkably similar to that of females (though in 1980 Antony and Jallon had not yet completed their work, so that the substances were not chemically analyzed). Normally when adult males first emerge from the pupa, they emit female-type odors; immature wildtype males therefore stimulate mature males to court (Tompkins, Hall and Hall, 1980). It is likely that *fru* blocks the male's normal post-developmental shut-down of female pheromone production that occurs along with initiation of courtship tendency and mating drive.

If female odors are necessary to induce male courtship, then it would be expected for males with an olfactory defect to exhibit less courtship than wildtype males and to be less efficient at mating. The smellblind (*sbl*) X-chromosome variant does not respond to female pheromones even at high concentrations when two males are tested together for a courtship index assay; *sbl* males do not respond to organic substances generally in an odor avoidance test, so their defect is a general odor sensory input deficiency (Hall *et al.*, 1980a, 1980b; Tompkins, Hall and Hall, 1980). Courtship by *sbl* males towards females is depressed to about half that of wildtype males. In females, the *sbl* variant does not respond to courtship by slowing down random movement which is the typical response of wildtype females to normal male courtship.

Siegel and Hall (1979) described a major factor in the conditioning of *D. melanogaster* males by prior sexual experience: males that have been paired with unreceptive fertilized females fail to court virgin females in a normal manner; in fact they will perform little or no courtship with virgin females until about three hours have passed. Experiments (Tompkins *et al.*, 1983) indicated that males had not been fatigued by the rejection of fertilized females because their general activity was unimpaired, nor were they sexually exhausted since they courted newly emerged males at a normal level. However, mutant males with general olfactory defects (*sbl* and *olf-C*, the latter being sex-linked and not allelic with the former) continued to court virgin females after experiencing fertilized-female rejection. Also wildtype males exposed to extracts from fertilized females will not court virgin females at anywhere near a normal level. Thus wildtype males appear to learn (in a broad sense) the odors emitted by fertilized females. In fact an antiaphrodisiac odor from such females is indicated. The relevant substance found in the fertilized female's reproductive tract is the ester *cis*-vaccenyl acetate which is hydrolyzed by esterase-6 to *cis*-vaccenyl alcohol (Mane, Tompkins and Richmond, 1983). Both the ester (produced in the male's reproductive tract) and its hydrolyzed product greatly lower the courtship behavior of males.

Finally by testing a number of single mutant strains that are defective in ability to be conditioned (mutants isolated on the basis of their inability to associate a chemical non-sexual odor with an electric shock), Tompkins *et al.* (1983) and Gailey, Jackson and Siegel (1984) found that aversive olfactory learning and experience-dependent courtship share common features. Their working model was stated as follows:

> 'Fertilized females present to courting males cues that have excitatory and aversive effects and that during courtship males learn to associate these cues, such that, when subsequently presented with only courtship-stimulating cues (by a virgin female), the courtship response is repressed in 'anticipation' of the aversive cue...'

Experience-dependent modification of courtship and olfactory conditioning by electric shock share common elements in the fly's learning process and both are under genetic control.

Visual signals

Light is necessary for mating in the majority of *Drosophila* species (Grossfield, 1971; Spieth, 1974a), though some species will mate in darkness at a depressed rate compared with the rate in light. Movement stops almost completely in darkness for most Hawaiian species. Light is essential for orientation and approach of prospective mates, particularly in those species that depend on complex movements in courtship, as do *D. subobscura*

(Pinsker and Doschek, 1979) and the picture-wing species from Hawaii (Spieth, 1974b, 1978).

For species that can mate in darkness (*D. melanogaster*, *D. simulans* and *D. pseudoobscura*, for example) the relative importance of visual stimuli to courtship and mating becomes a problem for analysis of discrimination signals and the information between prospective mates. Several mutants of *D. melanogaster* with defective vision show impairment of courtship by males, as summarized by Hall *et al.* (1980a). Males that are blind because of neurological defects (for example, no-receptor potential, *norpA* and optomotorblind, *omb*^{H31}) either misdirect their courtship actions or fail to respond to movement of other flies and thus do not orient toward moving females.

Males with impaired vision and pleiotropic morphological traits such as tan body (*t*), ebony body (*e*) and glass (*g*) have trouble tracking females. Also eye pigment mutants such as alleles of white (*w*) are affected similarly (see Burnet and Connolly, 1973; Pak, 1975; Matsumoto, O'Tousa and Pak, 1982). In turn, a female's movement, which is usually agitated when she enters an experimental chamber, tends to slow down after a latency period and when she is courted by a wildtype male, will come to a momentary stop followed by slow definite movements allowing the male to orient. Males with impaired vision take much longer than wildtype to bring the female to a stop and start her slow forward movement (Hall *et al.*, 1980a; Tompkins *et al.*, 1982). Just how important female movement is to the male's orientation and continued courtship was demonstrated (Tompkins *et al.*, 1982) by employing females of a temperature-sensitive paralytic mutant (*shibire*^{ts}) that cannot move when temperatures are over 25 °C. Males court these females vigorously in cool temperature but nearly stop courting when females are paralyzed.

In a multivariate analysis of *D. melanogaster* courtship elements and transmission of information between the sexes, Markow and Hanson (1981) found the female behaviors of preening and standing still to affect courtship by indicating a female's willingness to mate. *D. simulans* has much reduced mating in darkness and therefore must have more dependence on visual cues than its sibling species, *D. melanogaster*. Males of *D. simulans* spend more time orienting in front of females, vibrating around the female, often in arcs somewhat suggestive of Hawaiian species like *D. silvestris*.

Cook (1979, 1980) devised a revolving chamber which allowed a male either to track a stationary female on a slowly moving floor or to track a female tethered and rotated on the radius of a disc. Variation in the courtship around the female was shown to be limited by the velocity at which the male had to run in order to court the female. At lower velocities males did more circling to the side of the female and the mean distance of the male behind the female increased with velocity of revolving; the amount of this effect estimated the reciprocal feedback gain of the control system for tracking, and strain differences in these courtship variables were revealed. The proportion of wing extension during tracking varied widely, the lowest value being displayed by the strain that tracked the fastest; thus the outcome of courtship was dependent upon the relation between wing extension and tracking speed. Males oriented to the female's abdomen at close range (less than 4 mm) but more centrally to her body at greater distance. These experiments established the importance of visual stimuli for males' orientation toward females, and strain differences indicate genetic determination of tracking abilities.

How much visual stimulation is necessary for the female is far from clear. Probably for *D. melanogaster* vision plays a lesser role for the female than it does for *D. simulans* and other more light dependent species (*D. subobscura* and the picture-wing Hawaiian species) where mutual actions between the sexes must be observed continually during initiation and throughout courtship displays. Visual stimuli seem most important for the male's initiation activity. Movement of a slow and definite pace by the female attracts the male's 'attention' so that his behavior escalates to orientation and vibration.

Tactile signals

In most species, males usually tap other flies that are near and have slowed down random movement. A male taps by extending his prothoracic tarsus, and if the other fly is a female she often reciprocates by tapping with a mesothoracic leg. A receptive female may then give an acceptance response and orientation proceeds. If he taps a female of a different species or a non-receptive female, she will usually give a rejection response. Often the fly tapped is another male which may in turn respond by wing flicking (male to male rejection in *D. melanogaster*), and the tapping male usually desists from further courtship.

Stimuli from tapping signals could be entirely mechanical, chemosensory, or both. Hairs (setae) of the legs and body are sensory receptors, but in addition the tarsi, antennae, and several other structures possess sensilla (sensory pits or protrusions in the epidermis) for chemoreception (Fig. 1) (Bryant, 1978; Hodgkin and Bryant, 1978). In *D. pseudoobscura* and its sibling species, *D. persimilis*, tapping nearly always precedes orientation, while in *D. melanogaster* tapping is not as regular nor as persistent as in the *D. obscura* group species. It is apparent that tapping is a discrimination device for sex and for species recognition. Surgical removal of males' foretarsi often lifts restrictions of males toward courting of females of different species that are usually avoided (Spieth, 1974a).

Elaboration of complex tactile stimuli has been described for the Hawaiian species, in particular where tarsal and tibial setae are varied in size, number and shape. Not only in morphology but in their behavioral use of vibrating, striking or rubbing the female being courted, the variety is extraordinary

Fig. 1. Scanning electron micrograph of *Drosophila pseudoobscura* female's prothoracic third tarsal segment showing two sensilla pits, receptors of chemical stimuli. × 2000. *Photograph by C. C. Dapples, University of Illinois at Chicago, SEM Facility, Department of Biological Sciences*

when the array of species is compared (Spieth, 1978; Carson and Bryant, 1979; Spieth and Ringo, 1983). No doubt sexual selection is the predominant mechanism at work in perfecting these highly specific mate recognition systems.

DISCRIMINATION AMONG INTRASPECIFIC PROSPECTIVE MATES

From the extensive literature on *Drosophila* mating behavior that has accumulated over the past 30 years, it is evident that mating is seldom 'at random' between individuals in natural populations. Given any significant variation among prospective mates in their sensory abilities or sexual signals, as outlined in the previous sections, both competition among males for receptive females and choice by females favouring mates with 'proper signals'

become highly likely. It is the latter phenomenon especially that concerns us here for at least the following reasons:
1. The female controls whether mating will occur.
2. Most of the literature on choice of mates points to the female's choice as critical to the outcome.
3. We wish to know whether discriminatory mating leads to improvement in the efficiency of reproduction over non-discriminatory mating. (Discrimination by males among females will be discussed later.)

When two or more strains of males compete for female acceptance, a central problem to be solved is whether mating success can be attributed to relative competitive ability among the males or to choice by the females or to some combination of both. To some extent the former cause has been substantiated, both for mutant strains like *e* of *D. melanogaster* and for wildtype strains that differ in the qualities of their courtship elements from the normal (Spiess, 1970). Female choice, however, plays a considerable role as demonstrated from two lines of evidence:
1. 'Female choice' experiments in which the female's ability to sense differences among males has been tested; and
2. Female acceptance based on varying frequencies among courting males ('rare male mating advantage').

Female ability to sense male differences

A number of experiments with *D. melanogaster* in which virgin wildtype females were tested with two types of wildtype males have indicated that successful males tended to be associated with greater fertility. This outcome has important obvious consequences for the evolution of the mate recognition system. Markow and her colleagues (Markow, Quaid and Kerr, 1978; Long, Markow and Yaeger, 1980) found that virgin males were more successful than males that had just mated ('experienced') one hour previously or less in spite of equality in such factors as lag time before courtship and time spent in courtship. Also for males of different ages, the older (four days) were more successful than the younger (two days) in spite of absence of differences in courtship elements. In both cases the males preferentially chosen were the more productive of offspring.

If female choice is operating in these cases, we must assume that competing males are providing sensory cues detectable to the female; by contrast, in the absence of sensing ability, females would not be able to make choices, so that the outcome would depend on the males' relative courtship efficiency. Markow (1986) has conducted experiments with single females, either wildtype or mutant for sensory deficiencies, given two males of contrasting fertility that would compete for the female. Each of four types of female were tested as a single virgin confined with two males:

1. Wildtype.
2. *norpA* (blind).
3. *sbl* (smell deficiency).
4. *al th*(aristaless-thread antennae with auditory deficiency).

The two males (= treatments for the female strain) were of four combinations:
1. Canton-S (CS) versus Oregon-R (OR) both four-days old.
2. Four-days old versus two-days old, both CS strain.
3. Virgin males versus males that had mated one hour previously, CS strain four-days old
4. Random males four-days old from a freshly caught wild strain

When wildtype females were presented with competing males, treatment 1 CS males consistently mated twice as much as OR males and the data for treatments 2 and 3 confirmed earlier findings (above) that older virgin males mated more than younger experienced males. Also the 'winner' males produced significantly more progeny than the 'losers' (which were mated without choice to the same strain of female). Finally, for wildtype males (treatment 4) from a freshly caught strain, 'winner' males were significantly more fertile than 'loser' males. For the first three combinations of male competitors (treatments 1 to 3), there was no significant difference between competing males in courtship lag time, size or in time spent in courtship. In the case of random wildtype males, 'winners' were larger and courted ahead of 'losers'. For this last combination mating success probably was the result of male competitive ability rather than the female's sensory perception and choice. (Large size plays a role in male–male competition, as noted by Partridge and Farquhar, 1983.) By contrast, when the female was a mutant deprived of a principal sense perception, the outcome was changed in certain important cases: With blind (*norpA*) females, CS males versus OR males and four-day versus two-day old males were not significantly different (though the former mated more than the latter in each case), while for virgin versus experienced males there was no change from the wildtype female tests and also with random wildtype males the winners were still first to court and larger than losers. In marked contrast, with the *sbl* females (smell-blind), all advantages displayed by one type of male over another disappeared completely except that winners were those males that courted first and were larger, regardless of the type tested. Thus any differences between CS and OR males, between ages, and between virgin versus experienced condition of males could be attributed to odor detectable by wildtype but not by *sbl* females. Finally when using the *al th* auditory-deficient females, the results of all tests resembled the outcome when wildtype females were used.

According to Markow (1986):

'Using wildtype females, there was a strong and consistent advantage for males of a given treatment, whether it was strain, age, or mating status. That these advantages were dependent upon different olfactory cues is suggested by the change that occurs when males of various treatments competed for smell-blind females... In the absence of the females' ability to detect olfactory cues, male success appears to be dependent upon two correlated traits, larger body size and greater courtship propensity.'

Thus CS and OR males, old and young males, and virgin and mated males provided different odors to females. Further, with blind females (*norpA*) some of the CS advantage over OR males and some of the four-day old advantage over two-day old was lost; relevant to those changes a fine structure analysis of these types of males' courtship behavior by Markow and Hanson (1981) revealed some visible differences in the sequence of courtship elements between the male types.

In summary, males that display odor and visual differences to a female being courted bring the female into a receptive state such that she appears to prefer one type of male over another. Overall courtship intensity, however, also plays a role in male to male competition particularly when the signals to the female's sensory perception are partially nullified so that her ability to choose is reduced. Thus the female's central nervous system presumably responds to several messages, and no single factor among the cues received is so critical or so overriding that if deprived of that factor the female cannot mate. Thus choice by the female and more intensive courtship by the males each play a role in the final outcome.

Rare male advantage

In the general problem of attributing mating success among males to choice by females, demonstration of the rare male mating advantage has been interpreted as favorable evidence (Petit and Ehrman, 1969; Ehrman, 1978; Ehrman and Probber, 1978; Spiess, 1970, 1982a,b; Spiess and Kruckeberg, 1980). In spite of differences in mating activity among male strains, when two or more strains of males compete for female acceptance, either strain may gain an advantage by mating proportionally more as its frequency becomes low. Data supporting a behavioral basis for this phenomenon were summarized in my review (Spiess, 1982a) where arguments for and against were discussed. Here I shall restate briefly some of the points from that review and add the latest developments.

When experimentalists first described the proportional increase in mating success brought on by rarity (Petit in 1951–58, later by Ehrman and by Spiess in 1966 as cited by Spiess, 1970), it was logical to conclude that females must be able to evaluate cues from two or more kinds of courting males and to alter their receptivity in favor of the rarer type (once we determined that it was by varying the frequencies of males but not by varying frequencies of females that the advantage was demonstrated). However, when mathematical models of frequency dependent mating success were explored by O'Donald (1977, 1980), he pointed out that a rare male advantage did not require a behavioral change in the selective agent (virgin female's change in receptivity). In fact he showed that the only necessary ingredients for rare male advantage are constant preferences among a sufficient number of females (α = proportion preferring type A males and β = proportion preferring type B males) and input frequencies (u for A males and v for B males) such that the ratio $u/v < \alpha/\beta$. However, with less restrictive and more realistic conditions, for example if female preferences are conditioned by their encounters with each type of male (frequency dependent 'preferences'), similar outcomes of rare male advantage can be demonstrated. In fact, O'Donald (1980) described numerous models that agree in varying degrees with observed results. The interpretation for those data remained for a behavioral analysis to resolve.

A reasonable behavioral basis for the rare male advantage was forthcoming from experiments in which my students and I tested single females of *D. melanogaster* with two types of males easily recognized by the observer because of an eye color difference due to a single allelic substitution at the brown (*bw*) locus with scarlet (*st*) homozygous (Spiess and Schwer, 1978; Spiess and Kruckeberg, 1980; Spiess, 1982b). What proved to be the most important observation was uncovered as we asked the following question: is the rare male successful irrespective of the sequence of courtship experienced by the female? The answer was most clear: the female tends to mate about 80% of the time with the type of male that is not the first to court, irrespective of the input ratio of male types. Thus the female appears to learn (in a broad sense) signals from the first type of male that courts her; since her initial response is refusal and on the average she takes about five bouts of courtship before accepting a male, she usually samples more than one male's courtship. When two types of male are experienced by the female in unequal numbers, the more common type is likely to court first and then be rejected. Thus when several pairs of flies are tested, the rarer type male gains an advantage.

In these experiments with eye color mutants, flies were deficient in visual acuity; and males differing in their *bw* genotype displayed noticeable courtship differences including general activity, ability to follow the female, courtship sequence duration and wing vibration duration (Spiess and Kruckeberg, 1980). Whether they also displayed any odor differences was not tested, but since *bw* is a pteridine inhibitor there are several biochemical pathways that could be affected in the mutant flies. Whatever the recognition cues, females that discriminate against the type of male first to court must be distinguishing those two types by one or more signals emitted by the males. A randomly mating fraction of females must either be unable to distinguish between the types of males or have such a low threshold that they will accept any type of male that emits a broad range of signals for the species. Thus it is our working hypothesis that we can distinguish discriminatory females (from non-discriminatory) by allowing females to be courted by two or more types of males, observing four or more bouts of courtship and rejection of the type male that was first to court. A female that accepts the male that is first to court after his initial three or four courtships we may define as having a low threshold, and presumably she is mating with less discrimination (thus more or less randomly); that is if no evidence of the female's rejection is observed, we may assume that she has less discriminating ability or takes less time to exercise that ability than a female that does display rejection.

As for a mathematical model for this behavior, O'Donald (1983) pointed out that our data appear to illustrate a case of 'frequency-dependent expression of preference', in which a female encounters a group of males but having been habituated to or conditioned against the phenotype of the first male she has met, she mates preferentially with another phenotype, provided that it is encountered within a limited number of additional courtships. If preference parameters for females (α and β) are conditional upon frequency of encounter and represent the fraction of females (d) that are discriminatory, as defined above, and if input frequencies of males are u for type A and v for type B, then probabilities for matings, P(A) and P(B) by A and B type males respectively, will be given by the following (if $\alpha = \beta = d$):

$$P(A) = u + d(1 - 2u)$$
$$P(B) = v + d(1 - 2v)$$

as given by O'Donald (1983) and by Spiess (1982a).

Additional observations extending the evidence for female discrimination against first-to-court males have been described for wildtype *Drosophila* in three species where genetically distinct types of males from polymorphic populations could be tested (Spiess, 1982a). In *D. silvestris* of Hawaii, single females were tested alternatively first to a male from a population on one side of the island and second to a male from the opposite side, these males differing in possession of an extra row of tibial bristles that come into play during the male's 'head under wings' approach to a receptive female (Carson and Bryant, 1979; Spiess and Carson, 1981). Females with discrimination ability rejected the first-to-court male by a ratio of nearly 3:1. In *D. persimilis*, tests involving homokaryotype strains for Klamath versus Mendocino arrangements of the third chromosome indicated that discriminating females mated to the second type of courting male at a rate of 80%. In *D. melanogaster*, males from strains that had been selected for high and low sternopleural bristle number were tested with single females both from the H and L strains and females from three partially inbred wildtype strains (LS, CS, and OR). Females discriminated against the type of male first-to-court at about 70% except for the last two wild strains (CS and OR) whose females mated randomly (Spiess, 1982a, Table 4).

Subsequently, further tests with *D. melanogaster* have continued to indicate the phenomenon of female discrimination against the first-to-court type male:
1. Strains derived from a natural population polymorphic for allozymes of esterase-6 (Est-6 Fast, Est-6 Slow, and Est-6 Null) contributed by Professor R. C. Richmond, Indiana University in 1982, were tested in our laboratory (Spiess and Stubblefield, unpublished). Each Est-6 homozygous genotype (F, S and N) female was placed in a glass vial with a pair of males (either F versus S, F versus N or S versus N) and courtship sequence and mating recorded. With 90 tests (30 for each female type) in which the female was courted at least once by each male, two-thirds of the matings were by the male that was second to court (homogeneous results over the three types of female). Richmond and his colleagues (Gromko, Gilbert, Sheehan and others) have found that allozyme variants of esterase-6 can influence repeat mating by females and sperm precedence after multiple matings; it appears that esterase-6 is synthesized mainly in the male reproductive tract and its activity is to hydrolyze a lipid found in the male seminal fluid after it is transferred to the female during mating. The allozymes Est-6F and Est-6S have different substrate specificities and certain kinetic parameter differences; thus they could have a male pheromonal influence and be detectable as distinct odors by females (Richmond and Senior, 1981; Mane, Tepper and Richmond, 1983; Gromko, Gilbert and Richmond, 1984).
2. Techniques similar to those employed by Kruckeberg (1978), Spiess and Kruckeberg (1980) and Spiess (1982b) with eye color genotypes of 'red' (bw^{75}/bw^{75}; st/st) and 'orange' (bw/bw^{75}; st/st) were repeated with single females tested with two males, one of each genotype. However, an age difference between competing males was made a critical factor (Spiess and Salazar, 1983). One difference between our earlier (1978–80) and later (1980–82) tests that might account for some of the mating success differences between those tests was age: in earlier tests flies were aged five days past eclosion and red and orange males mated nearly equally, while in later tests flies were aged only three days and orange males mated at about half the frequency of five-day old males. Experiments by Long, Markow and Yaeger (1980) indicated that age increased mating success. We supposed that if red-eyed males had a sexual advantage over orange during the first two or three days of adulthood, that advantage might diminish as both types aged to five days. Eye pigments of these mutants darken within four to five days and difference in visual ability could be minimized thereby. Alternatively, whatever factor females detect in older males (see section above) such as an odor difference or general courtship activity may make them more successful. Thus flies were either matured equally to an age of three days when tested (experiment one) or matured for two days in the case of red-eyed males but three days for orange-eyed

Table 2. Numbers of males mating as a function of first-to-court type and age. (Reproduced from Spiess and Salazar, 1983.) X^2 = Contingency chi-square with one degree of freedom. R = red eye, O = orange eye.

A. Experiment 1: matings with all males three days old

Female	Male first to court	Male mated R	Male mated O	X^2	P
R	R	12	15	5.9	0.015
	O	18	5		
O	R	16	11	4.7	0.04
	O	20	3		
Total		66	34		

B. Experiment 2: matings with red males two days old and orange males three days old

Female	Male first to court	Male mated R	Male mated O	X^2	P
R	R	8	17	13.9	≪0.01
	O	21	4		
O	R	11	20	13.6	≪0.01
	O	19	3		
Total		59	44		

males (experiment two). Testing was done with one red and one orange male introduced into a glass vial followed by a virgin female and timing of courtship was begun. Tables 2A and 2B present the data for experiments one and two respectively for those matings in which both types of male courted (about 70% of the trials). In agreement with previous results, females accepted the male second to court preferentially in both experiments. Associations are significantly negative between first-to-court and mating. However, that effect is more marked when the orange male is a day older than the red (experiment two) than when both types are the same age (experiment one); in addition, orange males mated about 10% more in experiment two than in experiment one, a significant difference and a level comparable to that in our earlier tests with five-day old flies.

3. The experiments with two genotypes of eye color based on the brown (*bw*, *bw*[75]) pair of alleles and homozygous for scarlet (*st*) were extended to three genotypes of males competing for acceptance by single females

(Bowbal, 1984; Spiess and Bowbal, 1987). Techniques were similar to those used before, but eye color genotypes were the segregants of the brown locus, 'red' bw^{75}/bw^{75}; st/st, 'orange' bw^{75}/bw; st/st, and 'white' bw/bw; st/st. Either three males each of two types or two males each of three types were first placed in a glass vial followed by the female. Courtship sequences and times were recorded. There were twelve different combinations tested: three combinations of two male types plus one combination of three male types by each of three eye color female types; a minimum of 50 trials for each combination were recorded in which each eye color type male had courted at least once and a mating had occurred within 30 mins. (Actually over 2,000 females were tested since several had a low mating threshold with only one male courting and mating; unless a female has experienced courtship from more than one male type, it cannot be said that she has had an opportunity to be discriminating.) In brief, overall mating success favored males with the most pteridine: red males achieved the most matings (52–58), being about twice as successful as orange (29–33) and four times as successful as white males (13–15). However, the negative association between first courtship and mating success was maintained both for the trials with two types and for trials with three types of males. Tables 3A, 3B and 3C present the pertinent overall data arranged according to combinations of males in their first courtship order and mating. Table 3A presents results similar to those already observed with two competing types of male, as in references cited above, with the added white eye genotype that tends to be more successful whenever it is not the first type to court. When three types of males are present, one, two or three types may court before the female becomes receptive; and the data given in Tables 3B and 3C only include the trials where all three types courted. (For further details on remaining trials, see Bowbal, 1984.) When all three types of males court, possible associations are between just the first type of male to court and the type that mates, between the first and second courting types combined and the mating type, between the second type alone and the type that mates, and so forth. From Table 3B it is evident that there remains a negative association between the type of male first to court and the male that mates; in fact if the average mating frequency for each type of male is compared with the row by row frequencies, we note that the latter frequencies are higher than their average for those trials in which that type of male was not the first to court (with the exception of red males mating, 49%, when white males were first to court). No significant association was found between either the second male to court or second type of male to court and the mated type. However, when the data of Table 3B are partitioned into combinations of the first two types of males courting, as in Table 3C, the type of male that mated was more often in favor of the type which the female had not yet

Table 3. Numbers of males mating as a function of first-to-court type. (Data from Bowbal, 1984; Spiess and Bowbal, 1987). R = red eye, O = orange eye, W = white eye.

A. Trials in which two eye-color type males courted, pooled over all types of females

		Male mated				Male mated				Male mated		
		R	O	Σ		R	W	Σ		O	W	Σ
	R	49	36	85	R	53	24	77	O	33	37	70
First to court	O	67	19	86	W	83	15	98	W	67	19	86
	Σ	116	55	171	Σ	136	39	175	Σ	100	56	156
		Contingency $X^2 = 8.0$				$X^2 = 6.3$				$X^2 = 15.9$		
		$P = 0.005$				$P = 0.01$				$P < 0.001$		

B. Trials in which three eye-color type males courted, pooled over females. Percentages within each row (first-to-court male) in parentheses

		Male mated				
		R	O	W	Σ	
	R	32 (43.2)	30 (40.6)	12 (16.2)	74	$X^2 = 10.1$
First to court	O	41 (64.1)	12 (18.7)	11 (17.2)	64	d.f. = 4
	W	28 (49.1)	23 (40.4)	6 (10.5)	57	$P = 0.04$
	Σ	101 (51.8)	65 (33.3)	29 (14.9)	195	

C. Trials from B (above) partitioned by the combined first two types of males to court. Bold type indicates types that mated but not in the first two types to court

		Male mated				
		R	O	W	Σ	
	R O	17	7	7	31	
	R W	15	**23**	5	43	
First two types	O R	18	8	7	33	$X^2 = 21.85$
to court	O W	**23**	4	4	31	d.f. = 10
	W R	10	**14**	2	26	$P = 0.02$
	W O	**18**	9	4	31	
	Σ	101	65	29	195	

experienced among the first two courting types. Note that in the row by row combinations of courtship the highest frequencies (bold type) of mating for each type of male are those that the female had not yet experienced after two types had courted. Thus the female discriminates not only against the first type of male that courts but also against the combination of first two types.

Thus we find considerable evidence that a virgin *Drosophila* female samples courting males' signals. Because her first response is generally to reject males' advances for at least a few courtship bouts, she has the opportunity to sample various male genotypes until her threshold against mating is lowered. Her acceptance thereafter is based on signals different from those that she experienced first in courtship, yet the signals must be within her species recognition range for acceptance of the male. This behavioral evidence for discrimination ability by the female leads to the logical significance of the rare male mating advantage. When diverse types of males court a group of females, if a significant proportion of the females discriminate among prospective mates in the way just described, the rare male advantage will be expressed. As pointed out elsewhere (Spiess, 1982a), arguments that try to find evolutionary reasons for females' preference toward rare males 'put the cart before the horse' and divert our attention from the point of measuring female discrimination ability.

Criticisms of rare male mating advantage

In my review (Spiess, 1982a) some of the criticisms directed at experimental design and interpretations of the rare male effect were discussed, particularly cases investigated by workers who did not observe expression of the effect or where the effect was induced by wing clipping in houseflies (Bryant, Kence and Kimball, 1980) or where *Drosophila* wildtype males were selected from tops or bottoms of storage vials (Markow, 1980). From experiments (Spiess, 1982b) in which genetic backgrounds were made substantially homozygous or heterozygous, I had concluded that a female's discrimination in favor of a male that differs from the male first to court is best expressed when she is heterozygous. Thus either extreme homozygosity (isogenicity) in a female or lack of difference between males in their courtship elements or both would be expected to reduce or eliminate the rare male effect. Males distinguishable to a female by their general activity (as in Markow's tops and bottoms of storage vials experiment) or by their relative debilitation (wing clipping in houseflies noted by Bryant, Kence and Kimball) would be expected to display a rare male effect.

Much commentary has followed the report by Bryant and co-workers on the debilitation of males by wing clipping. This marking of flies was implicated as leading to a spurious rare male mating advantage in the trials where the marked flies were common and the unmarked rare. Those authors implied that such a detrimental effect in *Drosophila* experiments could account for the rare male advantage even when marked and unmarked flies were alternated, as the technique has usually been applied; thus a systematic bias in the experimental design was implicated as creating an artifact. Of course what might produce an artifact for houseflies that do not display elaborate courtship does not necessarily mean that *Drosophila* species would perform in the same manner. Counterarguments to the Bryant, Kence and Kimball (1980) paper can be summarized as follows:

1. If wing clipping were detrimental for males of the common strain, thus benefiting the rare unmarked strain in mating tests, then the reverse

situation (clipping the rare strain males) should cause the advantage to disappear. (Incidentally as a criticism of the Bryant, Kence and Kimball paper I should point out that those authors omitted to make a comparison for each strain across the test ratios when that strain was the marked strain, or conversely when that strain was the unmarked; the reader is not given information on whether any strain(s) when marked would have any advantage at the 4:16 ratio, for example, compared with the same strain when marked at the equality or 16:4 ratio.) Leonard and Ehrman (1983) used AR and CH inversion strains of *D. pseudoobscura*, clipping only the rare type males in mating tests with 50 flies per chamber. The rare male advantage was demonstrated at about the same level as that of previous tests where clipping had been alternated between strains.

2. Anderson (1987) has reanalyzed data collected by him with Ehrman previously. All replicate mating chambers indicated an average small decline in mating frequency (16%) for clipped males compared with unclipped. However, by applying a formula devised by Kence (1981) to estimate mating success for a given debilitating effect of clipping, Anderson showed the amount of advantage gained by unclipped flies is negligible, only 0.8% relative to the common (clipped) males, an advantage too small to play any prominent role in the observed large increases in mating success at low frequencies.

3. In a progressive removal of portions of the males' wings with *D. melanogaster* (Canton-S strain), Robertson (1982a) found that male mating success did not decrease with amounts of wing clipped until a large part (more than half) of the wing was removed. These results contrast with earlier work of Ewing (1964) who did find a decrease in mating success with amount of wing removed.

4. Knoppien (1984) has commented on the possible artifactual (Bryant, Kence and Kimball, 1980) effect of wing clipping and on the formula Kence (1981) used to show the overall mating success induced by marking of males. First, Knoppien indicated an unjustified bias in Kence's formula that weights equally the number of matings by rare and common strains when marked. In fact, the number of matings by rare males will be greater on the average when the rare strain is marked than when the common strain is marked, when observation time is fixed. Knoppien revised Kence's formula and pointed out that no bias would result from alternate marking of wings. Second, Knoppien showed several incorrect applications by Bryant's group of their computer simulation model to experiments done in my laboratory in which marking of flies was not done (our 'red' and 'orange' eye experiments cited above) as well as to several experiments with flies marked by wing clipping done in other laboratories as well as mine. He concludes that any artifact of rare male advantage caused by wing clipping will be less important than suggested by Bryant,

Kence and Kimball. Knoppien (1985) has reviewed these issues further with similar general conclusions.

5. While I have never detected any detriment to mating by marking of flies (poking a hole through the wing near the distal margin in our *D. persimilis* experiments previously), the Bryant paper controversy stimulated me to assess the marking of various wildtype strains of *D. melanogaster* in some of our recent behavioral tests. Spiess and Wilke (1984) employed strains that had been selected disruptively for high (H) and low (L) sternopleural bristle number as well as unselected laboratory strains (Canton-S, Oregon-R and Lausanne-S) in order to check for female discrimination between rival males and their courtship sequence (Spiess, 1982a). Each mating test was done with a virgin female aged five days together with one marked male and one unmarked male. Marking was done by poking a hole with an insect pin through the wing margin distally between the second and third longitudinal veins; often the wing margin was torn in the process. In alternate tests the H or L male was marked for cases numbers 1–5 in Table 4. Cases numbers 6 and 7 were 'null tests' in which both sexes were of the same strain (LS), unselected for bristle number and one male was marked, the other not. As indicated in Table 4, none of the cases displayed significant departure from the null hypothesis (no difference between mating success of marked and unmarked males). Case number 6 is somewhat in the direction of more success for the unmarked male ($X^2 = 3.0$, $P = 0.08$, n.s.), but case number 7 with the identical strain shows the opposite result. So I conclude that for these strains a small hole poked through one wing or a torn wing margin has no significant effect on the male's mating success.

Table 4. Number of matings for males marked or unmarked in rivalry, one of each with a single female per glass vial.

Case number	Strain[1] female	Strain male	Number of matings[2]	
			Marked male	Unmarked male
1	L	H or L	29	21
2	H	H or L	23	27
3	CS	H or L	25	25
4	OrR	H or L	27	23
5	LS	L or L	22	28
6	LS	LS	30	45
7	LS	LS	29	21
Totals:			185	190

[1] L = low sternopleural bristle number selected subpopulation,
H = high sternopleural bristle number selected subpopulation,
CS = Canton-Special, OrR = Oregon-R, LS = Lausanne-Special wild type strains, from Spiess and Wilke (1984).
[2] N = 50 females tested for each case except for number 6 in which N = 75.

Although it is difficult to prove a negative, these counterarguments to the statements that the marking of flies induces an artifact in experiments on the rare male advantage in my opinion show those statements to be without merit in the case of *Drosophila*. Houseflies, lacking elaborate courtship, may be different; as Leonard and Ehrman (1983) have pointed out, when testing species that have not been carefully checked for changes in behavior with some amount of mutilation, it is hazardous to draw conclusions or cast doubts on work with organisms that have been checked.

Finally, criticism of Spiess and Kruckeberg's results with 'red' and 'orange' mutant strains of *D. melanogaster* has come from Partridge and Gardner (1983) who failed to replicate our results, even though they tried to replicate our conditions in their laboratory at the University of Edinburgh. I have reserved detailed counterargument for another paper (Spiess and Bowbal, 1987). However, a brief mention of some factors that could have contributed to their different results are as follows: First, they used carbon dioxide as anesthesia while we used ether. Gilbert (1981) has pointed out that CO_2 can have many effects if given to young flies, and particularly that CO_2 treatment tends to increase females' latency to mounting and decrease copula duration relative to ether treatment. Thus CO_2 used in collecting virgins may have long term effects on female reproductive responses. Second, they maintained their cultures by backcrossing heterozygous females (bw^{75}/bw; st/st) to homozygous (bw^{75}/bw^{75}; st/st) males. In our laboratory we have consistently avoided that backcross since it is known that the bw locus is a complex cistron and the alleles bw^{75} and bw are recombinable in females while in males crossing over is very much reduced or non-existent. Third, we have observed sporadic slight changes in eye color that generally bring about behavioral variation in the outcome of these tests, sometimes accompanied by depression of fertility, so that we have found it necessary to 'weed out' variants that disturb the behavior of the two phenotypes we use as markers. Fourth, and perhaps most important, the Edinburgh authors did not seem to be aware of the contribution to solving this problem made by experiments with genetic background (Spiess, 1982b). In our laboratory the 'red' and 'orange' males tend to court at nearly equal frequencies (55% red: 45% orange) and to mate at about 40% for orange males when their genetic background is heterogeneous; but with increased inbreeding or with 'mutants off the shelf', orange males mated at a much lower frequency (20–25%). Partridge and Gardner observed this low rate of mating by orange males, but they said nothing about controlling genetic background. Thus it appears their failure to verify the rare male effect or the Spiess and Kruckeberg results may be largely due to use of comparatively homozygous strains. Resolution of these discrepancies can only be brought about by more thorough analysis of the genetic basis for female discrimination ability in these flies.

Populational demonstration of rare male mating advantage

Variation in mating success among males of numerous genotypes has been shown often to be a major component of fitness and thus critical to genotypic change in populations of *Drosophila* (Prout, 1971; Anderson and Watanabe, 1974; Anderson and McGuire, 1978). When their frequencies become low, many genotypes tend to increase their mating success, as reviewed above. Part of the criticism of experiments measuring this rare male advantage has been leveled at possible biases introduced by marking (above section) or by laboratory conditions, but recent critics may have lost sight of the fact that earlier population studies demonstrated maintenance of polymorphic equilibria attributable to the advantage gained by rarity (e.g. Ehrman *et al.*, 1965). More recently the rare male advantage in large populations either in laboratory or natural conditions has been demonstrated sufficiently to establish it beyond the level of a technical artifact.

1. Ehrman (1970) released into a large space (2,655 cu ft) 2,000 *D. pseudoobscura* with a 4:1 ratio of orange-eyed mutant:wildtype or the reverse ratio and then allowed the flies two days to be recaptured in baited traps. All released flies were virgin and the sexes were equal in number. Wildtype flies were recovered in greater proportion (46%) than orange-eyed (30%), and 86-90% of the females were inseminated. When wildtype males were rare, they enjoyed an enormous advantage, having inseminated more than 65% of the females. Also orange-eyed males when rare mated with 38% of the females, a proportion nearly twice their input frequency (20%), while wildtype males when common mated at 10% less than their input frequency. A balance between the two phenotypes would occur if these frequency-dependent mating values were the only factors affecting their proportions.
2. Anderson and Brown (1984) designed a population experiment using ST and CH chromosomal arrangements of *D. pseudoobscura* (Mather, California strains), specifically to avoid the possible artifactual bias from marking flies as was mentioned above. Strains carrying ST and CH arrangements were distinguished by amylase allozyme variants ($Amy^{1.00}$ for ST and $Amy^{0.84}$ for CH), and all flies were progeny of crosses between strains so that genetic heterozygosity was encouraged. Each of five population cages contained 500 virgin females of ST/ST karyotype followed a few hours later by 1,000 males divided among three karyotypes, ST/ST, ST/CH and CH/CH at various frequencies ranging from high CH (95%) low ST (5%) through 50% each to opposite frequencies. After one day for mating, flies were vacuumed from the cage, and females were isolated into separate food vials for progeny determination and thus inference of the male parent's karyotype. (In addition, since ST and CH arrangements are

part of a balanced polymorphism, both in natural and in experimental populations, the same strains used in the mating cages were also set up in two long-term populations to confirm their balanced state under laboratory conditions.) Both ST/ST and CH/CH males displayed a significant mating advantage when rare (30% input frequency producing the best advantage) in comparison with their mating success at midrange frequencies: about a three-fold increase for CH/CH and a six-fold increase for ST/ST. The heterokaryotype ST/CH never achieved a measurable rare advantage, though in three out of five populations it was either superior (cages I and IV) or about equal (cage III) to ST/ST in mating success. The authors point out that although CH/CH males were usually less successful in mating than ST males, their success was best at frequencies typical of populations at equilibrium in the laboratory (70–80% ST, 20–30% CH). Thus the male mating success is likely to be a major contributor to an overall balancing selection over a wide range of frequencies.

3. Demonstration of mating success for specific genotypes (or karyotypes) in natural populations is difficult, but Salceda and Anderson (1987) were able to ascertain karyotypes of wild *D. pseudoobscura* from a natural population sufficiently to judge whether matings in the wild were frequency dependent with rare male advantage or simply occurring in proportion to input frequencies. They sampled a population from Amecameca on the southern edge of the Valley of Mexico in 1977. Wild flies were separated by sex in the field. Females inseminated before capture were isolated and laid eggs, each female in a separate food vial; then salivary gland chromosomes of their progenies were analyzed. After two weeks, males of ST/ST karyotype were added to each female's vial to give an opportunity for remating. Since the ST arrangement is not found in Central Mexico populations of *D. pseudoobscura*, examination of larval progeny from the laboratory matings served to diagnose the female's karyotype. Also, wild males were mated to ST/ST females, so that by examining larval progeny each male could be diagnosed. Thus nearly all captured wild flies were identified as to karyotype, and the progenies of wild females that they would have produced in nature allowed inference of types of males that had mated in the wild population. The chromosomal arrangement Cuernavaca (CU) was most frequent in the sampled population at about 60%, with Tree Line (TL) next at 30%, followed by Estes Park (EP) at 8%, and six other rare arrangements; all were equally represented in the two sexes. Changes in chromosome frequency owing to male mating success were calculated by comparing frequencies in adult males with those in the chromosomes they contributed to their offspring. Such changes were sizable and provided evidence that male mating success is a major component of selection on these chromosomal arrangements in the natural population. Most significant were the facts that the most

common arrangement (CU) showed a drop in frequency by more than 8% between the adult males and the chromosomes contributed from wild males to the next generation while the rarer arrangements all showed increases from about 1.5% to 4% in the same comparison. The authors estimated that the total frequency of chromosomes contributed by rare male karyotypes was 24% and that the relative mating success of rare male karyotypes was 1.8 times that of the common karyotypes in that population. Thus these data give substantial evidence for asserting the reality of the rare male mating advantage in nature.

STAGES IN DISCRIMINATION LEADING TOWARD SEXUAL ISOLATION

It is logical to assume that, in the evolution of two or more mate recognition systems (sexually isolated populations) out of a single system, an early stage must first include selection for genetic control of discriminatory ability. Once such ability is gained in both sexes or at least in the sex that controls whether a mating will occur, preferences for specific mate signals may become fixed; that is, highly specific signals must be finely tuned for recognition and response between potential mates. Positive assortative mating must then be established for efficiency of species reproduction and for lessening or preventing the mixing of populational adapted genomes. These steps in the evolution of ethological isolation may well be rapid and require few genetic changes.

Discriminatory ability among phenotypes for distinguishing prospective mates within strains of a species has hardly been genetically analyzed because of the technical problem of defining and recognizing such phenotypes when determining whether features of strains are behavioral, physiological, or both. The experiments cited above under the rare male advantage represent a major attempt at solving that problem. On the other hand, tests for discrimination against interspecies crossing seem easier and are numerous including genetic or chromosomal analysis of the discrimination trait (see Spieth and Ringo, 1983, for a summary). Generally, artificial selection to improve discrimination, for example, between sibling species, has been successful (for example, Kessler, 1966, for *D. pseudoobscura* versus *D. persimilis*; Eoff, 1975, 1977, for *D. melanogaster* versus *D. simulans*). Thus genetic variation controlling mate recognition signals is abundantly available to maintain or to improve discrimination between closely related species that often occur sympatrically. In addition, tests for incipient isolation among geographically distinct natural populations of single species have been made by at least nine research groups (summarized by Spieth and Ringo, 1983) and have demonstrated that behavioral elements have diverged among discrete populations of many species, making incipient isolation likely to evolve when gene flow between populations is reduced. Nevertheless, it is the emergence of assortative mating tendencies within a population's gene pool that we need to study for a genetic basis leading toward ethological isolation.

Assortative mating tendencies within strains and the origin of isolation

Genetic potential for assortative mating has been demonstrated by increasing its expression under various selection techniques, as documented for example by the following:
1. In 1962–70 Thoday and Gibson (cited in Spiess and Wilke, 1984) selected disruptively for bristle number in *D. melanogaster* and found that high and low subpopulations became reproductively isolated.
2. Crossley (1974) marked two strains with recessive non-allelic mutants, selected against wildtype hybrids, and achieved significant response by an increase in pre-mating assortative tendencies within the strains.

Attempts to repeat Thoday and Gibson's experiments by Scharloo in 1970 and by Barker and Cummins in 1969 (cited in Spiess and Wilke, 1984), while producing changes in mating behavior and slight temporary assortative tendencies, reverted in later generations with no significant emergence of assortative mating. The earlier achievement could not be considered a general result.

Another attempt at selecting disruptively for assortative mating coupled with bristle number was made (Spiess and Wilke, 1984) using two inbred strains and their hybrids of *D. melanogaster*. The strains differed in the number of sternopleural bristles. The method was thought to be an improved technique for assortative selection. Although disruptive selection for the morphological trait over 16 generations was effective in producing a bimodal distribution, no significant or clearcut assortative tendencies emerged within the subpopulations of high and low bristle number. From the distribution of bristle numbers among mated pairs in the last generations of selection, it was apparent that mating may not have been strictly random; in fact the distributions seemed dissected with three or four scattered low peaks among the mated pair bristle numbers, perhaps indicating a mixture of assortative and random mating. In fact there was a one-sided assortative tendency for low × low parents but high bristle number females mated randomly. We concluded that failure to achieve a positive response in both subpopulations could have arisen from:
1. Lack of genetic potential variation in the base population (considered unlikely because preliminary testing had indicated a slight but significant correlation in bristle number for mating pairs).
2. Genetic variation structured in such a way that the selection procedure was not effective in making it available.

We discussed theoretical models by Felsenstein (1981) and by Sved (1981) indicating that some initial linkage disequilibrium between assortative-mating genes and the marker loci under selection is a necessary ingredient for progress. If there had been a slight disequilibrium initially, we postulated that it probably broke down or was not enhanced sufficiently for a positive response. Thus attainment of assortative mating for morphologically distinct types within a population requires a genetic structure of a rather special sort which has yet to be described in any detail.

A significant assortative tendency can arise during divergence of populations for adaptive resource utilization. Dodd (1984) selected eight *D. pseudoobscura* laboratory populations for their utilization and preference for either starch or maltose (four populations each). It had been shown by Powell and Andjelković (1983) that populations reared on starch or on maltose differ in their frequencies of α-amylase (*Amy*) alleles and the patterns of amylase activity in the flies' midgut; the populations could be said to have adapted to the two media. Dodd tested larvae from these populations for habitat choice and found that larvae did spend more time in the medium to which they had been adapted when given a choice of equal portions of starch or maltose medium. Of greater interest, however, were Dodd's experiments on mating preferences of adults from these two resource-adapted populations. Flies to be tested from either starch-adapted or maltose-adapted populations were reared on standard cornmeal–molasses–agar medium for a generation and aged, and sexes were separated and stored on standard medium for three to six days as virgins. Twelve pairs from each resource-type of population were introduced (females first) into observation chambers and matings recorded for an hour or more. Marking of flies was done by wing clipping with populations alternated. A minimum of 50 matings were recorded for each combination of starch × maltose paired populations. Of the 16 starch × maltose population combinations, eleven were significantly assortative with isolation indices ranging from 0.26 to 0.49 [where isolation index = (homogamic matings − heterogamic matings)/total matings], while for the remaining combinations all indices were positive but ranging from 0.18 to 0.24. As pointed out by the author, this assortative tendency could have arisen as a 'byproduct of the [resource] adaptation... or it could be a result of bottlenecks in population size when the populations were first established.' If the latter were the case, then testing for assortative matings among the four starch-adapted populations or among the four maltose-adapted populations should indicate as many positive isolation indices as when the two resource-adapted populations were tested against each other. In contrast with the latter tests, none of the within-resource tests deviated significantly from random mating, the indices not being significantly different from zero. Clearly the isolation was only between the starch- versus maltose-adapted populations and had arisen during their respective adaptations, not as a result of going through any founder effect (i.e. bottleneck) of population size reduction. Finally in checking for any post-mating isolation factor, such as a drop in viability or fertility following crossing between the two resource-adapted populations, Dodd found no evidence of any loss in those fitness properties in population hybrids. Thus assortative mating evolved while the populations were diverging for resource preference. It would be interesting indeed to look further into the genetic basis of the assortative behavior and its association with the amylase locus and any determination of the habitat choice mechanism.

Further evidence that mate recognition within a population can evolve as 'accidental divergence' between lines was provided by Powell (1978) when he tested Carson's (1968, 1975) founder-flush speciation theory with laboratory populations of *D. pseudoobscura*. A founder event by just a few individuals or by a single gravid female entering a habitat suitable for survival and expansion of progeny can establish new genetic combinations with frequencies different from the founder's ancestral population. Furthermore, a series of such founder-flush events can increase the probability for establishment of new populations that may attain reproductive isolation (assortative tendencies) that separate them from their ancestral population and from each other. Powell derived twelve subpopulations from a large, genetically diverse, expanded population by choosing twelve single pairs for establishment of each subpopulation. After allowing those progenies to expand, he tested pairs of subpopulations for reproductive isolation, but just one pair out of 45 pairs tested indicated significant assortative mating. Eight of the twelve subpopulations were then put through an additional series of founder-flush cycles by allowing just a single founder pair ('bottleneck') to carry on each subpopulation, then allowing expansion ('flush'), and continuing (again a single pair, expansion, and so forth). Finally more than a year after the last bottleneck, all pairs of subpopulations, including the original control large population (duplicated) and eight inbred lines (not put through a founder-flush-crash cycle) were tested for mating preferences by direct observation: twelve pairs of flies from each of two subpopulations were tested together. Out of 36 combinations of the eight founder subpopulations plus the original control, eleven expressed significant assortative mating, with three of the eight subpopulations most highly isolated. Tests six months later with the most isolated lines confirmed the maintenance of the same isolation level, that is the ethological change was not transitory. No assortative mating was detected among either the original control sublines, the inbred control sublines, or between the original and the founder-flush subpopulations. No post-mating isolation was found. Powell pointed out that when a founder event occurs followed by population growth and recombination, genetic control of mate recognition may become perturbed; but during population expansion, slightly different behaviors may well become established and lead to sufficient divergence to make assortative preferences emerge if sympatry ensues.

'Accidental divergence' has been recorded for other species also, particularly when opportunities for population expansion following a 'bottleneck' occurred during maintenance of a laboratory line. Notable cases are those recorded by Arita and Kaneshiro (1979) for two strains of *D. adiastola* and by Ahearn (1980) for strains of *D. silvestris*, both Hawaiian picture-wing species in which one strain freshly collected from a natural population was tested for assortative tendencies with a strain that had been kept under laboratory conditions for a long period with 'bottlenecks' at times during laboratory maintenance. Mating tests were done by 'male choice' technique, one male observed with two females, one from the male's strain and one from the opposite strain. In both sets of experiments, females from the strain most recently collected in nature ('ancestral') tended to reject males from the laboratory ('derived') strain while 'derived' females readily accepted both types of male. Thus the assortment was not achieved mutually. Unfortunately the authors do not indicate whether the males in any way discriminated between the two females, that is, whether there was preferential courtship towards either female by the male. (For further discussion of these results and some unfounded criticisms by others, see Giddings and Templeton, 1983; their deductions concerning the model by which mating preference may be used to deduce the direction of evolution—Kaneshiro's model—lie beyond the scope of this chapter.)

It appears rather ironic that of the several attempts at selecting for assortative tendencies from single large populations such as those cited at the beginning of this section, many failed to achieve a significant response, while accidental divergence seems to occur commonly, brought about either by possible 'hitch-hiking' with resource adaptation (as observed by Dodd) or by the founder-flush-crash cycle technique (as observed by Powell). Thus seemingly we display our ignorance of the genetic factors that bring about assortative mating from a population that is large, genetically heterozygous at an enormous number of loci, yet does not often appear to be selectable for the behavioral steps that can lead to ethological isolation between subpopulations. Indications from such circumstantial evidence suggest that mate recognition systems have a genetic structure consisting of such tightly associated components that selection within populations under ordinary conditions of propagation cannot evoke separation into two or more assortative systems. A 'genetic revolution' such as could be achieved through accidental and sudden alterations in the gene pool followed by expansion of numbers under new conditions does seem more likely to evoke the separation. Nevertheless, in view of our ignorance of genotypes controlling mate recognition, it is not surprising that we have so little information about how it evolves. We need to describe in genetic detail what stages are necessary for such crucially important evolutionary events.

Interspecies discrimination

The literature describing mating tests between ethologically isolated populations, separate species, is very large; Spieth and Ringo (1983) have reviewed much of the experimental work, most of which will not be summarized again here. However, certain points need to be made and one or two examples given for completeness.

In testing strains of allopatric populations or separate laboratory populations for the emergence of partial isolation, most authors either show evidence or make a tacit assumption that *Drosophila* males do not discriminate among strains of females, that is, they randomly court any female. It is logical to infer, however, that if emergence of isolation necessitates formation of assortative tendencies, then males should acquire discrimination ability, and if given a choice of female types they should court conspecific females preferentially. Indeed discrimination by both sexes would reinforce the isolation between closely related sympatric species. Wood and Ringo (1980) have reviewed the experimental literature on the isolation between *D. melanogaster* and *D. simulans*, as well as between other pairs of sibling species, with the general conclusion that males indeed show considerable discriminatory ability when it comes to distinguishing other species females from their own. I have similarly observed male discrimination in tests (unpublished data) with *D. pseudoobscura* and *D. persimilis* derived from a pair of sympatric natural populations of the two species: using two females × six males (three males of each species), courtships observed were only by the males conspecific with the females while the heterospecific males did not court; however, if given no choice and only one species of male is present, that male will court the heterospecific female but be rejected in nearly all tests.

Robertson (1983) has observed details of courtship and mating between *D. mauritiana* (endemic to the island of Mauritius) and its sibling species *D. simulans* and *D. melanogaster*. The exotic island species resembles *D. simulans* more than it does *D. melanogaster* in the male's courtship wing display, wing vibration songs and complex movements. Visual signals are apparently important for male *D. mauritiana* for courtship initiation since it rarely mates in the dark. When *D. melanogaster* and *D. mauritiana* are paired, males of one species with females of the other, almost no courtship is initiated by either species of male. Between *D. melanogaster* and *D. simulans*, the males of the former species hardly discriminate against females of the latter species, but the latter's males are strong discriminators against the former species' females. In contrast, *D. simulans* and *D. mauritiana* court each other interspecifically with little hesitation, though *D. simulans* females resist the other species' males. Robertson presents evidence that a male's tapping with foretarsi allows him to detect species differences and thus provides him with discriminative information. Apparently, as evidenced by differential responses of *D. melanogaster* males, *D. mauritiana* females differ from *D. simulans* females in the chemical stimuli received by these males from tapping and close proximity, but the males of the latter two species do not respond to that difference. Thus interspecies discrimination is largely a male's function at initial encounter, though continuation of courtship and final acceptance by the female is the decisive factor in mating. We may generalize that female discrimination and choice of mate seem to function predominantly among conspecific pairs, while male discrimination comes into play when a prospective mate belongs to a species different from his own. Thus the male's discrimination provides additional insurance against interspecific error, supplementing the defenses of the female.

SUMMARY AND DISCUSSION

Drosophila is described as a model animal for exploring behavioral and genetic factors controlling discrimination ability and mate recognition. Technical advances during the past two decades have gained precision in measuring male courtship signals and female responses or female signals and male responses. Acoustic variation (wing vibration in males predominantly), odor differences (chemical signals emitted by both sexes), visual stimuli and tactile stimuli have been monitored in sufficient species to give us insight into the sequence of courtship factors and the opportunities for discrimination. In most species, normal courtship begins when a male orients to a moving fly and taps with his foretarsus. Visual and tactile stimuli allow the male to discriminate at least the sex and species of the fly approached. Pheromones (odor) from the courted fly and appropriate signals indicating receptivity of a female escalates his courtship, bringing about his vibration of one or both wings toward the female and continuation of courtship, leading toward copulation. As for the female, she may detect a male's odor before he orients; but his activity in following her, his vibration (song), odor, and persistence all play a sequential role in reducing her receptivity threshold. Thus signals are provided to the female that allow her to discriminate and finally to accept a male.

Variation in acoustic signals has been recorded in about 40 species of *Drosophila*. Distinct features in pulse song and sine song characterize genotypes, populations and species in ways that indicate acoustic differences to be major factors for discrimination by courted females. The chemistry of pheromones, now known for a few species, has emphasized that female aphrodisiac substances are polymorphic in some populations. These substances, generally unsaturated long chain hydrocarbons localized on the fly's cuticle, are detected by males by tapping and by sensory receptors of the antennae. Male-specific compounds, consisting of shorter chains than female substances, contribute to signals needed for female discrimination and receptivity. Also, male compounds are transferred to females during mating, and they inhibit courtship by other males following insemination. Visual stimuli, important in early stages of courtship for a male's tracking of female movement, lead to attempted copulation when the female's movement ceases and her wings spread slightly. Males that can keep the female in view are far more successful in mating than blind or eye-mutant males.

Female choice and male intensity of courtship each play a role in acceptance of a mate. Males display visual, auditory and odor differences to the courted female, and she often appears to prefer one type of male over another. Her central nervous system apparently responds to a complex combination of messages. When two or more types of male are courting, females whose threshold prevents them from accepting the first male to court tend to require several courtship bouts from more than one male before accepting one. This hesitancy allows the female to receive diverse signals; thus she samples the courting males' signals, and often she accepts a male whose phenotype (combination of signals) is different from that of the male who was first to court her (provided the phenotype is within the normal range for species recognition). Such learning by the female leads to an outcrossing (disassortative) mating system among a polymorphic group of genotypes (phenotypes); preferential acceptance of the rare type of male results. While this rare male effect has been criticized for various reasons, evidence for its existence includes not only laboratory tests with flies that have been marked artificially and with easily recognized mutant types but also populational studies in the laboratory and in nature. Thus male mating success depends not only on specific courtship signals but also on factors of relative abundance, relative familiarity of females with diverse male signals, and the resulting discrimination ability of females.

Because of *Drosophila* females' tendency to sample courting males and to mate with rare types, it is perhaps not surprising that emergence of positive assortative mating leading to sexual isolation within a breeding unit has been difficult to achieve by the usual methods of artificial selection. It is ironic that divergence of mate recognition-response systems (positive assortment) has come about perhaps more by accident (founder-flush-crash cycles) or by 'hitch-hiking' to a resource adaptation, than by concerted selection methods. On the other hand, it is quite likely that critical factors in evolving separate recognition-response systems have not yet been taken into account when attempting to select for them. A critical step may well be discrimination by males when beginning courtship; at least interspecies tests indicate that male discrimination of conspecific from heterospecific females is a common factor helping to prevent errors in mating between closely related species. Both sexes must be selected not only for acceptance of their own type (positive assortative ability) but also for rejection of the wrong type (other species or other genetic combinations separated by prevention of gene flow).

DEDICATION

Dedicated to Hampton L. Carson, in honor of his seventieth year and his accomplishments in evolutionary genetics and behavior.

LITERATURE CITED

Ahearn, J. N. 1980. Evolution of behavioral reproductive isolation in a laboratory stock of *Drosophila silvestris*. *Experientia*, **36**, 63–64.
Anderson, W. W. 1987. The effect of wing clipping on male mating success. *Behavior Genetics*, in press.
Anderson, W. W. and C. J. Brown. 1984. A test for rare male mating advantage with *Drosophila pseudoobscura* karyotypes. *Genetics*, **107**, 577–589.
Anderson, W. W. and P. R. McGuire. 1978. Mating pattern and mating success of *Drosophila pseudoobscura* karyotypes in large experimental populations. *Evolution*, **32**, 416–423.
Anderson, W. W. and T. K. Watanabe. 1974. Selection by fertility in *Drosophila pseudoobscura*. *Genetics*, **77**, 559–564.
Antony, C. and J.-M. Jallon. 1982. The chemical basis for sex recognition in *Drosophila melanogaster*. *Journal of Insect Physiology*, **28**, 873–880.
Arita, L. H. and K. Y. Kaneshiro. 1979. Ethological isolation between two stocks of *Drosophila adiastola* Hardy. *Proceedings of the Hawaiian Entomological Society*, **13**, 31–34.
Averhoff, W. W. and R. H. Richardson. 1974. Pheromonal control of mating patterns in *Drosophila melanogaster*. *Behavior Genetics*, **4**, 207–225.
Averhoff, W. W. and R. H. Richardson. 1976. Multiple pheromone system controlling mating in *Drosophila melanogaster*. *Proceedings of the National Academy of Sciences, USA*, **73** 591–593.
Bateson, P. 1983. Optimal outbreeding, in Bateson, P. ed., *Mate choice*, Cambridge University Press, Cambridge, England, pp. 257–277.
Bennet-Clark, H. C. 1971. Acoustics of insect song. *Nature*, **234**, 255–259.
Bennet-Clark, H. C. 1975. Acoustics and the behaviour of *Drosophila*. *Verhandlung Deutsche Zoologisches Gesellschaft*, **1975**, 18–28.
Bennet-Clark, H. C., M. Dow, A. W. Ewing, A. Manning and F. von Schilcher. 1976. Courtship stimuli in *Drosophila melanogaster*. *Behavior Genetics*, **6**, 93–95.
Bennet-Clark, H. C. and A. W. Ewing. 1969. Pulse interval as a critical parameter in the courtship song of *Drosophila melanogaster*. *Animal Behaviour*, **17**, 755–759.
Bennet-Clark, H. C., A. W. Ewing and A. Manning. 1973. The persistence of courtship stimulation in *Drosophila melanogaster*. *Behavioural Biology*, **8**, 763–769.
Bennet-Clark, H. C., Y. Leroy and L. Tsacas. 1980. Species and sex-specific songs and courtship behaviour in the genus *Zaprionus* (Diptera—Drosophilidae). *Animal Behaviour*, **28**, 230–255.
Bowbal, D. A. 1984. *Association between courtship sequence and mating success in certain eye color mutants of* Drosophila melanogaster, Master of Science Thesis, University of Illinois at Chicago.
Bryant, E. H. 1979. Inbreeding and heterogamic mating: an alternative to Averhoff and Richardson. *Behavior Genetics*, **9**, 249–256.

Bryant, E. H., A. Kence and K. T. Kimball. 1980. A rare-male advantage in the housefly induced by wing-clipping and some general considerations for *Drosophila*. *Genetics*, **96**, 975–993.

Bryant, P. J. 1978. Pattern formation in imaginal discs, in Ashburner M. and T. R. F. Wright, eds., *The genetics and biology of* Drosophila, Academic Press, New York, Vol. 2c, pp. 229–335.

Burnet, B. and K. Connolly. 1973. The visual component in the courtship of *Drosophila melanogaster*. *Experientia*, **29**, 488–489.

Burnet, B. and K. Connolly. 1974. Activity and sexual behaviour in *Drosophila melanogaster*, in van Abeelen, J. H. F. ed., *The genetics of behaviour*, American Elsevier, New York, pp. 201–258.

Burnet, B., L. Eastwood and K. Connolly. 1977. The courtship song of male *Drosophila* lacking artistae. *Animal Behaviour*, **25**, 460–464.

Carson, H. L. 1968. The population flush and its genetic consequences, in Lewontin, R. C. ed., *Population biology and evolution*, Syracuse University Press, Syracuse, New York, pp. 123–137.

Carson, H. L. 1975. Genetics of speciation. *American Naturalist*, **109**, 83–92.

Carson, H. L. and P. J. Bryant. 1979. Change in a secondary sexual character as evidence of incipient speciation in *Drosophila silvestris*. *Proceedings of the National Academy of Sciences, USA*, **76**, 1929–1932.

Chang, H.-C. and D. D. Miller. 1978. Courtship and mating sounds in the species of the *Drosophila affinis* subgroup. *Evolution*, **32**, 540.

Cook, R. M. 1973a. Courtship processing in *Drosophila melanogaster*. I. Selection for receptivity to wingless males. II. An adaptation to selection for receptivity to wingless males. *Animal Behaviour*, **21**, 338–358.

Cook, R. M. 1973b. Physiological factors in the courtship processing of *Drosophila melanogaster*. *Journal of Insect Physiology*, **19**, 397–406.

Cook, R. M. 1979. The courtship tracking of *Drosophila melanogaster*. *Biological Cybernetics*, **34**, 91–106.

Cook, R. M. 1980. The extent of visual control in the courtship tracking of *Drosophila melanogaster*. *Biological Cybernetics*, **37**, 41–51.

Cowling, D. E. and B. Burnet. 1981. Courtship songs and genetic control of their acoustic characteristics in sibling species of the *Drosophila melanogaster* subgroup. *Animal Behaviour*, **29**, 924–935.

Crossley, S. A. 1974. Changes in mating behavior produced by selection for ethological isolation between ebony and vestigial mutants of *Drosophila melanogaster*. *Evolution*, **28**, 631–647.

Crossley, S. A. and J. McDonald. 1979. The stability of *Drosophila melanogaster* courtship across matings. *Animal Behaviour*, **27**, 1041–1047.

Dodd, D. M. B. 1984. *Behavioral correlates of the adaptive divergence of* Drosophila *populations*, Doctoral Dissertation, Yale University, New Haven, Connecticut.

Eastwood, L. and B. Burnet. 1979. Variation in the importance of acoustic stimuli in *Drosophila melanogaster* courtship. *Experientia*, **35**, 1159–1160.

Ehrman, L. 1970. A release experiment testing the mating advantage of rare *Drosophila* males. *Behavioral Science*, **15**, 363–365.

Ehrman, L. 1972. A factor influencing the rare male mating advantage in *Drosophila*. *Behavior Genetics*, **2**, 69–78.

Ehrman, L. 1978. Sexual behavior, in Ashburner, M. and T. R. F. Wright, eds., *The genetics and biology of* Drosophila, Academic Press, New York, Vol. 2b, pp. 127–180.

Ehrman, L. and J. Probber. 1978. Rare *Drosophila* males: the mysterious matter of choice. *American Scientist*, **66**, 216–222.

Ehrman, L., B. Spassky, O. Pavlovsky and T. Dobzhansky. 1965. Sexual selection, geotaxis, and chromosomal polymorphism in experimental populations of *Drosophila pseudoobscura*. *Evolution*, **19**, 337–346.
Eoff, M. 1975. Artificial selection in *Drosophila melanogaster* females for increased and decreased sexual isolation from *Drosophila simulans* males. *American Naturalist*, **109**, 225–229.
Eoff, M. 1977. Artificial selection in *Drosophila simulans* males for increased and decreased sexual isolation from *Drosophila melanogaster* females. *American Naturalist*, **111**, 259–266.
Ewing, A. W. 1964. The influence of wing area on the courtship behaviour of *Drosophila melanogaster*. *Animal Behaviour*, **12**, 316–320.
Ewing, A. W. 1969. The genetic basis of sound production in *Drosophila pseudoobscura* and *Drosophila persimilis*. *Animal Behaviour*, **17**, 555–560.
Ewing, A. W. 1977a. Communication in the Diptera, in Sebeok, T. A., ed., *How animals communicate*, Indiana University Press, Bloomington, Indiana, pp. 403–417.
Ewing, A. W. 1977b. The neuromuscular basis of courtship song in *Drosophila*: the role of the indirect flight muscles. *Journal of Comparative Physiology*, **119**, 249–265.
Ewing, A. W. 1978. The antenna of *Drosophila* as a 'love song' receptor. *Physiological Entomology*, **3**, 33–36.
Ewing, A. W. 1979. Complex courtship songs in the *Drosophila funebris* species group: escape from an evolutionary bottleneck. *Animal Behaviour*, **27**, 343–349.
Ewing, A. W. and H. C. Bennet-Clark. 1968. The courtship songs of *Drosophila*. *Behaviour*, **31**, 288–301.
Ewing, A. W. and A. Manning. 1963. The effect of exogenous scent on the mating of *Drosophila melanogaster*. *Animal Behaviour*, **11**, 596–598.
Felsenstein, J. 1981. Skepticism towards Santa Rosalia, or why are there so few kinds of animals? *Evolution*, **35**, 124–138.
Gailey, D., F. Jackson and R. Siegel. 1984. Conditioning mutations in *Drosophila melanogaster* affect an experience-dependent behavioral modification in courting males. *Genetics*, **106**, 613–623.
Giddings, L. V. and A. R. Templeton. 1983. Behavioral phylogenics and the direction of evolution. *Science*, **220**, 372–378.
Gilbert, D. G. 1981. Effects of CO_2 versus ether on two mating behavior components of *Drosophila melanogaster*. *Drosophila Information Service*, **56**, 45–46.
Gromko, M. H., D. G. Gilbert and R. C. Richmond. 1984. Sperm transfer and use in the multiple mating system of *Drosophila*, in Smith, R. L. ed., *Sperm competition and the evolution of animal mating systems*, Academic Press, New York, pp. 371–426.
Grossfield, J. 1971. Geographic distribution and light-dependent behavior in *Drosophila*. *Proceedings of the National Academy of Sciences, USA*, **68**, 2669–2673.
Hall, J. C., R. W. Siegel, L. Tompkins and C. P. Kyriacou. 1980a. Neurogenetics of courtship in *Drosophila*. *Stadler Symposium (University of Missouri)*, **12**, 43–82.
Hall, J. C., L. Tompkins, C. P. Kyriacou, R. W. Siegel, F. von Schilcher and R. J. Greenspan. 1980b. Higher behavior in *Drosophila* analyzed with mutations that disrupt the structure and function of the nervous system, in Siddiqi, O., P. Babu, L. M. Hall and J. C. Hall, eds., *Development and neurobiology of* Drosophila, Basic Life Sciences, Plenum Press, New York, Vol. 16, pp. 425–455.
Hodgkin, N. M. and P. J. Bryant. 1978. Scanning electron microscopy of the adult of *Drosophila melanogaster*, in Ashburner, M. and T. R. F. Wright, eds., *The genetics and biology of* Drosophila, Academic Press, New York, Vol. 2c, pp. 337–358.

Hoikkala, A., S. Lakovaara and E. Romppainen. 1982. Mating behavior and male courtship sounds in the *Drosophila virilis* group, in Lakovaara, S., ed., *Advances in genetics, development, and evolution of* Drosophila, Plenum Press, New York, pp. 407–421.

Hoikkala, A. and J. Lumme. 1984. Genetic control of the difference in male courtship sound between *Drosophila virilis* and *Drosophila lummei*. *Behavior Genetics*, **14**, 257–268.

Ikeda, H., H. Idoji and I. Takabatake. 1981. Intraspecific variations in the thresholds of females responsiveness for auditory stimuli emitted by the male in *Drosophila mercatorum*. *Zoological Magazine (Japan)*, **90**, 325–332.

Ikeda, H. and O. Maruo. 1982. Directional selection for pulse repetition rate of the courtship sound and correlated responses occurring in several characters in *Drosophila mercatorum*. *Japanese Journal of Genetics*, **57**, 241–258.

Ikeda, H., I. Takabatake and N. Sawada. 1980. Variation in courtship sounds among three geographical strains of *Drosophila mercatorum*. *Behavior Genetics*, **10**, 361–375.

Jallon, J.-M. 1984. A few chemical words exchanged by *Drosophila* during courtship and mating. *Behavior Genetics*, **14**, 441–478.

Jallon, J.-M. and Y. Hotta. 1979. Genetic and behavioral studies of female sex appeal in *Drosophila*. *Behavior Genetics*, **9**, 257–275.

Kence, A. 1981. The rare-male advantage in *Drosophila*: a possible source of bias in experimental design. *American Naturalist*, **117**, 1027–1028.

Kessler, S. 1966. Selection for and against ethological isolation between *Drosophila pseudoobscura* and *Drosophila persimilis*. *Evolution*, **20**, 634–645.

Knoppien, P. 1984. The rare male mating advantage: an artifact caused by marking procedures? *American Naturalist*, **123**, 862–866.

Knoppien, P. 1985. Rare male mating advantage: A review. *Biological Reviews*, **60**, 81–117.

Kruckeberg, J. F. 1978. *A behavioral basis for minority male mating advantage of certain eye color mutants of* Drosophila melanogaster. Master of Science Thesis, University of Illinois at Chicago.

Leonard, J. and L. Ehrman. 1976. Recognition and sexual selection in *Drosophila*. *Science*, **193**, 693–695.

Leonard, J. and L. Ehrman. 1983. Does the rare male advantage result from faulty experimental design? *Genetics*, **104**, 713–716.

Leonard, J., L. Ehrman and A. Pruzan. 1974. Pheromones as a means of genetic control of behavior. *Annual Review of Genetics*, **8**, 179–193.

Leonard, J., L. Ehrman and M. Schorsch. 1974. Bioassay of a *Drosophila* pheromone influencing sexual selection. *Nature*, **250**, 261–262.

Long, C. E., T. A. Markow and P. Yaeger. 1980. Relative male age, fertility, and competitive mating success in *Drosophila melanogaster*. *Behavior Genetics*, **10**, 163–170.

Mane, S. D., C. S. Tepper and R. C. Richmond. 1983. Studies of esterase-6 in *Drosophila melanogaster*. XIII. Purification and characterization of the two major isozymes. *Biochemical Genetics*, **21**, 1019–1040.

Mane, S. D., L. Tompkins and R. C. Richmond. 1983. Male esterase-6 catalyzes the synthesis of a sex pheromone in *Drosophila melanogaster* females. *Science*, **222**, 419–421.

Markow, T. A. 1980. Rare male advantages among *Drosophila* of the same laboratory strain. *Behavior Genetics*, **10**, 553–556.

Markow, T. A. 1986. Genetics and sensory basis of sexual selection in *Drosophila*, in Huettl, M., ed., *Genetics of invertebrate behavior*, Plenum Press, New York, in press.

Markow, T. A. and S. J. Hanson. 1981. Multivariate analysis of *Drosophila* courtship. *Proceedings of the National Academy of Sciences, USA*, **78**, 430–434.

Markow, T. A., M. Quaid and S. Kerr. 1978. Male mating experience and competitive courtship success in *Drosophila melanogaster*. *Nature*, **276**, 821–822.

Matsumoto, H., J. E. O'Tousa and W. L. Pak. 1982. Light-induced modification of *Drosophila* retinal polypeptides *in vivo*. *Science*, **217**, 839–841.

McDonald, J. 1979. Genetic analysis of lines selected for wing vibration in *Drosophila melanogaster*. *Behavior Genetics*, **9**, 579–584.

Miller, D. D., R. B. Goldstein and R. A. Patty. 1975. Semispecies of *Drosophila athabasca* distinguishable by male courtship sounds. *Evolution*, **29**, 531–544.

O'Donald, P. 1977. The mating advantage of rare males in models of sexual selection. *Nature*, **267**, 151–154.

O'Donald, P. 1980. *Genetic models of sexual selection*. Cambridge University Press, Cambridge, England.

O'Donald, P. 1983. Do female flies choose their mates? A comment. *American Naturalist*, **122**, 413–416.

Pak, W. L. 1975. Mutations affecting the vision of *Drosophila melanogaster*, in King, R. C. ed., *Handbook of genetics*, Plenum Press, New York, Vol. 3, pp. 703–733.

Parsons, P. A. 1973. *Behavioral and ecological genetics: a study in* Drosophila, Oxford University Press, Oxford, England.

Partridge, L. and M. Farquhar. 1983. Lifetime mating success of male fruit flies (*Drosophila melanogaster*) is related to their size. *Animal Behaviour*, **31**, 871–877.

Partridge, L. and A. Gardner. 1983, Failure to replicate the results of an experiment on the rare male effect in *Drosophila melanogaster*. *American Naturalist*, **122**, 422–427.

Paterson, H. E. 1980. A comment on 'mate recognition systems'. *Evolution*, **34**, 330–331.

Petit, C. 1972. Qualitative aspects of genetics and environment in the determination of behavior, in Ehrman, L., G. S. Omenn and E. Caspari, eds., *Genetics, environment, and behavior*, Academic Press, New York, pp. 27–47.

Petit, C. and L. Ehrman. 1969. Sexual selection in *Drosophila*. *Evolutionary Biology*, **3**, 177–223.

Pinsker, W. and E. Doschek. 1979. On the role of light in the mating behavior of *Drosophila subobscura*. *Zeitschrift Naturforschung*, **34C**, 1253–1260.

Powell, J. R. 1978. The founder-flush speciation theory: an experimental approach. *Evolution*, **32**, 465–474.

Powell, J. and M. Andjelković. 1983. Population genetics of *Drosophila* amylases. IV. Selection in laboratory populations maintained on different carbohydrates. *Genetics*, **103**, 675–689.

Powell, J. and L. Morton. 1979. Inbreeding and mating patterns in *Drosophila pseudoobscura*. *Behavior Genetics*, **9**, 425–429.

Prout, T. 1971. The relation between fitness components and population prediction in *Drosophila*. I. The estimation of fitness components. II. Population prediction. *Genetics*, **68**, 127–167.

Richmond, R. C. and A. Senior. 1981. Studies of esterase-6 in *Drosophila melanogaster*. IX. Kinetics of transfer to females, decay in females, and male recovery. *Journal of Insect Physiology*, **27**, 849–853.

Robertson, H. M. 1982a. Female courtship summation in *Drosophila melanogaster*. *Animal Behaviour*, **30**, 1105–1117.
Robertson, H. M. 1982b. Female preference for normal or wingless males in *Drosophila melanogaster*. *Drosophila Information Service*, **58**, 127.
Robertson, H. M. 1983. Mating behavior and the evolution of *Drosophila mauritiana*. *Evolution*, **37**, 1283–1293.
Salceda, V. M. and W. W. Anderson. 1987. Rare male mating advantage in a natural population of *Drosophila*. *Proceedings of the National Academy of Sciences, USA*, in press.
Shorey, H. H. 1962. Nature of the sound produced by *Drosophila melanogaster* during courtship. *Science*, **137**, 677–678.
Shorey, H. H. and R. J. Bartell. 1970. Role of a volatile female sex pheromone in stimulating male courtship behaviour in *Drosophila melanogaster*. *Animal Behaviour*, **18**, 159–164.
Siegel, R. W. and J. C. Hall. 1979. Conditional responses in the courtship patterns of normal and mutant *Drosophila*. *Proceedings of the National Academy of Sciences, USA*, **76**, 3430–3434.
Siegel, R. W., J. C. Hall, D. A. Gailey and C. P. Kyriacou. 1984. Genetic elements of courtship in *Drosophila*: mosaics and learning mutants. *Behavior Genetics*, **14**, 383–410.
Sloane, C. and E. B. Spiess. 1971. Stimulation of male courtship behavior by female 'odor' in *Drosophila pseudoobscura*. *Drosophila Information Service*, **46**, 53.
Spiess, E. B. 1970. Mating propensity and its genetic basis in *Drosophila*, in Hecht, M. K. and W. C. Steere, eds., *Essays in evolution and genetics in honor of Theodosius Dobzhansky*, Appleton-Century-Crofts, New York, pp. 315–379.
Spiess, E. B. 1982a. Do female flies choose their mates? *American Naturalist*, **119**, 675–693.
Spiess, E. B. 1982b. Minority mating advantage of certain eye color mutants of *Drosophila melanogaster*. III. Female discrimination and genetic background. *Behavior Genetics*, **12**, 209–221.
Spiess, E. B. and D. A. Bowbal. 1987. Minority mating advantage of certain eye color mutants of *Drosophila melanogaster*. IV. Three genotypes and female discrimination. *Behavior Genetics*, in press.
Spiess, E. B. and H. L. Carson. 1981. Sexual selection in *Drosophila silvestris* of Hawaii. *Proceedings of the National Academy of Sciences, USA*, **78**, 3088–3092.
Spiess, E. B. and J. F. Kruckeberg. 1980. Minority advantage of certain eye color mutants of *Drosophila melanogaster*. II. A behavioral basis. *American Naturalist*, **115**, 307–327.
Spiess, E. B. and L. I. Salazar. 1983. Age of males as a factor in female mate choice in *Drosophila melanogaster*. *Drosophila Information Service*, **59**, 124–125.
Spiess, E. B. and W. A. Schwer. 1978. Minority mating advantage of certain eye colour mutants of *Drosophila melanogaster*. I. Multiple choice and single female tests. *Behavior Genetics*, **8**, 155–168.
Spiess, E. B. and C. M. Wilke. 1984. Still another attempt to achieve assortative mating by disruptive selection in *Drosophila*. *Evolution*, **38**, 505–515.
Spieth, H. T. 1968. Evolutionary implications of sexual behavior in *Drosophila*. *Evolutionary Biology*, **2**, 157–193.
Spieth, H. T. 1974a. Courtship behavior in *Drosophila*. *Annual Review of Entomology*, **19**, 385–405.
Spieth, H. T. 1974b. Mating behavior and evolution of the Hawaiian *Drosophila*, in White, M. J. D., ed., *Genetic mechanisms of speciation in insects*, Australia and New Zealand Book Co., Sydney, pp. 94–101.

Spieth, H. T. 1978. Courtship patterns and evolution of the *Drosophila adiastola* and *planitibia* species subgroups. *Evolution*, **32**, 435–451.
Spieth, H. T. and J. M. Ringo, 1983. Mating behavior and sexual isolation in *Drosophila*, in Ashburner, M., H. L. Carson and J. N. Thompson, Jr., eds., *The genetics and biology of* Drosophila, Academic Press, New York, Vol. 3c, pp. 223–284.
Sturtevant, A. H. 1915. Experiments on sex recognition and the problem of sexual selection in *Drosophila*. *Journal of Animal Behaviour*, **5**, 351–366.
Sved, J. A. 1981. A two-sex polygamic model for the evolution of premating isolation. I. Deterministic theory for natural populations. II. Computer simulation of experimental selection procedures. *Genetics*, **97**, 197–235.
Tompkins, L. 1984. Genetic analysis of sex appeal in *Drosophila*. *Behavior Genetics*, **14**, 411–440.
Tompkins, L., A. C. Gross, J. C. Hall, D. A. Gailey and R. W. Siegel. 1982. The role of female movement in the sexual behavior of *Drosophila melanogaster*. *Behavior Genetics*, **12**, 295–307.
Tompkins, L., J. C. Hall and L. M. Hall. 1980. Courtship-stimulating volatile compounds from normal and mutant *Drosophila*. *Journal of Insect Physiology*, **26**, 689–697.
Tompkins, L. and J. C. Hall. 1981a. The different effects on courtship of volatile compounds from mated and virgin *Drosophila* females. *Journal of Insect Physiology*, **27**, 17–21.
Tompkins, L. and J. C. Hall. 1981b. *Drosophila* males produce a pheromone which inhibits courtship. *Zeitschrift Naturforschung*, **36C**, 694–696.
Tompkins, L., R. W. Siegel, D. A. Gailey and J. C. Hall. 1983. Conditional courtship in *Drosophila* and its mediation by association of chemical cues, *Behavior Genetics*, **13**, 565–578.
van den Berg, M. J., G. Thomas, H. Hendriks and W. van Delden. 1984. A reexamination of the negative assortative mating phenomenon and its underlying mechanism in *Drosophila melanogaster*. *Behavior Genetics*, **14**, 45–61.
Venard, R. 1980. Attractants in the courtship behavior of *Drosophila melanogaster*, in Siddiqi, O., P. Babu, L. M. Hall and J. C. Hall, eds., *Development and Neurobiology of* Drosophila, *Basic Life Sciences*, Plenum Press, New York, Vol. 16, pp. 457–465.
Venard, R. and J.-M. Jallon. 1980. Evidence for an aphrodisiac pheromone of female *Drosophila*. *Experientia*, **36**, 211–212.
von Schilcher, F. 1976a. The role of auditory stimuli in the courtship of *Drosophila melanogaster*. *Animal Behaviour*, **24**, 18–26.
von Schilcher, F. 1976b. The function of pulse song and sine song in the courtship of *Drosophila melanogaster*. *Animal Behaviour*, **24**, 622–625.
von Schilcher, F. 1977. A mutation which changes courtship song in *Drosophila melanogaster*. *Behavior Genetics*, **7**, 251–259.
von Schilcher, F. and A Manning. 1975. Courtship song and mating speed in hybrids between *Drosophila melanogaster* and *Drosophila simulans*. *Behavior Genetics*, **5**, 395–404.
Wood, D. F. and J. M. Ringo. 1980. Male mating discrimination in *Drosophila melanogaster*, *Drosophila simulans*, and their hybrids. *Evolution*, **34**, 320–329.
Wood, D. F. and J. M. Ringo. 1982. Artificial selection for altered male wing display in *Drosophila simulans*. *Behavior Genetics*, **12**, 449–458.

CHAPTER 6

Kin Recognition in Subsocial Arthropods, in Particular in the Desert Isopod Hemilepistus reaumuri

K. E. LINSENMAIR
Zoologisches Institut (III) der Universität Würzburg, Röntgenring 10, D-8700 Würzburg, West Germany

INTRODUCTION

Kin recognition has been documented rarely within the huge number of non-eusocial arthropods. This rarity most probably reflects the factual situation, despite our poor knowledge of most species. The large majority apparently lacks social behavior other than courtship and mating. Moreover, where we find, for example, extended brood care, ecological conditions usually do not engender selection for a kin recognition system, since exclusivity is guaranteed in other ways and there is little danger of mistaking foreign conspecifics for kin. Kin recognition has been demonstrated in the isopod genus *Hemilepistus* (Marikovsky, 1969; Linsenmair and Linsenmair, 1971; Linsenmair, 1972, 1975, 1979, 1984, 1985; Schneider, 1971, 1975) and in the genus *Porcellio* (Linsenmair, 1979, 1984, unpublished) and besides these woodlice, in the roach *Cryptocercus* (Seelinger and Seelinger, 1983). Of these only the desert isopod, *Hemilepistus reaumuri* (Fig. 1), has been thoroughly investigated. I shall concentrate on the highly evolved recognition system of this species and present an account of some of the results (mainly unpublished) obtained in field and laboratory work during recent years.

Hemilepistus reaumuri belongs to a small group of unusual isopods which have invaded xeric habitats. This species and probably the other eight in the genus (for taxonomy see Lincoln, 1970) are ecologically important within their habitats. In large arid areas in North Africa and Asia they are the

Fig. 1. The desert isopod *Hemilepistus reamuri*. The burrow guard in the foreground has just left the den to admit its food-carrying partner. An immature isopod, about seven weeks old, is leaving the entrance to forage

major macrofaunal herbivores and detritivores, having considerable impact on their ecosystems not only through decomposition of organic matter but also through soil turnover by burrowing and consumption of large amounts of substrate, thus facilitating soil erosion and desalination (Shachak and Yair, 1984). Although *Hemilepistus* is, for an isopod, physiologically and morphologically well adapted to living on land (Linsenmair, 1975: Coenen-Stass, 1981; Hoese, 1982), its success is only comprehensible through an appreciation of its peculiar behavioral adaptations, especially its social behavior.

Many species of terrestrial isopods aggregate in refuges with favorable microclimates (Takeda, 1984). To these aggregations any conspecific is admitted. In the desert isopod, *H. reaumuri*, families dissolve during February and March after hibernation, their members usually dispersing tens to hundreds of meters from their birthplaces before settling down. Weather conditions in early spring restrict activity to a few warm hours in the early afternoon. During unfavorable times, unsettled individuals congregate in retreats. There, no control over admittance is exerted. This non-discriminatory phase may last from eight days to six weeks; at all other times *H. reaumuri* lives in strictly closed social units. First, monogamous pairs are formed, in which the

Fig. 2. The entrance of a *Hemilepistus* burrow surrounded by the fecal embankment

partners know one another individually. These then give rise to cooperative family units consisting of one set of parents and their progeny (a single brood, since females of *H. reaumuri* are semelparous (Warburg, Linsenmair and Bercovitz, 1984). Foreign conspecifics that appear inside the 'fecal embankment' at the burrow entrance (Fig. 2), are, after inspection with the antennae, attacked and driven off by the burrow-owning parents and by offspring that are at least six weeks old. If alien isopods are met outside the fecal embankment, they are either attacked or avoided. Wherever and whenever members of a family meet, they neither show mutual aggression nor mutual avoidance, provided the family has not yet dissolved and has not been experimentally manipulated.

These observations show that:
1. The families are closed units.
2. The ability to distinguish kin from non-kin is not site-dependent and is not restricted to a particular behavioral context.

How then do desert woodlice identify kin? Most of this chapter deals with this question.

Readers desiring a general account of *Hemilepistus* ecology and kin-related behavior may turn to the Discussion and Summary at the end of this chapter before reading the experimental results.

METHODS

Maintenance

For breeding, the isopods were kept in containers consisting of a plastic pipe (diameter 10–15 cm, height 25 cm) filled with clay, in which the isopods constructed their burrows. On top was placed an open plastic box (20 × 20 × 8.5 cm) containing a thin layer of dry clay or gypsum. During times of above-ground activity, some family members stayed inside this box and could be tested there.

For practical reasons (see below) most experiments were performed in another apparatus (the standard container), consisting of two plastic boxes connected to one another by a narrow opening. One container (10 × 10 × 6 cm) was kept dark and humid, the other (20 × 10 × 6 cm) open and dry. The floors of these containers were covered with a 1–2 cm layer of clay, too shallow to permit the construction of permanent burrows.

Test procedures

To test the reactions of a control group of isopods toward a conspecific, the latter was held with forceps and presented to the test animals in such a way that they could touch the individual on its tergites with their second antennae. Thus mutual chemical contamination could be avoided; it is otherwise unavoidable when putting an unrestrained individual into a group. Many field and laboratory experiments, however, determined whether or not manipulated animals were admitted to the family burrow. Here, the possibility of contamination was always considered. Extracts of possible recognition pheromones (discriminators), hemolymph and exuvial fluid were either applied to the tergites of test animals or to the bead-like ends of glass rods (diameter 5 mm). When dry, these glass rods were held close to the antennae of test animals. As controls, clean glass rods (and others treated with pure solvent) were presented.

For ready reference, several terms are explained here:
1. *Signature*. An individual's own chemical recognition characteristics, consisting of those discriminators (recognition pheromones) produced by the individual itself, or of these plus materials derived from other individuals.

2. *Badge*. The recognition pheromones of a natural family or experimental group, consisting of a mixture of the signatures of the group members. It is made up of all those components of the individuals' discriminators that are present in family or group members above a threshold of detectability or response.
3. *Contamination*. Transfer of discriminators among alien individuals and from isopods to their surroundings (for example burrow walls, fecal pellets, etc.) by natural contact or by artificial manipulation.
4. *Alienation*. An individual becomes unacceptable to its own family or group members by acquisition or loss of signature or badge elements.
5. *Intermolts*. Individuals that are between molts.

OBSERVATIONS AND EXPERIMENTAL INVESTIGATIONS ON SIBSHIPS OF *H. REAUMURI*

The family badge: evidence for its chemical nature

An individual must be touched with the 'apical cones' on the second antennae to be identified. The apical cones are complex sensory organs composed of mechanosensitive, olfactory, and mainly contact chemical receptors (Seelinger, 1977, 1983; Krempien, 1983; Holdich, 1984, for other species). Their amputation or reversible elimination incapacitates *Hemilepistus* for discriminating foreigners from kin. In the role of burrow guards, isopods deprived of their apical cones either defend their dens indiscriminately against both family members and strangers or they admit every conspecific regardless of relatedness; in all other situations they never show aggression.

Although scanning electron micrographs revealed no morphological features that might be parts of a recognition system, numerous findings clearly demonstrate the chemical basis for such a system. These findings include the following:
1. With suitable solvents, extracts can be obtained that:
    (a) when applied to a neutral carrier elicit aggression like a live foreign conspecific; and
    (b) when applied to members of another family, alienate them entirely from their own kin.
2. Intensive direct body contacts between members of different families usually lead to long-lasting and strong alienations.
3. *Hemilepistus* contaminate the substrate with family-specific marks wherever they aggregate.

Thus, if glass rods are put into the maintenance boxes, these quickly become contaminated, and when they are presented to members of another family, the effects (Fig. 3A) are similar to those of transferring a live isopod.

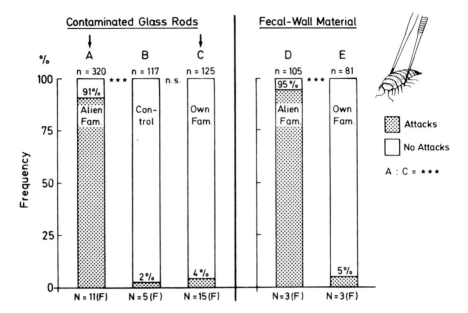

Fig. 3. Discriminator contamination of glass rods and feces. Glass rods placed on the bottom of the maintenance containers for 48 h were presented to alien (A) and to contaminating (C) families. B: Reactions to clean glass rods presented to five of the eleven alien families tested in experiment A. D, E: Fifteen individuals from each of three different families were placed singly into small vials and covered with material from either alien or their own fecal embankment for 5 min (while turning and shaking the vials) and then presented to their kin.
Here, and in all following figures, n = number of individual responses observed; percentage figures inside columns = frequency of attacks; *: $p \leq 0.05$; **: $p \leq 0.01$; ***: $p \leq 0.001$; n.s.: $p > 0.05$. N(F) = number of families tested

Conversely, when presented to the contaminating family (Fig. 3C) the rods release no aggression. The substrate around the entrance, in particular the fecal embankment material, can be used for experimental alienation of foreign individuals but not of family members (Figs. 3D and 3E). These results can easily be understood if we assume that transferable chemical 'signatures' (Beecher, 1982) or 'discriminators' (Hölldobler and Michener, 1980) varying among families are involved.

The extent of interfamilial variability

Preliminary observations

Temporarily and locally *Hemilepistus* lives in dense populations. During the pair formation phase more than 20 inhabited burrows per m^2 may be found in some North African habitats and some months later up to 14 families may still dwell there, each with an average of about 80 young. Field observations of many thousands of encounters between members of different pairs at and near their burrow entrances and, later, between members of different families inside their respective fecal embankments, have never produced evidence of mistaken identification. Alien desert isopods are always denied entry into an inhabited burrow, even at highest densities. Variability of the chemical signatures and family badges, therefore, must be high and these discriminating substances must be rather stable over time and must withstand environmental influences in order to preserve their individual- and family-specific characteristics. It would otherwise be difficult to imagine how they could guarantee exclusivity of family membership for over half a year in spite of high burrow densities with neighboring families exposed to similar environmental conditions, frequent encounters between members of neighboring families, and repeated orientation mistakes leading to numerous attempts to enter the wrong burrow.

Experimental investigations

Unless otherwise stated, results described in this chapter concern young (< 8 months old) isopods. Parents differ in many respects (see below and K. E. Linsenmair, unpublished). To scrutinize our field observations on interfamily variability (and to investigate additional problems) more than 50,000 laboratory tests were carried out, involving in every case a reciprocal exchange of several individuals between pairs of families. No two families have ever been found to tolerate one another (with the exception of laboratory lines in the third generation of brother-sister pairings). Very rarely, however, we observed unilateral toleration between natural families.

In the most extensive of several comparable tests designed to clarify the extent of interfamily badge variability among neighbors in their natural habitat, 78 families were examined for signature coincidence by systematic exchange of juveniles which were five to eight months old and more or less full-grown. These families were excavated in the field in an area of 50×50 m. In 4,826 of 4,830 non-repetitive exchanges we observed unambiguous aggression. There were four combinations in which the young (> 10) from one family (A) were not attacked by the members of the other family (B), but in which individuals from the accepting family (B) were always attacked when presented to members of family A.

In these experiments the isopods could not be expected to display their highest level of discriminative acuity. Two reasons for this are:
1. Only reactions of young toward young were evaluated, whereas parents are usually the most aggressive and the most discriminating individuals

within a family. Exclusion of parents from the tests results in an underestimate of the interfamily variability.
2. The individuals most critical in judging signatures are burrow guards checking individuals trying to enter the den. But maintaining families for practical reasons (see below) in containers without burrows eliminates this role.

These two points indicate that the small misidentification rate of about 0.1 percent does not prove that an alien badge cannot be differentiated from that of a test subject's own family. The outcome of these exchange experiments corroborates the conclusions from field observations; family badges of *H. reaumuri* are extremely variable.

What variables influence aggressive behavior?

In the exchange experiments, distinct intrafamily differences in aggressive behavior were observed. The following three points deserve closer examination.
1. Among juveniles the readiness for reacting aggressively against alien conspecifics is neither site- nor age-independent and is, furthermore, influenced by current and/or preceding behavior.
2. Siblings may, in spite of identical environmental conditions, react differently towards the same alien isopod.
3. The aggressive reactions of individuals from one family to members of different foreign families may differ greatly, and for long periods reproducibly, in the average frequency and intensity of attacks.

Where and when are aliens attacked under natural conditions?

Parents with progeny not yet half-grown usually attacked alien conspecifics wherever they met with them, even when carrying food on their way back from foraging excursions (the food being dropped and usually lost). Half-grown or older young sometimes attacked aliens during foraging trips (mainly on their way out—see Discussion and Summary), but in the majority of encounters (64% of 159 observations) they showed avoidance behavior, especially when feeding or carrying food items.

For large (three-quarters- to full-grown) juveniles the highest level of aggression toward aliens (96% of 80 individuals reacting aggressively in at least three consecutive trials) was restricted to the fecal embankment area of the family burrow. Half- to full-grown young (more than 10 mm in length) attacked strange conspecifics irrespective of whether they had just left the family burrow (while cleaning it or starting to forage) or had entered the fecal embankment upon returning from feeding. Smaller young (less than 10 mm in length) behaved differently. Those under 5–6 mm never ($n > 100$)

attacked aliens, but ran away, in particular when encountering strange adults. Those 6–9 mm long were regularly aggressive only inside the fecal embankment area and, additionally, only when leaving their burrow (81% of 57 young tested). Just outside the fecal embankment on an outward trip only 11% of 101 were aggressive. On returning they normally showed avoidance behavior, even when touching the same foreigner as before at the same place inside the embankment (then only 4% of 76 individuals were aggressive). Of 45 on their way back, none was aggressive outside the fecal embankment. Which conditions enhance or reduce the aggressiveness of young family members?

We have long known (Linsenmair and Linsenmair, 1971; Linsenmair, 1972) that the burrow walls and the fecal embankment area are marked with the family-specific badge. This could be one factor responsible for the site-specificity of aggressive behavior in young isopods, but it cannot be solely responsible in individuals not yet half-grown. To explain why these young behave differently on the same spot when departing and returning, an additional factor must be sought. It is possible that contacts with family members might be the decisive proximate factor here. To test this hypothesis families must be manipulated, which could have profound effects on their aggressiveness.

Effects of experimental manipulations and maintenance on aggressiveness

Using families living in their natural burrows (or in laboratory containers) with an excursion range of a few meters around the entrance, data collection would be hampered for the following reasons.

1. Only individuals staying within the fecal embankment could be used for reliable tests. During activity phases, often lasting not over 2–4 h per day, at best a small minority of the population would be testable.
2. Outside the activity phases only the burrow guards are accessible for tests and, as a consequence of the peculiar spatial situation and of the guard's special behavior (see Linsenmair and Linsenmair, 1971), most experiments could not be performed.
3. With rare exceptions, molting desert isopods never leave their dens. Therefore, all color marks are lost during molting. Since molting takes place in two steps, separated by 18–28 h for half-grown and older *Hemilepistus*, the isopods are, when observable, easy to keep track of individually during moltings by remarking the newly molted posterior halves before the anterior hemiexuvia are shed. This is a precondition for numerous essential experiments.

Hence, great advantages are gained from maintaining families and experimentally assembled groups in containers without burrows, where all members are continuously accessible.

As all our tests have demonstrated, most families kept for prolonged periods in the standard containers showed no lowered aggressiveness in comparison with controls living under natural conditions in their burrows and tested inside the fecal embankment. Excavating families in late summer causes the strongest disturbances; burrows may then reach depths of about 90 cm and a length (including side-branches) of 2 m, and they often are situated in hard ground. Such families ($n = 30$) were used to determine whether removing isopods from their dens and placing them in small standard plastic boxes and transferring them within 2 h to a laboratory environment reduced their aggressiveness for a significant length of time. In spite of the massive disturbance, the members of all families examined under these conditions regained their previous level of aggressiveness within 1–2 h and many individuals within less than 30 min. This level then remained constant for many weeks if climatic and maintenance conditions were not materially changed.

Does a family's own badge stimulate aggressiveness?

In the first set of experiments, whole families ($n = 10$) or a group of siblings (> 20 individuals per group, $n = 25$ groups) were transferred to new boxes and substrates and were tested immediately afterwards with single alien conspecifics. No significant differences in the readiness to attack alien conspecifics could be detected when comparing groups placed either on a substrate consisting of material from their own fecal embankments (taken from the field directly before excavating or trapping the family) or on neutral soil not contaminated with badge substances of any family. In the first situation 80% of 98 individuals and in the second 83% of 93 individuals reacted aggressively in encounters with aliens. A significantly lower level of aggression resulted if the substrate consisted of discharge from an alien burrow (65% of 88 tested individuals responding aggressively).

The last finding demonstrates that chemical properties of the substrate can influence the level of aggression. Why did the neutral soil not produce a similar effect? Judging by the behavior of isopods outside the fecal embankment in the field, reduced aggressiveness was to be expected. One possible explanation is that because a whole family or a group of siblings was transferred, frequent contacts with family members might have compensated for any aggression-reducing effects of neutral soil.

If this interpretation is correct, then clear evidence would be expected from displacing single isopods. Such evidence was obtained (Figs. 4A and 4B). Single isopods placed on neutral soil behaved less aggressively than similarly treated sibling groups or whole families. Within seconds of transfer, eight of the transferees (of a total of 26 from five families) strongly attacked alien isopods after the first three to six antennal contacts. But then they suddenly

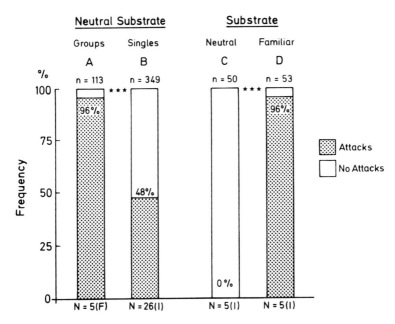

Fig. 4. Frequency of attacks in encounters with alien conspecifics following transfer of individuals or groups to different substrates. A–C, responses immediately upon transfer to a neutral substrate; D, following transfer to a substrate contaminated with the badge of their own family (see text for further details). N(I) = number of individuals used (originating from five families)

stopped attacking and often showed escape behavior after touching an alien, as did other individuals earlier.

In a further experiment (Figs. 4C and 4D) some of those isopods that did not react aggressively after being displaced to a neutral substrate were placed in solitary confinement for 2 h and then retested on neutral soil. They displayed an even higher degree of 'anxiety' (i.e. a more intensive escape behavior) after contacting unfamiliar conspecifics. However, if these isopods were returned to their family container, from which all siblings and the parents had been removed shortly before, in 96% of all encounters with a foreigner, attacks were released. Evidently, they recognized their surroundings, most probably by contamination of the substrate with their family's badge, and at the same time regained their aggressiveness. (All tests took place during the first 5 min after replacement and the young used in this experiment were nearly full-grown.)

These results suggest plausible explanations of the why and when of aggressive acts against aliens under natural circumstances. The fact that young desert isopods defend mainly their burrow and the fecal embankment area is clearly an effect of the family badge. Perception of this badge by touching kin and/or the contaminated substrate increases the motivation to attack aliens; even short periods (about 10 s with small young) without renewed perception of the family-specific badge lowers the level of aggression.

These results also show that keeping families in small containers should be a suitable means of inducing and maintaining a high level of aggressive motivation, since family members live in mutual contact on a substrate heavily contaminated by family-specific chemicals. Thus these conditions conform in important respects to those of the family burrow and the fecal embankment area.

The above conclusions are valid for the numerous North African populations we know of (from western Egypt to eastern Algeria), but do not apply to progeny of pairs originating in the Negev and bred in the laboratory. Even at their burrow entrances fully-grown Negev juveniles ($n > 20$ families) showed a considerably reduced aggressiveness toward alien conspecifics in comparison with the progeny of parents from North African stock. If transferred to the standard maintenance boxes, most Negev families lost their aggressiveness toward non-kin entirely. It remains an open question whether this represents normal field behavior or is a laboratory artifact, since we have not investigated the behavior of free-living, fully-grown juveniles in the Negev. Apart from this peculiarity, there are a few others in the Negev population, e.g. a skewed sex ratio. However, the Negev *Hemilepistus* is not a separate species (our breeding results, see also Lincoln, 1970).

Intrafamilial variability in aggressiveness

Within families a striking interindividual variability in the frequency and intensity of aggressive reactions towards identical aliens is often observed. What are the reasons for these differences?

1. *Molts and reduced vitality.* The variability is caused, in part, by molting. Isopods either do not attack at all or only with reduced frequency and intensity for 12–24 h before shedding the posterior hemiexuvium, and for about 24 h after discarding the anterior part. Under field conditions, such isopods do not leave the burrow and do not guard the den. Thus they would not normally encounter alien conspecifics. In other cases unaggressive isopods were either heavily infested with parasites (nematodes and larvae of acanthocephalans) or were ill from unknown causes, and most died within a few days. These factors do not explain lack of aggressiveness of many other individuals. Other causes must be hypothesized, e.g. different levels of 'attentiveness' and different 'judgements' of alien badges.

2. *'Attentiveness' and the recognition of aliens.* Tests conducted in the standard containers at different times of the day demonstrated the existence of a pronounced cyclic variation in the readiness of family members to respond aggressively toward alien conspecifics. The changes correspond well with the activity cycle. Isopods which are, or would be, active outside their burrows show a low threshold, readily responding aggressively. The same individuals, however, will often not react to contacts with the same aliens during the resting hours. A trivial partial explanation is that some isopods are not disposed to touch aliens presented to them. But after a light push, a considerable number will inspect a foreigner thoroughly, yet still not show aggression. There is great intrafamilial variability in this respect, but nearly always there are a few individuals that remain attentive and critical. Guards are recruited from this set of family members, as becomes apparent if one allows a family (with its members individually marked) to move into a burrow. Nearly always, the parents belong (sometimes solely) to this permanently alert group, provided they are not molting and not yet senile. A few individuals may quickly activate the 'sleepers' by excitedly running about and bumping against other family members. Thereafter, brief antennal contacts with the same alien may release strong attacks by individuals that previously did not react aggressively despite extensive touching.

The cyclic changes in the readiness to react aggressively upon meeting the bearer of a deviating badge are well correlated with the changing probability of encountering an alien. During phases of inactivity the guard is the only individual that may occasionally come into contact with alien conspecifics and it is always fully alert during these periods.

3. *Variability in response to aliens.* Most alien badges release strong attacks. In testing members of numerous sibships in one test family, particular foreign badges are always found that release less frequent and less intensive attacks than the majority. In presenting individuals which release only weak attacks, striking interindividual differences within the test family can also sometimes be found. This form of variability in response to similar stimuli is not due to differences in arousal but points to 'divergences of views' between close kin. Parents may be strikingly different from their progeny, or the greatest differences may occur between a small group and the large majority of siblings. For example, the parents, or a small sibling group, may react by attacking at each encounter, whereas other family members do not attack, or their initially weak aggression quickly vanishes.

Here, the badges of the two families evidently resemble each other. In this situation mainly guards, or individuals ready to act as guards, are motivated to inspect the aliens thoroughly enough to detect and respond to the differences. All others 'see' no deviations or at least do not respond to them and therefore behave peacefully (see Discussion and Summary).

The origin and nature of the family badge

There is no doubt that the characteristics used for identifying kin and non-kin are chemical in nature. Are these substances acquired from the surroundings and/or are they influenced by the food spectrum?

Are environmental odors a source of interfamily variability?

Environmental odors, defined as substances which are not produced by the bearer or its conspecifics but are acquired via adsorption from the surroundings, have often been assumed to provide the basis for individual or colony signatures of eusocial Hymenoptera (e.g. Ribbands, 1953, 1965; Wilson, 1971; Michener, 1974). But in recent years a growing body of evidence supports the assumption that genetically based sources of variability may be more important in these insects than was previously supposed (Kukuk *et al.*, 1977; Greenberg, 1979; Breed, 1983; for summaries see Chapters 7 and 8).

In a series of experiments performed at the beginning of our investigations of *Hemilepistus* (Linsenmair, 1972) we were able to demonstrate that environmental odors probably do not cause high interfamily badge variability. These earlier findings have been supplemented by numerous additional experiments, but only a few results will be mentioned here.

If environmental odors play an important role in the generation and maintenance of the interfamily variability of badges, it should be possible to reduce and finally to eliminate such acquired variability by keeping different families for long periods under identical living and feeding conditions. With increasing duration of equal treatment, interfamily aggression should diminish and the frequency of misidentifications in encounters with alien conspecifics should increase. During the last 15 years we have excavated (or trapped) in the field and transferred to the laboratory more than 2,000 *Hemilepistus* families. Large numbers of these families were kept under identical conditions (concerning maintenance boxes, soil, nourishment, temperature, air and soil humidity and light/dark cycle) in close proximity to one another for several months. These families were collected at different times after the young had left their mother's brood pouches. Thus, for at least a few hours, the young isopods had been exposed to uncontrolled conditions outside the brood pouches, which could have had some influence on the acquisition of the badge. This rather improbable influence was eliminated by transferring pairs, before delivery of their young, from the field to the laboratory where many hundreds successfully raised their progeny. Finally, and most crucially, in several thousand cases, parents that produced successful broods in the laboratory originated from the second to the fifth generations bred under identical conditions in captivity.

The laboratory conditions specified above should be ideal for equalizing family badges if, in fact, environmental odors are essential components of badges. However, in several thousand exchange experiments there was no evidence of reduced accuracy in distinguishing family members from aliens. This suggests that the interfamily variability produced under the homogeneous laboratory conditions does not differ recognizably from that found under natural conditions.

These results, however, demonstrate only that differing environmental odors are not a prerequisite for achieving high variability and reliable specificity of family badges. In addition to an endogenous source, there might still be a second, environmental, source of recognition substances that enhance the diversity and specificity of the signatures. This question was addressed by dividing families and feeding the subgroups differently or by placing them on different soil for periods of several weeks. Feeding experiments yielded no evidence that nourishment influences specificity (see Linsenmair, 1972). Experiments with different substrates also failed to demonstrate significant effects. However, as later experiments occasionally gave somewhat ambiguous results, they were repeated.

Four small experimental groups, each consisting of five individuals, were separated from their respective families (n = six families). Two experimental groups (called controls) and the much larger remainder of the family (family group) were maintained on soil from the vicinity of the natural burrow from which the families were excavated. The other two experimental groups were kept on soil from other parts of Tunisia and mainly from the surroundings of Regensburg and Würzburg in West Germany. After being separated for four to 18 days, members of the experimental groups were temporarily returned to their kin in the family group.

The outcome of this experiment confirms our previous results. Within 18 days a few individuals became slightly alienated, but these were equally distributed between the experimental animals maintained on different soils (5% of a total of 228 encounters led to attacks) and the controls kept on the same soil as the rest of the family (7% aggressive reactions among 211 observed encounters).

Since acquisition of substrate-borne substances might be possible only by newly molted isopods, individuals (n = 110) were separated from their families shortly before molting and half were placed singly in containers with their original substrate, half with a different substrate. As in the previous experiment, no statistical difference between groups could be found. In a total of 1,645 encounters with the 55 individuals that molted on different soil, 15% resulted in attacks; in the control groups the corresponding figure was 17% (n = 1,528 observed contacts). Thus substances acquired directly from the substrate or indirectly via the digestive tract through substrate feeding (soil is always eaten in great amounts, see Shachak, 1980; Coenen-Stass, 1981; Shachak and Yair, 1984; K. E. Linsenmair, in preparation) could not be shown to change the family badge.

Finally, the investigation was taken a step further by directly besmearing family members ($n = 20$) with different soils. This led only to prolonged tactile inspections until a more or less uncontaminated piece of cuticle was found.

Despite these numerous negative results, there remains the possibility that under special conditions, environmental odors may play a significant role. Most probably the 'Regensburg-phenomenon' has to be explained on this basis. In numerous laboratory experiments, involving several hundred isopods from more than 100 families, water from different sources (distilled and bidistilled, as well as tap water) almost always changed the character of the family badge, if applied directly, e.g. by spreading a single drop over the tergites or briefly submerging the subjects. After drying, the great majority (> 90%) of treated individuals released very strong attacks. In control experiments in the field, in the laboratory in Tunisia and in Würzburg, these results could not be confirmed with respect to either the regularity or to the strength of the alienating influence of water. We have to assume that unknown substances in the maintenance rooms in Regensburg, most probably entering via the central air conditioning, somehow changed the badges.

Weak effects of water were shown under both field and laboratory conditions at places other than Regensburg (22% of 203 individuals tested released temporary aggression, but were finally admitted to the family burrow). Therefore, the above findings must be classified as laboratory artifacts in a quantitative but not qualitative sense. Whatever their causes, these phenomena demonstrate that exogenous factors may alter badges, although they seem to play an insignificant role under natural conditions. Thus in conclusion, all findings support the hypothesis of a chiefly genetic basis for the discriminators used by *Hemilepistus* for signatures and badges.

Where are the discriminators produced?

For discrimination to occur, an isopod must be touched with the apical antennal cones somewhere on its cuticle. The discriminators are distributed over the whole surface, as many different experiments with live and dead isopods, body parts and exuvia have shown. In active isopods a brief (< 1 s) and localized (< 1 mm^2) contact is sufficient for a reliable classification as kin or non-kin. Lightly contaminated family members, as well as aliens differing only slightly in their badges, are usually investigated intensively (i.e. > 10 s at the entrance of a burrow; an inspection may last for many minutes before a 'decision' to accept or repulse them is made).

Since discriminators are evidently uniformly distributed over the whole body surface, a narrowly localized production site would require either a specific wiping behavior, or the substances (or a solvent) should possess good surface spreading qualities. There is no indication that any of the known gland fields (on epimeres) (Holdich, 1984) or large glands (in the uropods) produce secretions with alienating properties. Indeed, such a function can be definitely excluded for the uropod secretions as interfamily exchanges of these secretions never ($n = 30$ individuals) provoked aggression.

In isolated, molting *Hemilepistus* no form of behavior that would cause active spreading of discriminators occurred during continuous observations, yet individuals tested some hours after shedding a hemiexuvium gave no indications of not having uniformly distributed discriminators.

The spreading potential of the badge substances was investigated first with extracts obtained with different solvents from alien conspecifics. These were applied in small amounts but high concentrations to narrowly confined areas of the experimental animal's dorsal surface. Secondly, discriminators were transferred directly between individuals of different families by wiping large areas of the tergites of one animal onto narrow tergal parts of the tested individual.

The experimental isopods were returned to their families shortly after contamination; they were then separated and again presented at intervals. The site of antennal inspection could be influenced by the manner in which the isopods to be checked were held with forceps and presented to their kin, and the actual point of contact was observed. Between tests, these contaminated isopods were kept in isolation without a burrow. Thus, distribution of the transferred substances via contacts with conspecifics or by wiping along burrow walls was prevented.

All the results demonstrated that a surface-spreading potential of locally applied discriminators exists (Figs. 5 and 6), but is limited, as it never extended the response to the transferred alien substances more than one to four mm.

Spreading may play a role in the mixing of discriminators over small areas and may explain why individuals that were only weakly contaminated with alien discriminators sometimes became acceptable to their kin again within 6–24 h (after contamination) and why the contaminated areas lost some of their aggression-eliciting efficiency within 24 h.

A single localized source for production and delivery of the badge substances is not likely but two possibilities merit further investigation:
1. Badge substances might be produced in numerous small glands distributed over the whole cuticle;
2. they might be synthesized within the body and dispersed in hemolymph, reaching the surface via cuticular canals.

If the second hypothesis is true, hemolymph should elicit aggression either directly or after being extracted and applied to glass rods. Moreover, if foreign hemolymph is used to besmear family members, it should have alienating properties and it should produce a smaller effect or none if wiped on a donor's sibling.

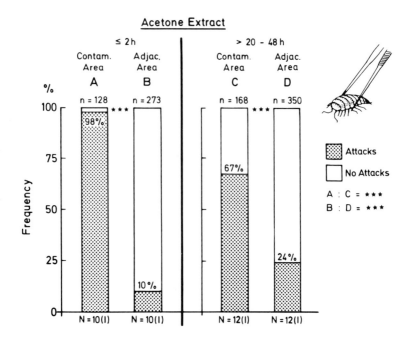

Fig. 5. Spreading potential of experimentally applied discriminator extracts on the cuticular surface. Application of 0.1 ml highly concentrated exuvial acetone extract to the dorsomedial part of the sixth or seventh tergite (A). Adjacent area (B) = lateral parts of the same tergite and adjoining halves of the next posterior and anterior tergite respectively

Hemolymph was obtained by cutting off an antenna, the emerging drop being quickly removed before it could spread over the cuticle. It was then applied directly to a glass rod, or distributed over the posterior tergites of an experimental animal, or dispersed in a solution of acetone/water or methanol/water (proportions, 80:20 in either case, with one drop of hemolymph/ml solution) and tested with glass rods. Both hemolymph and the solutions were tested only after they had dried, since fresh hemolymph induced cannibalistic behavior.

The results of these experiments are summarized in Fig. 7. Apart from some ambiguous results in experiments performed in late autumn with fully-grown young (which are not included in Fig.7), the outcome was clear. In the majority of tests both hemolymph extracts and crude hemolymph elicited aggression by members of foreign families (Figs. 7A and 7B).

Transfers of hemolymph between alien family members (via besmearing) also yielded clear results: alienation was consistent and often very strong (Fig. 7D). In exchanges of hemolymph between family members it became immediately obvious that sibling hemolymph is less effective in its alienating properties (Fig. 7E), but the expectation that intrafamily transfer would cause no badge alteration was met only in ten of 21 individuals, each taken from a different family. It is still an open question as to why 52% of intrafamilial hemolymph transfers resulted in alienation, albeit slight in most cases. Possibly chemically differing precursors as well as discriminator end-products are sometimes present in the hemolymph, or a larger spectrum of potential discriminators is present from which only a restricted number are released to the cuticular surface by means of selective transport mechanisms. This is pure speculation; we need more information about what substances are used as discriminators before definite answers can be given to the questions of where and how the substances are produced and how they reach the surface.

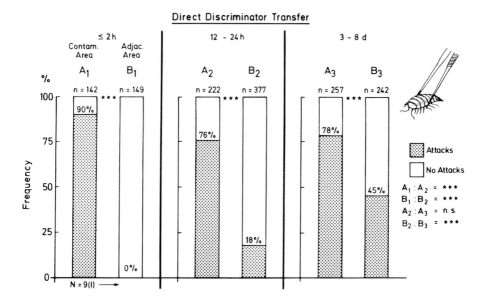

Fig. 6. Spreading potential of discriminators in relation to time after contamination. Direct transfer of recognition substances between aliens was to the same tergites as in Fig. 5 (for method see text). Comparison of the results presented in Fig. 6 with those of Fig. 5 shows that extraction does not change the surface-spreading properties of discriminators

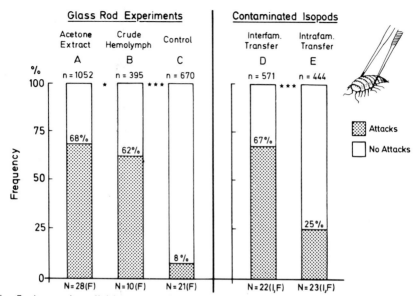

Fig. 7. Aggression-eliciting properties of acetone-extracted or crude hemolymph tested in different ways in alien (A, B, D) and own family (E). In the control experiment (C) responses toward either clean glass rods or those treated with pure solvents (after drying) are summarized. N(I,F) = number of families from which at least one individual was tested

The multi-component hypothesis of the family badge

Two findings mentioned so far lead to the postulate that family badges are blends of several components.

1. Because no two identical badges have been discovered, we are compelled to assume the existence of thousands, at the very least, of discernible family-specific discriminator patterns. In fact, attaining interfamilial variability with 10^6 or more different badges should cause no special problems if blends of several substances are always involved (see Discussion and Summary). Such variability could not be achieved with a single chemical compound for each badge.
2. The observed (and reproducible) graded aggressive responses toward members of different alien families, extending in the extreme to unilateral acceptance, are easy to interpret within the framework of a multi-component model. The reason for a low level of aggression could be that an alien family conforms in most of its components to the badge of the test family. For unilateral acceptance the accepted family may have a badge consisting only of components that are also present in the signature of the accepting family, whereas the latter possesses at least one additional component, allowing the accepted family to identify members of the accepting family as alien.

Intrafamilial signature variability and the common family badge

Family members recognize each other and are able to distinguish their kin from an enormous number of aliens. Theoretically, this performance could result, at one extreme, from every individual knowing all other family members individually. At the other extreme, if there were no intrafamilial variability, it would suffice for an individual to become acquainted with only one family member in order to recognize all its nearest kin. The reality may lie somewhere between these two extremes.

Are the family members individually recognized?

To answer this question, adoption experiments were performed. Parents were separated from their own young and kept with 15–30 young of another family. When, 3–35 days after the forced adoption, these parents were introduced for the first time to siblings of their adoptive young that they had never met before, six out of seven of the sibling groups released no aggression by the foster parents (Linsenmair, 1972). These experiments, in which young six to eight weeks old and families with few progeny were used, were repeated with young from large families (> 80 individuals) six to seven months old, in order to vary sibship sizes. The results and further explanations are given in Fig. 8.

The outcome of all adoption experiments compels us to discard the hypothesis that individual characteristics alone are the basis for recognizing kin. If this assumption were correct, all unfamiliar siblings of the foster young should have been rejected. On the other hand, the results do not support the second extreme hypothesis (no intrafamilial variability) either, since in the second series of experiments, a considerable number of the unfamiliar siblings from the main groups were not accepted by the foster parents, especially by those cohabiting with small groups of siblings that could not represent the whole chemical diversity of a sibship. Even if there were not exceptions to the prediction that members of the separated sibling group would not be attacked, the second hypothesis would need support from additional evidence before acceptance.

Do all family members produce identical signatures?

In addition to adoption experiments, there are other considerations that render the assumption of no intrafamilial variation in signatures *a priori* improbable. Firstly, *H. reaumuri* shows normal bisexual reproduction. Emigration of the young, following the dissolving of families at the end of the hibernation period, thoroughly mixes the population. Therefore, we have to assume that members of more than 99% of all mated pairs come from

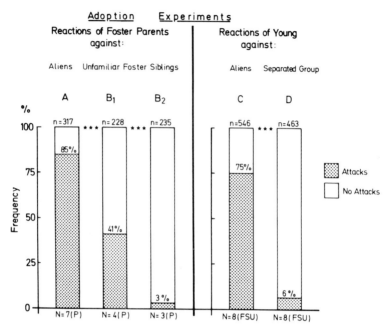

Fig. 8. Investigation of intrafamilial signature variability through adoption experiments. Alien parents (three pairs and a single adult female) were forced to adopt different numbers of foster young from four large sibships (> 80 siblings each). Two pairs (B_1) adopted relatively small groups of 14 and 21 siblings respectively. In B_2 the single female and the third pair accepted larger groups (28 and 41 individuals). The main group of young of each sibship was kept separately. For the first time on the third day after adoption and subsequently six times until day 45, siblings of the main groups were presented to the foster parents. The frequency of attacks released by these siblings of the foster young, previously not met by the tested parents, did not change after the sixth day. The four adults that had adopted small groups (B_1) were significantly more aggressive than the B_2 parents, but significantly less so than toward foreign conspecifics (A; concerning these reactions, B_1 and B_2 parents did not differ). Siblings exchanged between the two subunits of each sibship elicited few attacks (D), whereas alien young were always attacked with equal probability in both subunits. This experiment was performed close to the time of the hibernation phase; overall aggressiveness was therefore reduced. N(P) = number of parents used; FSU = family subunit

different families possessing different badges (Linsenmair, 1984, 1985). No evidence points to sex linkage of the genes that are responsible for the production of discriminators. However, with normal autosomal inheritance and the presumption of a multilocus gene system with several alleles per locus (see Linsenmair, 1985), variability among siblings with respect to discriminator patterns is to be expected.

A further consideration is that a paired female that loses her male normally pairs anew with a second, or even a third, male until shortly before her parturial molt. A conservative estimate, based on data gathered in the field, is that at least 5–10% of all pairs that successfully raise young include a female paired with a second male. The only pairs considered for the above estimate were those in which the replacement of the first male occurred after it had already performed the long copulations during which the bulk of sperm is transferred and which usually terminate receptivity in females not separated from their first male for more than a few hours (see Linsenmair, 1979, 1984).

By using X-ray irradiated males either as the first or the second mate and by determining the proportions of normally developed embryos in the marsupial mancas shortly before hatching, we demonstrated that no perfect sperm clumping occurs. In 60% of all ($n = 126$) cases, both males fathered some of the progeny (K. E. Linsenmair, in preparation). It is obvious that in such families an extra source of signature variability exists. Young fathered by the second male should add new components to the discriminator spectrum of the family, augmenting the intrafamilial heterogeneity. We conclude that uniformity among all family members in the genes responsible for discriminator production is highly improbable.

Is the intrafamilial variability in discriminators secondarily reduced?

Given the postulated genetic basis for considerable intrafamilial variability in individual discriminators, a common family badge could be achieved by a secondary adjustment. One of several mechanisms would be to confine the signature substances produced within the family to the smallest common denominator of all its members. The question of whether such a process takes place can be tackled in different ways.

Individuals might match their discriminator production to their badge environment. If this were a continuous process and not a single event fixed within a short critical period, amputations of the apical antennal cones might cause changes in the produced spectrum of recognition substances. No evidence favors this possibility; in amputating the apical cones of more than 300 individuals from more than 25 families, kept singly or in groups for up to six weeks before being tested, we never found indications of increased alienations compared with controls consisting of identically treated but intact animals.

Another method involved the long-term integration (4–8 weeks) of single intact individuals into strange families. In preliminary experiments we found that the adopting family and the family from which the integrated individuals were taken showed strong mutual aggression. During the experimental period the integrated isopods molted one to several times. Shortly before a further molt, they were separated from their foster families, tested in their genetic

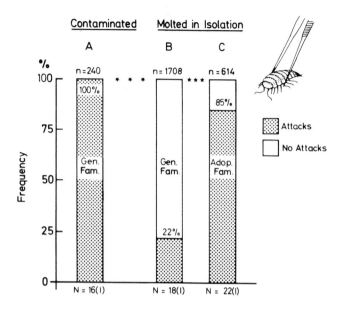

Fig. 9. Effects of molting on acceptance of individuals to their genetic and foster families. Individuals that had been integrated for weeks into alien families were presented before molting to their genetic families (A), and after molting in isolation, both to their genetic families (B) and to their previous foster families (C)

families and then, during the molt, put into solitary confinement. Thereafter, at least 24 h after molting, they were presented first to their genetic and then to their adoptive siblings or vice versa.

As was to be expected from previous findings, the integrated, still unmolted isopods released severe aggression when tested against members of their genetic families (Fig. 9A). This demonstrates beyond any doubt that the signatures of these isopods had changed for some reason connected with the alien badge environment. However, after molting in isolation, the majority of previously heavily attacked individuals were peacefully accepted in their genetic families, but mostly treated very aggressively by members of their foster families (Fig. 9B and 9C). This result clearly contradicts the hypothesis of the integrated animal's adjustment in discriminator production to a changed chemical environment. Obviously, integration results in contamination with alien substances, not a changed production. Two points deserve closer inspection: Firstly, only 43% of all molted individuals were accepted by their genetic kin without aggression ($n = 56$ individuals, from which one series of 38 was not tested in its adoptive family; these were not included in Fig. 9). The rest were attacked, although in most cases only mildly and by a minority of the control siblings. Secondly, 15% of the contacts between the newly molted isopods and their adoptive 'siblings' did not lead to attacks, whereas in encounters with aliens a lower percentage ($< 10\%$) of contacts were expected to be non-aggressive at that time in the families used.

Do these two points show that, although slowly and imperfectly, some adjustment nevertheless can take place? This question will be answered below. For the moment we conclude that the hypothesis of a secondary, exact and rapid adjustment by matching the individual discriminator production with properties of the badge environment lacks support.

Is a common badge achieved by mixing individual discriminators?

Early experiments (Linsenmair, 1972) demonstrated that keeping large groups of approximately equal-sized individuals from different families together in small containers for several days resulted in alienation of all individuals from their respective families. Later experiments showed that by crowding isopods together, e.g. in a small petri dish (6.5 cm) for 24–72 h, one strange individual can contaminate a maximum of 10–20 siblings, rendering them entirely unacceptable to their uncontaminated kin for days. By intensive and frequent body contacts with aliens (using small containers with 20–30 isopods, adding a single member of a test family, and then shaking and turning the container), not more than 10 s are needed for sustained alienation of an individual. To cause alienation it also is sufficient to rub two isopods from different families two to five times back to back while exerting slight pressure.

All these experiments, performed thousands of times in different contexts with identical results, show that during direct body contacts discriminators are always exchanged. Members of families, living in their burrows in close association with one another and with frequent direct body contacts, therefore, cannot avoid almost continuous exchange of discriminators. Such exchange should result in a thorough mingling of all recognition compounds produced by the members of the family, thus harmonizing individual differences, to some extent at least, and producing the family badge.

Are discriminators transferred only by direct contact or also as a vapor?

As long-term controls reveal, strong alienations resulting from body contacts with foreigners are not short episodes, but may last for weeks. Most such individuals, kept solitarily, are not acceptable before their next molt. This enduring alienating efficacy of strange discriminator substances points to low volatility of the compounds and/or their firm adsorption onto the cuticle. Properties of extracts unambiguously support the assumption of low volatility; only three such properties will be mentioned here.

1. Solutions may be heated to 90 °C and still no discriminators appear in the distilled solvent in amounts detectable by bioassay; they remain without loss of efficacy in the residue.
2. Effective extracts on glass rods do not lose their efficacy at room temperature, when openly stored, for many days or weeks.
3. Exuvia retain their aggression-releasing properties for more than half a year when kept at temperatures of up to 30 °C.

These findings make it highly probable that exchanges of discriminators as vapors would require considerable time and would be effective only among large numbers of isopods in small containers. As a test, small plastic boxes were divided into halves by a double wire screen with a mesh width that did not allow the smallest (half-grown) individuals to pass through, and that reliably prevented mutual touching, even with the tips of fully extended antennae. Air flow between the two parts of the container was certainly not appreciably impeded. Members of two different families, or of a family and an extreme mixed group (each individual taken from a different family), were placed on either side of the screen. Groups were used that had displayed a high degree of mutual aggression in preliminary experiments. A relatively large subunit of each pure family was kept isolated from aliens to serve as controls. Members of each group, after they had spent 10–45 days within the divided containers, were tested with the control group of their genetic family, as well as by mutual exchange between the adjacently-living groups.

Should test isopods be accepted without aggression by their kin, a discriminator exchange via the vapor phase in detectable amounts would have to be excluded. If, on the other hand, a complete exchange of discriminators were achieved by diffusion, members of adjacent groups should develop mutual tolerance. The results were unambiguous. No alteration of badges could be detected during a period of more than four weeks; members of test groups never gave any indications of being contaminated when returned to their respective control groups ($n = 14$ families). Slight aggression between family subunits separated for more than 20 days was observed in four cases, but this was certainly not caused by discriminator transfer (see below). Exchanges between groups living in adjacent compartments always released aggressive reactions at intensities not distinguishable from those of the preliminary tests.

Could it be that the properties of the cuticle change during the molting cycle, so that only molted individuals adsorb alien discriminators from the vapor phase? During the experiment just described, most individuals (n > 30) molted more than once. These isopods were tested frequently at different times after molting, but no differences from intermolt animals could be detected.

To be on the safe side we performed additional experiments. Aliens ($n = 18$) integrated into several families and fully accepted there, were placed, shortly before molting, in two stainless steel wire cages ($2 \times 2 \times 2$ cm) of fine mesh, preventing direct contacts between the caged individuals and others outside, but allowing free air circulation. These cages were placed in the standard maintenance containers of the respective integrating families. Within these boxes, lacking a burrow, the wire cages were favorite places for isopods to congregate. Most of the time, therefore, these cages were densely covered with isopods. Thus, ideal conditions for indirect transfer of volatile badge substances during molting were guaranteed. However, as in all previous experiments, no evidence was obtained of such acquisition of alien recognition substances. All the aliens, tolerated before the molt, were severely attacked afterwards.

One may conclude from these results that an exchange of badge substances requires direct body contacts. Adsorption of discriminators from the vapor phase does not occur in above-threshold amounts. Only by exchanging and mixing discriminators via direct contacts could individual differences in badge production be harmonized.

Do family members produce individually differing discriminator patterns?

As has been demonstrated above, a *Hemilepistus*, when it molts, discards along with its old cuticle the discriminators acquired from conspecifics. Moreover, environmental odors normally do not contribute to badge variability, and the badge environment perceived via the apical cones does not influence the spectrum of substances produced. Separating isopods from their families and maintaining them in solitary confinement during molts, therefore, should be an appropriate means of obtaining individuals carrying only their own self-produced discriminators.

In an experiment described above (see section 'Is the intrafamilial variability in discriminators secondarily reduced?') many individuals (57% of 56) that had been integrated into foster families for weeks were not unaggressively accepted by their genetic kin, after molting in isolation. On the other hand, some of the molted isopods elicited less aggression in their adoptive families than strangers normally do (see above and Fig. 9C). The obvious question is whether these results indicate the beginning of an adjustment to a changed badge environment, or alternatively demonstrate intrafamilial variability, and point to learning processes by the adoptive family.

If the postulate of an incipient adjustment is correct, one might expect that those individuals eliciting attacks by their natural families (indicating a deviation in their badge from the norm of the family) should be the ones releasing the least aggression in their previous foster families. Individuals accepted by their genetic kin without aggression obviously did not change their original badge. Therefore, upon replacement in their foster families, these molted isopods should be as fiercely attacked as in the preliminary tests preceding all experiments.

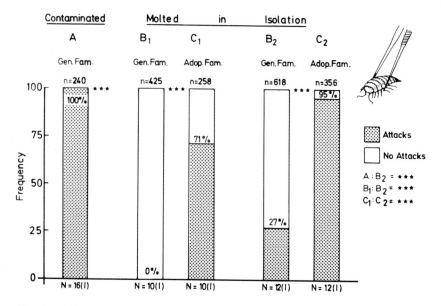

Fig. 10. Effects of molting on acceptance. Further analysis of a part of the results presented in Fig. 9: responses of the genetic and adoptive families toward identical groups of molted individuals (for details, see text)

The results obtained do not support the assumption of an adjustment. The same three individuals that elicited no, or almost no, aggression in their foster families, (column C_1, Fig. 10) belonged to the group of molted isopods that released no attacks by their genetic kin (B_1). On the other hand, without a single exception, all those treated as more or less alien by their genetic families (B_2) also encountered strong aggression from their adoptive families (C_2).

Before considering this as final proof of the existence of intrafamilial variability and before definitely rejecting the adjustment hypothesis, yet another explanation for the above outcome must be examined. Could a slow, imperfect and individually variable adjustment have accidentally produced these results?

If any form of slow adjustment influenced badge production, then individuals alienated a few hours before their molt by brief contacts with foreign conspecifics should differ from newly molted isopods that had been integrated into foster families for many weeks. Single isopods ($n = 20$) whose next molt was imminent (within 24–48 h) were each placed in a group of 30–40 siblings from an alien family or in an equal-sized extreme mixed group for five or 15 min. These test subjects became strongly alienated and did not differ significantly in this respect from siblings living for weeks in foster families (96% attacked in 288 encounters compared with 98% in 195 encounters in the latter). The reactions of kin towards members of the two experimental groups differed significantly after molting, however. Of the formerly integrated isopods, 22% were attacked (Fig. 9B), compared with only 12% in 724 encounters for the briefly contaminated individuals. Does this result support the adjustment hypothesis? Before examining this, we have to ask whether the fact that in both groups more than 10% of all encounters released aggression (in each group a minimum of about 40% of the individuals became at least slightly alienated) was caused by an after-effect of contamination or by intrafamilial variability. Varying contamination time and thereby contamination strength produced no corresponding variation in the frequency of post-molt alienations. Contamination times were 30 s to 1 min, 5 min, 15 min, and 4–6 h respectively ($n > 30$ individuals in each case).

These results indicate that for all isopods, including those that were heavily contaminated, foreign substances were, as postulated above, lost by molting or were at least diluted below the threshold of detectability. Thus, differences in the extent of contamination ought not to cause any differences in the outcome of the above experiments. If contamination with foreign substances does not survive molts, it could not have been responsible for the failure of some contaminated individuals to regain their acceptability following a molt. If intrafamilial signature variability is the answer to this issue, then isolating normal, i.e. uncontaminated family members for molting should yield an incidence of molt-caused alienations similar to that found in the previous experiments.

To test this, isopods were separated from their families 0–48 h before they started to molt and placed singly into identical small containers. Up to six days after they had shed their anterior hemiexuvia, they were returned to their kin. Since in addition to molting, the separation time (three to eight days) could be responsible for alienation, family members at an intermolt stage were, in a number of cases, isolated in parallel with molting siblings under identical conditions and returned as controls either immediately before or after their molted siblings. Molted isopods ($n = 280$), returned to their families after isolation times of three to six days, released attacks in 11% of all encounters ($n = 7,141$), compared with only 0.4% (two of 501 observed contacts) for the unmolted controls ($n = 55$). Comparing the incidence of molt-caused alienations of uncontaminated individuals (12%) with that of isopods contaminated shortly before their molt (11%) demonstrates that the null hypothesis of a uniform reaction cannot be rejected.

Families differed in regard to the incidence of alienations following molts in isolation (Linsenmair, 1985). In the majority of families most molted members were not attacked when returned, especially if not isolated for more than six days. In a minority of families, however, far more than 50% of molted members were treated aggressively, with striking interindividual differences in the intensity, frequency and duration of aggression. These findings clearly support the hypothesis of intrafamilial variability in signatures.

How does separation time influence molt-induced alienations?

The explanations presented so far do not answer the question of why molted individuals that had been integrated for a prolonged period into alien families released a significantly higher incidence of aggression than did isopods isolated shortly before their molt. Is the adjustment hypothesis yet to be saved? An obvious difference between the two groups is the mean separation time. While the integrated isopods were separated for weeks, all the others were absent from their families for only three to eight days. Is this difference crucial?

Prolonging the isolation of uncontaminated molting individuals without changing any other external condition caused more individuals to become alienated and attacks on them to occur at a higher frequency (Fig. 11). However, this increase in aggression occurred over a period of three to twelve days only under our experimental conditions. Uncontaminated molting family members isolated for seven to twelve days released aggression with the same probability as did the integrated animals (Fig. 9B) after a total separation time exceeding four weeks.

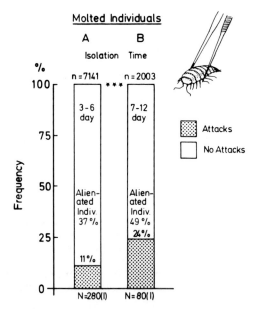

Fig. 11. Frequency of attacks released in encounters with molted family members in relation to increasing isolation times from three to twelve days. In A, 37% of all molted individuals released at least light attacks in some of their family members; in B this figure is 49%

What is the nature of this temporal factor? How does it strengthen the alienation effects of molts? The following two hypotheses require examination.
1. In the days following a molt, further modifications of the signature might take place, e.g. an increasing concentration of individual components could enhance the unfamiliarity of the signature.
2. If in normal families interindividual variability persisted despite extensive intermingling of discriminators, then prolonged separation could cause individual peculiarities of separated individuals to be gradually forgotten by the rest of the family, thereby causing increased aggression with increasing isolation time.

These two hypotheses are not mutually exclusive; both mechanisms could act simultaneously.

The following favors hypothesis 2: if changes after the completion of molting are decisive, as is suggested in hypothesis 1, then the critical quantity would not be the total separation time, but only the time spent in isolation after molting. In fact, the data collected so far indicate that the total time is the more important variable.

Since separating intermolt juveniles for six to ten days leads to alienation only in exceptional cases, the temporal effects are evidently based on molt-caused changes that are somehow strengthened with increasing separation time. A possible explanation within the framework of hypothesis 2 is as follows: during a short separation the individual traits are remembered by family members and the molted individual would be categorized as belonging to the family even though its signature now deviates from the common family badge through the loss of acquired discriminators. During a prolonged absence, however, the individual-specific information cannot be renewed and is gradually lost from memory of the family members. Are these assumptions supported by experimental evidence?

Memory and recognition of badges

Family badges may be forgotten completely and rather rapidly. If three-quarters- to fully-grown intermolt young are put into solitary confinement for periods exceeding eight days, then a considerable number react upon return to their family as if they had been transferred to an alien family. Initially, they behave aggressively, but after several contacts (three to more than ten), they often behave 'anxiously.' If they attacked kin only if they themselves released aggression, it could be concluded that their badges change during isolation and that these isopods rely on their own signatures as a reference for a phenotype-matching mechanism of discrimination (Sherman and Holmes, 1985; Discussion and Summary; Chapters 2, 3, etc.).

Such a correlation does not exist; attacks by the returned isopods during initial encounters are independent of how they are treated by their kin. This alone, however, does not definitely disprove the assumption of a changed badge as reference for comparisons. Changes leading to unilateral acceptance could be invoked here, but such a hypothesis is improbable because in many hundreds of instances in which separated individuals had either been forcibly alienated by contamination or molts had resulted in such alienations, no indications of more rapid development of aggression (during the first eight days of isolation) towards bearers of the family badge have been found. Rather, the time course of this process (in isolated animals) seems to be independent of the animal's own signature; it depends solely on the duration of separation from kin and thus points to a process of forgetting.

Integration of a group of foreign siblings into a foster family causes their badge to change rapidly. Under crowded conditions, alienation requires only a matter of seconds, as has been demonstrated. Consequently, after the first contacts, integrated isopods could no longer perceive their original family badge. Nevertheless they will accept uncontaminated genetic siblings for a prolonged period. Eventually, however, the same badge releases aggression (see below and Fig. 20).

These findings show that an individual signature may differ considerably from the actual common badge of a group, yet not induce aggression. On the other hand, the identical signature will, after a certain period, be forgotten and change its significance for the group members. It will then be classified as alien and thereupon will release aggression.

Additional and particularly clear indications that learned signatures are forgotten if refreshment of the memory is prevented were obtained through experiments with parents. The progeny of large families (> 80 young) were divided into two equal groups maintained in separate boxes under identical conditions. Either both parents were placed in one group or the female and male in separate groups. Under these conditions the common badge of the young in both groups remained unchanged for weeks; siblings of both groups could be mutually exchanged without any recognizable effect. In a ratio of at least 20 young to one adult, parents did not contribute to the common family badge. On the other hand, if parents were exchanged between groups after a separation for six to ten days, violent attacks usually resulted ($n >$ 100 tested parents, K. E. Linsenmair, in preparation).

Parents evidently possess a peculiar status within the family and probably bear a particular cue identifying them as adults. They seem to produce less transferable discriminators, but otherwise there are no indications of fundamentally deviating signatures (K. E. Linsenmair, 1984, in preparation). The interesting point here is that parents are obviously integrated into family communities on the basis of some form of individual recognition, which functions only if the parents are encountered more or less regularly by their young. Without new encounters, a parent's individual signature traits are reliably recalled for not more than four to seven days by its own young; thereafter they are usually forgotten. These statements are based on results obtained with three-quarters- to fully-grown young.

Long-term integration of young alien isopods into foster families (not more than one alien per 25 family members) gave identical results to those described above. As in the case of the parents, these aliens did not contribute to the common badge (see below). If, following a prolonged period of integration, either a group of integrating siblings or the integrated aliens were separated at an intermolt stage and then returned, the reactions depended on the duration of separation. After short separation (one to two days) nothing striking happened, but beyond that time the formerly integrated animals were attacked with a frequency that increased with increasing duration of isolation. After an integration time of eleven to twelve days a three to five day separation resulted in 24% attacks (of 277 observed encounters with seven integrated aliens). Prolonging the separation time to nine days, after the same integration period, led to a highly significant increase in the frequency of aggressive acts (to 87% of 151 contacts between five previously integrated aliens and their respective adoptive families). All these individuals did not molt during isolation and the discriminators acquired from the adoptive family were certainly not lost within that time (they remained unacceptable for their genetic families). These findings demonstrate that there are no important differences between the results obtained in experiments with parents and those with young that deviate considerably in their discriminator set from that of the host family. In both cases the deviation persists despite heavy contamination with the common badge of their young or adoptive siblings respectively.

In summing up the results presented so far, the following are of special importance:
1. The hypothesis of intrafamilial variability in badge production, postulated on theoretical grounds, has been firmly established by unambiguous experimental evidence.
2. Intragroup differences in signature production are not harmonized entirely by the exchange and mixing of all recognition substances fabricated by members of this group. To bear the smallest common denominator, however, apparently does not guarantee acceptance, since additional substances can completely cancel the character of the signature. This became evident in the experiments with isolated parents and isolated, previously integrated and fully accepted, aliens, as well as in experiments in which short contacts by individuals with foreigners caused alienations. All these isopods bear the common badge and additional components besides.
3. Isopods become acquainted with the characteristics of deviating conspecifics, and this learning, rather than a complete badge adjustment, is prerequisite for the toleration of deviating family members.

4. Deviating patterns are remembered only for a limited time, so that the stored information must be renewed by more or less frequent contacts. Consequently, in categorizing the signature of a conspecific as alien or familiar, it has not only to be matched with a common, family-specific discriminator set, but in addition, with learned deviations from this common badge. Since remembering and forgetting of such learned deviations obviously can influence decisions fundamentally, problems in the interpretation of experimental results arise which have always to be kept in mind. For instance, if a family member is attacked upon return after prolonged isolation, this may indicate that its chemical attributes have changed. But it could indicate that its chemical characteristics, deviating somewhat from the norm of the family, have been forgotten, although the individual itself has remained unchanged. On the other hand, when a molted individual is not aggressively treated upon its return, this is not a reliable demonstration of unchanged badge. The deviation could have been learned, and not necessarily via this individual only (see below).

The extent of intrafamilial variability

We have seen that parents' signatures deviate from the common badge of their progeny and that this can easily be demonstrated by separation and exchange experiments. Exchanging young between parts of partitioned and separately maintained families or returning non-molting juveniles isolated for six to ten days, however, does not usually lead to aggressive behavior in those siblings and parents to which the transferred or isolated young were presented. This could indicate either that by interindividual exchange of discriminators, all variability among siblings has been harmonized as long as molts do not disturb this balanced state, or that persistent and (in principle) detectable differences are less pronounced or are valued differently in siblings as compared to parents (or integrated aliens).

Experiments with young, molted in isolation, yielded clear evidence for intrafamilial variability. Nevertheless this method is not suitable for obtaining a definite solution to the problem of the extent of sibling variability. We do not know which molt-induced deviations are tolerated on the basis of learning and which alterations go too far, thereby releasing aggression.

The problems caused by learning cannot always be circumvented by extending the separation times of the molting individuals to more than ten days, although that time is normally long enough to guarantee forgetting of those features of a family badge that disappear with the isolation of the individual concerned. The reason: if in a family there are a number of individuals resembling one another in their signatures, then isolation of one or a few individuals would not eliminate their 'type' of discriminators from the family. Thus, these separated individuals could remain acceptable through their doubles even when isolated for long periods. Then, despite considerable variability, aggressive behavior could be largely suppressed and one might assume a non-existent degree of homogeneity.

If siblings vary considerably in their signatures, and if intermingling of discriminators within a family does not lead to complete harmonization, the adoption experiments described above are to be interpreted in the following ways. Acceptability would not only presuppose conformity to the smallest common denominator of recognition substances in both groups, but would also require that the adopted group should possess the spectrum of more or less all discriminators, including those produced only by a small number of family members. Changing the form of adoption experiments could answer the question about the extent of sibling variability.

Adoption experiments

Foreign parents were forced by the usual procedure into close contact with all the young (> 50 siblings, half- to almost fully-grown) of another pair ($n = 7$). Thus most transferable recognition substances of the alien parents were distributed among the young, and thereby, because of the large number of young, diluted below the threshold of response. After having been confined together for 4–6 h, all but five young were removed from the foster parents and maintained in a separate box. Since the parents' transferable substances had been diluted, and since adult individuals produce new transferable discriminators slowly and in small quantities, they cannot be expected to influence the adoptive young's discriminator patterns (K. E. Linsenmair, in preparation). The first exchange experiments between the two groups were performed six to ten days later. Toward siblings transferred from the large to the small group, the foster parents nearly always reacted aggressively (Fig. 12A). But when these reactions toward siblings of the main group were compared with those toward arbitrarily chosen young from other alien families, a significant difference was found (Figs. 12A and 12B).

This result demonstrated that in all seven progenies, the entire spectrum of discriminator patterns found in the main group of siblings was not represented to its full extent by five individuals. Since all the young remaining with the foster parents molted during the six to ten days before the test, either these young already varied individually at the moment of their separation, or variations among them developed quickly afterwards.

Before drawing definite conclusions, an important objection must be considered: do these results, despite the experimental procedure used and contrary to our expectation based on many control experiments (K. E. Linsenmair, in preparation), not in reality point to nothing more than a contamination of the five adoptive young by the foster parents, which thus were prevented from becoming acquainted with the pure family badge of these young? This

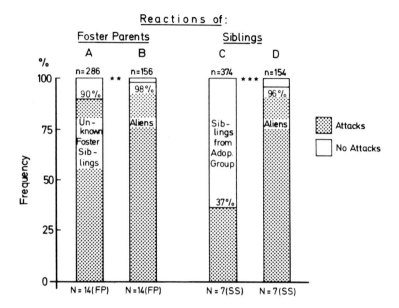

Fig. 12. Responses of foster parents to unfamiliar siblings of their five adoptive young (A) and to alien conspecifics (B). C and D: Responses of the young in the main group of family members toward their five adopted siblings and toward aliens. (FP) = foster parents; (SS) = sibships

possibility could be discarded if all five adoptive young in all seven cases had been accepted in their respective main groups of siblings without releasing aggression, but this was the case for only two-thirds of them (Fig. 12C and 12D). Although it is improbable that any contamination via adoptive parents occurred (K. E. Linsenmair, in preparation), it cannot be excluded with certainty in all instances. However, since the majority of individuals transferred from the adopted to the main groups did not elicit aggression, most foster parents (ten of 14) had had an opportunity to learn a badge unchanged by contamination. These ten adults were not less aggressive towards siblings from the main groups than were the rest. Therefore, the above objection is negated.

This adoption experiment was continued for several more weeks with the number of adoptive young reduced from five to three. Subsequent tests were performed during the fifth and sixth weeks after assembly of the groups. By then all adoptive young had molted at least twice and those possibly contaminated had certainly lost all alien discriminators which they might have worn since their early close contacts with their foster parents (and, of course, those acquired in their own families). New contamination by discriminators of the foster parents can be excluded with certainty (K. E. Linsenmair, in preparation).

In the following experiments young from the main group of siblings were isolated for molting and returned at the earliest one day after its termination, but in most cases not until nine days later. The reactions toward them of both the foster parents and other siblings in the main group were then recorded. Of 39 molted siblings, 28 were treated by the foster parents as entirely alien, but six were not attacked and five released only comparatively mild attacks. All reactions together differed highly significantly from those displayed toward foreigners (in 320 observed encounters 75% elicited attacks, whereas in 132 controls with aliens presented to the foster parents, 92% released aggression). Of those returned to the main group of their siblings, the majority (25 of 43) did not elicit the slightest aggression.

These findings corroborate the results obtained in the previous experiments, as is shown by the following.

1. Three young did not represent the complete spectrum of substances or patterns used within the entire group of siblings. Learning of their limited spectrum leads to rejection of most pure individual signatures of siblings by the foster parents.
2. On the other hand, not all individual signatures from molted unfamiliar siblings released aggression. This finding points to a restricted variability which, of course, is to be expected in groups of mainly full-siblings.
3. The rather long isolation times (see above) caused a large number of molted individuals to be attacked in the main group upon their replacement. The finding that the foster parents accepted only 15% of the molted young without aggression, whereas 58% were accepted in the main sibling group, indicates that there were individual differences among those unaggressively treated by their siblings in the main group. Since all individuals accepted by the foster parents were also accepted in the main group but not vice versa, the group of 58% cannot have been homogeneous with regard to their signatures. In normal isolation experiments this heterogeneity would not have been detected.

Integration experiments

To demonstrate that both parents differ in their signatures from their young as well as from each other (K. E. Linsenmair, in preparation), a suitable experimental procedure was to divide their progeny into halves, to allocate one parent to each half, and later to exchange them. This procedure should also be appropriate for determining whether arbitrarily chosen siblings normally differ with regard to their individual signatures.

Families were divided into halves and one individual or a few siblings from a different family were introduced into each half. After various integration times these subjects were exchanged between subunits. From the results of previous experiments it was expected that these integrated individuals would become heavily contaminated with their foster family's discriminators, but usually without their own signatures becoming entirely masked. Hence these integrated aliens should not lose their individuality completely. They should remain deviant, and therefore identifiable members of their adoptive groups. If the siblings integrated into the two half-families varied individually, exchanges should release aggression. There are many possible reasons why existing differences might be obscured or totally masked. Although this type of experiment is suitable for revealing differences, therefore, it is not suitable for disclosing high degrees of resemblance. Results: upon exchange, 56% of all integrated individuals were attacked ($n = 50$ isopods, i.e. 13 groups of two siblings each and six groups of four siblings each). This outcome clearly supports the hypothesis of a rather high intrafamilial variability, but as expected, there were also cases of high similarity between siblings under these circumstances.

Discrimination by partners in incestuous pairs

A further method of obtaining reliable information about the extent of intrafamilial variability presents itself in pair formation. Since the family badge is not a barrier to incest, sister-brother pairs are easy to form under experimental conditions (see K. E. Linsenmair, 1985, in preparation).

All families utilized had lived for months without contamination by alien conspecifics. Immediately after hibernation males and females were separated and kept as single-sex groups for a further eight to twelve days before a female, and 24 h later a brother, were put into a pipe container having a short artificial burrow. At various times afterwards exchanges of siblings were made within each sex. Not more than 30 min passed between separation of the partner and its replacement by a sibling. In pairs that had lived together for 24 h (the first hours after pair formation do not concern us here, see Linsenmair, 1984), a separation time of less than two to four hours never loosened the pair bond. This implies, among other things, that the partner remaining in the burrow would not accept an alien mate without resistance for considerably more than an hour. Only the familiar partner was, after being touched with the antennae, admitted at once, i.e. if removed and put back after 10–30 min, the average admittance time was 153 ± 120 s ($n = 84$). If an exchanged conspecific was allowed entry into the burrow in less than ten times the average time taken to admit genuine paired partners, then a mistake in identification was considered to have occurred.

To admit a foreigner inadvertently is usually a costly error. If an alien is permitted to enter the lower part of the burrow, it will usually expel the owner within a few minutes. The former owner will not then be allowed to re-enter without passing through a complete admittance ritual (cf., Linsenmair and Linsenmair, 1971; Linsenmair, 1985). During this procedure, which often lasts for hours, the isopod is in danger of being driven off by a competitor.

During the first three days the majority of exchanged siblings were obviously misclassified (19 of 30 were admitted). Already, however, 37% of the burrow owners demonstrated by their strong defense behavior that they did not consider the siblings to be their mates. The proportion of individuals that could discriminate increased with the time the pairs had cohabited. In tests performed after the third day, 86% of 97 exchanged siblings were not accepted.

These results show that the interindividual variability among siblings may be very high, at least at the end of hibernation. It is conceivable that, as a consequence of the low activity during hibernation, the exchange of discriminators might be drastically reduced, leading to a strengthening of individual characteristics, but the results do not point to a fundamental change with regard to the discriminators an individual produces during this or any other stage of its life. (This does not exclude the possibility of signatures being influenced by different situations, developmental stages, and sex-specific pheromones.)

What badge features vary among siblings?

The exchange experiments with pairs of siblings suggest that complete conformity of the discriminator patterns of several siblings might be the exception rather than the rule. The question of the nature of the intrafamilial variability may now be addressed.

Molt-induced deviations could be caused in the following ways:
1. A lack of important components in the signature of molted isopods compared with the common badge and its tolerated deviations could release aggression. If badges are blends, with no single individual contributing all components that characterize a family, a loss of badge substances acquired from other family members must always take place at a molt.
2. Individually contributed components may not match the common badge quantitatively.
3. Individuals could differ qualitatively in producing one or several compounds which only they bear in above-threshold concentrations. This could cause a strengthening of alienation already induced by loss of components acquired from other family members.

Although there is clearly a high intrafamilial variability, we must nevertheless assume that there is considerable overlap in the discriminators produced within a family as a consequence of the high degree of genetic relatedness (mostly full-siblings). It is to be expected that in molting, different siblings lose different parts of the overall badge produced by the family, but through close contact they should make up their deficiencies to a considerable extent. Therefore, assembling groups of siblings, molted in solitary confinement, could lead to a partial resolution of our problem.

Firstly, from a number ($n = 14$) of pure families showing a rather good molt synchronization among their members, siblings whose molts were imminent ($n = 143$) were separated on three to six consecutive days and kept singly. They remained in isolation until there were at least five that had completed molting 24 h earlier. Thereafter, these molted individuals were presented to their families. Of 143 such molted individuals, 43% were treated entirely non-aggressively, whereas 24% released slight, and 33% relatively strong attacks (altogether 19% attacks in 2,264 observed encounters).

Following this test molted siblings of each family (five to 14 per family, but 28 in one large family) were at first crowded together for four hours and then assembled in a narrow container. From time to time (minimally one day and maximally five weeks after assembling the partial families), individuals were tested, either by presenting them to their siblings in the main group or, in the reverse experiment, by determining their reactions toward members of the main group of their siblings. In comparing the reactions of families toward their molted members before and after the assembly of subunits, three principal responses were observed, but in different proportions (Fig. 13).

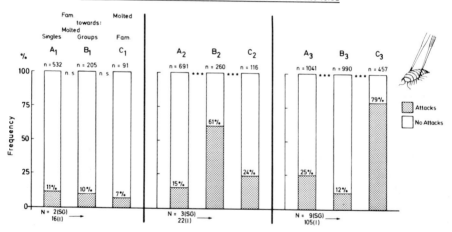

Fig. 13. Interactions between siblings molted in isolation and the main groups of their family members. Siblings molted in solitary confinement were individually tested in their families ($A_{1,2,3}$) and thereafter several molted siblings were assembled to form a family subunit. Then at different times, these were again presented to their kin in the main groups ($B_{1,2,3}$). In $C_{1,2,3}$ relations were reversed, with the main group's members presented to the molted individuals. (SG) = sibling groups

1. In two cases no statistical differences were found (Figs. $13A_1$ and $13B_1$).
2. In three families a strengthening of the aggression-eliciting properties of the molted individuals was observed (Figs. $13A_2$ and $13B_2$).
3. In the majority of families (nine of 14) the individuals released fewer attacks after assembly (Figs. $13A_3$ and $13B_3$), which was to be expected if individuals lost, through molting, important parts of the common discriminator spectrum, but re-acquired these, more or less, through exchanges with their siblings.

Concerning the two families in which no effect could be detected, intrafamilial variability was evidently less pronounced than in other families.

If molted individuals differ from all other family members in fabricating unique recognition substances, then memory of these particular discriminators is a prerequisite for their continued acceptability. In prolonging separation time from a more or less large number of family members, a formerly accepted component will be entirely forgotten, causing the unfamiliarity of its producer to increase. Thus, taking temporal effects into account and assuming that at least one and perhaps two or more individuals producing unique components were included in the subunit, the increasingly aggressive reactions in three of the families are not difficult to interpret.

Further clarification may be expected from the reverse experiments, i.e. from tests of the reactions of members of the molt groups toward their siblings from the main groups (see Fig. 13C). In two of the three cases in which the molted individuals (after assembly) released stronger aggression in the main group, members of the main group were treated less aggressively in the molt group (Fig. $13C_2$). This points to an asymmetry in resemblance, with the members of the main group being classified as more similar to those of the molt group by the latter than in the reverse experiment where the members of the main group showed stronger aggression upon contacting individuals from the molt group. This outcome is in accord with the hypothesis that in the molt group individuals were included that produced recognition substances lacking in the main group. One of the strongly deviating molt groups of ten individuals was then crowded with ten siblings from the main group, with the result that the latter ten were also alienated without the molt-group individuals being less aggressively treated. This result further supports the suggestion of a qualitative deviation by production of additional substances in the molt group.

Provided we were right in assuming that the unchanged and low aggression in two of the families before and after group formation was an indication of low variability among the members of these families, we predicted that the isopods would not behave differently in the reverse experiment. This prediction was fulfilled (Fig. $13C_1$).

Reversing the exchange procedure in category three (the majority group) of our families also gave a clear answer. The molt group was more aggressive toward the members of the main group than vice versa (Fig. 13C$_3$). Many experimental results have indicated that additional substances are more aggression-enhancing than is the lack of components. This result is not surprising, because most molt groups comprised only 15–25% of the family members, so that the probability was high that the main group possessed a more complex pattern of recognition substances than did the smaller molt groups.

Quantitative differences: do they influence the specificity?

Molts should cause both qualitative and quantitative alterations in the badge structure, but the latter have been almost completely excluded from the discussions thus far. What role do they play? Is it possible for some of the results (e.g. in the previous section) that were explained in terms of qualitative differences, to be interpreted in quantitative terms?

By thorough rubbing of the cuticle with chemically cleaned cotton-wool and extracting this with different solvents, discriminator solutions may be obtained. These release strong aggression when applied to dummies. Extracts from individuals of a family release less aggression within their own than in alien families (Figs. 14A and 14B), but they always release some attacks. This result demonstrates that the substances removed do not wholly match the complete family badge. The fact that effective extracts can be obtained by rubbing discriminators off shows that the badges can be altered in a quantitative manner.

Rubbing off and rinsing off discriminators

If absolute concentrations play an important role in discrimination, then rubbing off discriminators ought to lead to alienations. However, in 83% of all isopods treated in the described manner ($n = 41$), not the slightest indication of any aggressive response to them could be detected and in the remaining cases only light aggression by family members occurred (4% attacks in 987 observed encounters). Provided only the relative amounts of the different discriminator substances were important, it seems that the rubbed individuals lost all components in the same proportions as those present in the unchanged badge. This assumption, however, conflicts with the unequivocal result of tests with extracts, which released aggression in the donor family, although weak and not elicited from all members. This outcome is most plausibly interpreted by the assumption that rubbing does not yield the total spectrum of discriminators in an above-threshold concentration. (Probably acquired substances are more easily removed than are self-produced components—see below.) This assumption suggests that not only were absolute concentrations changed by the procedure, but also that the relative propor-

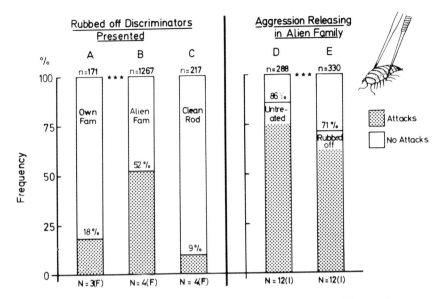

Fig. 14. Partial removal of discriminators by surface rubbing. Glass rod experiments with extracts from cotton-wool that had been used to rub the tergites of ten siblings from each of three families. Tests in alien (B) and own (A) families and controls with clean rods (C). E: Effects of rubbing off discriminators on the aggression-releasing efficiency of test isopods presented to alien families in comparison to the responses released by untreated siblings (of the isopods used in E) in the same test families (D)

tions of the badge components were changed. Nevertheless, if alteration of the badges on the isopods does not lead to alienation, it must be concluded that the specificity of a badge is mainly determined by its spectrum of discriminators and not by concentrations that are above the threshold of detection.

The few instances in which rubbed individuals were aggressively treated could possibly demonstrate that important components were reduced to below the threshold of perception. Quantitative reduction in above-threshold concentration does not cause the badge to lose its character; it remains familiar for members of the family and remains alien for others. But results with aliens show (Figs. 14D and 14E) that the badge's aggression-releasing efficiency is lowered by thorough rubbing. This reduced efficiency becomes obvious only in brief contacts on the rubbed tergites. Upon long and extensive tactile inspections the aliens were always attacked as fiercely as before their treatment.

Similar results were obtained by rinsing isopods with distilled water. Apart from the case previously referred to as the 'Regensburg phenomenon,' the majority of isopods (78% of 203) originating from pure families were not alienated at all through this procedure. But aggression-releasing solutions were obtained by submerging *Hemilepistus* for 2 s only (2 ml per individual). This finding clearly demonstrates quantitative changes in discriminator concentrations by rinsing. When these solutions were tested using the glass-rod method and only one drop of the original solution, the following results were obtained: Only slight aggression, if any, was engendered in the donor families (13% attacks in 207 encounters in four families); in alien families, however, the rods were significantly more effective, releasing aggression in 40% of all (n = 633) encounters. These results provide further evidence for our view that quantitative changes are not crucial with respect to family specificity.

Assuming that despite the aforementioned negative results, varying concentrations do, in fact, change the character of a badge, then dilution of solutions should change the responses to extracts. For example, in the intrafamily test the extract could, and plausibly should, become more alien with increasing dilution as it deviates more and more from the concentration to be found on the surface of the average family member. Of course, this ends where all substances become diluted below threshold.

In alien families, homogeneous reactions would not be expected. Rather, variable results should be obtained, since relatively important discriminators should be more concentrated or more diluted than in the family badge used as a reference for matching. However, the experimental evidence gained in thousands of test series in the context of our chemical analysis of *Hemilepistus*' discriminators (Esswein, 1985; K. E. Linsenmair et al., in preparation), refutes these suggestions. As a rule, diluting an extract leads to a decrease in its aggression-inducing effects, but usually not in a linear fashion, as the effectiveness often remains unchanged within a range of two to three orders of magnitude.

Mixing experiments

Finally, mixing experiments also support the view that quantitative differences do not contribute to badge specificity. In numerous trials we found that a single foreigner forced into close contact with siblings from one family may alienate between ten and 20 individuals. These alienated family members sometimes release as fierce attacks as does a single sibling after being crowded with 20 or 50 members from the same alien family. This result does not prove that quantitative differences do not change badge characteristics, but if this form of variability were crucial, we should have found, depending on the amount of the transferred alien substances, non-homogeneous reactions. This, however, was not observed in trials with several hundred alienated individuals. With quantitative increases in contamination, either the aggression level remained constant or it too increased.

In observing exactly what happens with increasing contamination we found that:
1. The latency time between the start of antennal inspection and attacking was reduced, i.e. with higher concentrations a faster decision was possible.
2. The probability of attack became more and more independent of the area touched on the body of the alienated individual. This clearly indicates that greater contamination, through prolonged and intensified body contacts with a large number of strange isopods, leads to a more uniform distribution over the body of acquired substances in above-threshold concentrations. Brief or less intensive contacts with strangers do not result in homogeneous distribution of alien substances. Taking into account the relatively low potential for surface spreading of the discriminators, this outcome was expected. The weaker the contamination, the more patches, randomly chosen for tactile inspection, should be free of alien substances and correspondingly the frequency of attacks should be lower. On other areas the quantity of substances present would not suffice to allow an immediate decision and more information could be gained only in prolonged and spatially more extensive tactile inspection, causing an increased latency time.

In summing up this section we may conclude that above the threshold of detection, differences in concentration of discriminators do not change the specificity and therefore do not contribute to the variability. Quantities, of course, play an important role wherever thresholds are crossed in one or the other direction and they influence the speed and reliability of recognition performances.

THE EXTREME MIXED GROUP (EMG) AS AN EXPERIMENTAL FAMILY MODEL

Many results demonstrate the existence of intrafamily variability in badge production. As we have seen, the extent of this variability is not easily demonstrated by behavioral experiments alone. It is nevertheless beyond doubt that individuals within natural families may deviate as strongly from the norm of the family in their discriminator production as an alien conspecific. Furthermore, the assumption also seems to be justified that within those groups of siblings in which behavioral experiments point to close resemblance in discriminator spectra, most members will nevertheless vary sufficiently to be individually identifiable under certain conditions.

To overcome the problem of variability within families in our experiments, we formed artificial groups in which we could be certain that every individual differed from all others in its self-produced signature. This was achieved by assembling extreme mixed groups (EMGs) with every member originating from a different family. If such individuals would merge to form family-like communities, they would be suitable for testing some of the hypotheses suggested above and for asking further questions. A few selected aspects only will be treated here.

We knew already that one of the necessary preconditions for the formation of EMGs was fulfilled. *Hemilepistus* is permanently capable of learning small as well as radical alterations in its badge environment. This continuous preparedness for learning allows, under natural conditions, constant adjustment, a fine tuning, to intrafamilial changes caused by deaths and by moltings (see below). Thus the first crucial question here was not whether a new badge could be learned and, after a time, be comparable to that of a natural group, but whether the high complexity of the badge, to which more than 100 individuals (each from a different family) sometimes contributed, would lead to an overloaded system, thereby inducing its total breakdown.

The formation of EMGs

When single individuals each from a different family were placed together in standard containers, only short-lived aggression lasting a few seconds, if any, arose. The animals, which under other conditions would have treated one another with extreme aggression, calmed down with surprising rapidity. One to two hours after they had been placed in the container (all containers without a burrow) they huddled together like a natural family. One crucial property that characterizes every normal family was never found at this early stage, namely, aggression toward aliens from any source.

An obvious question was whether this lack of aggression was caused by the changed badge environment. The following procedure seemed likely to provide clear answers through elimination of dissimilarities among individuals that might have been caused by their different histories. From families that were excavated and transferred to standard maintenance boxes several days previously, single isopods were isolated shortly before molting. After molting, most of them were united in groups consisting of siblings only (groups of ten to 28 each, $n = 9$); in two cases EMGs were formed (15 animals each). The results are summarized in Fig. 15. Three to seven minutes after the individuals were assembled in groups they were tested for the first time. Thereafter, they were tested repeatedly for several hours. Those forming sibships attacked an alien presented to them in an overwhelming number of encounters (Figs. $15A_1$ and $15A_2$). The fact that a significantly higher level of aggression was found in these groups in tests repeated 4–72 h after group assembly may be explained as follows:
1. In order not to exceed a separation time of six days and not to form groups with less than ten animals, it was impossible to use only isopods that had terminated their molts at least 24 h before the test; others that had molted less than 24 h before had to be used too. These initially

unagressive, 'anxious', individuals became aggressive during the next 24 h.
2. The new substrate always induced some members upon first contact to feed on it for a prolonged period and these were often not disposed to touch the test isopods held up to their antennae.

The behavior of the individuals forming small EMGs was different in that initially they were non-aggressive, but slowly these too regained their aggressiveness (Figs. $15B_1–B_3$). One further fact is striking and deserves closer analysis. Introduced into pure families, EMG members were especially good stimuli for releasing aggression, whereas in the mixed groups, members of pure families were themselves particularly effective in eliciting this response (Figs. 15C and 15D).

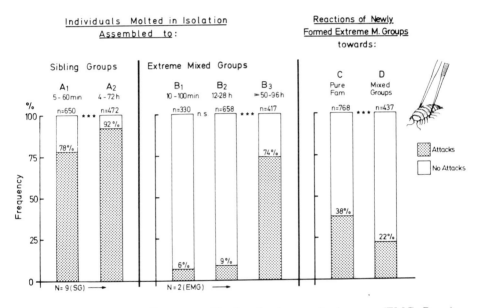

Fig. 15. Responses of sibling groups ($A_{1,2}$) and extreme mixed groups (EMG: $B_{1,2,3}$) toward alien conspecifics in relation to time after assembly. C, D: Frequency of attack behavior of the EMG members in encounters with pure family members (C) or individuals taken from other mixed groups (D) during the first 96 h after assembly

Usually the EMGs were much larger than the two groups used above and consisted of intermolt animals collected in the field during an activity phase. It was necessary, therefore, to determine whether the size of the group influenced the regaining of aggression and whether intermolt isopods behaved distinctively. The results of experiments concerning time until renewal of aggression toward pure family members in relation to group size are presented in Fig. 16.

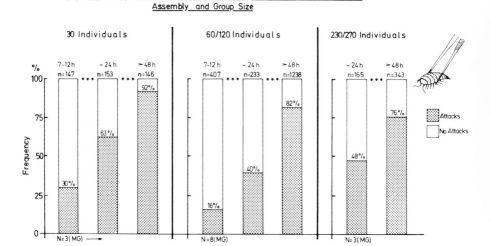

Fig. 16. Responses of EMGs toward aliens from pure families in relation to the variable indicated

In EMGs of intermolt animals (three-quarters- to fully-grown) aggression against members of alien pure families was regained somewhat faster than in the above-mentioned EMGs consisting of newly molted isopods. During the first four hours hardly any aggression could be elicited. However, 7–12 h later a few individuals reacted with nearly normal aggression in all groups. The number of aggressive group members then increased until about 48 h after group formation, at which time a level was attained that persisted for weeks and sometimes months if external conditions remained unchanged. The size of the group influenced its aggressiveness. At all times groups of only 30 individuals proved more ready than larger groups to attack alien family members. Groups of 60 and 110–120 individuals did not differ statistically. Still larger groups (>230 members) were, in the final state, always the least aggressive.

The essential result of these experiments is that new mixed groups can be assembled at any time, as long as the number of differing individuals does not greatly exceed the upper size limit of a natural family (about 140–150 members). For such a group it is on average only a matter of about two days before it becomes a new unit with its own identity. After this short period, the mixture of isopods shows some of the most striking properties of natural families: mutual tolerance (but with restrictions during the first few days with respect to molting members, see below), cooperation, and aggressiveness toward non-group members. This general statement does not apply without restrictions to the very young (< 8 mm long) and to the grown young in late autumn taken from the field into the laboratory.

Simple theoretical considerations, involving Mendelian genetics, lead one to expect that even though interindividual variability within a family may be high and many gene loci may influence discriminator production, the variability should be less than that in an EMG of the same size. As obvious and plausible as this assumption of greater variability may be, it is desirable to demonstrate it directly.

Are badges of EMGs more complex than those of pure families?

Small EMGs of newly molted individuals reacted more aggressively against members of pure families than against individuals from other EMGs. This may point to difficulties that isopods could have in discriminating reliably between increasingly complex badges. Is this an effect of the increasing number of discriminator substances associated with an increasing number of integrated individuals? Do patterns of large (> 100) EMGs eventually approach the total discriminator spectrum present in the population?

So far the number of these substances can only be guessed at (see Discussion and Summary), but assuming that the number is not very high, that the usual substances are more or less evenly distributed among badges, and that badges are more or less evenly distributed among families in a limited part of the deme, then it should not require an astronomically large number of individuals to reach a situation in which equal-sized samples from the same population (but not the same families) would contain similar discriminator spectra. Above a certain mean number of individuals, all groups should develop similar group-specific badges and discrimination between these groups should become difficult.

On the other hand, the difficulty of differentiating complex mixed groups could also be based on problems of processing increasingly complex chemical information. Possibly, before badges become qualitatively similar, the neural machinery of the isopod's brain, consisting of only 10,000 neurons, of which not more than 6,000 may be involved in the central processing of chemical stimuli (Krempien, 1983), could become overloaded. One may imagine that badges must contain a rather low number of components to permit exact matching.

Before pursuing this question, we must examine what role group size and the duration of togetherness play in the behavior of EMGs composed of intermolt individuals taken directly from the field. In a first series of experiments, tests of these variables were repeated during the first few days after group formation, and the results obtained were in full accord with those achieved with small mixed groups of newly molted individuals. Independently of the size, in all EMGs tested (30 ± 5, 60 ± 10, 105 ± 5, 250 ± 20), members of pure families released more aggression than members of any of the EMGs (Fig. 17). When these experiments were repeated 10–20 days (and occasionally up to two months) later, similar results were obtained.

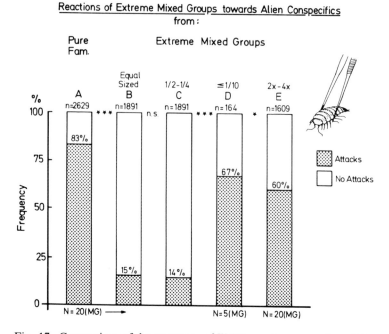

Fig. 17. Comparison of the responses of EMG members more than 48 h after assembly, toward aliens from pure families (A) and from other EMGs of differing sizes (B–E)

Close analysis of the results (Fig. 17) revealed distinct differences, dependent upon the size of the EMG, with respect to the reactions toward members of other mixed groups. Those individuals originating from large EMGs always released significantly more attacks than members of units only one-half or one-quarter of the size. When the size was further reduced, the aggression-releasing efficiency increased. This was expected, since for all EMGs with 25 or more members, individuals taken from pure families were always the best to elicit aggression. The fact that, when members of two mixed groups of differing complexity were exchanged, individuals from the more complex groups regularly released the higher level of aggression, clearly demonstrates that additional components are more important in the matching process than the absence of some components.

In several exchange experiments, involving EMGs of 200 or more individuals from the same population, exchanged individuals (intermolt juveniles) were never treated as aliens. Are the discriminator patterns too complex for matching because of problems with the peripheral acquisition and/or central processing of the chemical information, or are the two badges too similar because almost all discriminators present in the population are represented?

Exchange experiments between EMGs consisting of > 200 individuals assembled from families originating from different populations (separated by a few hundred to over 2,000 km), point to the second hypothesis without excluding the first. Members of these groups (from four different populations) treated one another as alien. The complexity of their patterns was certainly of the same high level. The differences in these patterns could have been caused by different discriminators in separated populations or by an identical set of substances with differing frequencies. The relevant information is lacking, but it seems probable that both factors are involved.

In conclusion, according to expectations, there are essential differences between natural family badges and those generated by an EMG. We have to assume that the latter always comprise more components, other factors such as group size being equal, thereby giving rise to differentiation problems. The natural family badge most probably represents only a relatively narrow, but interfamilially variable, section of the whole spectrum of qualitatively distinguishable discriminators. Thus the problem of too great an overlap at the family-typic level of badge complexity is avoided.

Consequences of high badge complexity in mixed groups

In the previous sections it has been mentioned several times that the best stimuli for inducing strong aggression by pure families are the badges from EMGs. If this is an effect of the large number of components which are never to be found together in a single natural badge, then members of any family should become alienated on close contact with individuals from an EMG, but the reverse experiment should not cause the same effect. When using 20–30 individuals from a large EMG in the usual contamination procedure, every pure family member (⩾ 5,000 individuals) thus contaminated became strongly alienated within 30–60 s.

In mixing families, there are rare cases in which members of one family remain unchanged after contacts with individuals from a second family, i.e. they release no aggression when returned to their control siblings, whereas in the reciprocal tests the individuals of the second family become alienated. This is always so in combinations of two families showing unilateral acceptance. Despite normally efficient aggression-releasing properties, a family may, in combination with a certain other family, totally lack alienating potential. This suggests that the badge of the alienated family consists of a discriminator subset which comprises only components of the pattern of the alienating family. In contrast, the badge of the latter contains at least one additional recognition substance. Experiments with EMGs were designed to test this hypothesis. Assuming that an EMG with more than 100 individuals contains most discriminators present in the population, one may postulate that the members of pure families are usually attacked, not because they possess substances not found in the badge of the EMG, but because their patterns consist of much simpler subsets of the EMG pattern.

If this is indeed the usual situation, then trying to alienate members of EMGs via close contact with a large number of siblings from a natural family should not be very effective, because most members of the EMG would possess those discriminator substances forming the badge of the pure family. Therefore, the members of the EMG would change their badge character, as has been shown above, and by keeping the contamination time short, as usual, we can be sure that all the other recognition substances in the EMG badge are not diluted below the threshold of detection through transfer and distribution among members of the pure family.

The results of these experiments are represented in Fig. 18; using contamination times of one to ten minutes (and sometimes even hours), intensively attacked individuals ($n = 30$ per family) of pure families (Fig. 18A) were not able to alienate members of each of eight EMGs (C) which, in the reverse experiment, caused profound alterations (D). Thus our hypothesis was substantiated.

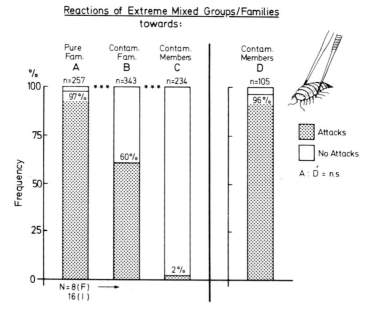

Fig. 18. Contamination experiments with EMGs and pure families. At first pure family members were tested in the EMGs (A); 30 individuals of each of the eight pure families were then crowded together with ten EMGs and tested again with the EMG (B) and in their genetic family (D). C shows the responses toward 16 EMG members (two of each of eight EMGs) used for contaminating the pure family members in their own group

The trials in which attempts were made to mask signatures of alien individuals by massive contamination at the family level failed in most cases. Although a ratio of > 25 siblings to one alien was beyond the alienating capacity of the latter, the alien nevertheless remained identifiable as such. A hypothetical explanation is as follows:
1. Both families differed qualitatively by producing discriminators not present in the other's badge.
2. Self-produced substances on an individual were somehow protected against dilution beneath the detection threshold and against being completely masked by alien discriminators.

If the attacks released by pure family members in EMGs are normally due to missing substances, then it should be possible to adjust these individuals by contamination, as missing substances should always be transferable. Fig. 18B points clearly to the validity of this hypothesis, which has been unequivocally supported in hundreds of individual trials in various experiments. After contamination by isopods from an EMG, the efficacy of previously pure family members to release aggressive behavior in EMGs was rapidly reduced and they became entirely acceptable if crowded together with 20–30 individuals (from an EMG of at least 100 members) for a prolonged period (24–60 h). This outcome favors the hypothesis that large EMG's contain an almost complete spectrum of all discriminators present in the population.

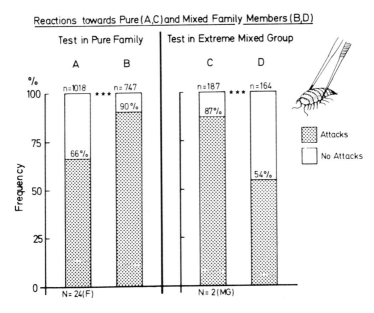

Fig. 19. Consequences of mixing members of families releasing weak attacks in pure families or in mixed groups (for details see text)

When members of a large number of different families were exchanged, a few combinations were always found in which individuals from one family released less frequent and less intensive aggression in another particular family. In mixing members of the two or three families eliciting the weakest attacks among those tested, and then again presenting these contaminated individuals to members of the same test family, a highly significant increase in their aggression-stimulating efficiency was always observed (Fig. 19A and 19B). Presumably in the pure state these families released comparatively low aggression because their badges did not differ much from the badge of the test family. Furthermore, according to what we know about the high degree of interfamilial variability, it was expected that families, although treated identically, would hardly ever differ in the same manner with respect to single components included in or absent from their respective family badges. Therefore, in mixing, the number of alien components should increase in the experimentally generated new badge, rendering it more foreign and thus releasing higher aggression. Exactly this has been found as the general outcome. When this procedure was repeated with EMGs as test units, using the three pure families that released the least aggression and mixing their badges by forcing some of their members into close bodily contact, the opposite result was obtained (Figs. 19C and 19D). The efficiency of the mixed badge was significantly lower than that of the respective pure badges.

We can interpret this result within the framework of our understanding of the recognition system of *Hemilepistus*. Those pure family badges that released less aggression in EMGs than the majority of family badges, most probably consisted of an unusually complex blend of recognition substances, deviating to a lesser extent from the highly complex pattern of the mixed test group. Forming a new badge by mixing several particularly complex family badges should yield an especially complex product, which would deviate even less from the reference badge than each of the complex family mixtures alone, and should therefore be less effective in releasing attacks.

Missing recognition substances as causes for badge strangeness

Many of the above results indicate that absence of substances may be the sole reason for categorizing a badge as alien. Mixing experiments should provide a suitable means of further examining this issue. Here, no EMGs were used, as their discriminator patterns could be too complex to be differentiated reliably. Instead, 'mixed families', consisting of rather large numbers of siblings (10–50) from two or a few families, were employed. One aspect is of particular interest, namely the reactions of members of mixed families toward their own and toward the pure, uncontaminated kin of their adoptive siblings.

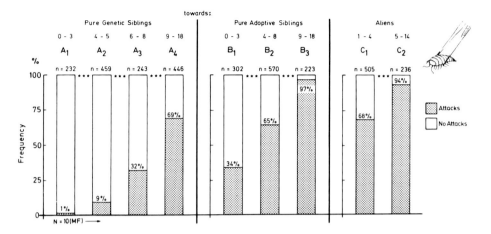

Fig. 20. Effect of time together on responses of mixed family members toward pure genetic and pure adoptive siblings. Two to four groups of siblings from different families were used to form a mixed family (MF) and groups were distinguished by color-marking their members. Siblings from pure family subunits were then presented to their integrated kin and to the members of other integrated groups. The numbers above the letters for the columns indicate days after group assembly

Where members of two or more families were assembled to form a new community, genetic siblings would usually have had a chance to become acquainted with the pure badges of their adoptive siblings only within the first seconds of the enforced contacts. Within a few seconds badges change to such an extent that they are unacceptable to the pure kin in the rest of each family. This means (since a few seconds are not sufficient, according to all our results, to learn a new discriminator pattern) that the mixed family members become acquainted only with the mixed product and not the unchanged original pattern of the adoptive siblings. Additionally, it is henceforth impossible for any member of the integrated group to meet a genetic sibling with the intact original family badge. As already mentioned, pure siblings are normally accepted for several days without any aggression by the majority of their genetic kin within a mixed family, but thereafter gradually become alien (Fig. 20A_1–A_4). To their adoptive siblings they are mostly only acceptable for the first two or three days. Thereafter, the level of aggressiveness is significantly higher and rises much more steeply during the following days than for genetic siblings, but remains lower than toward aliens until about the ninth day (Figs. 20B_1–B_3 and 20C). These results demonstrate that an absence of substances alone can induce aggression. A fit of at least 50% on average of the discriminator substances expected to be contributed from either pure family (in the case of two mixed families) is insufficient to render individuals in mixed families permanently acceptable to one another.

The learning of new badges

All manipulations leading to changed badges from the integration of a single alien into a family to the assembly of a large EMG prove that *Hemilepistus* is able to learn altered or entirely new badges. The system still seems to function well even with badges a great deal more complex than a natural family could ever have. As we have seen, an individual transferred to a new badge environment will need some time before it demonstrates the effects of having learned the new badge. After a drastic change in the badge environment, attacks can at first be elicited only with strongly deviating discriminator patterns; in a small complex mixed family, aggression is only elicited by presenting members of large EMGs, and within such EMGs only pure family badges lead to aggressive responses. Obviously at first, following a drastic badge alteration, only a crude discriminatory ability exists, which becomes more and more subtle within a few days.

Two to four days after the formation of EMGs, when strong attacks against aliens from pure families can be elicited, such stimulation sometimes leads to an outbreak of intragroup aggression involving either general mutual fighting or attacks aimed at particular individuals. This demonstrates that the regaining of aggressiveness does not require as a precondition familiarity with and the full acceptance of the complete new badge environment.

When young isopods two to six months of age were used to form EMGs, intragroup aggression normally disappeared after the third to fifth day following assembly. By then the new badge had obviously been learned. As has been demonstrated (Fig. 16), the regaining of aggression in EMGs depends to some extent upon the number of members. If not complete, then at least partial familiarity with the new badge is probably necessary before an isopod can begin to behave as it would in its natural family. Does the time needed to achieve this partial acquaintance vary greatly with complexity of the new socio-chemical environment?

Chemical complexity of a new group and the regaining of aggressiveness

As in many other contexts, the high intra- and interfamilial variability in *Hemilepistus* behavior must be stressed. One example concerns a mixed family consisting of three groups of 18 young each, from three different families. Although the original families had an identical rearing history, had been reared as pure families, and were indistinguishable in their behavior toward alien conspecifics, their respective members reacted very differently after integration (Fig. 21). The reasons for these differences are unknown.

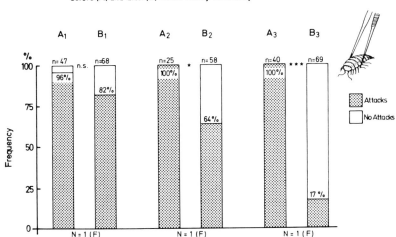

Fig. 21. An example of the common interfamilial variability in behavior of *Hemilepistus* (see text)

In mixed families composed of two equally large groups (> 25 individuals from each family) a rapid recovery of aggressiveness was observed. By the usual method, all isopods were crowded together for on average 5 h ($n = 4$ mixed families). They were then placed in a new container and tested a few minutes later for the first time. With 92% aggressive reactions ($n = 98$ encounters with aliens) there was no statistically significant difference as compared to 98% attacks by the original family members before group assembly. Thus, these mixed families differed significantly from more complex mixed families or groups in the rapid recovery of aggressiveness. The more complex groups did not reach a comparable level for at least 48 h. It is particularly noteworthy that within these mixtures of two families, no mutual attacks among juveniles could normally be released 5–7 h after the initial crowding together.

Mixed families composed of equal parts of three or four progenies always showed reduced aggressiveness compared with their pure siblings in the first 36 h after assembly. Thus, after 24–28 h, in five such families, only 45–60% of encounters with aliens led to attacks, whereas in the original family the lower limit was 95%. They resemble, with respect to the time needed to regain their highest level of aggressiveness, EMGs having a membership of about 110 individuals.

In additional experiments, not presented in detail here, mixed families were composed of older (> 6 months) individuals. Although the recovery of aggressiveness was considerably slower and never reached the original level, similar results were obtained. Here, comparisons between mixed groups of three, four, and six families showed that up to five days after their formation, the least and the most complex groups differed significantly, with the latter showing the lower level of aggression.

All these results demonstrate that the speed with which a high level of aggressiveness is regained is negatively correlated with group complexity. The particularly rapid learning of their new badge by mixtures of two families is probably best interpreted as a process of re-adjusting, whereas in more complex mixed families, and especially in EMGs, an entirely new learning must occur. If this explanation is correct, the time difference may not be caused solely by differences in learning tasks, but may be at least partly attributable to differences in the motivation to learn. A desert woodlouse that encounters an alien badge environment under natural conditions should strive to leave it and not try to learn the new discriminator pattern. Only when it is unable to escape the strange environment and therefore is also unable to avoid manifold contacts with strangers for a prolonged period, should it learn the new pattern (see Linsenmair, 1984, for similar problems in becoming acquainted with potential mates during pair formation). On the other hand, within the familiar environment among family members, it would be adaptive to learn to recognize changed badges rapidly, e.g. in the context of molting (see below), and therefore a particular preparedness for learning under these circumstances should exist.

In a mixture of two families with an equal number of members from each, what is saved from the original badges may be just sufficient to enhance the motivation for learning. There are a number of experimental results that support this view, but at present the validity of the assumption is uncertain.

Besides the complexity or strangeness of a new badge, the temporal recovery of aggressiveness by the members of an experimentally assembled group is also influenced by the aggressive behavior of other group members. The recovery is slowed considerably by intragroup aggression (details will not be given here).

THE FAMILY BADGE, INTERINDIVIDUAL VARIABILITY AND THE MOLTING PROBLEM

The mixed character of the family badge necessarily causes discriminator losses through molting. When members of a large number of pure families were isolated during molts, more than 50% were not attacked upon their replacement. Obviously alterations to their signatures do not surpass tolerable limits, although in a large minority the molt-induced changes led to at least light attacks. We assume that in short-term isolations (60–100 h) the loss of too many components of the family badge induces the aggression. In prolonged separations further problems arise as a consequence of forgetting individual signature traits.

To study the problems caused by discriminator losses through molting, mixed groups were used. In such units interindividual differences in the self-produced signatures are more pronounced than in the average pure family, and we know beforehand in which individuals molts should cause major deviations.

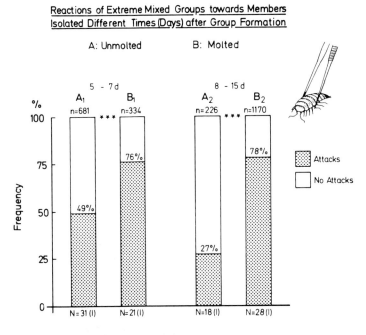

Fig. 22. Effects of isolation without ($A_{1,2}$) or with molting ($B_{1,2}$) in members of EMGs in relation to time together in EMGs. Individuals of both groups were kept in solitary confinement under identical conditions. B individuals molting in isolation were not returned for at least 24 h after molting had terminated

Results obtained by returning to their previous foster families aliens molted in isolation that had been integrated singly into foster families for prolonged periods, have already been presented (Figs. 9 and 10). The majority became alienated and were repulsed by their foster families. Temporary isolation of molting members of EMGs yielded similar results. All individuals tested 36–96 h after molting ($n = 60$) were attacked in their groups. Control experiments demonstrated that isolating intermolt members from an EMG may lead to alienations without molts, especially when members were removed within the first seven days after group formation (Fig. 22). Responses of EMG members to intermolts (when these were returned after temporary isolation) compared with responses to members of the same EMG's which molted in isolation, show that molting always causes more fundamental alterations than isolation without molting for a similar period.

Molt-induced badge alterations and protection of deviating family members

One urgent question so far avoided now has to be answered. How can individuals deviating in their self-produced discriminator patterns survive molts within their own families? The newly molted *Hemilepistus* can hardly move; it is extremely vulnerable, and can survive only if not exposed to any serious attack. An injury that causes hemolymph leakage often instantly releases cannibalism even in a pure, unmanipulated family. If members of mixed families or EMGs that molt within their artificial units were treated like members isolated during their molt and returned more than 24 h after the shedding of the second hemiexuvia, then at least 95% of the newly molted isopods would be killed.

Are molting individuals protected against aggression by other group members? Integration of aliens that were about to molt into foster families gave a clear answer to this question. All foreigners ($n = 30$) that were integrated singly into pure foster families and that had been accepted in their unmolted state, were killed if they molted within 36 h. EMG members that molted during the fourth to eighth day after group assembly had a 74% chance of survival (222 of 292 individuals). Isopods that molted later than this had even better prospects of surviving. Only 5% of those that molted on the ninth day or later were killed (116 of 2,396 members of EMGs). Single aliens, integrated into pure families eight days before molting, survived in more than 96% of all observed cases ($n \gg 200$ molts).

The striking dependence of the tolerance on integration time suggests that learning crucially influences the acceptance or rejection behavior. Before examining this possibility we should interpret the difference in response that depends on whether the integrated isopod molts within its group or in isolation. At times when 95% of all individuals within extreme mixed groups survive a molt without suffering a single serious attack, 100% of all those members of the same unit that molt in isolation are heavily attacked when returned 24–48 h after termination of the molt.

Is the molting individual protected through site choice?

A molting *Hemilepistus* retreats to the lower parts of the family den. There it comes into contact only with individuals going to rest or already resting. As previously stressed, most isopods behave very differently toward the same discriminator pattern when active and when at rest. Data not presented in detail here demonstrate that both alienated family members and aliens, adjusted to some extent through close contacts with a large number of members of a test family, may be tolerated by the resting group, whereas they may be fiercely attacked if encountered during the phase of above-ground activity. This situation-dependent behavior strengthens other mechanisms, but the complete protection of molting members cannot rest solely on it, as is demonstrated by the following observations:

1. Protection against life-threatening attacks during the resting phase is not at all perfect. Individual family members may react differently, and some may attack and bite. For the molting individual it is then irrelevant that the majority showed peaceful behavior.
2. In pure families and in mixed groups established for more than eight days, members may molt during activity times without the protection of a burrow and yet risk almost no attacks. Why then, are group members that are returned more than 24 h after their molt not equally well protected? Is the protection limited to a relatively short period after shedding the exuvia?

Are newly molted individuals protected in the short term?

In the experiments so far described, individuals that molted in isolation were returned at the earliest 24 h after they had stripped off their second (anterior) hemiexuvia. How do EMGs that have lived together for a minimum of twelve days react toward members returned to them at different times after molting in isolation? (Note: in these and all comparable experiments, test individuals were, as usual, held with forceps and presented in such a way that only the newly molted half of the body could be touched by group members.) The outcome is presented in Fig. 23. It confirms results obtained with more heterogeneous units (see Linsenmair, 1985): the aggression-releasing effect rose steeply with increasing isolation time during the first 48 h after the molt.

The attacks during the first 12 h were usually light. Upon touching a newly molted individual, in particular during the first 6 h, clear signs of inhibition were recognizable. Often, isopods immediately stopped moving forward, then withdrew their antennae rapidly while retreating a few steps.

Obviously the newly molted individuals possessed inhibitory properties. The question as to whether this temporary protection is enjoyed by all newly molted isopods, including those transferred to alien groups, can be answered unequivocally in the negative (see Linsenmair, 1984, 1985). Although in most encounters, arranged within the first 30 min after discarding the exuvia, alien controls showed signs of inhibition, this did not protect the still soft and extremely vulnerable animal. Sooner or later within the first few minutes of the test procedure, every isopod, including the very newly molted, was attacked.

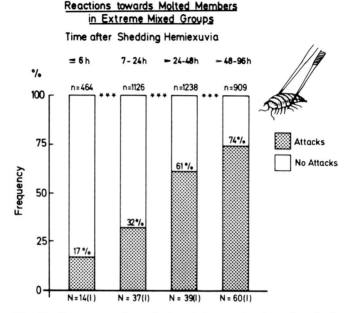

Fig. 23. Frequency of attacks in relation to the time after shedding hemiexuvia. The EMGs in this experiment consisted of 60–120 members

In all circumstances it was obvious that the newly molted alien released less aggression during the first hours than it did later. Are the inhibitory properties alone responsible for this reduced aggression, or are low discriminator concentrations also responsible?

Do newly molted individuals carry smaller amounts of discriminators?

If discriminator substances are secreted only after shedding the exuvia, and if this is a slow process, then low discriminator concentration could be partly responsible for the reduced aggression-stimulating properties of newly molted aliens and alienated group members. Low, but above threshold, concentrations, though not changing the specificity of a badge, diminish the probability of attacks, as has been shown above.

Three methods were used to examine whether discriminators are already present when the hemiexuvia are discarded and whether discriminator concentration is lower than in the average intermolt individual.
1. Exuvial fluid was applied to glass rods and tested in different alien families and in the family from which the molting individual originated. Additionally, exuvia were broken open and presented to the controls in such a way that they could only touch the inner side.

2. At specific times after having shed the hemiexuvia, individuals were used to alienate members of a control family by rubbing their newly molted body half on the posterior tergites of the intermolt test animal.
3. At the moment when the old cuticle was shed, a few isopods were killed; they were presented at different times afterwards to alien conspecifics.

The results of the glass rod experiments are summarized in Figs. 24A and 24B. The most important findings were as follows:
1. The exuvial fluid obviously contained discriminators. The fluid and its dried remains were mostly acceptable within the family, but were attacked in about every second encounter by aliens.
2. Compared with the outer side of the exuvia, the exuvial fluid had a much lower aggression-releasing efficiency in alien families (Figs. 24B and 24D); in the majority of encounters the test animals reacted by retreating, or at least by withdrawing their antennae.
3. The aggression-releasing efficiency of the besmeared glass rods increased within a short period if they were left lying exposed at 25 °C and about 30% RH. Obviously inhibitory properties that had been transferred to the glass rods were relatively quickly lost.

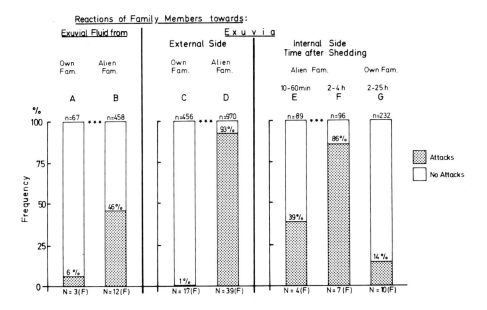

Fig. 24. Aggression-releasing efficiency of exuvial fluid applied to glass rods (A, B). The exuvia used for testing responses to their outer surfaces (C, D) had been shed within the previous six hours. For testing responses to inner sides of exuvia (E–G), these were broken open and presented in a way that excluded contacts with the outer surfaces

Comparable results were achieved with the inner sides of exuvia tested in alien families. Here too the aggression-releasing effect increased considerably with the drying of the exuvial fluid (Figs. 24E and 24F) and during the following two to five hours. In tests of exuvia within the families to which the molting individuals belonged, no differences from the results of the comparable glass rod experiments were detected during the first two hours. In later tests (at 2–25 h) the inner sides of exuvia of molted individuals that had been accepted without aggression now released attacks in 14% of all contacts (Fig. 24G). However, with equally old exuvia from group members that released aggression upon their return, a highly significant difference was found—74% attacks in 110 trials in four different families. Even greater aggression was elicited by the producers of these hemiexuvia (which had been kept in solitary confinement since before the onset of molting) when they were presented to their group members two days after molting (93% of contacts elicited attacks, $n = 53$).

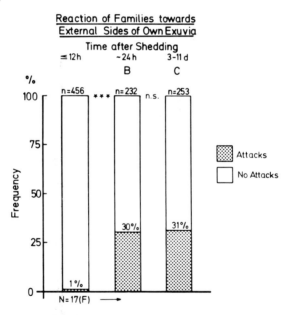

Fig. 25. Responses to outer surfaces of exuvia. The time effects were not dependent upon how exuvia were stored. Whether the exuvia were kept in a closed vial, at high or low RH, or left exposed to laboratory conditions, did not recognizably influence results

Concerning the unexpected aggression-releasing properties of the inner sides of exuvia from family members, the same phenomenon was demonstrated for the outer sides also (Fig. 25). Outer surfaces of exuvia of family members, accepted after molts in isolation, were treated like these family members themselves for about 12 h, after which they became alienated to a limited extent. They clearly retained much similarity to the badge of the living members, as can be seen from the results of presenting alien outer sides of exuvia (Fig. 24D). What causes the partial alienation of the exuvia from the family is still unknown.

According to these results, the exuvial fluid already contains discriminators and the lowered aggression-releasing effect rests upon transferable inhibitory properties that quickly fade. But quantitative differences could also play a role, since inner sides always remained less effective in stimulating attacks by foreigners than outer sides, or than the live producers themselves (alienated through the molt) when isolated for about 48 h after molting and then returned to their kin.

Rubbing two intermolt *Hemilepistus* of different families back to back a few times reliably caused alienation of both individuals. When using a newly molted (0–10 min) foreigner to alienate an intermolt animal, it was not possible to exert the same pressure as with two intermolt isopods. To compensate for this the newly molted individual was not rubbed the usual three to five times, but 15–50 times against a small tergal area of the test isopod. On presenting the test isopod to its kin, no aggression could be released for the first 20 min, but often clear retreating behavior. However, isolating these intermolt isopods singly and testing them at intervals after the rubbing procedure showed that a transfer of alien recognition substances had taken place (Figs. 26A–26C). Nevertheless, the level of aggression achieved with intermolt isopods was never attained.

This outcome also shows that discriminators are present directly after the shedding of exuvia. Both the fact that alienation becomes recognizable only after an interval, and the frequently observed inhibitory reactions, suggest a transferable inhibitory principle.

Comparisons between the aggression-releasing effect of the alienated intermolt family member at the time of its replacement and that of the newly molted foreign individual used for the alienation, yielded highly significant differences, with the newly molted one being the more effective in the test family (see below). Obviously, only small amounts of discriminators of the newly molted were transferred by rubbing to the intermolt member of the test family. The reason could have been either that only small amounts are present on the new surface or that the mechanical properties of the still soft cuticle impede the transfer. In additional rubbing experiments, pressure was exerted on the newly molted isopod just below the maximum mechanical stress that could be tolerated and the rubbing was continued for several minutes. The results indicated that the softness of the cuticle plays an important part in reducing the amount of discriminators transferred, since now the alienated intermolts did not differ significantly in their aggression-releasing

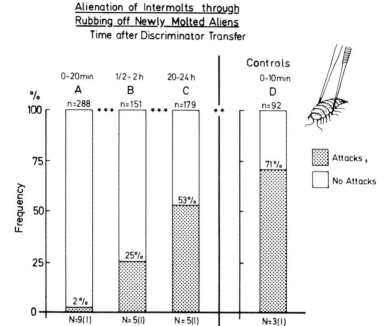

Fig. 26. Effects of rubbing material from newly molted isopods in relation to time after contamination. Between tests, contaminated intermolt isopods were kept in isolation. In the control experiment, three intermolt *Hemilepistus* that had been alienated by rubbing a normal intermolt alien three to five times against their posterior tergites were presented

properties from the newly molted individuals themselves; 48% of 195 trials led to attacks in the former case and 50% of 189 in the latter. However, with normal intermolt isopods even light rubbing between strangers causes far more effective alienation. Therefore, it is reasonable to assume that a newly molted isopod possesses less recognition substances which, in addition, are probably tightly bound to the epicuticle.

Yet another method was used to determine whether newly molted individuals carry discriminators, namely killing isopods immediately after shedding of exuvia. When tested, surprising results were obtained; none of the individuals killed ($n = 15$) ever released attacks, regardless of how they were treated, whether left exposed and allowed to dry, kept at high RH, or deep-frozen. Additional experiments are needed to clarify these results.

Possible consequences of small discriminator quantities after molting

Besides the fact that an individual deviating from the family norm may reduce its risk of being attacked by possessing only small amounts of discriminators, another important point must also be considered. Both normal intermolts and just hardened, newly molted siblings (14 sibling pairs from 14 different families) were lightly alienated, in the usual way but with small numbers of aliens, large vials and short contact times. Upon return to their families, the intermolts released aggression in 33% of 493 encounters, whereas the newly molted individuals released aggression in 70% of 483 trials. Thus the acquisition of recognition substances from conspecifics is apparently facilitated in newly molted *Hemilepistus*. There are in addition hints of a more enduring alienation in the molted siblings. If these observations can be confirmed, the most plausible explanation would be that in addition to the presumed tight binding of self-produced discriminators to the cuticle, a limited quantity of other recognition substances, acquired from family members, can be adsorbed in an equally effective manner. This could also prevent some of the acquired substances from being rapidly dissolved and falling below the threshold of detection.

The nature of the inhibitory principle

As the aforementioned experimental results show, the inhibitory properties are not inseparably bound to the newly molted cuticle. The inhibitory effects are transferable to normal intermolt animals and to inanimate objects also. It is not yet known whether specific inhibitory, and relatively volatile substances are involved or whether chemical properties of the discriminators are somehow changed. Possibly, the results of the tests with isopods killed at the moment of discarding the exuvia point in the latter direction, but without additional data further speculation is futile.

Does the two-stage molt help to protect a deviating family member?

Since molting takes place in two distinct steps, could it not be that a still fully accepted unmolted anterior body half prevents attacks against the newly molted posterior half with its changed discriminator pattern and, *mutatis mutandis*, that the same holds true for the hind end, which could, by the time the anterior body half is molted, have already adjusted to the family norm partly by acquisition of missing substances? All relevant experiments unambiguously demonstrate that this possibility does not offer sufficient protection. The main cause is that normally only a limited area is touched during antennal inspection. It depends on where the contact occurs, but usually this is located either on the posterior or the anterior half of the body and hardly ever at the borderline between the two. Even when this happens by chance or by experimental manipulation, no reliable inhibition results. In the majority of observed cases with either simultaneous or successive contacts, the aggression-eliciting stimuli were dominant. In eight newly molted isopods, 43 more or less simultaneous contacts of both body parts were induced and, in 34 cases attacks were released. For individuals deviating only slightly from the common badge, however, which are often extensively inspected both spatially and temporally, the information provided might help to prevent aggression.

Learning of individual traits as an additional factor protecting newly molted isopods

It has been demonstrated that a newly molted isopod:
1. Possesses aggression inhibitors.
2. Has its own discriminators quantitatively reduced, thereby lowering their attack-releasing properties.
3. Retreats to parts of the burrow where other group members react less aggressively on contact with a deviant badge.

Yet these three factors together do not provide reliable insurance against dangerous attacks. Under most circumstances protection becomes perfect only after the members of the family or mixed group have had the opportunity to live with the deviating animal for a period of at least six days. This necessity has nothing to do with a change in the production of discriminators or with a complete adjustment of the individual's badge through discriminator exchange. Rather, one must assume that a period of cohabitation is necessary to allow group members to become familiar with the chemical peculiarities of the deviant. How can such a recognition system, apparently requiring an ability to learn to recognize deviant signatures individually, function when family badges are mixtures of the individually produced components? Is it only a matter of learning each individual component of the family badge? When integrating aliens into families, new components must often be added to the existing family pattern borne by all members. Do only these new components which become part of the new badge have to be learned, and if so, is any arbitrary subset of all discriminators in the family badge tolerated if it is coupled with the signal 'newly molted'?

Considering integration experiments in which aliens were inserted into families in a proportion of one to > 20, we note that learning to recognize all components of the common family or group badge cannot be the complete solution to the problem, because the additional discriminators contributed by the integrated animals are, on being exchanged, diluted below the response threshold. They do not show up in the family badge, which we define as the smallest common denominator of all above-threshold components. We have never found an instance in which > 20 siblings could be alienated by a single foreign individual. The quantity of individual discriminators produced and transferred is naturally limited, and permanent losses to the surroundings (and through molting) are unavoidable, thus preventing accumulation of above-threshold concentrations of all discriminator substances in larger groups. Consequently we must postulate that in a large family, which may consist of up to 140 members, only those substances that are produced by several or many individuals will form the family badge. All recognition compounds produced by a single isopod or by only a few individuals remain 'personal' components playing no role in the family badge. Evidently the producers of exclusively personal components would be greatly endangered by molting if only the components of the common badge were learned, since the addition of substances to the common badge via contamination usually leads to aggressive behavior by the family members. We have recently found that single pure compounds isolated from discriminator extracts and applied in synthetic form, may alter a badge (K. E. Linsenmair et al., unpublished).

How, then, is a deviant family member protected when it molts? The crucial point, which has already been stressed in the context of integration experiments, is that the individual traits are never entirely lost. Upon contacts with foreigners the integrated isopods, in fact, become alienated quickly, but complete masking by the common family badge was never observed in all those alien individuals that produced substances not contained in the pattern of their integrating group (which can be checked by mixing experiments; see above). If complete masking of personal substances or their dilution below the threshold of detection on the producer itself were possible, then *Hemilepistus* could not have developed the present recognition system.

Even where differences in badges of normal intermolt members of separated groups are not detectable upon exchange, as is often the case in large EMGs composed of a similar, but not identical, spectrum of original families (see above), the learning of personal components seems to be important. When members of such EMGs were separated for molting 15 days or more after group assembly and then, 6–20 h after molting, returned to their original group, and also to a second group that did not differ recognizably in its group badge, they were more frequently attacked in the second group (see Linsenmair, 1985). This result demonstrates that two EMGs may resemble one another in their common badges to such an extent that mutual tolerance results, and yet they may consist of members differing in their personal signatures. Obviously some deviations are tolerable and do not elicit aggressive behavior as long as they are embedded in the common pattern. Clearly, however, these deviations are detectable upon close inspection and are learned.

As many experiments and incidental observations have shown, an alien that is integrated into a family may be fully accepted there and not even attacked when the family is made highly aggressive, and yet it will be killed while molting during the first two to four days after integration.

Whereas the family badge can be perceived by touching any member of the group, the detection of personal components requires direct contact with the producer. In a large group it certainly does not happen that each member touches every other individual during every activity period. Furthermore, it is just as certain that a number of contacts will be necessary to identify a deviant conspecific reliably. From this perspective, then, the fact that learning to recognize a deviant group member's personal traits requires a relatively long integration time is easy to understand.

In many cases where molt-caused deviations can be demonstrated, learning alone guarantees reliable protection. But there are numerous cases in which the inhibitory properties of a newly molted individual are also needed to protect it, at least during the times of highest vulnerability. The inhibitors may attain their greatest importance where family members lose many important badge components during molting. There is no indication that isopods can differentiate between produced and acquired discriminators, even though there may be differences in the way they are bound to the cuticle. Therefore, it is difficult to imagine how an isopod could learn which substances will be lost by its molting sibling. Here, the signal 'I am newly molted' offered at the right place, namely inside the family burrow, could help to identify this individual as a member of the group.

> 'Memory plays a central role in all identification responses, and this is not the least of the reasons why it is difficult to judge the extent of intrafamily variability by using the indirect methods of behavioral experiments. It is therefore impossible to estimate the demands individual and kin recognition make on the learning abilities of the family members. We do not yet know how many components the family badges contain and how many other discriminators may be 'concealed', to be learned only through direct contacts with the individual concerned. But there is considerable evidence that very complex combinations of characteristics, including many components that do not appear in the group badge, can be learned.' (Linsenmair, 1985)

The regaining of lost badge components

Has a family member already re-acquired all lost substances composing the common badge by the time inhibition is no longer effective? This does not seem probable. In families divided into subunits for prolonged periods (> 14 days) and in mutually tolerant large EMGs, the following has often been observed: exchanging individuals between groups does not lead to the slightest aggression of group members toward each other with one exception.

Members of one group that had undergone molts in their group and never suffered any attacks there were picked out by the exchanged members and treated aggressively. In a considerable number of incidents the attacked isopods had molted 24–48 h before. That the majority of these attacked individuals lost their aggression-releasing properties during the next 24–48 h supports the view that they had been attacked on the grounds of missing components. Usually, however, the fact that a molted member had not yet re-acquired all lost components was not evident; this became obvious only upon exchange experiments following the partitioning of the unit. These findings apply to laboratory conditions with groups kept in containers without burrows. Very probably, within burrows the acquisition of missing discriminators is accelerated, since newly molted isopods display a particularly high level of activity as soon as the cuticle is somewhat hardened, rubbing themselves against family members and the burrow walls.

The problem of losing the presumed inhibitory substances before the badge is completely restored could also be solved by learning. If a subset of recognition substances is presented together with the molt signal, this might be especially conducive to learning. If this hypothesis is correct, we should expect the learning process to be effective, to require few contacts and to be assisted by special behavior on the part of the molted animal. For example, after the first hardening of its cuticle, and while its inhibitors are still effective, it might seek locations that facilitate contacts with all members of the group. This is supported by observations, but further investigation is needed for full evaluation.

THE PRIMARY RECOGNITION OF OFFSPRING

As we have seen, alien conspecifics are in many instances attacked and never accepted. In particular, parents always display very aggressive behavior. Small alien young are generally not only driven off from the vicinity of the family burrow by adults, but hunted and eaten if caught. This holds true for all adults, irrespective of their stage within the reproductive cycle. Whether a female has or has not already delivered her young, is still carrying marsupial mancas or eggs, or has not yet performed her parturial molt, does not change either her cannibalistic tendencies or those of her mate. But if one is prepared to eat a neighbor's children, one must be sure of being able to recognize one's own (see Linsenmair, 1984).

What protects young against cannibalism by their parents?

The first idea that springs to mind is that toleration of young may be site-specific. Those young would then be accepted which appear first in the home burrow. This hypothesis can be refuted however. Placing alien offspring, delivered from the brood pouch more than 48 h previously, into the lower parts of a burrow in which the female is about to deliver her own progeny, inevitably leads to the cannibalism of all the alien young by both the female and male ($n = 20$ pairs).

Do parents somehow recognize their progeny at the time of delivery before coming into contact with them? When parents ($n = 18$ pairs) were separated from their young immediately after their release and then reunited with them 24–70 h later, they cannibalized them without exception, regardless of where they met—in a neutral burrow, in their own burrow, or outside a burrow. But when isolated pairs of adults ($n = 40$) were brought into contact with alien young delivered from the brood pouch of their mother not more than six hours earlier, the great majority ($n = 34$ adults) showed no aggressive behavior no matter where they met them, even in an alien burrow or a small container previously inhabited by foreign conspecifics. Six adults showed some aggression initially, but attacks were only light and quickly ceased; no killing occurred. There were no sex-specific differences; adult males and females acted alike. These results clearly demonstrate that the parents' own young are neither recognized at 'birth', i.e. upon delivery from the brood pouch, nor are they site-specifically tolerated.

What is the nature of the aggression-inhibiting property?

Do the newly delivered young lack signatures? If this assumption is correct, newly released offspring should be unable to render older young, forced into close contact with them, alien, whereas they themselves should become alienated. The results obtained (Fig. 27, see also Linsenmair, 1984) contradict this hypothesis unequivocally. The older young, mixed for 35–100 min in small vials with newly released young (5–12 h after delivery, and in three recently performed experiments, 0.5–2 h afterwards) all became completely alienated and unacceptable to their parents. On the other hand, the newly released young remained, without a single exception (13 progenies tested with > 60 young in each), fully acceptable and entirely protected against attacks from their parents. Apparently the very young desert isopods already possess typical badges. But although these must be alien to both their genetic and foster parents, their aggression-eliciting efficacy is somehow suppressed.

This inhibition of aggressiveness reminds one of the situation with newly molted family members. Does *Hemilepistus* employ inhibitory substances under these circumstances too? Are the same inhibitors used as with newly molted isopods? A negative answer to the last question may be given rather easily. The inhibitors found at molting provide a reliable protection only if coupled with subsets of a previously learned discriminator set, including all personal components. The inhibitory properties of the newly delivered young, however, operate independently of previous learning, or rather, they exert their effects despite previous learning (see below). Furthermore, in bringing intermolt test animals into close contact with newly molted aliens or in using exuvial fluid transferred to glass rods or to other isopods, inhibitory effects were always evident (see above). But observations of parents with older young, alienated through contacts with the newly released progeny of another pair, never yielded any clear indication that inhibitory substances had been transferred.

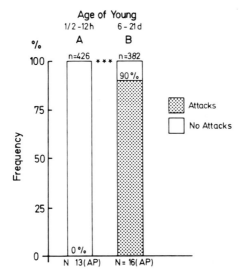

Fig. 27. Contamination experiments with newly delivered and older young. Two groups of the indicated ages were brought into close contact for 35–100 min and then returned to their respective parents. Given the small size of the young, presentation of test individuals in the usual way (by holding them with forceps) was not possible

These findings do not support the view that inhibition by newly molted individuals and by newly delivered young have identical bases. If the same substances are in fact employed, one or more other factors must also be involved. At the moment clear evidence of inhibitory chemical compounds on the newly born young is lacking.

Who is inhibited?

Which conspecifics are inhibited from attacking newly released young? No inhibition, or no sufficiently protective inhibition, could be detected in half- to fully-grown juveniles of *Hemilepistus*. The same holds true for adults that are not already parents or are not about to become parents. Parents with young not over four to six weeks old and adults that will become parents in the near future are inhibited. Concerning the latter, significant sex-specific behavioral differences can be observed, with males being considerably more tolerant toward alien newly delivered young than females (Fig. 28). This is not surprising, since males usually have much less information than females concerning the time of birth of their young (Linsenmair, 1984).

What are the prerequisites for males to accept newly released young? According to data collected so far, two requirements must be fulfilled.

1. Males must have copulated with a female to which they had been paired at least three days before the parturial molt.
2. They must have lived at least 14 days with the pregnant female. This female need not necessarily be the one with which the male copulated.

Of 30 males forced to become acquainted with an unfamiliar female exchanged for their own shortly after the parturial molt, none behaved aggressively toward the young not sired by himself. Among males that were forced, without having copulated, to live with pregnant females during the whole pregnancy phase, 23 of 25 did not accept the young of 'their' female, but cannibalized the whole brood.

Provided the two preconditions are met, males are very tolerant (Fig. $28A_1$–A_3). Not one of 26 males tested with young up to six hours old showed any aggressive reaction. Alien offspring delivered 6–24 h earlier were still accepted without attacks by 18 of 27 males, whereas only one of 22 males did not attack and try to cannibalize young that were older than one day (up to six days).

Females behave much less tolerantly than males toward alien newly released young. The majority of females tested (23 of 31) that did not deliver their own young within the ensuing 24 h, did not even accept young less than six hours old (Fig. $28B_1$). Only two of 35 females reacted peacefully toward alien young older than six hours (Fig. $28B_2$ and $28B_3$). On the other hand, females that released their own young within the next 24 h were significantly more tolerant. Seven of ten did not attack when encountering alien young less than five hours old (for more detailed figures, see Linsenmair, 1984). Complete tolerance of newly released foreign young by all females is restricted to the final 2–4 h prior to the release of their own progeny.

In adults that can be influenced by the inhibitory properties of newly delivered young, further effects besides the immediate suppression of aggression can be detected. The inhibitory properties not only protect the bearers

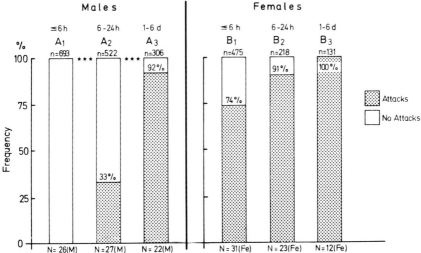

Fig. 28. Sex-specific tolerance of parents toward young. Males had copulated and were paired with females pregnant for more than 14 days. All females used were carrying eggs or marsupial mancas in their brood pouches

themselves, but can also induce changes in the behavior of adults that have spent 1–3 h with newborn young. These changes outlast the presence of the newborns. This may be demonstrated as follows: Adults that are prepared to accept newborn young (but that are not yet in contact with these) will always attack, catch, and cannibalize alien young aged two to ten days wherever they meet them. After spending 1–3 h with newly released young, however, the majority of such adults will behave completely benignly if they meet the same or any other older young (up to about three weeks of age). This is the only natural situation in which an alien intermolt badge, suddenly encountered, will be accepted unaggressively in the absence of inhibitors.

The inhibitors seem to interfere with learning. Even if the young are touched frequently, they can be successfully exchanged with alien progenies of about the same age for 18 and, in part, even up to 48 h after birth (Linsenmair, 1984). Does *Hemilepistus* need up to 48 h and many hundreds of antennal contacts to learn the signatures of its offspring, while in other situations, e.g. by changing a badge through contamination, or by forcibly integrating aliens into older sibships, much shorter periods and an order of magnitude fewer contacts are sufficient? This appears highly improbable.

It seems more likely that we have found another special behavioral adaptation to particular requirements of the recognition system, its probable value being as follows. The discriminators are distributed over the entire body, including the brood pouches. Hence, upon hatching the young cannot help but be contaminated with the recognition substances produced by their mother. If the mother deviates in her spectrum of discriminators from those of her offspring as is usual, then the signatures of the young may become materially changed via contamination. Large amounts of these discriminators will be lost only by molting, loss via evaporation being negligible owing to low volatility. All young molt during the first 12 h after delivery and one may assume that by the termination of this molt, most maternal discriminators have been lost. Subsequent contamination with parental recognition substances is unimportant, especially in the very young (K. E. Linsenmair, unpublished). After the molt-caused loss of maternal substances, parents can, and obviously do, learn the pure badge of their progeny. Assuming that the inhibitors of the newborn are substances resembling discriminators in their low volatility, their disappearence would indicate the stage at which the pure badge could be learned. But the mechanism whereby this retardation of learning is regulated may be quite different. This and many other questions await experimental elucidation.

KIN RECOGNITION IN OTHER SUBSOCIAL ARTHROPODS

To my knowledge, apart from *Hemilepistus* and *Porcellio*, comparable kin recognition has been experimentally demonstrated among subsocial arthropods only in *Cryptocercus* (Seelinger and Seelinger, 1983). It is usually considered to be a primitive cockroach, but may possibly be a primitive termite. Like *Hemilepistus*, *Cryptocercus punctulatus* lives in monogamous pairs and forms, together with its single brood, closed family communities. *Cryptocercus* dwells in the rotting wood of fallen logs and by eating the wood constructs rather complicated burrows consisting of a number of chambers connected by galleries. In laboratory experiments it has been demonstrated that these burrows are defended, mainly by the adults, against alien conspecifics, but the biological significance of this intraspecific aggressive behavior is still unclear. Alien adult or young *Cryptocercus* can be distinguished by an adult from its mate and from the pair's own progeny. Evidently, discriminators of an unknown chemical nature function as identifying cues.

During the first two instars, i.e. during their first year, *Cryptocercus* larvae are dependent on trophallaxis with symbiont transfer, but what advantages the closed unit offers its members during the following two to four years of association are as yet unclear. The Seelingers have speculated that it might be defense by the adults against predators of their young. Details of the recognition system are unknown and many questions concerning other aspects of the social behavior of these arthropods cannot be answered at present.

DISCUSSION AND SUMMARY

The exclusivity of families

To understand the life history pattern and many features of the social behavior that have evolved in *Hemilepistus*, the following points have to be taken into consideration.

1. Despite remarkable physiological and morphological adaptations to the abiotic conditions of its arid environment (Linsenmair, 1975; Coenen-Stass, 1981; Hoese, 1982), an adult *Hemilepistus* is hardly able to withstand the weather conditions that prevail above ground during the hot season for a single day or, under extreme conditions, for even a single hour. Small young are the most sensitive and will not survive the average midday weather outside their burrow for 5 min. For most of the year *Hemilepistus* is totally dependent on its burrow. Only inside dens do temperatures not exceed the critical maximum of 37 °C during the hottest months and relative humidity does not drop below 100% at the time of the progeny's delivery, or later on, below the critical minimum of about 92–95% for the half- to fully-grown individuals.
2. *Hemilepistus* lacks effective adaptations for digging, so that the construction of a new burrow is a slow process and is limited by climatic conditions to spring. During the entire summer and autumn *Hemilepistus* would have no chance of digging a new, sufficiently deep burrow in the hard, dry soil. Therefore the timely securing of a burrow is the basic prerequisite for survival and reproduction.
3. The fact that no new burrow can be excavated in summer prevents a family from dissolving until long after the time at which the young could otherwise care for themselves. One of the remarkable North African subsocial *Porcellio* species, *P. albinus*, which itself merits the name 'desert isopod' since its wide distribution includes the central Saharan sand seas, lives partly in the same habitat as *H. reaumuri*. It is exposed to identical climatic conditions and does not differ either in the food it eats or in various other basic features of its ecology, yet it has developed a different life history pattern. Females are iteroparous and multivoltine, brood care (provided only by the female) is confined to a much shorter period and families dissolve when the young are half-grown. All these differences are probably largely due to a single factor—the type of soil in which they live. While *Hemilepistus* is able to excavate its burrow only in solid soil, the digging behavior of *P. albinus*, and of a number of closely related species, is exclusively adapted to light sandy soils. A *Porcellio* will need, at worst, 1–5 h to reach a protective depth even during the summer months (Linsenmair, 1979, 1984, unpublished).
4. In *Hemilepistus* there is strong competition for burrows, which constitute

a valuable resource; continuous guarding is necessary to retain possession of a burrow. This requires the continuous presence of an isopod in the burrow entrance, which prevents it from doing anything else from early spring until late autumn. *Hemilepistus* has solved the problem of continuous burrow defense in an elegant way. The adults form monogamous cooperative pairs and later a strictly closed family community with the young of their single brood. Under natural conditions these family communities have never been observed to admit a single alien conspecific. Why are they so exclusive?

That parents do not allow entry to alien adults as long as their young are still small is easy to understand—foreign adults are cannibalistic and can exterminate a brood of small young in a short time. Furthermore, since the young are for some weeks fully dependent upon their parents' brood caring activities (guarding, feeding, providing the necessary microclimate by excavating and cleaning the burrow) and on mutual cooperation, it is also easy to comprehend that during this phase all young conspecifics are also repulsed. The society has to protect itself from both intentional and accidental parasitism, in order not to expend costly effort without adequate direct or indirect reproductive compensation. But why is this control not somewhat relaxed later in the year, when every family member forages mainly for itself and food provisioning is less important?

Dens are initially deepened by digging, but this activity more or less ceases two to three weeks after delivery of the young. Further deepening is achieved mainly by eating the substrate and expelling it in the form of fecal pellets. There is a close correlation between the number of family members and the speed of this form of burrow construction. In the hottest and driest North African habitats a family not comprising at least 30–35 individuals by mid-June will not survive the next two months, because it cannot deepen its burrow fast enough (see also Shachak, 1980). Such families should either try to gain admittance to families in possession of suitable burrows, or they themselves should permit entry of alien conspecifics. However, neither type of behavior has ever been observed. Why do families not accept alien conspecifics?

The young desert isopods are 'altricial' nestlings for the first 10–20 days of their lives. If brood care behavior were confined to this period, as in one of the subsocial Canarian isopods of the genus *Porcellio* (K. E. Linsenmair, in preparation), it would be sufficient for parents to recognize their burrow and their mate and to deter every other conspecific that tries to enter the den. There would be no necessity to recognize their own progeny. However, when young desert isopods begin to leave the den regularly to forage, they are still dependent for weeks on parental care and for months on mutual cooperation among themselves. In addition, burrow densities are sometimes extremely high with entrances occasionally separated by only a few centimeters. Therefore, the problem of distinguishing kin from non-kin cannot be solved through recognition of a special site and a corresponding tolerance of young found at that site. To maintain exclusivity a true and site-independent recognition mechanism is necessary. While some social insects have genetically specified discriminators (Chapters 7, 8), others have solved this problem partly by using body odors acquired from the environment (see Wilson, 1971; Hölldobler and Michener, 1980). In *Hemilepistus*, in which at peak population densities more than ten families, i.e. up to about 1,000 individuals, may dwell on the same square meter and have foraging areas that overlap more or less completely, one cannot imagine how such a system could allow accurate discrimination. The genetically based signature system that has evolved in *Hemilepistus* solves this problem in a highly effective way. It permits the production of an extremely large number of different badges by using blends of substances which are extraordinarily resistant to most external influences. This resistance, especially to badge equalization via substances adsorbed from the substrate or from the food, results in a strict exclusivity even among nearest neighbors.

One external influence has to be avoided at all costs and this is direct body contacts with alien conspecifics. During such contacts discriminators are exchanged, normally causing alienations of all parties concerned. From this standpoint the recognition system of *Hemilepistus* is vulnerable, since additional components usually change a badge fundamentally. Despite the fact that a contaminated individual often still carries all family-specific discriminators, it will be classified as alien and rejected. Acceptance of foreigners into a family burrow would not only result in a profound alteration of the discriminator pattern of family members coming into contact with the aliens, but also initially in changes in the accepted isopods' badges. There would therefore be a hiatus in badge efficiency until a new common badge was formed, and until all the family members become acquainted with the individual traits of the newly accepted conspecifics, no admittance control could be exerted. To give up admittance control for a few days would lead to an entire breakdown of the system, at least under conditions of high population densities. Although most isopods avoid the fecal embankment of an alien family burrow, orientation errors (Hoffmann, 1985), as well as overlapping fecal embankments of neighboring families, result in frequent attempts by foreigners to intrude into a particular den. With admittance of the first alien, either control of access would have to be abandoned or the risk would have to be taken that in re-establishing control, some family members would also be rejected. Since these are both poor options, one may expect families to remain exclusive for as long as some control over access to the family burrow is necessary. Since each burrow offers protection (against heat, dryness, etc.) to only a limited number of inhabitants and the food resources are limited within the restricted foraging range (Linsenmair, 1972, 1979), limitation of access to burrows is necessary.

Thus, from the time a new family is founded, from pair formation until the end of hibernation, no free access to a burrow should be permitted. On the other hand, neither ecological nor sociobiological considerations would forbid the integration of a few foreigners into a family at certain stages. There are, in fact, circumstances in which permitting access to a limited number of alien conspecifics could be life-saving, but the recognition system does not allow this to happen. The system functions according to an all-or-none principle. For as long as limitation of access, based on the individual- and family-specific recognition system, is, on average, advantageous compared with no control at all, the community must remain exclusive.

This system also explains, in part, the pronounced aggression displayed toward alien conspecifics far from the burrow, which otherwise would hardly be understandable. Such behavior makes no sense in an ecological context, since foraging territories do not exist. Additionally, many attacks that may be observed in the vicinity of the burrow seem to be superfluous, given that only forcible entry into a burrow needs to be prevented; most foreigners retreat voluntarily from the fecal embankment area before they reach the entrance. These aggressive acts are most probably (and in attacks far from the burrow, certainly) not aimed at defending the burrow or any other resource, but they serve to prevent unintentional contacts which could lead to dangerous alienations. Usually the attacked individual runs away, thereby reducing the risk of coming into contact with an alien conspecific. Merely avoiding an alien would probably be less effective in preventing casual contacts.

Recognition of an alien presupposes antennal contact, and since this requires that animals come into close proximity, involuntary contacts may sometimes occur, especially during feeding on a favorite piece of food. At high population densities one may observe after every activity phase, at up to about 1% of family burrows, individuals which, judging by their behavior, are members of the family unit, but are having difficulty in gaining admittance. They have obviously become somewhat alienated, presumably by too close contact with foreigners.

In this context it becomes understandable why some alien badges or slightly alienated family members release aggression in burrow guards only, and not in the rest of family members. For burrow guards, the decision on admitting or rejecting an individual should be taken on the basis of the conspecific's membership in the kingroup of the guard. If differences are detected between the family badge (including deviating signatures) and that of the individual striving for admission, it should be repulsed irrespective of whether it could, upon close contact, alienate family members. Away from the burrow, however, the alienation potential of conspecifics should be the decisive variable responsible for the reaction. Since badges that represent such a risk can probably be detected by superficial antennation, there is no necessity for thorough inspections away from the burrow. To attack an individual that represents no alienation risk is hardly adaptive; on the contrary, it may be disadvantageous for the following reason: not only being attacked, but also attacking, especially in fast chases (which frequently occur) may interfere with the idiothetic orientation mechanism (Hoffmann, 1984, 1985), thereby causing navigation problems on the way back to the family burrow. Such problems become greater with increasing distance from the burrow. They are of comparatively little importance within the fecal embankment or its near vicinity, and they are much smaller at the beginning of an excursion than at the same spot when returning from an extended trip. Mistakes that cause time-costly searching (see Hoffmann, 1978, 1983a,b) are more risky for small isopods owing to their lower resistance to adverse weather factors. These facts may explain why, especially in young isopods, regular attack behavior is restricted to the near vicinity of the burrow entrance and is enhanced by perceiving, or recent perception of, the family badge. They may also explain why, especially in the not yet half-grown young, there is a negative correlation between the readiness to attack and the distance already walked since leaving the burrow, and a positive correlation between body size and aggressiveness outside the fecal embankment. Since *Hemilepistus* does not cooperate in chasing aliens from the vicinity of the burrow, this form of 'feeling secure at home' can hardly influence the aggressive motivation and explain its dependence on the spatial and temporal conditions.

Interfamilial variability

The radius within which the members of a family forage can be as great as 6 m and is usually 1.5–4 m. Assuming an average foraging radius of 2.5 m and an already high but not extreme population density (five families per m^2), it is to be expected that members of about 100 families could come into the near vicinity of any particular burrow. An upper limit, for very high population densities (10–12 families per m^2) and extensive foraging excursions (up to a 6 m radius) due to food shortage caused by the density, would be in the order of 1,000 families, the members of which could make contact with inhabitants of a single burrow. Within a complete family there are, as a rule, at least three distinctly different signatures or badges: that of the male, the female and their progeny. These differ so greatly that familiarity with the identifying characteristics of one of the three does not result in acceptance of the other two. Moreover, within sibships individual variability through personal substances not showing up in the family badge is frequent. Thus the possible number of differing discriminator sets an individual could encounter at its burrow and mistake for a kin signature may increase by another order or, at an absolute maximum, another two orders of magnitude.

In accordance with the above considerations, the maximum number of different signatures a family member could encounter, none of which should be confused with those of its siblings and parents, is 10^5. This is probably an overestimate, but as we have not found a single case in which two families had the same badge, we can reasonably postulate a diversity of discriminator patterns sufficient to ensure that even in the most demanding circumstances misidentifications should remain rare exceptions. The occurrence of unilateral acceptance in 0.1% of the exchange experiments is probably an overestimate owing to the conditions of testing (see above). Thus it seems reasonable to assume that within a population the number of different discriminator patterns produced may considerably exceed 10^6. As we have seen, badge differences seem to be exclusively qualitative in nature with quantitative differences above the threshold of detection not contributing to either the variability or specificity of badges. How many single mutually distinguishable recognition substances would be necessary to produce, for example, 10^6 through 10^8 badges?

In our chemical analysis, after a number of purification steps (Hering, 1981; Esswein, 1985; K. E. Linsenmair et al., in preparation), eight fractions were obtained which differ considerably from each other in chemical properties of the compounds and which proved effective in eliciting aggression in our bioassay. (Previously only seven were thought to be active; see Linsenmair, 1985.) If we consider these eight fractions as eight independent badge dimensions (corresponding in a formal way to eight different gene loci with several alleles at each), we would not need more than an average of nine different compounds in each fraction (provided they lead to phenotypic differences) to generate a number of repetition-free permutations exceeding 100 million.

Such a system, requiring about 80 different recognition substances within the population, should be entirely feasible as far as production is concerned. Morover, it would not make excessive demands on the sensory capabilities of these arthropods, which orient to the surroundings chiefly by chemoreception and which therefore most probably were preadapted for the peripheral acquisition and central processing of chemical information.

The chemical compounds used in this system are strongly polar; have relatively low molecular weights; are extremely resistant to pH changes, oxidation and high temperatures; and have a low volatility. We have also just succeeded in identifying the first pure substances (K. E. Linsenmair et al., in preparation). These were obtained from extracted exuvia and they can alter badges, thereby releasing aggression. However, many investigations must yet be undertaken before a picture will emerge that can usefully be discussed.

Some Canarian and North African *Porcellio* species that have developed, convergently, a social behavior resembling in many respects that of *Hemilepistus* (Linsenmair, 1979, 1984, unpublished), mark their burrows with individual-specific secretions and in at least two species the young are recognized by their mother. The substances which these species carry on their cuticle are capable of alienating *Hemilepistus*. This possibly indicates, as do many other findings, that the discriminators are in part at least not exotic compounds, but may be drawn from the pool of intermediate metabolites that probably do not differ greatly in the two genera. The fact that hemolymph contains effective substances also favors this view.

Intrafamilial variability

As already discussed in detail, there are several reasons why a primary conformity among family members with respect to their individually-produced signatures is improbable. There are no indications of any secondary adjustment to increase intrafamily resemblance via changes in production. The only form of mutual adjustment results from passive exchange of substances during direct body contacts, giving rise to the family badge. In a large family the badge contains only those discriminators which are produced by several individuals. All compounds originating from one or only a few animals are, through exchange and mixing, diluted below the threshold of detection or at least of response and are therefore not effective components of the common badge. They nevertheless seem always to be effective components of the individual discriminator signature borne by the producer. Self-produced substances are apparently protected in a yet unknown way against dilution below the detection threshold, whereas components acquired from conspecifics can easily be diluted to inefficacy through contact with a large number of siblings.

All individuals producing subsets of the pattern representing the common badge, as well as those family members that chance to produce just this pattern, are equalized through intermingling of discriminators. To recognize all the components therefore, it would suffice to become thoroughly acquainted with a single individual. Any other individual producing components which are deviant in respect to the smallest common denominator would usually not be recognized as belonging to the family. The badges *Hemilepistus* has developed exhibit many properties of a Gestalt character (Crozier and Dix, 1979, Chapter 4). One of the striking features of the Gestalt is that it is far more vulnerable to addition than to subtraction of components. Therefore, to become acceptable and to remain so, the deviating features (i.e. the additional individual components) have to be learned, and this learning has to be repeated frequently in order not to forget the particular properties of the deviants. As our experiments with EMGs have demonstrated, unusually complex badges can also be learned relatively quickly. In discriminating among intermolt conspecifics, *Hemilepistus* employs a phenotype matching mechanism (Sherman and Holmes, 1985; Chapters 2 and 3).

As a reference the common badge is used, as well as the additionally learned deviations of personal signatures, but never the animal's own characteristics. It is still uncertain whether *Hemilepistus* learns, in addition to the common family badge, all additional substances produced in its group, and thereby accepts every permutation in excess of the unchanged basic common badge, or whether particular individual combinations are learned when group members deviate in more than one component. The data favor the latter hypothesis.

The molting problem

Frequent molting makes the *Hemilepistus* system extraordinarily complicated. All members molt at more or less short intervals and each time they lose all the discriminators acquired from conspecifics. Since family badges are always mixtures, components must be lost and this must interfere with the Gestalt of the badge. This conclusion is documented by the results of isolating family and mixed group members during molting.

As all integration and mixed group experiments have shown, becoming familiar with a deviating individual is a precondition for its acceptance in the newly molted state. By thorough inspection, a group member's deviating features can be recognized and learned. But there are no indications of an ability to recognize which components are produced by the individual itself and which are acquired from conspecifics. Therefore, no behavioral adjustment specific to the components lost at molting can occur. To avoid acceptance of a foreigner that happens to match one of the presumably many possible subsets of a complex family badge, additional safeguards are required. Under natural conditions a deviant family member is protected in several ways against attacks by its kin. During the most critical period in the first hour after molting, the individual, being almost immobile, is reliably protected through inhibitory properties. Because these properties are transferable to inanimate objects, as well as to intermolt conspecifics, we assume that special substances are involved, the chemical nature of which is still unknown. These substances display their full efficacy only if all members of the group have become familiar with possibly deviating personal traits of the molting individual during previous encounters. The inhibitory substances are relatively quickly lost. It is doubtful that a newly molted individual always contacts all of its kin while inhibitors are still effective, although its high activity following the partial hardening of the cuticle suggests such contacts. 'Resting' conspecifics often do not inspect an individual pushing against them. Furthermore, it is doubtful whether only one or a few contacts are sufficient to become fully acquainted with the pattern as altered by the molt, even though we assume that *Hemilepistus* is able to learn major changes very quickly.

While it crawls about, a newly molted isopod comes into close contact with its kin and with the burrow walls, thereby regaining some of the lost badge discriminators on the one hand, but continuously changing its badge pattern on the other, which hardly facilitates the learning task of its kin. The newly molted cuticle seems to possess particularly good adsorption properties, but despite this fact we cannot assume that complete renewal of the old badge is possible while inhibitors are still effective. Besides the reduction in aggression-eliciting properties of a badge associated with a low concentration of discriminators on the cuticle, additional factors come into play. These are tolerance of deviating individuals inside the burrow and the behavior of the molting individual in not leaving the burrow between the two molting stages and for a short time after the termination of molting. Thus it avoids the most critical examination, namely that of the burrow guard. Spending a prolonged period within the burrow after molting ensures that a maximum of lost discriminators will be regained and that other family members will have an opportunity to familiarize themselves with still deviant traits before a newly molted individual again leaves the protective family den.

This complicated recognition system, which makes high demands on the learning capacity of *Hemilepistus*, is somewhat simplified by the fact that different concentrations of discriminators are not employed as identifying and variability-enhancing criteria.

Primary recognition of offspring

The ability to recognize one's own kin allows the development of aggression, e.g. in the form of cannibalism, against foreign conspecifics. The emigration behavior of *Hemilepistus* that follows family dissolution in early spring makes it unlikely that any relative will dwell within the usual activity range of a family. Therefore it is not surprising that young aliens are caught by adults and cannibalized.

Because parents cannot be familiar beforehand with the badge their progeny will wear, the problem arises of how young are protected against attacks by their own parents. *Hemilepistus* has a mechanism that not only protects the offspring, but at the same time minimizes the risk of accidentally accepting alien young. Because the inhibitory properties of newly emerged young are effective without previous learning of a discriminator pattern, and because in mixing experiments the transfer of inhibitors to older young could not be demonstrated, we conclude that the inhibitors found on newborn isopods are not identical to those on molting individuals. Probably these substances are also responsible for the delayed learning of the discriminator patterns of the newborns, which we assume to be, partly at least, an adaptation to the unavoidable contamination of the young by maternal substances. These substances will in many cases profoundly change the character of the progeny's badge and in losing these through molting, a further alteration results.

Comparative comments

The recognition system of the modern *Hemilepistus* species, possibly already developed in their common ancestor, is ideally suited to guarantee family exclusivity under demanding conditions. On the other hand, it has also caused a number of problems that have resulted in the development of secondary adaptations. The present-day mechanisms found in *Hemilepistus*, therefore, represent a highly derived system. As far as can be judged from minimal information available on other species in the genus (Marikovsky, 1969; Schneider, 1971; K. E. Linsenmair, unpublished), all seem to conform to the *H. reaumuri* model in the essential features of their kin recognition behavior. Thus, information concerning its evolutionary development probably cannot be gathered by comparing the species of the genus. Convergent to *Hemilepistus*, a number of other isopod species living under similar ecological conditions have evolved subsocial behavior. In attempting to determine the steps in the evolution of the highly developed discriminating system of *H. reaumuri*, therefore, comparative studies of North African and Canarian *Porcellio* species may be useful (Linsenmair, 1979, 1984, unpublished).

ACKNOWLEDGMENTS

This study was supported by the Deutsche Forschungsgemeinschaft (Li 150/6–10, SF B4 'Sinnesleistungen: Anpassungen von Strukturen und Mechanismen'). Special thanks are due to my wife for her most valuable help during field work over many years. I am very grateful to the editors for their great efforts in improving my English.

LITERATURE CITED

Beecher, M. D. 1982. Signature systems and individual recognition. *American Zoologist*, **22**, 477–490.

Breed, M. D. 1983. Nestmate recognition in honey bees. *Animal Behaviour*, **31**, 86–91.

Coenen-Stass, D. 1981. Some aspects of the water balance of two desert woodlice, *Hemilepistus aphganicus* and *Hemilepistus reaumuri* (Crustacea, Isopoda, Oniscoidea). *Comparative Biochemistry and Physiology, (A)* **70**, 405–419.

Crozier, R. H. and M. W. Dix. 1979. Analysis of two genetic models for the innate components of colony odors in social Hymenoptera. *Behavioral Ecology and Sociobiology*, **4**, 217–224.

Esswein, U. 1985. *Über das chemische Familienabzeichen der Wüstenassel* Hemilepistus reaumuri. Dissertation der Naturwissenschaftlich-Mathematischen Gesamtfakultät der Universität Heidelberg.

Greenberg, L. 1979. Genetic component of bee odor in kin recognition. *Science*, **206**, 1095–1097.
Hering, W. 1981. *Zur chemischen Ökologie der Tunesischen Wüstenassel* Hemilepistus reaumuri. Dissertation der Naturwissenschaftlich-Mathematischen Gesamtfakultät der Universität Heidelberg.
Hoese, B. 1982. Morphologie und Evolution der Lungen bei den terrestrischen Isopoden (Crustacea, Isopoda, Oniscoidea). *Zoologische Jahrbücher, Abteilung für Anatomie und Ontogenie der Tiere.*, **107**, 396–442.
Hoffmann, G. 1978. *Experimentelle und theoretische Analyse eines adaptiven Orientierungsverhaltens: die 'optimale' Suche der Wüstenassel* Hemilepistus reaumuri *nach ihrer Höhle*. Dissertation des Fachbereichs Biologie der Universität Regensburg.
Hoffmann, G. 1983a. The random elements in the systematic search behavior of the desert isopod *Hemilepistus reaumuri*. *Behavioral Ecology and Sociobiology*, **13**, 81–92.
Hoffmann, G. 1983b. The search behavior of the desert isopod *Hemilepistus reaumuri* as compared with a systematic search. *Behavioral Ecology and Sociobiology*, **13**, 93–106.
Hoffmann, G. 1984. Orientation behavior of the desert woodlouse *Hemilepistus reaumuri*: adaptations to ecological and physiological problems. *Symposia of the Zoological Society of London*, **53**, 405–422.
Hoffmann, G. 1985. The influence of landmarks on the systematic search behavior of the desert isopod *Hemilepistus reaumuri*, I. Role of the landmark made by the animal; II. Problems with similar landmarks and their solution. *Behavioral Ecology and Sociobiology*, 325–334, 335–348.
Holdich, D. M. 1984. The cuticular surface of woodlice: A search for receptors. *Symposia of the Zoological Society of London*, **53**, 9–48.
Hölldobler, B. and C. D. Michener. 1980. Mechanisms of identification and discrimination in social Hymenoptera, in Markl, H. ed., *Evolution of social behavior: hypotheses and empirical tests*, Dahlem Konferenzen, Verlag Chemie, Weinheim and Deerfield Beech, Florida, pp. 35–58.
Krempien, W. 1983. *Die antennale Chemorezeption von* Hemilepistus reaumuri Audouin u. Savigny *(Crustacea, Isopoda, Oniscoidea)*. Dissertation der Fakultät für Biologie der Universität Würzburg.
Kukuk, P. F., M. D. Breed, A. Sobti and W. J. Bell. 1977. The contributions of kinship and conditioning to nest recognition in a primitively eusocial bee, *Lasioglossum zephyrum* (Hymenoptera: Halictidae).*Behavioral Ecology and Sociobiology*, **2**, 319–327.
Lincoln, R. J. 1970. A review of the species of *Hemilepistus* s. str. Budde-Lund, 1885 (Isopoda, Porcellionidae). *Bulletin of the British Museum (Natural History) Zoology*, **20**, 111–130.
Linsenmair, K. E. 1972. Die Bedeutung familienspezifischer 'Abzeichen' für den Familienzusammenhalt bei der sozialen Wüstenassel *Hemilepistus reaumuri* Audouin u. Savigny (Crustacea, Isopoda, Oniscoidea). *Zeitschrift für Tierpsychologie*, **31**, 131–162.
Linsenmair, K. E. 1975. Some adaptations of the desert woodlouse *Hemilepistus reaumuri* (Isopoda, Oniscoidea) to desert environment. *Verhandlungen der Gesellschaft für Ökologie, Erlangen*, **1975**, 183–185.
Linsenmair, K. E. 1979. Untersuchungen zur Soziobiologie der Wüstenassel *Hemilepistus reaumuri* und verwandter Isopodenarten (Isopoda, Oniscoidea): Paarbindung und Evolution der Monogamie. *Verhandlungen der Deutschen Zoologischen Gesellschaft*, **1979**, 60–72.

Linsenmair, K. E. 1984. Comparative studies on the social behaviour of the desert isopod *Hemilepistus reaumuri* and of a Canarian *Porcellio* species. *Symposia of the Zoological Society of London*, **53**, 423–453.
Linsenmair, K. E. 1985. Individual and family recognition in subsocial arthropods, in particular in the desert isopod *Hemilepistus reaumuri*, in Hölldobler, B. and M. Lindauer, eds., *Experimental behavioral ecology and sociobiology*, Symposium in memoriam K. v. Frisch. *Fortschritte der Zoologie*, **31**, 411–436.
Linsenmair, K. E. and C. Linsenmair. 1971. Paarbildung und Paarzusammenhalt bei der monogamen Wüstenassel *Hemilepistus reaumuri* (Crustacea, Isopoda, Oniscoidea). *Zeitschrift für Tierpsychologie*, **29**, 134–155.
Marikovsky, P. J. 1969. A contribution to the biology of *Hemilepistus rhinoceros* [in Russian]. *Zoologitcheskii Zournal*, **48**, 677–685.
Michener, C. D. 1974. *The social behavior of the bees: a comparative study*. Harvard University Press, Cambridge, Massachusetts.
Ribbands, C. R. 1953. *The behaviour and social life of honeybees*. Bee Research Association Ltd, London.
Ribbands, C. R. 1965. The role of recognition of comrades in the defence of social insect communities. *Symposia of the Zoological Society of London*, **14**, 159–168.
Schneider, P. 1971. Lebensweise und soziales Verhalten der Wüstenassel *Hemilepistus aphganicus* Borutzky 1958. *Zeitschrift für Tierpsychologie*, **29**, 121–133.
Schneider, P. 1975. Beitrag zur Biologie der afghanischen Wüstenassel *Hemilepistus aphganicus* Borutzky 1958 (Isopoda, Oniscoidea) Aktivitätsverlauf. *Zoologischer Anzeiger*, **195**, 155–170.
Seelinger, G. 1977. Der Antennenendzapfen der tunesischen Wüstenassel *Hemilepistus reaumuri*, ein komplexes Sinnesorgan (Crustacea, Isopoda). *Journal of Comparative Physiology*, **113**, 95–103.
Seelinger, G. 1983. Response characteristics and specificity of receptors in *Hemilepistus reaumuri* (Crustacea, Isopoda). *Journal of Comparative Physiology*, **152**, 219–229.
Seelinger, G. and U. Seelinger. 1983. On the social organization, alarm and fighting in the primitive cockroach *Cryptocercus punctulatus* Scudder. *Zeitschrift für Tierpsychologie*, **61**, 315–333.
Shachak, M. 1980. Energy allocation and life history strategy of the desert isopod *Hemilepistus reaumuri*. *Oecologia*, **45**, 404–413.
Shachak, M. and A. Yair. 1984. Population dynamics and role of *Hemilepistus reaumuri* (Audouin u. Savigny) in a desert ecosystem. *Symposia of the Zoological Society of London*, **53**, 295–314.
Sherman, P. W. and W. G. Holmes. 1985. Kin recognition: issues and evidence, in Hölldobler, B. and M. Lindauer, eds., *Experimental behavioral ecology and sociobiology*, Symposium in memoriam K. v. Frisch. *Fortschritte der Zoologie*, **31**, 437–460.
Takeda, N. 1984. The aggregation phenomenon in terrestrial isopods. *Symposia of the Zoological Society of London*, **53**, 381–404.
Warburg, M. R., K. E. Linsenmair, and K. Bercovitz. 1984. The effect of climate on the distribution and abundance of isopods. *Symposia of the Zoological Society of London*, **53**, 339–367.
Wilson, E. O. 1971. *The insect societies*. Harvard University Press, Cambridge, Massachusetts.

Kin Recognition in Animals
Edited by D. J. C. Fletcher and C. D. Michener
© 1987 John Wiley & Sons Ltd

CHAPTER 7

Kin Recognition in Primitively Eusocial Insects

CHARLES D. MICHENER
Departments of Entomology and of Systematics & Ecology, University of Kansas, Lawrence, Kansas 66045, USA.

and

BRIAN H. SMITH
Department of Entomology, University of Kansas, Lawrence, Kansas 66045, USA.

INTRODUCTION

Primitively eusocial insects have provided some of the best information available on kin recognition. They typically live in small colonies, such that all colony members can be distinctively marked and their interactions recorded. Their genealogical relationships are often knowable and can be related to behavioral observations. Above all they are small, common creatures, inexpensive to maintain in the laboratory and to find and observe in the field. We commend them to all who have been studying large, rare and expensive animals. Although only a tiny fraction of insect species are primitively eusocial, these are enough to keep many observers busy for centuries. The insects involved are certain bees and wasps.

Eusocial (or truly social) insects are those in which one or more mothers (queens) live in a colony with one or more non-reproductive or less reproductive individuals (workers). In the Hymenoptera, the workers are all females, and are ordinarily daughters of the queen (or queens). Thus the colonies consist of females only; in the following pages we always discuss females except where males are clearly specified. There is reproductive division of labor between the female castes, and frequently there is also division of labor relative to other functions.

Primitively eusocial species are those that:
1. Lack major morphological caste differences.
2. Usually have caste determination in the adult stage.
3. Usually live in rather small colonies (down to one queen and one worker).
4. Have colonies started not by a swarm (i.e. a migrating or dispersing colony or colony fragment) but by one or more gynes (i.e. queens and potential queens).
5. Have temporary (e.g. annual) colonies rather than perennial ones.

Semisocial colonies are similar to primitively eusocial ones, but consist of individuals of a single generation, commonly sisters, one of which is the queen. Such colonies may exist as stages in the life histories of the bees and wasps discussed here, or as laboratory artifacts. No differences between eusocial and semisocial colonies relevant to kin recognition are known except those that result from different genetic relationships among colony members. Since colonies of the species discussed below obligately become eusocial, and to simplify expression, they are called primitively eusocial species.

It has long been known that workers of highly eusocial Hymenoptera such as ants and honey bees often distinguish nestmates, which they accept, from other conspecifics, which they fight. As explained below, similar behavior is found in primitively eusocial colonies.

Certain terms and concepts require clarification for purposes of this chapter. Confusion in discussion of kin recognition has sometimes resulted from failure to distinguish clearly the recognized individual, i.e. the source of recognition cues, from the recognizing individual that responds to the cues. In many interactions, such as elaborate courtships, both participants play both roles. We use 'cue bearer' or 'recognized individual' (whether accepted, rejected, familiar or unfamiliar) for the one that is the source of recognition cues (pheromones, odors from the environment, behavior, sound, structure or whatever the cues may be). We use 'responding individual' for the one that does the recognizing, i.e. the one whose responses demonstrate that it has recognized the cues. In specific cases words like guard or intruder or even female and male usually make the meaning clear, but for discussion and generalization, general terms are needed.

In some cases a group of cue bearers may be alike in their recognition cues and from this standpoint constitute a 'homogeneous group' (Barrows, Bell and Michener, 1975) within which individuals are probably not usually discriminated (see also Chapter 3). On the other hand, a group of cue bearers not united by any common cue may be recognized because of prior experience of the responding individual who may discriminate, for example, familiar from unfamiliar cue bearers as individuals. Such individuals constitute a 'heterogeneous group' because of their dissimilar recognition cues.

There are several methods by which an animal may distinguish its kin from unrelated conspecifics or assess the degree of relatedness of individuals it encounters. While other chapters provide comprehensive accounts of these problems, the following comments seem most relevant to the insects discussed here: Among animals that recognize individual, familial, or other intraspecific differences, there may be a particular stage at which these differences are learned by the responding individual. The term imprinting is used for some such cases. This learning stage may occur in association with particular places, times, nests, etc., where related individuals are unlikely to be mixed with unrelated ones. Then the learned individual or group attributes will usually be those of kin, not because kin have features indicative of the relationship but because the individuals were at the right place at the right time. Much of the learning by *Lasioglossum* and *Polistes* discussed below appears to be of this type. (We define learning and memory broadly, to imply habituation and all forms of conditioning. The types of learning, Shepherd, 1983, Chapter 30, have not yet been worked out for the cases described below.)

Having experienced attributes of itself, its kin or other associates, an animal may use the information for recognition of the same or similar individuals. Lacy and Sherman (1983) and Holmes and Sherman (1983) make a case for 'phenotype matching', i.e. comparing phenotypes. They postulate that the responding individual, having learned recognition cues, matches the phenotype of an individual with the learned template of characteristics, e.g. of chemicals in a pheromone. The responding individual can assess the degree of relationship by the goodness of the match. The fact that in some cases, e.g. *Lasioglossum* guards at nest entrances, there are all levels of response to unfamiliar individuals from prompt acceptance to complete rejection, supports the phenotype matching idea because there can be different degrees of matching to the learned template. On the other hand, perhaps guards merely vary in their motivation or their ability to recognize odor differences. In *Polistes*, however, as noted in the Discussion below, there is no evidence of a continuum in responses and a threshold model has been proposed.

An important question is the extent to which the recognition cues are genetically specified, environmentally determined, or stochastic. In the first case, organisms which have close genealogical relationship will on the average have more similar recognition cues than organisms not so closely related. In the second case, cues derived from food, nest odors, or other individuals may be wholly or partly responsible for discrimination. Related conspecifics (cue bearers) are then not necessarily more similar than less related ones, although there could be a correlation between the cues and relationship because relatives may live near one another and therefore share foods and other environmental odor sources. In the third case, stochastic mediation of recognition cues (e.g. from genetic or developmental accidents, Rutter, 1984) may result in individual differences that others can learn. The recognition cues used by any species may have genetically specified, environmentally controlled, and stochastic components. Such components are difficult or impossible to segregate fully, but it may be possible to determine whether there is a major component.

For purposes of this chapter discrimination is primarily or wholly by chemical mixtures (as explained below). Such 'discriminating substances' can be of:
1. Environmental origin (e.g. directly from food).
2. Endogenous but ultimately of environmental origin (e.g. food eaten as a larva that influences odor of adults).
3. Genetically specified.

As noted in Chapter 2, Hölldobler and Michener (1980) used the synonymous terms 'discriminators' and 'recognition pheromones' for genetically specified odor signals that differ among the cue bearers of a population. Ordinarily discriminators, like other discriminating substances, must first be learned, after which recognition or discrimination may occur. Discriminators differ from other classes of pheromones (primers, releasers) in that responses vary widely according to prior experience (learning) of the responding individual. They are in a sense releasers but the behavior released can be diverse and depend entirely on learning. Thus responding individuals in the same physiological state can react to a discriminator in completely opposite ways depending on experience, e.g. one guard may aggressively respond to a conspecific intruder's discriminators while another in a different nest would readily admit the same individual to its nest.

In much of the experimental work discussed below, pupae or young adults taken from a field (or laboratory) nest were used to form a laboratory colony. Such individuals are probable sisters; for brevity we sometimes call them 'sisters'. We also call them 'kin' females, and the colony is a kin colony living in a kin nest. Conversely individuals of diverse origins are non-sisters and may form a non-kin colony. Unrelated individuals whose cues, as a result of experimental manipulations, have been learned as kin, are treated as kin (e.g. as colony members) and become 'pseudokin' or 'pseudosisters'.

In the following material we exemplify some of the approaches needed for studies of kin recognition and its mechanisms. Not all are practical with any one species. More detailed information, for example, on the controls for various experiments, is presented in the original papers cited. For background information the reader may consult general accounts of the life cycle and behavior of the sweat bee *Lasioglossum zephyrum* (Batra, 1966; Michener, 1974) and of *Polistes* wasps (West-Eberhard, 1969; Evans and West-Eberhard, 1970). Reviews of some of the topics discussed below are found in Michener (1982), Hölldobler and Michener (1980) and Gamboa, Reeve and Pfennig (1986).

Studies of the mating biology of *Lasioglossum zephyrum*, a bee also known as *Halictus zephyrus* or *Dialictus zephyrus*, showed that there is recognition of female kin or of familiar individuals by males. Such kin recognition by males would presumably be equally important in solitary ancestors of *L. zephyrum* (it belongs to the same subgenus as solitary species) and in other insects. We therefore explain first this presumably ancestral function of kin recognition by males of *L. zephyrum*. Except for the presumed use of the same female-produced discriminators, this nuptial behavior probably has nothing to do with the eusocial behavior of the females although it happens to have been studied in a eusocial species. Next we take up female nestmate recognition within the same species. We then examine discrimination of kin from non-kin in *Polistes* wasps whose eusocial organization is at a level similar to that of *L. zephyrum* but which have no close solitary relatives. Finally, we note what little is known of kin recognition in the more advanced primitively eusocial bees and wasps, *Bombus* and the Vespinae.

KIN RECOGNITION IN THE SWEAT BEE *LASIOGLOSSUM ZEPHYRUM*

Mate selection by males

Nests made and occupied by females are burrows in earthen banks, usually along streams or rivers. Such nest burrows are commonly aggregated, so that dozens to thousands are located in limited sections of bank, and similar nearby sections may lack nests. Males permanently leave the nest burrows soon after maturation and spend a great deal of time flying (patrolling) over the bank, each in a limited area but not a defended territory, for ranges of different males broadly overlap. In the presence of female odor, patrolling males pounce on any small, dark objects or shadows on the ground beneath (Barrows, 1976). They appear to be responding sexually although often they do not extend the genitalia or exhibit other copulatory behavior unless the target of a pounce turns out to be a female bee (see Fig. 5A). In the field most females reject mating attempts. Either they are already mated, or they are workers which usually do not mate. Therefore repeated unrewarded pounces by males are the rule, not the exception, at *L. zephyrum* nesting sites.

For laboratory elucidation of the field findings, Barrows (1975a) used cages containing about 20 males patrolling over a paper with several black ink dots 3 mm in diameter. He found that if a tethered female was introduced nearby, males pounced on the dots at increased frequencies. Moreover, if females were confined for 15–24 h in a small container with pieces of moist filter paper, and a piece of this treated paper was introduced instead of a living female, the result was similar. (Controls consisted of pieces of moist paper kept in a small container without bees.) Responses of males suggested that the treated papers had become impregnated with a female-produced pheromone that functioned both as an attractant and as an aphrodisiac. After about 1 min the responses of males waned, perhaps due to learning.

Barrows observed that after responses of males to the odor of a given female had waned, they were often rejuvenated when another female or its impregnated paper was introduced in the cage. He therefore suspected that individually different odors of females might be involved, and that males learned the different females' odors independently. In terms of our Introduction, the female is the cue bearer, producing the recognition cues, and the male is the responding individual.

If two impregnated papers from the same female are presented sequentially (5 min each) to a group of patrolling, caged males, each paper having remained in the container with the female until the time of presentation, waning of the responses of males ordinarily continues with the second paper (Fig. 1). Thus in 55 out of 61 trials of impregnated papers from different females, there were fewer pounces during the second five minutes of a test (second paper) than during the first ($p < 0.001$), in spite of the stimulating disturbance of removing the first paper and inserting the second (Barrows, 1975b; Barrows, Bell and Michener, 1975). This result cannot be plausibly explained by assuming that production of odor by a female decreases while she is in the cage with males or that she produces a repellent or that males mark females with repellents. We only have evidence for individually different female odors (i.e. discriminating substances) which males learn.

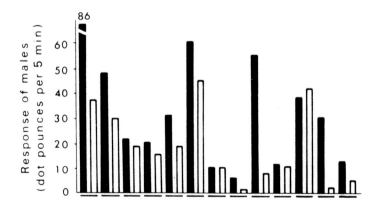

Fig. 1. One series of responses of males of *Lasioglossum zephyrum* to dots during presentation of female odors on pieces of paper. Black, first papers; white, second papers. Each short line below links results from two papers impregnated by one female and presented in sequence. *Reproduced from Barrows, Bell and Michener, 1975, with permission*

Males do not immediately forget the odor of a given female, even while odors of other females are being presented. When presentations of first papers were followed by 60–135 min of presentations of other female odors before the second papers were presented, responses to second papers were lower than to first papers ($N = 14$ trials, $p < 0.005$). One might suspect that the males became fatigued after long periods of unrewarded pouncing on dots, and that this could explain low pouncing rates in response to second papers. However, to control for this, a paper from another female, unfamiliar to the males (i.e. a first paper) was inserted in the cage immediately after each second paper was removed. Pouncing rates then increased again, as in responses to other first papers (Fig. 2). We do not know how long males remember the odor of a female; if their memories are similar to those of females (see below), they may do so for several days.

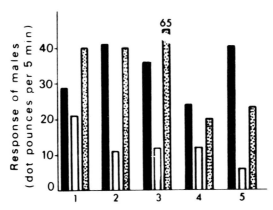

Fig. 2. One series of male *Lasioglossum zephyrum* responses to dots during first presentations of odors of individual females (black bars), second presentations 60–135 min later (open bars), and immediately afterwards, odors of different females as controls (stippled bars). *Reproduced from Barrows, 1975b, with permission*

The individually distinctive odors of females that males recognize are genetically specified. As will be explained more fully below, females retain their own odors when kept together rather than acquiring a mixed 'colony' odor (Greenberg, 1979; Buckle and Greenberg, 1981). Moreover, Smith (1983) showed that males seem to 'mistake' close relatives of a familiar bee for the familiar bee itself. He used small cages containing three males each, and introduced a female (a worker, which ordinarily refused to mate) into each cage for 10 min. Then, on removing the worker, he introduced another

female of known relation to the first and counted mating attempts during the next 2.5 min. Across seven degrees of relationship (coefficient of relationship, $r = 0–0.8$) there was a strong negative correlation between:
1. The number of mating attempts by the males with the second female.
2. Her genealogical relationship to the first female (Fig. 3).

As expected, reintroduction of the first female elicited little response (right side, Fig. 3) but close relatives of the first female also elicited relatively little response. Such results clearly indicate genetic specification of the recognition cues, which may therefore be called discriminators or recognition pheromones.

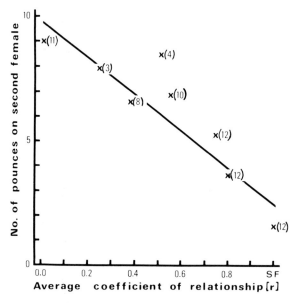

Fig. 3. Linear regression of the number of copulation attempts by males of *Lasioglossum zephyrum* elicited in 2.5 min by second females as a function of their relatedness to the initial females presented to the males; $r = 0$ = females presumed to be unrelated, i.e. obtained from sites separated by at least 10 km; $r = 0.75$ = outbred sisters; SF = the same (first) females reintroduced. The figure presents pooled data from three unrelated family lines; a statistical test for homogeneity of slopes of the separate lines yielded no significant difference ($p > 0.05$). Each female was tested separately. A total of 72 trials were performed and each point shown is weighted by the sample size for that value (shown adjacent to each point). *Reproduced from Smith, 1983, with permission*

To test further whether the discriminators produced by females are learned by males, Smith (1983) recorded the waning number of pounces by males during the 10 min exposures of first females to males. The resulting curve (Fig. 4) has the appearance of a typical learning or habituation curve (Alloway, 1972).

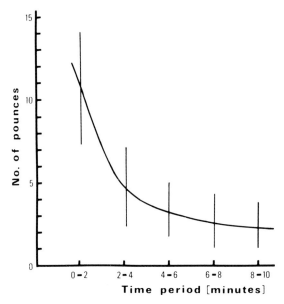

Fig. 4. Response curve for males of *Lasioglossum zephyrum* to a female during the initial 10 min period of exposure. The curve is drawn through the mean for each time period and the vertical lines represent standard deviations of the number of copulation attempts elicited by the female during each of the five successive 2 min intervals (n = 12 for each period).
Reproduced from Smith, 1983, with permission

Mating behavior such as we have described could affect the fitness of males. It would reduce wastage of time and energy by males trying repeatedly to mate with non-receptive females. Such females should be abundant in insect populations whose females mate only once or only a few times, and are especially abundant in eusocial insects whose workers commonly do not mate. Furthermore, the mechanism can presumably evolve to promote whatever level of inbreeding or outbreeding is optimal (Bateson, 1983), assuming that there are coadapted gene complexes. Meager data suggest that males of *L. zephyrum* patrol in the vicinity of the nests from which they emerge. We also believe that females establish new nests near their mothers' nests (Kukuk and Decelles, 1986). Thus populations are probably viscous, but the mating behavior described would tend to reduce close inbreeding. Males may first become familiar with discriminators of closely related females (perhaps nestmates) and thus would be likely to mate readily with more distantly related neighbors or with occasional unrelated transients.

Greenberg's (1982) data support this conjecture. He offered groups of adult males (brothers) in a small cage simultaneous choice between two black spots, behind one of which were hidden two of the males' sisters, and behind the other, two sisters unrelated to the males. The females were placed in position once daily for 15 min and the responses of the males observed. Greenberg reported a significant initial preference by males for their own sisters, but this conclusion is not supported by his Table 1. After several days, however, a preference was shown for non-sisters in all trials. Use of different sisters did not alter this finding. Presumably sisters commonly have similar but not identical odors; the males must generalize their sisters' odors, for they respond similarly to the odors of different sisters.

In field studies (B. H. Smith and W. T. Wcislo, pers. comm.) with females that had been killed by freezing and pinned in a nest aggregation where males were patrolling, there was a significant tendency for females from the same nest (presumably sisters) to be alike in their 'attractiveness' to males of a given population. However, in some cases when females from a particular nest were very attractive to males in one population, their sisters were unattractive, relative to females from other nests, when tested with males in a second population. We do not yet understand the meaning of such results but they do show that mate recognition by males of *L. zephyrum* as demonstrated in the laboratory (Smith, 1983) also occurs in the field, and is probably based on the genetic differences that exist among females. Females extracted in hexane were significantly less attractive than unextracted individuals, indicating, as do laboratory studies, that a pheromone is involved.

Why should males discriminate among potential mates? Would they not increase their fitness by trying to mate with every female encountered? An important factor may be the loss of time and energy that males would experience trying to mate with unreceptive females (e.g. workers, Fig. 5C) that return repeatedly to the same site and would thus be repeatedly pounced upon by the local patrolling males. In addition, there may be a negative effect of mating or of mating attempts on male longevity and consequently a limit to the number of copulations. If this speculation is substantiated, males cannot mate many times and should indeed select the 'best' females as mates rather than mating indiscriminantly. We have no evidence that this occurs, however, except presumably by mating preferentially with females having unfamiliar odors; such females may be on the average less closely related to a male than those having familiar odors, since the latter include the males' sisters and relatives that smell like the sisters.

Fig. 5. Sweat bees, *Lasioglossum zephyrum*. A. Attempted mating in the laboratory, male at right. B. A guard blocking a nest entrance in the field. C. A pollen-laden forager (worker) resting on a bank before entering its nest. D. Two females in their nest burrow in a laboratory observation nest; the lower one is probably the queen

Kin recognition by colonial females

Most studies of kin recognition in *Lasioglossum zephyrum* have dealt with interactions between females at the burrow entrances. In this primitively eusocial species one normally finds 2 to 20 females per nest. Behavioral intermediates between castes are common and some females can be placed as to caste only arbitrarily. One female is the queen, one or a few act as principal guards (Fig. 5B) at the nest entrance, others engage in nest construction and foraging, and a fourth group consists of unspecialized and largely inactive individuals (Brothers and Michener, 1974). The guard bee interacts with every bee attempting to enter the nest and, as has been shown in both field and laboratory observations, the guard will reject certain female bees that attempt to enter. Any female seems to be able to function as a guard, although guarding is the main activity of some. Patterns of agonistic behavior shown by guards at nest entrances are described and illustrated by Bell and Hawkins (1974). The guard can effectively prevent a foreign bee from entering the narrow burrow entrance; if it does get in or is artificially introduced, bees other than the guard often attack it (Bell *et al.*, 1974). In the following pages, the word guard is used for whatever individual was at the nest entrance when observations were made.

Since *L. zephyrum* is only weakly eusocial and is a close relative of solitary species, its social behavior is probably not old in evolutionary terms. The mechansism of kin recognition used in the eusocial context are therefore probably of recent origin and based in pre-existing mechanisms. The most parsimonious explanation of the origin of cues used by guarding bees to discriminate nestmates from foreign bees is that they are the same cues used by males to discriminate among females. Because young females stimulate male mating attempts, Kukuk *et al.* (1977) postulated that the sex pheromone discriminator is produced by females even when very young, serving as a basis for mate selection by males, and that it is different from the discriminator for nestmate recognition by females, which seems to function only after females are about two days old. Subsequent studies, reported below, make it seem probable that the two discriminating systems use the same products of the females. If true, this would explain the apparently life-long production of the sex pheromones by females, unlike many solitary bees which become unattractive to males after mating. According to this view, non-receptive females of *L. zephyrum* (workers and old mated individuals) are stimulating to males because they need the same pheromones in the context of the eusocial colony. The same pheromone system may also be used by both solitary and social bees and wasps for individual nest recognition (references in Hölldobler and Michener, 1980).

Except as otherwise indicated the tests described below were made in the laboratory, using observation nests of the type described by Michener and

Brothers (1971) as modified by Kamm (1974). A bee to be introduced to a guard was placed in a short piece of plastic tubing held in front of and nearly touching the nest entrance. Then the bee in the tubing was prodded from behind with a pipe cleaner or probe, and thus forced toward the guard. Reactions of the guard often began when the intruder was about 1 cm away, but sometimes were not noticeable until near contact. This contrasts to the male's approach to females which may begin at a distance of 10 cm or more.

In terms of our Introduction, the guard is the responding individual. The introduced bee (called intruder for simplicity and in recognition of the field situation where foreign bees intrude rather than being introduced) is the cue bearer or recognized individual.

Bell (1974) and Barrows, Bell and Michener (1975) sought to reduce the possibility of discrimination through cues of environmental origin by using bees. The guards react in the same way toward bees killed by freezing as toward living bees (Bell, 1974; Barrows, Bell and Michener, 1975). Therefore behavior and sound of the intruder are not necessary parts of the recognition cues. Bell (1974) examined possible roles for tactile, auditory and visual stimuli without obtaining positive results. Odors remain as the probable cues for acceptance or rejection, a view supported by the responses of males, probably to the same cues, which can be elicited by impregnated papers and are known to be genetically specified (see above). Detailed evidence that the odors to which guards respond are also discriminators was provided by Greenberg (1979) and is summarized below.

Bell (1974) and Barrows, Bell and Michener (1975) sought to reduce the possibility of discrimination through cues of environmental origin by using identical foods and nest materials. Bell also made experiments designed to determine whether certain environmental factors might influence acceptance. He imprisoned bees in nest entrances of others or in tubes containing feces and soil from another nest, transferred soil among nests, pumped air from one nest into another for 24 h, introduced foreign odors (peppermint), etc. In no case did he observe changes in anticipated behavior (acceptance of most nestmates, rejection of most foreign bees).

The only evidence of environmental influence on recognition is a study by Greenberg (in Buckle and Greenberg, 1981) who set up pairs of kin colonies (all bees in each pair of colonies were probable sisters) and fed one colony in each pair *Typha* pollen, the other a mixture of *Typha* pollen and pollen substitute. Guards reared on the mixture were more aggressive toward bees reared on *Typha* alone than were guards of either type reared on the same food as the introduced bee. Thus there does exist meager evidence of an environmental influence on recognition, although it is trivial compared to the evidence for discriminators.

When introduced into a foreign nest of any age, young adult females, two days old or less, are more readily accepted than older females (Bell *et al.*,

1974; Bell, 1974). Perhaps they have less discriminator than do older bees, or possess inhibitory cues (pheromonal or behavioral) that make guards less aggressive, but nest age is also involved in determining acceptance. It makes good sense that young adult females should be accepted by the colony members even though they might have odors unfamiliar to colony members. The females maturing over a period of months in a nest, although mostly daughters of the queen and sisters of the workers, will not be genetically identical (unless there has been much inbreeding). Some may be daughters of mated workers or of a deceased queen, so would be more different genetically. They must pass through a stage when they can be accepted in spite of having different discriminators from other colony members.

Indeed, many of the studies of *L. zephyrum* were made with artificially constituted semisocial colonies of mostly unrelated bees. A pool of field collected female pupae from many nests was the source of pupae for artificial nests; bees in the resulting non-kin colonies lived together and developed division of labor and castes, much as in normal eusocial colonies consisting of closely related bees.

Whether guards respond to homogeneous mixtures of combined discriminators from several nestmates forming a nest or colony odor, or to individual discriminators independently, has long been in doubt. Kukuk *et al.* (1977) examined this question using binary choice mazes in which both kin and non-kin females were tested as to frequency of entering their own empty nest versus empty tubes, their own versus a foreign empty nest, etc. They found differential responses to empty nests and one might conclude that the odors discovered by Bell (1974) and Barrows, Bell and Michener (1975) permeate the nest. It is difficult to explain the observation by Kukuk *et al.* (1977) that females from non-kin colonies, in contrast to those from kin colonies, tended to enter empty foreign nests faster than their own. However, if there is more fighting in non-kin colonies (B. H. Smith, in preparation), the recognition odors and the fighting may be associated to produce negative reinforcement. This hypothesis has not been tested.

Somewhat surprising in view of the observations of Kukuk *et al.* is the finding by Buckle and Greenberg (1981) that discriminators do not transfer among nestmates. Bees were kept in mixed colonies of six bees, two unrelated groups (X and Y) of three sisters in each colony. When bees (e.g. X bees) from such an X–Y mixed colony were introduced to the guard of a colony composed of their nestmates' sisters (Y bees only), they were rejected at the same rate as when bees from an unmixed X colony were introduced in a Y colony. (Responses to sisters are commonly similar; see below.) Thus the X bees in the mixed colony acquired nothing from their Y nestmates, either directly or indirectly from the nest, that influenced guard behavior at an unmixed nest of Y bees. A similar experiment showed that living with unrelated bees does not decrease the probability of being accepted by ones own sisters in a kin colony. One would expect that if odors permeate the nest, they would also be adsorbed on the bees, so that living with unrelated bees would influence acceptance rates as in *Bombus* (see below), but this is not the case. It may be, however, that the nest odors suggested by the experiments of Kukuk *et al.* (1977) are not the same as the discriminators.

Greenberg (1979) performed the definitive experiment establishing a major genetic component (overriding the other components described above) controlling the discriminators produced by female *L. zephyrum*. He reared several different genetic lines of *L. zephyrum* in a laboratory breeding program; some lines were brother–sister mated. He could introduce females to guard bees whose genealogical relationship by descent to the intruder was known. He made a total of 1,586 introductions to guards of bees never before seen by the guards. There were 14 different coefficients of relationship (r) of intruders to guards, from $r = 0$ (unrelated bees) to $r = 0.88$ (sisters in an inbred line). He found a strong positive correlation between r and the probability that the guard would admit the intruder (Fig. 6). The relation was highly significant even if one excluded cases where guards and intruders were sisters, reared in the same nest (although they had not been together as adults). In all other cases guards and intruders were reared in separate nests. Thus there seems to be no possibility that the positive correlation resulted from environmental influences.

Greenberg's study proved the genetic control of major components in the discriminators used by guards in distinguishing nestmates from non-nestmates. It was similar in this respect to Smith's later (1983) proof of genetic specification of the discriminators used by males in selecting mates. However, while guards accept females, on the average, more readily the closer the relation of the intruder to the colony members, males favor the more distantly related (or at least the less familiar) females.

The discriminators, then, are genetically specified; what about the responses of guards to them? The responding individual learns the cues of its associates and acts accordingly. Each of the colonies used by Greenberg consisted of sisters which were presumably reasonably similar in their discriminators. Yet as noted above, unrelated bees, if put together when young, can form a non-kin colony, i.e. of pseudosisters, and the guards accept them as nestmates, rejecting other bees. Thus close relationship is not necessary for bees to live together and form a eusocial colony. This is one of the best indications that, while the discriminators are genetically specified, they are learned by the responding bee, i.e. the guard.

Further evidence that guards learn odors of their nestmates comes from the study by Kukuk *et al.* (1977). They isolated all bees of non-kin colonies from one another in vials (with food) for one to twelve days. The guard was returned to its nest and after 24 h was tested by introducing nestmates and other isolated bees. The ability of the guard to admit nestmates and reject others varied with the duration of isolation. After one to three days of isolation 90–100% of nestmates were accepted but almost no foreign bees. Near the midpoint of the study about 50% of the nestmates were accepted and on some days, 30% of foreign bees were also accepted. By days eleven and twelve almost no bees were accepted, whether former nestmates or not. These results suggest a progressive loss of recognition of nestmates' odors, but of course isolation may have altered the motivation of the guards.

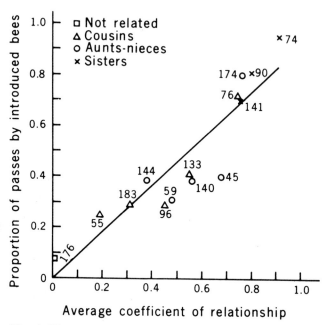

Fig. 6. Linear regression of frequencies with which introduced females of *Lasioglossum zephyrum* passed guarding bees, on the average coefficient of relationship between the bees. Each data point was weighted by the number of interactions for that relationship, shown next to each point. A total of 593 introduced bees was used to obtain the 1,586 interactions shown. If the sister data are excluded, the regression coefficient is still significant at $p < 0.001$. The data points show a better fit to a power function, but there are not enough points to show that it is significantly better. *Reproduced with permission from Greenberg, 1979, Science, **206**, 1095–1097. Copyright, 1979 by the American Association for the Advancement of Science*

Even in Greenberg's colonies consisting of sisters, and therefore relatively homogeneous, a certain number of 'mistakes' were made. About 8% of the bees unrelated to a guard (and its colony) were unexpectedly admitted (Fig. 6). This was not the result of errors—if such an admitted bee was removed and reintroduced, it would usually be accepted again. Some individuals must be odor mimics of others, even if not related. In a subsequent study, perhaps because of using laboratory strains, the percentage of bees unexpectedly admitted to nests rose to 20% or more (Buckle and Greenberg, 1981).

Greenberg's proof of genetic specification of discriminating odors opened the way for a series of further experiments which, like Greenberg's 1979 study, used the failure of guards to distinguish odors of close relatives. Thus guards probably mistake unfamiliar relatives of their nestmates for the nestmates themselves, i.e. they learn their nestmates' odors and then accept foreign bees with similar odors, usually relatives.

Buckle and Greenberg (1981) obtained further information about the learning of discriminators. They made four sets of three colonies each, six bees per colony. Each set consisted of a mixed colony of two groups of sisters (3X and 3Y as described above) and two kin colonies (6X and 6Y). (The letters X and Y represent source nests from different family lines.) Bees from the kin colonies, introduced to the mixed colonies, were accepted at almost exactly the same rate whether the guard was an X or a Y. This rate was much higher than the rate at which X bees from the kin colony were accepted at the Y kin colony. Thus the guard not only recognizes its nestmates but makes no distinction between its own sisters and sisters of its nestmates. It must have learned its nestmates' odors, regardless of whether they were its sisters or not.

The same authors made a similar experiment in which, however, 14 mixed colonies consisted of 5X and 1Y (one with 2X, 1Y). Most of the results were similar; when bees from a Y kin colony were introduced to X guards of the mixed colony, a moderate percentage (56% of 307 trials) were accepted. This shows considerable learning on the part of the X bees from their Y nestmates, the odd bees, even though there was only one in each mixed colony. The remarkable result, however, is that when the odd bees were guards, the rate of acceptance of their own sisters from kin colonies was low. Thus the odd bees accepted their own sisters at a lower rate than did their nestmates ($p < 0.005$). It is not that odd bees rejected all intruders, for they readily accepted their nestmates' sisters.

The odd bees seem to have learned their nestmates' odors, but do not use their own (see also the Discussion, below). Such findings support the idea (first based on the success of mixed colonies) that the learning occurs in the adult stage. If it occurs only at a particular time or context, one might explain the failure of odd bees to learn their own odors. For example, suppose learning occurs only when sitting near or passing other bees. An individual does not sit near or pass itself, and therefore would never learn its own odor.

The data reported above have to do with how females interact at nest entrances. To find out how the discriminators of *L. zephyrum* influence interactions among females within the nest (Fig. 5d), B. H. Smith (in preparation) set up 20 nests each containing six females, among which the within-colony coefficients of relationship varied from zero to 0.82. Ten kin colonies were each composed of sisters (average $r \geq 0.75$) and the remaining ten non-kin colonies of non-sisters (average r mostly 0, all $\ll 0.75$). Observations of activities and interactions were begun after establishment of colonial behavior, starting on the day that the first cell was completed (five to seven days after the first female was placed in the nest). When the genealogical relationship between queen and worker was low (i.e. in non-kin colonies), the workers engaged less often than in kin colonies in the more dangerous activities, such as guarding and foraging, and the interactions were more aggressive. The workers in non-kin colonies spent much more time in the areas of the nest where cells were being constructed and provisioned. This finding could be interpreted as increasing the chances for egg-laying by the workers. (Queens have been observed to destroy eggs laid by workers, Brothers and Michener, 1974.) Evidently relationship-correlated variables such as the discriminators which guards use at the nest entrance influence how a worker interacts with the queen, and in some way influence the activities of the worker even when the queen is not nearby.

Smith (1986) made a similar study of 22 laboratory colonies of five females each, eleven of the colonies consisting of sisters, the other eleven of bees from different and mostly unrelated parental colonies. On the average in the non-kin colonies, in comparison to the kin colonies, the queens were more aggressive toward workers (possibly for this reason there were more worker deaths), there were more workers with enlarged ovaries, and workers passed queens less readily in the nest burrows. Thus the level of relationship among colony members influenced both behavior and physiology. When the bees were unrelated, they acted more like solitary bees, the queen tending to lunge at others and the workers tending to be reproductive. Kin recognition may have been important for the evolution of social behavior in these insects, for the colonies of non-kin do not function as well as the kin colonies.

The origin and nature of discriminators

Various authors (Barrows, Bell and Michener, 1975; Michener, 1982) have suggested that the discriminators in *Lasioglossum zephyrum* as well as in other insects are mixtures of compounds, the relative abundance of which provides the distinctive features. Some known insect pheromones are mixtures of ten or more components (Blum, 1970). If a pheromone contains ten components (presumably coded for by at least ten alleles), and if four relative concentrations of each can be recognized, as seems well within the probable potentialities of bee sensory systems, then 4^{10} distinguishable combinations could exist. This is far more than needed to explain the findings for *L. zephyrum* (see Discussion).

Examining female bees for likely mixtures that might be the discriminators, Smith, Carlson and Frazier (1985) found that Dufour's gland, which is present only in females, contains a series of four macrocyclic lactones,

1. Octadecanolide,
2. Eicosanolide,
3. Docosanolide, and
4. Tetracosanolide,

together with monounsaturated homologues of each. Furthermore, a series of odd-carbon numbered saturated and unsaturated hydrocarbons and isopentenyl docosanoate were found in extracts of Dufour's gland and of whole female bees. Males tended to be more attracted to extracts of Dufour's glands than to extracts of heads, thoraces and abdominal terga, although extracts of these body parts were somewhat attractive.

The four lactones were synthesized. Bioassays were carried out in the field by placing small samples of the extract or of the synthetic compounds either separately or as mixtures on a bit of black velvet pinned 1.5 cm above the ground in a nest aggregation where males were patrolling. The number of approaches by males to within 5 cm of the treated velvet and the number of contacts were recorded. The numbers of contacts and contacts per approach were highest for the natural extract and next highest for mixtures of the four synthetic lactones in natural or other proportions. All these response rates were much higher than for one lactone alone or for other substances.

Although more work is needed, specifically bioassays with guards, it is probable that the lactones function as at least part of the discriminator. However, different females elicited widely different numbers of approaches, contacts, and contacts per approach by males, while male responses to the lactone mixture varied only with respect to the last two (B. H. Smith and W. T. Wcislo, pers. comm.). It is therefore probable that other cues also function in the recognition of females by males.

Further indications of the importance of the lactones as discriminators come from gas chromatographic studies of 'headspace' pheromone samples (washed from vials in which the live bee was kept for a few hours) from individual females of the same populations used by Smith and Wcislo. Principal components analysis of ratios among lactones of individual females showed that pairs of nestmates were more similar than random pairs. In addition, electrophoretic analysis of the same females by R. H. Crozier (in preparation) showed that nestmates were on the average more similar genetically than non-nestmates. These findings may be related to the variation in attractiveness among colonies of females.

KIN RECOGNITION IN *POLISTES* WASPS

Most of the information on kin recognition in wasps is based on temperate climate species of *Polistes*, and has been reviewed by Gamboa, Reeve and Pfennig (1986). Advantages, relative to *Lasioglossum*, for studies of kin recognition in *Polistes* are that:

1. Behavior on the nests is easily seen because the nests are fully exposed (Fig. 7);
2. Observations are easier because the wasps are large; and
3. The wasps are easy to manipulate so that their experiences prior to testing can be known.

Disadvantages are that there is no narrow nest entrance where guarding can be studied, and above all, that no one has yet learned how to reliably mate them in the laboratory, get them through the winter diapause and develop strains or individuals of known genealogical relationships.

Fig. 7. *Polistes fuscatus*. Two foundresses on a nest that is two days old. *Photograph by Dennis J. Hanser, courtesy of George J. Gamboa*

Mate selection by males of *Polistes fuscatus*

Mating in nature occurs away from nests at sites where males have territories or patrol foraging sites (Post and Jeanne, 1983a) and where both sexes congregate. Studies relevant to this topic, however, were carried out in the laboratory. Females (mostly or all young gynes) and males were reared by Larch and Gamboa (1981) from scattered nests (presumably unrelated colonies) and after ten to 30 days of isolation, were placed into separate boxes, one male, one of his sisters and an unrelated female in each box. There were 29 such triplets. Of the various activities recorded, none showed significant differences in frequencies of sibling and non-sibling interactions. Successful matings, however, were more frequent, although not significantly so, between unrelated individuals than between siblings. This observation suggests that in the natural environment there might be a bias toward outbreeding. In a differently designed experiment, however, using cages of ten males and ten gynes and determining mating success by dissections of gynes to look for sperm, Post and Jeanne (1983b) found no evidence of differential mating with siblings.

Negative evidence of recognition does not preclude the possibility that it exists, undemonstrated because of experimental design, wasp motivation, or other factors. Indeed it seems that prior experience of the female influences the outcome of experiments like those of Larch and Gamboa. Ryan and Gamboa (1986) modified the experimental design of Larch and Gamboa by using gynes that had never been exposed to conspecific males. Nestmates had a higher tolerance for one another's proximity (see below) but significantly fewer copulations than non-nestmates. Controls, in which gynes had been with males for two weeks, showed no difference in copulating frequency between nestmates and non-nestmates. Of course if outbreeding is desirable, it makes sense for a young gyne to mate preferentially with an unrelated male but if no such mating occurs, to mate later with any available male.

Kin recognition by females

Most studies of this topic have concerned gynes, i.e. potential queens (nest foundresses). In temperate species of *Polistes* gynes are produced in late summer or autumn. They hibernate, usually in clusters in protected places, and in spring build new nests, often near the nest where they matured the preceding summer, sometimes with two or more foundresses living together on a single nest.

The wasps tend to clump both on nests and in wintering sites, even when there are as few as two or three wasps. This tendency has been exploited in the experiments reported below. A study of clumping patterns during overwintering in *Polistes exclamans* suggests that at the time of clumping in autumn they recognize their nestmates (Allen, Schulze-Kellman and Gamboa, 1982). Putative gynes were taken from nests in autumn and placed in ten overwintering boxes, 20 wasps per box. Five of the boxes (kin boxes) each contained only sisters, five others (mixed boxes) each contained a mixed group of two different families of sisters from nests over 3 km apart. In the kin boxes the 20 wasps usually (65% of observations) formed a single clump. In the mixed boxes the wasps usually (54% of observations) formed two or more clumps. The mixed boxes had a significantly larger number of clumps than the kin boxes.

Although the wasps were not marked, these data suggest that since the wasps commonly overwinter in groups, they frequently associate with nestmates. There is no evidence that overwintering success is greater in a kin group than in a mixed group (although data should be collected on this point). Indeed there are no obvious reasons for the wasps' tendency to sort themselves into kin groups for overwintering. One possible explanation is that, because of aggression prior to overwintering, kin clusters might form more easily than non-kin clusters. Alternatively, kin clusters might perpetuate nestmate recognition so that in spring the overwintered wasps are more likely to reassemble to form multiple-foundress colonies of sisters. One could imagine that individuals passing the winter alone or with non-sisters lose the ability to recognize sisters or be recognized by sisters, and therefore the ability to join in a colony with sisters. Ross and Gamboa (1981), however, found that overwintered gynes of *Polistes metricus* are able to discriminate between sisters and non-sisters even after long isolation (74–99 days). Moreover, Post and Jeanne (1982) showed that nestmates of *P. fuscatus* recognize one another whether or not they overwintered together (see below).

The foundresses of *Polistes* are highly philopatric (Klahn, 1979; Noonan, 1981), showing a strong tendency to nest in the spring at or near the site where they matured the year before. Thus in Klahn's study of 80 gynes of *P. fuscatus* marked on their nests in autumn and seen nesting the following spring, 75 of the spring nests were on the buildings used by the same wasps in the autumn. Gynes that did not return to the previous year's nest site usually nested alone. Those that nested near the previous year's site were commonly joined by other wasps, almost always their sisters. In 25 foundress associations there were, of course, 25 original foundresses; these were joined by 34 marked individuals that were sisters of the wasps they joined, three marked non-sisters, and seven unmarked wasps. Philopatry appears to be important in bringing sisters together after hibernation and thus in promoting establishment of colonies of sisters.

Bornais et al. (1983) tested the tolerance of overwintered foundresses of *P. fuscatus* to sisters in the laboratory, where philopatry could not be involved because the original nestsites were not available to the wasps. For each replication of their experiment, they put three sisters from colony X in a box with three sisters from presumably unrelated colony Y, and recorded wasps resting close together before nesting as well as wasps starting nests. Before nesting, gynes preferentially paired with (i.e. rested close to) their sisters in contrast to forming random associations (p < 0.0001); that is, sisters are more tolerant of one another than of unrelated wasps. Moreover, of 18 foundress associations on nests started in the boxes, 16 consisted of sisters only. These data show that although philopatry may be important in drawing sisters to a particular vicinity, joining at least in the laboratory is mediated by nestmate recognition. An earlier study (Ross and Gamboa, 1981) reached similar conclusions for *P. metricus* although nest founding was not investigated. Interestingly, Pratte (1982) was not able to demonstrate nestmate recognition in *P. gallicus*.

Fig. 8. *Polistes fuscatus*. The triplet experimental design; the two marked adjacent females are sisters, the lone individual is unrelated to the other two. *Photograph by Dennis J. Hanser, courtesy of George J. Gamboa*

Several authors have sought information on the sources of the cues that lead to grouping of sister gynes. Post and Jeanne (1982) demonstrated that with *P. fuscatus* in cages, former nestmates preferentially associated and started nests with one another, regardless of whether they hibernated in clusters of their nestmates or in clusters containing unrelated wasps. Thus when no philopatry is possible, the nest of origin, not the makeup of the overwintering cluster, influences selection of associates in the spring. Shellman and Gamboa (1982) and Pfennig *et al.* (1983) placed additional emphasis on the importance of the natal nest. Both studies were made in autumn with prehibernation gynes in cages each containing a triplet of three wasps, two nestmates and a presumed unrelated non-nestmate. As in the Ross and Gamboa study, data were collected on behavioral interactions as well as on which individuals rested close to one another. For both *P. fuscatus* and *P. carolina* the results were similar. Exposure to the natal nest or nest fragments for as little as an hour or two immediately after emergence from the pupal cocoons was sufficient to cause, at a later time, sisters to show significantly higher tolerance for one another and to spend more time near one another, than non-sisters (Fig. 8). Discrimination was better, however, if gynes were on the nest for longer periods, e.g. five days. Exposure to adult wasps alone had no such effect and wasps removed from their natal nest immediately (within 30 s) after emergence did not discriminate nestmates from non-nestmates. Unrelated gynes of *P. fuscatus* exposed to different fragments of the same foreign nest became pseudosisters, i.e. they were significantly more tolerant of each other than unrelated wasps exposed to different foreign nests.

P. fuscatus workers, like gynes, can recognize sisters by cues learned at the natal nest (Pfennig, Reeve and Shellman, 1983). To modify these cues experimentally, nests of two gynes were exchanged about 20 days before eclosion of the first workers. Those workers (tested as the authors' 'reciprocal triplets') emerging on nests both of which were mixed products of the same two gynes, did not distinguish one another. They were all sisters or pseudosisters. However, when workers from such nests were tested (as the authors' 'non-reciprocal triplets') against workers from another nest, discrimination was evident. In all cases, workers emerging on the nests were not daughters of the gynes then on the nests.

These results show that soon after emergence as adults, wasps learn recognition cues from nests, nest fragments, or the immature stages therein. At the same time they acquire recognition cues. No other explanation could serve for the behavior of pseudosisters. The learning suggests imprinting, for it occurs quickly (in the young adult stage) and persists for 30 days or more for *P. fuscatus* (Pfennig *et al.*, 1983) or through 73–99 days of isolation for *P. metricus* (Ross and Gamboa, 1981). There are no data indicating waning of discriminatory ability, such as occurs with *Lasioglossum zephyrum* when bees are isolated for a week or ten days and guards appear to have forgotten their nestmates. It is likely that the cues are chemical, probably odors, and for simplicity are called odors below. Learning of recognition cues occurs in the presence of nests or nest fragments without adult wasps on them (except for the young learners) and only odors are likely to be transferred from nests to wasps using this experimental method.

As noted above, most tests in the *Polistes* studies have been based on mutual tolerance and on duration of resting in pairs or groups. It is therefore not possible to distinguish a cue bearer from a responding individual; both wasps in an interaction are probably playing both roles. The studies differ in this respect from those of *Lasioglossum zephyrum* where the female is the cue bearer and the male the responding individual, or where the intruder is the cue bearer and the guard the responding individual. The work to date on *Polistes* suggests that these wasps not only learn recognition cues but also acquire at least part of their recognition odors on the nest during a short period after emergence. Yet it seems unlikely that odor cues acquired during a brief period could be retained and remain effective for months.

Gamboa, *et al.* (1986) examined this problem and showed that non-nestmate gynes of *P. fuscatus* reared in the same laboratory environment (same room, food and nest-building materials) were more tolerant of each other than non-nestmate gynes from field nests. This finding supports the idea of an environmental source for the discriminating substances. These authors also believe that there is a genetic component involved, because even the laboratory reared gynes, after being off their nests for some days, preferentially paired with nestmates. The effect was small, however, and one cannot be certain of the uniformity of their nest environments. We tend toward the authors' interpretation but nest odors arising from fungi, bacteria, dead larvae, etc., are always possible and could be different for different nests even in the same room. In any case, recognition odors of environmental origin can overshadow any genetic component since initially the laboratory-reared non-sisters treated one another as sisters. Later, after being away from the nests, perhaps they lose the environmental odors and unmask the endogenous odors, which may mediate kin recognition after months off the nest.

The same authors believe that discriminating odors of environmental origin are partly acquired as brood and partly during the young adult stage. Wasps less than 2 h old already possess such odors. Thus the evidence indicates that a wasp's discriminating odors could consist of:
1. Materials acquired by young adults from the nest or brood.
2. Endogenous materials acquired by brood from food, secretions of adults etc., and thus possibly based in part on genetics of the adults that made the nest and reared the wasp.
3. Substances specified by the genes of the wasp itself (discriminators).

Various combinations of these three possible sources of recognition odors may be responsible for the behavior observed, and the combinations may change with age of the wasp. It is clear that the odors can have both heritable and environmental components, and that the latter have components acquired as brood and components acquired as adults. It is possible that the 'abdominal wagging' described by Gamboa and Dew (1981) involves deposition on the nest of discriminators that mediate some of the behavior described above. For example, part of the endogenous materials described in (2) above may be wiped onto nests from the sternal glands of females (Jeanne, Downing and Post, 1983).

Discrimination between kin and non-kin brood

The studies reported above indicate the importance of nests and brood in providing cues for kin recognition in *Polistes*. It is therefore not surprising that foreign nests and brood are recognized and sometimes destroyed. Klahn and Gamboa (1983) worked in the field with 150 similar nest boxes and marked gynes (foundresses) of *P. fuscatus* whose sister relationships were known. Before emergence of any workers, when gynes were off their nests, boxes and the contained nests (one per box) were switched, so that when the gynes returned, they found different nests than their own. Some switches were between sisters while others involved non-sisters, presumably unrelated individuals. In all 24 sister switches, the new nests were adopted, whereas twelve of 34 non-sisters deserted. Brood destruction (eating by the gyne, mostly of eggs and young larvae) occurred after six of 24 sister switches but 30 of 34 non-sister switches. In sister switches involving distant nests, returning females clearly distinguished a sister's nest from that of an unrelated foundress placed nearby.

These data suggest that the cues for nest or brood discrimination are heritable. Probably they are learned by the gynes. There was wide variation in responses, both in sister and non-sister switches, suggesting marked genetic variance in the cues, the responses, or both. Under natural circumstances ability to recognize and destroy foreign brood may be useful in multiple foundress colonies, in cases of usurpation, and in cases where workers lay eggs (Gervet, 1964). Destruction of eggs by the dominant individual, the queen, has been noted in all these cases. The cues that stimulate brood destruction could be the same substances, derived from brood or nest, that contribute to discrimination among adults.

Recognition of males by males in *Polistes fuscatus*

So far as we know this is the only insect in which male recognition of male nestmates has been reported. Ryan, Forbes and Gamboa (1984) found no evidence that males recognize their brothers. Shellman-Reeve and Gamboa

(1985), however, using slightly different methods and younger wasps, found that in caged triplets (two nestmates and an unrelated male) association of brothers was more frequent than random, and in part of the study such associations were of longer duration than associations of non-brothers. It is hard to say what function an ability of males to discriminate among other males might have. It seems likely that these authors have detected the general ability to recognize nestmates (or siblings) and that it functions for both sexes. Since males spend some time on the nests and participate in some colony activities, they probably need the same abilities as females to recognize and to be recognized by nestmates. Moreover, male *Polistes* sometimes aggregate at mating sites and recognition of brothers might play a role there.

KIN RECOGNITION IN BUMBLE BEES AND VESPINE WASPS

Certain bees and wasps are primitively eusocial but live in relatively large colonies and have marked differences in caste size. Bumble bees exemplify such forms and their social level, between the majority of primitively eusocial bees and the highly eusocial bees, has been documented statistically by Michener (1974). Vespine wasps occupy about the same position.

Free (1958) gave an excellent account of nest and colony defense in bumble bees. Nestmates and foreign bees were discriminated when anesthetized as well as when active, suggesting that discrimination is by odor. Moreover, bees caged for 1–2 h 5 cm above the cells of a foreign colony engendered attacks when returned to their own colony, suggesting that a discriminating substance had been adsorbed into the body surfaces. Such bees, however, were not readily accepted into the foreign colony either; they had neither fully acquired the discriminating cues of that colony nor lost those of their own.

Among vespine wasps, a study of *Vespula maculifrons* indicated that gynes mate preferentially with their nestmates when tested under laboratory conditions (Ross, 1983). Each gyne was given a choice between a sibling and a non-sibling male; 21 out of 29 gynes that mated were inseminated by siblings. This result is strikingly different from that reported above for *Polistes fuscatus*.

Recognition among females of *Dolichovespula maculata* was investigated by Ryan, Cornell and Gamboa (1986). They used triplet techniques similar to those described above for *Polistes* studies. Unlike *Polistes*, even gynes isolated at emergence showed higher mean tolerance to sisters than to non-sisters and spent more time paired with sisters than with non-sisters. Because the gynes, unlike those of *Polistes*, often spend some time, even hours, in their opened cocoons before coming out, they may learn and acquire the nest odor before actual emergence. If this is true, the system could be similar to that of *Polistes*.

DISCUSSION

The discrimination system

Getz (1982) modeled recognition systems perhaps applicable to *L. zephyrum* guards, using three modes of recognition: genotype recognition, foreign-label rejection and habituated-label acceptance (see Chapter 4). These modes are, respectively, acceptance of a female if her labeling genotype is represented in the group (i.e. colony), if all her labeling alleles are represented in the group, or if at least one of her labeling alleles is represented in the group. Making the simplifying assumption of random mating, he found that the recognition system of *L. zephyrum* as described by Greenberg (1979) could be modeled best by foreign-label rejection with alleles (approximately nine) distributed equally over three to four loci. Surprisingly, a point is reached at which an increase in the number of alleles and/or loci coding for the discriminators would decrease the efficiency of the recognition mechanism for exclusion of non-kin. Also, as expected, increasing the size of the colony increased the probability that sisters which had not previously been met would be rejected; increasing the colony size increases the sampling of the possible genotypes within a family group.

Buckle and Greenberg (1981), Gamboa *et al.* (1986), and Linsenmair (Chapter 6) have put forward evidence that an individual bee, wasp or isopod does not use its own odors (discriminators or other odors). Getz, however, believes that the data from the 'odd bee' experiments are not incompatible with the idea that guards use their own discriminators as well as those of their nestmates. Buckle and Greenberg report that in one case out of 14 repetitions of the 'odd bee' experiment, the guarding female who was the only representative in the nest from her family group, accepted all of her own sisters, whom she had never met, as well as sisters of her nestmates. Getz points out that, given the predicted number of alleles and loci from his model, the probability is moderately high (34%) that in one of 14 repetitions, a single recognition allele at each locus would be fixed in all members of a family, resulting in completely homogeneous discriminator genotypes within the family. Buckle and Greenberg's data therefore could be explained if guards, having experienced their own odors, recognize their sisters' discriminators only in the case where, by chance, all members of a family group possess the same discriminators.

Regardless of whether or not guarding females of *L. zephyrum* can use their own odors, it seems that their nestmates obtain more information from the guarding female's discriminators than the guarding female does. There would be no other explanation for the fact that the odd bee's nestmates accepted the odd bee's sisters at a much higher rate than did the odd bee herself.

The influence, if any, of the variance of discriminators in a colony on guards' actions should be studied. For example, would the odd bee respond differently if associated with bees of two or more diverse origins instead of with a group of sisters among whom the variance was presumably low? Colonies of unrelated bees, where each bee is an 'odd bee', have been used in various studies but data on acceptance of other bees do not exist. Mortality in the formation of such colonies (B. H. Smith, in preparation) suggests that there may be an upper limit to the tolerated variance.

Gamboa et al. (1986) believe that in *Polistes fuscatus* the recognition odors of diverse sources are not additive in their effects on tolerance. They therefore suggest a cue-similarity threshold model of recognition; depending on the similarity between the perceived cue and a learned template, a cue bearer is either tolerated (accepted) or not. There is no continuum with the degree of tolerance varying with degree of similarity. In *L. zephyrum* there appear to be all degrees of toleration of intruders by nest guards from rapid acceptance through delayed acceptance, moderate agonism, etc., to total rejection. Unfortunately such data have not been analyzed; the studies have concerned the final result, i.e. whether the intruder was admitted or rejected.

More work needs to be done on the recognition mechanisms and the ways in which discriminators are learned (i.e. habituation, sensitization, etc.). Once the chemistry of the discriminators is better known, experiments can be performed by adding material to females in order to study the effects on acceptance rates. In addition, nests of single females reared in isolation should be established; if females in these nests accept close kin more readily than non-kin, then females must be able to use their own discriminators. Getz (1982) also proposes some future experiments.

Comparative remarks

Since *Lasioglossum* and *Polistes* are only distantly related Hymenoptera, a reasonable hypothesis is that the features in common between these two insects are also characteristic of hundreds of species of other primitively eusocial bees and wasps. The influence of kin recognition on mating may be relevant to thousands of species, solitary as well as social.

In both *L. zephyrum* and *Polistes*, kin recognition appears to involve odors. In *L. zephyrum* genetically specified odors are important. In *Polistes* such odors exist and may be important. Odors of environmental origin appear to be of trivial importance in *L. zephyrum* but are important in *Polistes*. In neither is there evidence that behavior, sound or other means of communication serve for recognition. In both *Lasioglossum* and *Polistes* the cue bearer secretes the cues as discriminators or gets them from the nest environment. In both, the responding individual learns the cues of its nestmates and rejects or accepts accordingly.

Unfortunately, because of the differences in the habits and life histories of the two insects, studies of their kin recognition have followed largely different paths. For lack of information, therefore, many comparisons that would be interesting to make are not possible at this time. In *L. zephyrum* emphasis has been placed on the genetically specified recognition substances. Behaviors used to demonstrate such recognition are mostly mating activities and interactions of nest guards with other females. Environmental components in recognition substances have been demonstrated only once but should be studied further in the light of the *Polistes* data. In *Polistes* a genetic component of recognition substances has only recently been reported; emphasis has been on recognition odors derived from the nest and brood. The extent to which those odors are themselves specified by genotypes of larvae or of the queen requires further study. Kin recognition has been demonstrated to play a role in mating under some conditions; mating under more or less natural circumstances has been little investigated. Behavior used to experimentally demonstrate kin recognition is mostly differential clumping of sisters and unrelated adults in cages. Thus while kin recognition is well established in *Polistes*, the field conditions under which it functions are even more in need of study than in *L. (D.) zephyrum*.

For *Lasioglossum* it is known that caste can be recognized even among unfamiliar and unrelated bees (Breed, Silverman and Bell, 1978). The relation of this finding if any to the kin recognition mechanisms should be investigated; the whole matter of caste recognition should be examined for *Polistes*.

There is a little information for both *L. zephyrum* (Greenberg, 1981) and *Bombus* (Benest, 1976) on interactions among workers at artificial food sources, i.e. away from both mating sites and nests. In both species sisters were not agonistic while non-sisters frequently were. Similar agonism is not seen on flowers and it may be that the responses beside a container of honey-water, for example, are more akin to those that occur in nests—acceptance of nestmates and attack on strangers. No data about agonism at rich food sources have been obtained for *Polistes*.

In conclusion, we have presented a wealth of data showing that primitively eusocial wasps and bees discriminate among kin. They use partially different mechanisms of identifying kin. Differences in behavior between *Lasioglossum* and *Polistes* may be due in part to different conditions during development which exist in natural environments as well as in experiments. It would be interesting, through experimental manipulation, to test the plasticity of some of the mechanisms described for learning of kin odors in bees and wasps, and thus to find out to what extent the differences result from expression of different behavioral potentials of similar systems and to what extent they represent phylogenetic divergence.

ACKNOWLEDGMENTS

C. D. Michener acknowledges the work of his many students, assistants and associates whose publications are cited herein. From the Literature Cited, the names are Barrows, Batra, Bell, Breed, Brothers, Buckle, Decelles, Gamboa, Greenberg, Hawkins, Kamm, Kukuk, Richards, N. Ross, Smith and Sobti. A series of National Science Foundation grants facilitated the work. NSF grant BNS82–00651 made the preparation of this chapter practical and also supported work of B. H. Smith. Other work of B. H. Smith that contributed to preparation of this chapter was supported by the American-Scandinavian Foundation and the German Academic Exchange Service (DAAD).

George J. Gamboa was good enough to read an early version of the manuscript and to help, particularly in preparation of the section on *Polistes*, by providing prepublication versions of several important papers. Michael D. Breed made useful suggestions, particularly for the section on *Lasioglossum*. W. T. Wcislo and D. J. C. Fletcher read the manuscript and offered important suggestions, although they may not agree with everything we have said. Joetta Weaver helped in manuscript preparation and editorial matters.

LITERATURE CITED

Allen, J. L., K. Schulze-Kellman and G. J. Gamboa. 1982. Clumping patterns during overwintering in the paper wasp, *Polistes exclamans*: effect of relatedness. *Journal of the Kansas Entomological Society*, **55**, 97–100.

Alloway, T. M. 1972. Learning and memory in insects. *Annual Review of Entomology*, **17**, 43–56.

Barrows, E. M. 1975a. Mating behavior in halictine bees (Hymenoptera: Halictidae): III. Copulatory behavior and olfactory communication. *Insectes Sociaux*, **22**, 307–331.[1]

Barrows, E. M. 1975b. Individually distinctive odors in an invertebrate. *Behavioral Biology*, **15**, 57–64.

Barrows, E. M. 1976. Mating behavior in halictine bees (Hymenoptera: Halictidae): I. Patrolling and age-specific behavior in males. *Journal of the Kansas Entomological Society*, **49**, 105–119.

Barrows, E. M., W. J. Bell and C. D. Michener. 1975. Individual odor differences and their social functions in insects. *Proceedings of the National Academy of Sciences, USA*, **72**, 2824–2828.

Bateson, P. 1983. Opitmal outbreeding, in Bateson, P. ed., *Mate choice*, Cambridge University Press, Cambridge, England and New York, pp. 257–277.

Batra, S. W. T. 1966. The life cycle and behavior of the primitively social bee *Lasioglossum zephyrum*. *University of Kansas Science Bulletin*, **46**, 359–423.

Bell, W. J. 1974. Recognition of resident and non-resident individuals in intraspecific nest defense of a primitively eusocial halictine bee. *Journal of Comparative Physiology*, **93**, 195–202.

Bell, W. J., M. D. Breed, K. W. Richards and C. D. Michener. 1974. Social, stimulatory and motivational factors involved in intraspecific nest defense of a primitively eusocial halictine bee. *Journal of Comparative Physiology*, **93**, 173–181.

Bell, W. J. and W. A. Hawkins. 1974. Patterns of intraspecific agonistic interactions involved in nest defense of a primitively eusocial halictine bee. *Journal of Comparative Physiology*, **93**, 183–193.
Benest, G. 1976. Relations interspécifiques et intraspécifiques entre butineuses de *Bombus* sp. et d'*Apis mellifica* L. *Apidologie*, **7**, 113–127.
Blum, M. S. 1970. The chemical basis of insect socialilty, in Beroza, M. ed., *Chemicals controlling insect behavior*, Academic Press, New York and London, pp. 61–94.
Bornais, K. M., C. M. Larch, G. J. Gamboa and R. B. Daily. 1983. Nestmate discrimination among laboratory overwintered foundresses of the paper wasp *Polistes fuscatus*. *Canadian Entomologist*, **115**, 655–658.
Breed, M. D., J. M. Silverman and W. J. Bell. 1978. Agonistic behavior, social interactions, and behavioral specialization in a primitively eusocial bee. *Insectes Sociaux*, **26**, 351–364.
Brothers, D. J. and C. D. Michener. 1974. Interactions in colonies of primitively social bees, III. Ethometry of division of labor in *Lasioglossum zephyrum*. *Journal of Comparative Physiology*, **90**, 129–168.
Buckle, G. R. and L. Greenberg. 1981. Nestmate recognition in sweat bees (*Lasioglossum zephyrum*): Does an individual recognize its own odour or only odours of its nestmates? *Animal Behaviour*, **29**, 802–809.
Evans, H. E. and M. J. West-Eberhard. 1970. *The wasps*. University of Michigan Press, Ann Arbor, Michigan.
Free, J. B. 1958. The defense of bumblebee colonies. *Behaviour*, **12**, 233–242.
Gamboa, G. J. and H. E. Dew. 1981. Intracolonial communication by body oscillations in the paper wasp *Polistes metricus*. *Insectes Sociaux*, **28**, 13–26.
Gamboa, G. J., H. K. Reeve, I. D. Ferguson and T. L. Wacker. 1986. Nestmate recognition in social wasps: the origin and acquisition of recognition odours. *Animal Behaviour*, **34**, 685–695.
Gamboa, G. J., H. K. Reeve and D. W. Pfennig. 1986. The evolution and ontogeny of nestmate recognition in social wasps. *Annual Review of Entomology*, **31**, 431–435.
Gervet, J. 1964. Le comportement d'oophagie différentielle chez *Polistes gallicus* L. *Insectes Sociaux*, **11**, 343–382.
Getz, W. M. 1982. An analysis of learned kin recognition in Hymenoptera. *Journal of Theoretical Biology*, **99**, 585–597.
Greenberg, L. 1979. Genetic component of bee odor in kin recognition. *Science*, **206**, 1095–1097.
Greenberg, L. 1981. *The function of heritable odors in kin and nestmate recognition in a sweat bee*, Lasioglossum zephyrum. Ph.D. Dissertation, University of Kansas, Lawrence.
Greenberg, L. 1982. Persistent habituation to female odor by male sweat bees (*Lasioglossum zephyrum*). *Journal of the Kansas Entomological Society*, **55**, 525–531.
Hölldobler, B. and C. D. Michener. 1980. Mechanisms of identification and discrimination in social Hymenoptera, in Markl, H. ed., *Evolution of social behavior: hypotheses and empirical tests*, Dahlem Konferenzen, Verlag Chemie, Weinheim and Deerfield Beach, Florida, pp. 35–58.
Holmes, W. G. and P. W. Sherman. 1983. Kin recognition in animals. *American Scientist*, **71**, 46–55.
Jeanne, R. L., H. A. Downing and D. C. Post. 1983. Morphology and function of sternal glands in polistine wasps. *Zoomorphology*, **103**, 149–164.
Kamm, D. R. 1974. Effects of temperature, day length and number of adults on sizes of cells and offspring in a primitively social bee. *Journal of the Kansas Entomological Society*, **47**, 8–18.

Klahn, J. E. 1979. Philopatric and non-philopatric foundress associations in the social wasp *Polistes fuscatus*. *Behavioral Ecology and Sociobiology*, **5**, 417–424.

Klahn, J. E. and G. J. Gamboa. 1983. Social wasps: discrimination between kin and non-kin brood. *Science*, **221**, 482–484.

Kukuk, P. F., M. D. Breed, A. Sobti and W. J. Bell. 1977. The contributions of kinship and conditioning to nest recognition and colony member recognition in a primitively eusocial bee, *Lasioglossum zephyrum*. *Behavioral Ecology and Sociobiology*, **2**, 319–327.

Kukuk, P. F. and P. Decelles. 1986. Behavioral evidence of population structure in *Lasioglossum zephyrum* (Hymenoptera: Halictidae): female dispersion patterns. *Behavioral Ecology and Sociobiology*, **19**, 233–239.

Lacy, R. C. and P. W. Sherman. 1983. Kin recognition by phenotype matching. *American Naturalist*, **121**, 489–512.

Larch, C. M. and G. J. Gamboa. 1981. Investigation of mating preference for nestmates in the paper wasp *Polistes fuscatus*. *Journal of the Kansas Entomological Society*, **54**, 811–814.

Michener, C. D. 1974. *The social behavior of the bees*. Harvard University Press, Cambridge, Massachusetts.

Michener, C. D. 1982. Early stages in insect social evolution: individual and family odor differences and their functions. *Bulletin of the Entomological Society of America*, **28**, 7–11.

Michener, C. D. and D. J. Brothers. 1971. A simplified observation nest for burrowing bees. *Journal of the Kansas Entomological Society*, **44**, 236–239.

Noonan, K. M. 1981. Individual strategies of inclusive-fitness-maximizing in *Polistes fuscatus* foundresses, in Alexander, R. D. and D. W. Tinkle, eds., *Natural selection and social behavior: research and theory*, Chiron Press, New York, pp. 18–44.

Pfennig, D. W., G. J. Gamboa, H. K. Reeve, J. Shellman-Reeve and I. D. Ferguson. 1983. The mechanism of nestmate discrimination in social wasps (*Polistes*, (Hymenoptera: Vespidae). *Behavioral Ecology and Sociobiology*, **13**, 299–305.

Pfennig, D. W., H. K. Reeve and J. S. Shellman. 1983. Learned component of nestmate discrimination in workers of a social wasp, *Polistes fuscatus*. *Animal Behaviour*, **31**, 412–416.

Post, D. C. and R. L. Jeanne. 1982. Recognition of former nestmates during colony founding in the social wasp *Polistes fuscatus*. *Behavioral Ecology and Sociobiology*, **11**, 283–285.

Post, D. C. and R. L. Jeanne. 1983a. Male reproductive behavior of the social wasp *Polistes fuscatus*. *Zeitschrift für Tierpsychologie*, **62**, 157–171.

Post, D. C. and R. L. Jeanne. 1983b. Relatedness and mate selection in *Polistes fuscatus*. *Animal Behaviour*, **31**, 1260–1261.

Pratte, M. 1982. Relations anterieures et association de fondation chez *Polistes gallicus* L. *Insectes Sociaux*, **29**, 352–357.

Ross, K. G. 1983. Laboratory studies of the mating biology of the eastern yellowjacket, *Vespula maculifrons*. *Journal of the Kansas Entomological Society*, **56**, 523–537.

Ross, N. M. and G. J. Gamboa. 1981. Nestmate discrimination in social wasps (*Polistes metricus*, Hymenoptera: Vespidae). *Behavioral Ecology and Sociobiology*, **9**, 163–165.

Rutter, W. J. 1984. Molecular genetics and individuality, in Fox, S. W. ed., *Individuality and determinism*, Plenum Press, New York, pp. 61–76.

Ryan, R. E., T. J. Cornell and G. J. Gamboa. 1986. Nestmate recognition in the bald-faced hornet, *Dolichovespula maculata*. *Zeitschrift für Tierpsychologie*, **69**, 19–26.

Ryan, R. E., G. C. Forbes and G. J. Gamboa. 1984. Male social wasps fail to recognize their brothers. *Journal of the Kansas Entomological Society*, **57**, 105–110.

Ryan, R. E. and G. J. Gamoba, 1986. Nestmate recognition between males and gynes of the social wasp *Polistes fuscatus*. *Annals of the Entomological Society of America*, **79**, 572–575.

Shellman, J. S. and G. J. Gamboa. 1982. Nestmate discrimination in social wasps: the role of exposure to nest and nestmates (*Polistes fuscatus*, Hymenoptera, Vespidae). *Behavioral Ecology and Sociobiology*, **11**, 51–53.

Shellman-Reeve, J. and G. J. Gamboa, 1985. Male social wasps (*Polistes fuscatus*, Hymenoptera, Vespidae) recognize their male nestmates. *Animal Behaviour*, **33**, 331–333.

Shepherd, G. M. 1983. *Neurobiology*. Oxford University Press, Oxford, England and New York.

Smith, B. H. 1983. Recognition of female kin by male bees through olfactory signals. *Proceedings of the National Academy of Sciences, USA*, **80**, 4551–4553.

Smith, B. H. 1986. Effects of genealogical relationship and colony age on the dominance hierarchy in the primitively eusocial bee *Lasioglossum zephyrum*. *Animal Behaviour*, in press.

Smith, B. H., R. G. Carlson and J. Frazier. 1985. Identification and bioassay of macrocyclic lactone sex pheromone of the halictine bee *Lasioglossum zephyrum*. *Journal of Chemical Ecology*, **11**, 1447–1456.

West-Eberhard, M. J. 1969. The social biology of polistine wasps. *Miscellaneous Publications of the Museum of Zoology, University of Michigan*, **140**, 1–101.

ADDENDUM

For an account of nestmate recognition between males and gynes of *Polistes*, see Ryan and Gamboa, 1986.

Kin Recognition in Animals
Edited by D. J. C. Fletcher and C. D. Michener
© 1987 John Wiley & Sons Ltd

CHAPTER 8

Kin Recognition in Highly Eusocial Insects

MICHAEL D. BREED and BETH BENNETT
Department of Environmental, Population and Organismic Biology, Campus Box 334, The University of Colorado, Boulder, Colorado 80309, USA.

GENERAL CONSIDERATIONS

Mechanisms of kin recognition in the highly eusocial insects are among the most complex in the animal kingdom. Features such as intracolonial competition, multiple mating by queens, multiple queens in colonies and large colony size serve to complicate the evolutionary, genetic and behavioral factors that normally determine the nature of recognition systems. In this chapter we review some of the important features of highly eusocial insects that influence recognition systems, discuss theoretical approaches to predictions concerning the systems, and summarize the available data on kin recognition in the highly eusocial termites, bees and ants.

A highly eusocial insect species is one in which the mother (the queen) and her daughters (workers) (in termites a king and male workers are present) live in a colony in which there is reproductive division of labor and clear morphological and physiological differences among castes (Michener, 1974). All ants (Formicidae) and termites (Isoptera) are highly eusocial, but within the bees (Apoidea) and wasps (Vespidae) there is a gradation from solitary and primitively eusocial to highly eusocial species. Recognition systems in highly eusocial vespids are unstudied. The little known about such systems in the less highly eusocial vespids is summarized along with material on primitively eusocial vespids by Michener and Smith (Chapter 7). In most cases highly eusocial insect colonies survive for several years. In highly eusocial bees and some highly eusocial wasps, colonies are founded by swarms, while in many ant and termite species colonies are founded by lone reproductives. The termites differ from the other groups in having a diplo-diploid mode of sex determination; the hymenopteran social insects are haplo-diploid, a property that leads to special predictions concerning the advantages of serving as a worker in a colony (Hamilton, 1964; Chapter 4).

New workers and reproductive female offspring (gynes) in colonies of highly eusocial insects are normally produced from eggs laid by the queen(s) but male offspring often result from eggs laid by workers. There is considerable interspecific variation in whether the queen or workers lay male-producing eggs (Wilson, 1971). In general, competition for reproduction (dominance hierarchies, supersedures by previously lower ranking individuals) is less overtly apparent than in primitively eusocial insects or is absent.

Recognition systems in highly eusocial insects

Two research themes have emerged in the work on recognition in highly eusocial insects. These are:
1. Testing whether the cues used in recognition have an environmental or endogenous (to the colony) source.
2. Determining if there is a correlation between phenotype and genotype for those cues that are endogenous.

As this review shows, most of the work on these questions has been done on bees and ants; there is a paucity of information on recognition systems in termites.

The major adaptive context of recognition is thought to be the defense of the colony or its resources. In many cases, exclusion of non-nestmate conspecifics is a key component of such defense. In addition, an important focus in the study of recognition systems of ants has been colony specificity of territorial marking or trail pheromones.

It follows, then, that the most commonly studied behavioral aspect of recognition in highly eusocial insects is agonism. An individual perceives the presence of a conspecific, assesses that individual, and either treats that individual as a nestmate (i.e. tolerantly) or attacks it, using whatever tactics are available (e.g. biting, stinging).

There are many other recognition contexts involving within-colony behavior that might not be expressed as overt agonism. These contexts are less well studied; the one that has received the most attention has been preferential rearing of related brood as reproductives. Also possible is differential food flow among workers based on phenotypic or genetic similarity, preferential feeding of certain queens in polygynous situations, genetic segregation during swarming processes (colony multiplication) and choice of virgin queens.

Recognition templates

To discriminate between a nestmate and a non-nestmate a social insect must have information concerning the attributes (identities) of nestmates (template) and the identities of unknown individuals in the current encounter. A 'decision rule' concerning the magnitude of discordance between the template and the encountered individual will be used to determine appropriate subsequent behavior. Two major types of templates have been proposed. In allelic matching models (Crozier and Dix, 1979; Getz, 1982) comparisons are made to determine whether specific alleles are the same in the encountered individual and in self or nestmates. Information is thus, either internally derived (self-knowledge) or learned from surrounding individuals. Second, our probabilistic models rely on the fact that recognition characteristics may be continuously variable; consequently alleles are not represented as discrete states in the phenotype. In probabilistic models the nature of the decision rule is determined by the physical characteristics of the recognition cues, internal parameters such as degree of tolerance for deviation from the template, and the physiology of the sensory system.

THEORETICAL CONSIDERATIONS

Colony structure, reproductive output, and intercolonial competition

Several features of eusociality have a significant impact on predictions concerning recognition systems. Foremost are:
1. That workers will behave in a manner that maintains the genetic integrity of the colony (i.e. prevent intrusion of unrelated reproductives).
2. That workers, while they may not reproduce directly, may influence the genetic nature of the reproductive output of the colony.

The former is particularly directed to the reproductive members of the colony; because the fitness of the worker is entirely or almost entirely invested in the queen's offspring (and, in termites, the king, in most cases), it is critical that there be a high genetic correlation between the worker and its associated reproductive(s). This correlation exists because of familial relationship (a worker is ordinarily the offspring of its queen) and is maintained by the prevention of usurpation, supersedure or cleptoparasitism by less related individuals.

The latter (2) leads to a similar worker viewpoint concerning the reproductive output of the colony; the interests of the worker (fitness) are best served if the gynes and males produced have a high genetic correlation with the worker. The most interesting aspect of reproductive output is that workers may have the opportunity to play an active role in choosing which larvae or nymphs to rear as reproductives; there may be conflict among workers as to which are chosen. Such conflict should be most likely if the queen has mated more than once or if the colony is polygynous. If recognition phenotypes are genetically specified then genetic information available to workers would allow behavioral decisions consistent with kin selection.

Many social insect species occur under conditions in which colonies compete for limited food or other resources. Intercolonial competition may occur at resource sites. Moreover, the nest, food stores or brood may be a resource that requires defense from raiding conspecifics. Characteristics that can be used in discriminating nestmates from non-nestmates may be important in the maintenance of feeding territories and the exclusion of conspecific raiders from the nest.

Models for decision rules in recognition

Getz (1982) and Crozier and Dix (1979) have presented allelic models for decision rules (reviewed in detail in Chapter 4). These models may provide the basis for testing hypotheses concerning the number of alleles and loci involved in determining the nature of discrimination. The following discussion is centered on a number of behavioral issues influencing the recognition system.

One model (Crozier and Dix, 1979), the Individualistic model, requires one shared allele at each locus for acceptance into a colony. Another, the Gestalt model, postulates pheromonal transfer among workers, resulting in a mixture of odors that is unique to the colony. The Gestalt model is parsimonious in that it requires relatively few loci and alleles at each locus for discrimination among neighboring colonies (see Table 1 in Crozier and Dix, 1979). Getz (1982) extended the analysis of the Individualistic model by subdividing ways in which such discriminations might be made. He proposed that discrimination may be made on the basis of complete similarity of genotype ('genotype discrimination'), alleles that are not represented in the colony ('foreign-label rejection'), or common alleles ('habituated-label acceptance').

As Crozier and Dix (1979) point out, these models are limited by the fact that the available evidence leads to hypotheses that the recognition system operates on dosages (quantities of a large number of different pheromones) rather than simple presence or absence (Hölldobler and Michener, 1980). Also possible are systems in which the pheromonal integrity of each individual is maintained, rather than the 'Gestalt' mixing of discriminators.

Alternative models that deal with the dosage and mixing problems are:
1. The average quantity per individual of each component of the recognition pheromone is learned by colony mates and an individual tolerates a certain level of deviation from these averages ('mean template'; in the context of nestmate discrimination it gives a result equivalent to the Gestalt model).
2. A series of pheromonal profiles for different genetic subsets of a colony are learned, perhaps corresponding to different patrilines or different matrilines ('multiple template').

The mean template model has the disadvantage of failing under conditions of high intracolonial genetic diversity, as the colony mean will approach the population mean. It also does not provide information used in possible intracolonial conflicts among genetic groups. The multiple template model fails to take into account genetic variation within patrilines or matrilines. A more realistic conceptualization could combine the two in a 'multiple-mean template'. In this case an animal would perceive that the surrounding animals belong in groups, then group them, and finally calculate a mean template for each group. This deals with the difficulties of the two separate models, but requires a complicated learning process, as individual recognition phenotypes would have to be learned, sorted, and averaged into similiarity groups.

An alternative to the mean template and multiple template models invokes deviations from the assumptions made above. Individuals in a large, genetically diverse, colony may use a 'common feature' template. This is derived either from dietary or metabolic product odors that are ubiquitous in the colony or from pheromones produced by a single individual (e.g. the queen) and that are spread throughout the colony. Under these circumstances only one template needs to be learned; in one case it contains no genetic information, in the other it contains information concerning the maternal contribution to the colony but none concerning intracolonial genetic diversity.

From another point of view the discriminating individual is faced with a dichotomous choice. Either it can use its own characteristics as a template (as in the Individualistic model) or it can form a template based on its perception of the characteristics of the surrounding individuals. If self-characteristics are used in forming the template, then the template provides an accurate measure for decisions that affect inclusive fitness considerations, but is not very useful for defense of a colony resource, such as the nest, that is shared by a group of related, but genetically diverse, individuals. A mean or Gestalt template based on some combination of the phenotypes of surrounding individuals would be quite effective in the context of nest defense but useless in 'differences of opinion' between intracolonial genetic groups over allocation of resources. This is particularly true if the characteristics of the surrounding individuals are derived from the environment, and consequently contain no genetic information. The possibility that an individual uses different templates in different contexts, or a mixture of templates simultaneously, should not be ignored.

One assumption that underlines the foregoing discussion is that recognition cues are learned after adult emergence from self or from surrounding individuals. Learning of surrounding individuals is clearly supported in many species by the available data. Learning of self is difficult, if not impossible, to distinguish from 'recognition alleles' (Holmes and Sherman, 1983). We view recognition alleles as improbable, and have written this discussion based on the assumption that any self-knowledge is gained via a learning mechanism that is similar to that used in learning the cues of surrounding individuals. If a way could be devised to discriminate experimentally between self-learning and recognition alleles, it should be applied to studies of social insects.

Cheating

Cheating a recognition system can occur by a number of mechanisms. One possibility is the mimicry of recognition signals by unrelated individuals; such mimicry might be used by nest parasites or robbers to gain entry into a colony. Indeed, perhaps the most surprising fact is that cheating involving mimicry of nestmate recognition signals (the 'green-beard', Dawkins, 1976) is not more evident.

In most cases templates for recognition of relatives are learned and exploitation of this fact facilitates slavery. Young individuals (recently emerged adults in Hymenoptera) learn the identity of nestmates regardless of whether those nestmates are their relatives or are unrelated individuals who usurped the nest or stole larvae or pupae from another colony.

There are a number of reasons why recognition systems of highly eusocial insects are susceptible to cheating. Foremost among these are the factors making colonies of many species genetically diverse (e.g. polyandry, polygyny). This diversity selects for the use of learned templates, at least in certain contexts. Young individuals that have limited experience with colony mates are in danger of learning the 'wrong' template and consequently expending their labor for an unrelated individual.

TERMITES

Little is known about recognition systems in termites. Thorne (1982) reviews the older literature on this subject. Discrimination of non-nestmates and aggressive responses toward non-nestmates have been observed in a number of species, including *Reticulitermes hesperus* (Pickens, 1934), *Nasutitermes ripperti* (Andrews, 1911), and *Coptotermes acinaciformis* (Howick and Creffield, 1980). Clément (1978, 1982) has documented interspecific aggression in the genus *Reticulitermes*; the same cues may be used in intraspecific encounters among workers from colonies.

THE HONEY BEE

The highly eusocial bees are in the family Apidae. All species members of the genus *Apis* are highly eusocial but the recognition system has been investigated only for the honey bee, *Apis mellifera*. The several hundred species of the tropical subfamily, the Meliponinae, or stingless bees, are highly eusocial, but there has been little recent work on recognition in this group (see Michener, 1974).

Much of our knowledge of recognition mechanisms in highly eusocial insects comes from the honey bee, which has provided an excellent model for understanding the impact of eusociality on recognition mechanisms and vice versa. The honey bee is monogynous, and under normal conditions worker reproduction is completely inhibited, so that nearly all male and all female eggs are laid by the queen. Queens mate repeatedly (Laidlaw and Page, 1984) so that a honey bee colony is a mixture of half- and full-sisters (Page, Kimsey and Laidlaw, 1984). This genetic complexity may affect the learning of recognition templates and also may lead to conflict among genetic groups in the colony.

Discrimination among adults has been hypothesized in the contexts of worker recognition of queens (Boch and Morse, 1974; Ambrose, Morse and Boch, 1979; Boch and Morse, 1981; Breed, 1981), colony defense (nest entrance) (Kalmus and Ribbands, 1952; Ribbands, 1954), segregation of genetic groups during swarming (Getz, Brückner and Parisian, 1982), and recognition of larvae to be reared into queens (Page and Erickson, 1984; Breed, Velthuis and Robinson, 1984).

Boch and Morse (1974, 1979) demonstrated that honey bee queens can be individually discriminated by workers. Breed (1981) extended this finding by showing that the recognition cues of the queens persisted in the absence of environmental odor cues. He further argued that the cues were genotypically correlated, because responses of workers to sister queens were more likely to be identical than those to non-sister queens. This result was later confirmed independently by Boch and Morse (1982), although they previously reported (Boch and Morse, 1981) that artificial odors applied to a queen could modify workers' recognition responses to that queen.

Nest defense is probably the best understood context for recognition in honey bees. Bees active at or near the colony entrance are guards that encounter entering individuals. These guards display a characteristic posture (Ribbands, 1954) and attack—bite, sting or drag from the entrance—bees that are identified by guards by their flight pattern or by their odor. The population of guards at any one time is small (approximately 150 bees in an average midsummer colony) and is composed of bees slightly younger than foragers (M. D. Breed and A. Moore, unpublished). Only a small proportion of the workers in any age cohort become guards (M. D. Breed and A. Moore, unpublished) and they guard for a short period of time (average is less than two days). It is somewhat surprising to us that behavior requiring specialized knowledge of nestmate identity is not more persistent. After serving as guards bees become foragers.

Early investigations of nestmate recognition by workers showed that recognition cues could be modified by or based upon food odors brought into the nest (Kalmus and Ribbands, 1952). Recent laboratory investigations by Breed (1983) and Getz and Smith (1983) have shown, however, that in the laboratory and in the absence of environmental cues bees are still able to make discriminations. The cues used in this circumstance are genotypically correlated. Breed (1983) found that nestmates (full- and half-sisters) could be discriminated from non-nestmates, while Getz and Smith (1983) found that workers reared with only full-sisters could then discriminate full- from half-sisters.

Subsequently Breed, Butler and Stiller (1985) tested the hypothesis that individual bees in groups of mixed parentage would treat preferentially non-group members to which they were genetically related. A group of workers from two parental colonies, A and B, was assembled. A worker from colony A which was not a member of that group was then introduced into the group and behavioral interactions were recorded. The hypothesis was that the A bee would be more attacked by B bees than by A bees. This hypothesis was not supported by the data. Introduced bees were equally likely to be attacked by related and unrelated individuals.

Analysis of feeding interactions between workers in these groups indicated, however, that introduced bees engaged in significantly more feeding interactions with sisters than with non-sisters (Breed, Butler and Stiller, 1985). This result is supported by results from two normally functioning observation colonies (P. Frumhoff, personal communication). In this case two phenotypically color-marked patrilines (each the product of the same queen) were used. He found differential feeding and grooming within patrilines. Because bees in these groups had equal experience with patrilineal and non-patrilineal sisters, this result supports the hypothesis that bees are able to learn and use their own recognition characteristics. Noonan (1986a) found that workers preferentially feed worker larvae of their own patriline.

C. Evers (personal communication) has used a different approach to test the same hypothesis. In queenless colonies workers interact aggressively; this aggression is apparently associated with competition for access to a reproductive role (some workers lay male eggs under this circumstance). She found that workers are more likely to direct their aggression toward half-sisters than full-sisters.

As has previously been pointed out (Breed, 1983), the discovery that worker honey bees produce and use genotypically correlated cues does not negate previous evidence concerning environmentally derived cues. Bees can and do use whatever cues are available. Further experiments will be required to test the relative importance of each type of cue in nestmate recognition.

The preferential feeding of sisters is difficult to interpret in the context of natural selection, because differential feeding interactions among sterile workers probably have little or no effect on the reproductive contribution to the next generation. There are, however, three situations in which workers can influence the genetic contribution to the next generation; these are the determination of which bees leave the colony as a swarm, the choice of larvae to rear into queens, and drone rearing in laying-worker colonies.

Getz, Brückner and Parisian (1982) hypothesized that segregation on the basis of genetic relatedness would take place at the time of swarming. This would be due to the diversity of patrilines in the colony; workers that are half-sisters to the new queen would leave the colony with their mother, while bees that are full-sisters to the new queen would remain with her. This hypothesis was tested by collecting two swarms which had come from colonies of two phenotypically color-marked patrilines. In case A the original colony was 74% cordovan (one of the color genotypes); the swarm was 79% cordovan and the colony remnant was 64%. When only young bees (determined by wing wear) were considered the segregation was slightly better; 80% of the young bees were cordovan in the swarm. Case B followed the same pattern. The segregation of genotypes was statistically significant, but is obviously relatively incomplete. These data do suggest the possibility that individuals within the colony use a self-template or have a special mechanism for learning a template based on their own genetic subset of the colony (the multiple-mean template model).

Page and Erickson (1986a) studied the acceptance of virgin queens by workers. As more than one queen is reared under swarming or emergency queen-rearing conditions, workers may exert choice during the process of queen acceptance. Workers were much more likely to bite virgin queens with which they had a genetic relationship of 0.25 than queens with which the relationship was 0.75. The correlation between bites and coefficient of relatedness was highly significant. The referent for comparison, when the workers had no previous experience with queens, was the other workers in the group. When workers were exposed serially to virgin queens, they used the first queen as the referent. Page and Erickson (1986a) conclude that in natural colonies workers are likely to accept the first virgin queen to emerge, regardless of the patriline membership of that queen. In genotypically mixed colonies they found that the genotype of new queens was independent of the ratio of worker genotypes when brood ratios were held constant. In fact, only 28% of 32 colonies allowed more than one queen to emerge; the remainder were destroyed prior to emergence; this precludes post-emergence selection in most cases.

Discrimination among queen larva phenotypes is probably the most critical situation in which worker behavior could affect the genetic structure of the next generation. Because honey bee colonies produce only one or a few propagules (swarms) annually, the investment in each is large. There is contradictory evidence concerning discrimination of larval genotypes in selection of larvae for queen rearing; Page and Erickson (1984) report positive results while Breed, Velthuis and Robinson (1984) obtained a negative finding in similar experiments. It has been suggested (K. Visscher, personal communication) that the genotypes of eggs or newly emerged larvae can be discriminated while larvae of the age normally used for commercial queen rearing cannot be discriminated. Noonan (1986a) found that workers preferentially feed queen larvae of their own patriline. Larvae were grafted into queen cells in observation hives containing two worker patrilines discriminable to the observer by color phenotypes. Workers showed significant preferences for larvae of their own patriline for total visits to the queen cells and for feeding and maintenance visits. Similar preferences were found in the feeding of worker larvae (Noonan, 1986b). However, Page and Erickson (1986b) showed that under actual emergency queen rearing conditions the frequency of genotypes reared into queens reflected the frequency of larval genotypes in the colony rather than the frequency of worker genotypes. Thus the data indicate that workers can make discriminations, but it is highly questionable whether the preferences that result from those discriminations actually affect the outcome of queen rearing.

Yadava (1970) and Yadava and Smith (1971a,b,c) performed a series of experiments on the factors involved in the release of aggression in workers by queens. Of their studies the most illuminating is one in which they demonstrated that the mandibular glands of queens are the source of substances responsible for releasing aggressive behavior by workers toward queens (Yadava and Smith, 1971b). By removing the glands they eliminated aggressive responses of workers towards queens. Coating certain worker bees with glandular contents from queens elicited aggressive behavior toward them.

Crewe (1982) studied the mandibular gland secretions of queens of different honey bee races. He focused on 8, 9 and 10 carbon keto- and hydroxyacids from the glands of queens of *Apis mellifera mellifera*, *Apis mellifera capensis* and *Apis mellifera adansonii*. The relative percentages of these acids allowed him to cluster the queens in a way that corresponded with the race of the queens. This sort of study was performed on inbred lines, Ka, Yd and Gk, of *Apis mellifera* maintained in North America. He found that Ka bees were easily classified but that Yd and Gk mandibular products resulted in the misclassification of some individuals. Ka bees are the most inbred, so the decreased variation is not surprising. The data support the hypothesis that some recognition cues in honey bees are correlated with genotype. The source of the cues appears to be the mandibular glands and the chemistry of these glands supports the multiple component pheromone hypothesis (Hölldobler and Michener, 1980).

ANTS

Kin recognition in ants—overview

All ants (family Formicidae) are highly eusocial. The ability to discriminate nestmates from non-nestmates is nearly ubiquitous in ants, with only polygynous (multiply-queened) species lacking (to varying degrees) the ability to make such discriminations (Wilson, 1971; Brian, 1983). Nestmate recognition abilities were recognized by early myrmecologists (Wheeler, 1900) but only recently have experimental approaches been applied to understanding these systems.

In ants, nestmate recognition has most often been assayed as agonism in the context of nest defense. Agonism, which is fairly stereotypic in ants, occurs when individuals are recognized as alien to the colony (Wilson, 1971). Biting, seizing and dragging may sometimes result in injury to the attacked individual (Wallis, 1962). More severe attack includes stinging; in the subfamilies without stings gaster flexion occurs (Sudd, 1967). The typical response is that the foreign worker is attacked and rejected or even killed. This pattern is seen in primitive ants such as *Myrmica* spp. (Haskins and Haskins, 1950) as well as in more advanced species such as *Formica fusca* (Wallis, 1962) and *Myrmica rubra* (DeVroey and Pasteels, 1978). Agonism is most frequently seen when foreign individuals attempt to enter a nest. Consequently, experimental designs which introduce unfamiliar and unrelated test individuals at or near the nest entrance are most likely to elicit an aggressive reaction. Most experimental designs, however, utilize neutral arenas and may therefore underestimate the likelihood of an agonistic response to non-nestmates.

Demonstrations of discrimination in situations not involving agonism are rare. *Nothomyrmecia macrops* foragers preferentially associate with nestmates in the laboratory, even though non-nestmates do not elicit aggression (Hölldobler and Taylor, 1983). Although brood is typically accepted when introduced into unrelated colonies, selective brood care, so that related brood receive more attention, has been documented in *Cataglyphis cursor* (Lenoir, Isingrini and Nowbahari, 1982; Lenoir, 1984).

Genetic relationships within colonies

The amount of intracolonial genetic variance may correlate with both the efficacy of kin recognition systems and the type of template used to effect discrimination. In colonies which are relatively homogeneous genetically, individuals are expected to discriminate consistently between nestmates and unrelated individuals. Under this condition, a mean template may apply. In colonies with high levels of genetic diversity, nestmates are not always closely related and nestmate recognition systems are predicted to be less efficient; a multiple template or multiple-mean template may be used in such colonies. Here we review the available data on genetic variation within colonies of ants.

In monogynous colonies (with a single functional reproductive female) workers are either full- or half-siblings. If the queen mates only once, the workers are all full-siblings (average $\bar{r} = 0.75$) but if she mates more than once, they are a mixture of full- and half-siblings ($\bar{r} = 0.5$). In polygynous colonies, if queens are sisters, their female offspring will share almost as many genes by descent as full-siblings ($\bar{r}$ slightly less than 0.75). At the other extreme the offspring of unrelated queens will be minimally if at all related ($\bar{r}$ approaches zero). Queens in some polygynous colonies appear to stay in separate physical locations in the nest (Wilson, 1974), suggesting a nonrandom genetic structure within the colony, as worker cohorts in different parts of the nest might comprise sibships.

Solenopsis invicta, the North American fire ant, typically forms monogynous colonies (Tschinkel and Howard, 1978), although recently, a polygynous form of this species has been reported (Fletcher *et al.*, 1980). Intranidal (within-nest) relatedness in the monogynous type is high, approximately 0.75 between females; electrophoretic data indicate queens are singly inseminated (Ross and Fletcher, 1985). Queens in polygynous nests are also singly inseminated although relatedness among queens and between queens and workers in these nests is not significantly different from zero (Ross and Fletcher, 1985). In Australian populations of the *Rhytidoponera impressa* group studied by Ward (1980, 1983), two colony types commonly occur sympatrically: one is monogynous with fertilized queens, the second is usually polygynous, with laying workers rather than queens. In the monogynous colonies, the queen lays all eggs, both workers and males. Electrophoretic estimates of genetic relatedness are 0.70 between workers and 0.20 between workers and males, agreeing well with predictions of intracolonial relatedness for singly inseminated monogynous Hymenoptera. In contrast, non-laying workers in the polygynous colonies are related by 0.30 and to males by 0.16.

Estimates of b, the regression coefficient of relatedness (Pamilo and Crozier, 1982) obtained from allele frequency data in polygynous colonies range from a high of 0.42 in *Formica sanguinea* (Pamilo and Varvio-Aho, 1979), to 0.33 in *Conomyrma bicolor* (Berkelhamer, 1984), to approximately 0.1 in *Rhytidoponera mayri* (Crozier, Pamilo and Crozier, 1984) and *Myrmecia pilosula* (Craig and Crozier, 1979); all of these are significantly different from zero. Pearson (1983) found that b in some colonies of *Myrmica rubra* did not differ significantly from zero. Although intranidal relatedness in these species varies greatly there have been no studies to determine whether differential behavior occurs among worker groups of differing relatedness.

Workers from neighboring nests in a viscous population may be more highly related than workers from a single polygynous colony. This appears to be the case in *Rhytidoponera mayri* where worker relatedness between neighboring nests averaged 0.054 ± 0.033 (Crozier, Pamilo and Crozier, 1984); this is greater than that found within some nests of *M. rubra* (Pearson, 1983). Although nestmates are not invariably relatives, nestmate recognition systems serve to identify as kin (often pseudokin) individuals normally found in the same colony.

Addition of queens

Queen recruitment, a process whereby newly inseminated gynes are accepted by colonies that are not necessarily their natal colonies, may result in decreased intranidal relatedness among workers in either polygynous or monogynous colonies. It is not known whether queens are recruited in monogynous species of ants; if queen recruitment occurs, worker offspring of different queens will not be related if the queens were not related. In polygynous ants queen recruitment results in dilution of intranidal relatedness if recruitment queens were unrelated (Elmes, 1984). In colonies of the polygynous acacia ant, *Pseudomyrmex venefica*, Janzen (1973) noted that while recruitment of fertilized females is common, only females from the same tree branch are allowed to enter a nest thorn. Queen recruitment has been documented in *Solenopsis invicta* (Tschinkel and Howard, 1978) and *Myrmica* spp. (Elmes, 1984), although it is difficult to determine how long supernumerary queens are retained or whether they become reproductive.

Sources of recognition cues

Individual and environmental sources

Early myrmecologists hypothesized that both environmental and genetic factors are involved in recognition cues. Lubbock (1878) transferred *Formica fusca* pupae to a second *F. fusca* colony. These individuals were returned to their natal colony as adults and accepted. When subsequently related to the colony which reared them they were attacked. The initial acceptance by their genetic sisters seems to indicate an endogenous rather than an environmental odor source, although the hostility later evinced by the foster sisters suggests an acquired component. Fielde (1904), working with *Camponotus*, *Stenamma* and *Crematogaster*, reported that workers separated from their natal colony as immatures (brood) invariably clustered around their own queens, and were accepted by their older siblings after periods of up to two years. Goetsch (1957) maintained split colonies of *Messor* spp. on different diets and found extreme hostility between former nestmates after several months.

Endogenously derived recognition cues could be produced by queen(s) or workers or both. If the queen produces the odor, nestmate recognition will be highly reliable in colonies with one ('common feature' template) or closely related queens. This type of model would also apply in small colonies where all individual recognition phenotypes were learned. If queen(s) produce recognition cues, a relatively small number of recognition signatures must be learned, and few errors in nestmate discrimination are expected.

In *Nothomyrmecia macrops*, perhaps the most primitive living ant (Taylor, 1978), queen-produced cues may be used to label nestmates. Hölldobler and

Taylor (1983) found that workers from a queenless colony eventually fused with a neighboring colony, whereas workers from queenright colonies attempting to enter a foreign nest were attacked.

Haskins and Haskins (1950) split a colony of *Myrmecia tarsata*, one half maintained with the queen, and the other half was queenless. After a separation of four months, the two halves were reunited. Both groups accepted the queen, but the workers were never reconciled, suggesting that the influence of the queen contributes to nestmate recognition in this species. A colony of *M. nigrocincta* was separated in the same way, and upon recombination of the two halves, the queen and her group of workers were killed (Haskins and Haskins, 1950), suggesting either a rapid decay of memory or a temporal change in recognition cues.

Queenright colonies of *Odontomachus bauri* reject alien conspecifics, whereas colonies maintained in the laboratory without a queen eventually accept non-nestmates (Jaffé and Marcuse, 1983). Jaffé and Marcuse (1983) suggested the state of colony reproduction (i.e. amount or type of brood) might influence aggression toward non-nestmates in this species, although a simpler explanation might be that queen-produced cues are used to label nestmates.

Schneirla (1958, 1971) attributed recognition cues to the queen in many of the surface living Dorylines (army ants) he studied. Queenless colonies of *Eciton*, *Aenictus* and *Neivamyrmex* fuse with conspecific queenright colonies after 16 or more hours without a queen. The queen's retinue is most resistant to loss of discriminatory ability; unlike most other workers in the colony, they will continue to reject foreign workers for a long period after loss of the queen. Workers of these species accept foreign individuals which have been exposed to the odor of their own queen but reject those recently in contact with an alien queen. Workers appear to lose the odor(s) produced by the queen shortly after she is removed, thus allowing their acceptance by other colonies.

Carlin and Hölldobler (1983) formed mixed species nests consisting of workers of several *Camponotus* species reared together from the brood stage. As adults, these individuals rejected unfamiliar genetic sisters and accepted ants of the species with which they had been reared. These results were interpreted as indicating that the queen is the source of nestmate recognition cues.

Further experimentation (Carlin and Hölldobler, 1986) revealed that sources of nestmate recognition cues in the *Camponotus* species studied are hierarchical with queens providing the most important label. Groups of unrelated workers were maintained in separate nests and alternately exposed to the same queen. Under these rearing conditions non-kin are not significantly more aggressive toward one another than related workers. Further, this effect of queen-produced cues held even when the queen and workers were of different species. In queenright colonies, changing dietary cues had no effect on acceptance of unfamiliar sisters. Sisters maintained with different queens responded aggressively to one another whereas non-sisters exposed separately to the same queen were not more aggressive toward one another than sisters reared together.

Workers removed from their natal colonies and subsequently returned often elicit quantitatively different behavioral responses than workers which have not been removed. This difference could be due to loss of queen-produced cues (although decay of cues produced by other sources is also possible), and these studies are reviewed here.

Formica fusca workers removed from their nests and returned after varying lengths of time (>1 min–24 h) were licked significantly more often than control workers ($p < 0.001$); licking is a behavior found in agonistic contexts (Wallis, 1962). Non-nestmates elicited the greatest number of aggressive responses, separated workers an intermediate number, while ants within the nest elicited virtually no aggression from nestmates. DeVroey and Pasteels (1978) separated *Myrmica rubra* workers for two years from their nestmates. Upon their return, separated nestmates were licked more frequently than either nestmates or non-nestmates. Non-nestmates were more likely to elicit an agonistic response such as seizure or dragging than were separated nestmates. LeMoli and Parmigiani (1982) separated workers of *F. lugubris* from their natal colonies and after a week returned them either to their own colony or a distant one (separated by a mountain valley). Intracolonial introductions in the field resulted in significantly less antennal investigation than did heterocolonial introductions ($p < 0.002$, $n = 48$). Similar results were obtained in pairwise encounters staged in the laboratory ($p < 0.03$, $n = 55$ pairs).

Fielde (1904), using single field collected colonies of *Camponotus pennsylvanicus* and *Stenamma fulvum*, separated brood from the queens; when reunited later in life, the offspring invariably recognized their queens; they clustered around and groomed their mothers. Foreign queens were attacked. Worker offspring of the same queen sometimes rejected older siblings, although the older workers never failed to recognize their younger sisters. Fielde (1904, 1905) interpreted these results to mean that the characteristic odor of workers changed throughout their lives such that older individuals remember the colony-specific odor typical of younger workers whereas younger workers had no way of knowing the odor of their older siblings unless previously exposed to it. On the other hand, the odor of the queen did not appear to change over her lifespan as evinced by the success of reintroductions of the queen to offspring separated from her at the brood stage. If worker-produced cues do change over time, queen-produced cues would provide long-term colony-specificity.

Mintzer (1982) controlled environmental variation among colonies of *Pseudomyrmex ferruginea*, a species that inhabits thorns of certain species of *Acacia*, by providing nest thorns cloned from a single acacia. Workers were introduced near nest entrances; foreign conspecifics were rapidly rejected and usually killed. Worker brood from single queens was separated into four groups, two fostered to unrelated gynes and two reared without a queen. Siblings reared apart were reintroduced (two nestmates presented with one non-nestmate) on unoccupied *Acacia* branches one month after eclosion. Aggressive interactions between sisters were rare. Hostility between unrelated ants from both queenless and queenright treatments was much higher. These results suggest an endogenous source for recognition cues which do not originate with the queen.

Workers of *Pseudomyrmex ferruginea* from inbred lines were introduced to one another on nest thorns (Mintzer and Vinson, 1985). Workers from these colonies shared a substantial amount of genetic variation; however, in reciprocal transfers, different rates of rejection were noted, even though the genetic relationship between them was symmetrical. Such variation supports an Individualistic rather than a Gestalt model (Crozier and Dix, 1979) with multiple alleles coding for recognition labels (Mintzer and Vinson, 1985).

Wallis (1962), working with *Formica fusca*, separated ants from the same natal colony and maintained them in separate nest boxes, one queenless, one queenright. He found no aggression between workers when they were reintroduced and concluded that the queen did not contribute to worker recognition. It was not mentioned whether these workers were brood at the time of separation, or if they had in fact been exposed as adults to the odor of their queen. Such exposure might invalidate the conclusion that queen-produced cues were not used, as memory of this information may be persistent. In order to test this conclusion fully it would be necessary to test sister workers which were separated in the brood stage and reared with different queens as well as in a queenless group.

Worker pupae of *Camponotus vagus* were separated from their natal colonies just prior to eclosion (Morel, 1982). They were removed from their cocoons, reared in isolation, and subsequently introduced to adult workers from their natal colony. Both the young workers and their older sisters sometimes exhibited agonistic responses to one another. Morel suggests these results indicate the necessity of contact with older workers in order to establish 'normal behavior'. Alternatively, if cues associated with nestmates are not learned early in life, all conspecifics may be treated as non-nestmates.

Workers of *Rhytidoponera metallica* removed from their natal colony in the brood stage and returned as adults after approximately six months were unequivocally accepted (Haskins and Haskins, 1979). Workers maintained in the laboratory for four years were tested with workers from the same field site (presumably the same colony), and again there was complete compatibility. These results suggest that recognition cues in this species are derived from the workers themselves, as the laboratory colonies were queenless, and further, that recognition cues do not change over time.

Worker produced cues are not as important as cues derived from queens in *Camponotus* spp. (Carlin and Hölldobler, 1986). In experimental colonies —queenless groups of workers or mixed nests in which queens and workers were unrelated—information correlated with the worker genotype affected nestmate recognition. Sisters maintained with different, unrelated queens ($n = 210$) were less aggressive toward one another than unrelated workers ($n = 50$) reared with different queens ($p < 0.001$). In queenless groups fed the same diet, unrelated workers ($n = 223$) were significantly more aggressive toward one another than were sisters ($n = 227$) ($p < 0.05$).

Stuart (1985; personal communication) cross fostered pupae to create mixed colonies of *Leptothorax ambiguus* and *L. longispinosus* to investigate the sources of cues used in nestmate recognition. Individual workers reared in isolation were nearly always accepted by their natal colony (2.3% attacked, $n = 44$) whereas isolates introduced to foreign nests were attacked 84.1% of the time ($n = 44$). This indicates that workers produce colony specific cues. When workers (adoptees) reared by heterospecifics were returned to their natal nests, initial attack rates were high (56.5%, $n = 23$, not significantly different from individuals introduced to unrelated workers). Further observation, however, indicated that a larger proportion of these adoptees were later accepted by their sisters (60.9%, compared to 21.7% introduced to unrelated colonies, $p < 0.05$) and fewer were killed (8.7% versus 69.6%, $p < 0.001$). Related workers from the parental colony were introduced to the mixed group containing their unfamiliar sibling; the response of the mixed group was similar to that seen when the cross-fostered individual was returned to its natal nest except that the initial attack rate was lower. Workers are clearly capable of recognizing unfamiliar relatives; however, the cues are altered somewhat by information, presumably chemical, transferred from other, unrelated workers.

The efficacy of the label transfer was elucidated in a series of reciprocal cross-fosterings between paired colonies. In introductions of workers of both lineages (to their natal nest and the natal nest of their foster siblings), initial attack rate was slightly higher toward the unrelated foster sibling ($p < 0.05$, $n = 52$). Upon further observation, however, it was found that there were no significant differences in the proportions of true siblings versus foster siblings adopted or killed. These findings indicate that *Leptothorax* spp. are utilizing a Gestalt (Crozier and Dix, 1979) type of recognition label. The presence or absence of a queen did not affect the outcome of the experiments described here. Changes in the ratio of adoptee to resident ants strongly influenced the outcome of later trials. Thus in mixed colonies with few workers from a given parental colony, when those individuals are returned to their natal nest, they were more likely to be rejected ($p < 0.01$). This result further supports a Gestalt mechanism. Similar variations in the ratios of mixed colonies of *Camponotus* spp. had no effect on acceptance by unfamiliar siblings (Carlin and Hölldobler, 1986), indicating that the possibility of label transfer among workers varies among ant species.

Environmental influences on nestmate recognition have not been well studied. In the cases reviewed here, these effects are variable and rarely override genetic similarity. Wallis (1962) separated groups of *Formica fusca* workers from the same colony and maintained them on different diets for seven months. Slightly elevated levels of aggression were directed towards sisters eating different foods, but significantly higher levels of aggression were consistently directed toward non-nestmates ($p < 0.001$, $n = 16$ transfers). Elevated levels of aggresssion towards both sisters and non-nestmates were noted under conditions of increased temperatures (20–22 °C compared to 17–19 °C) and food deprivation. These results are important, but should be considered preliminary due to the small sample size: a single colony was separated and twelve to 16 test introductions made between groups.

Jutsum, Saunders and Cherrett (1979) manipulated factors affecting nestmate recognition in the leaf cutting ant, *Acromyrmex octospinosus*, in both laboratory and field studies. Workers from unrelated laboratory colonies were introduced to one another in a neutral arena. Between colonies maintained on identical diets, no intercolonial aggression was seen whereas former nestmates from split colonies maintained on different diets demonstrated high levels of investigation and non-injurious aggression. In field introductions, individuals with booty were placed at foreign nest entrances. These ants were less likely to elicit aggressive responses than were workers without vegetation. Wilson and Hölldobler split two laboratory colonies of *Atta cephalotes*, also a leaf cutting ant, and maintained each half on different forage. After six months, within-treatment introductions did not differ from between diet transfers (E. O. Wilson, personal communication).

Mintzer (1982) controlled for environmental variation, feeding all colonies on the same diet and providing nest thorns from a single *Acacia* individual. Nonetheless, he found uniformly high rejection rates for foreign workers of *Pseudomyrmex ferruginea*. Similarly, Jaffé and Sanchez (1984) maintained distinct colonies of *Camponotus rufipes* under identical conditions in the laboratory, and found no diminution of intercolonial hostility.

Diet alters recognition cues in *Camponotus* spp. only if queen-produced information is not available (Carlin and Hölldobler, 1986). Separated sisters maintained with different queens were equally aggressive toward one another regardless of whether they received the same or different diets. In queenless groups of separated sisters, different dietary treatments significantly ($p < 0.05$, $n = 228$) increased moderate aggression. The same treatment did not significantly increase aggression (which was already high; $n = 225$) between queenless groups of non-sisters. Further, in queenless groups of sisters from different natal nests maintained on the same diet, there was no significant reduction in aggression in between group transfers ($n = 223$).

These results indicate that while workers are capable of perceiving environmental effects on recognition cues, these effects are not as important as the endogenous information generated by the queen and the workers themselves.

Hangartner, Reichson and Wilson (1970) maintained laboratory colonies of *Pogonomyrmex badius* on identical diets. Groups of 30 workers were offered vials of their own nest material, sterile sand, and nest material from another colony. In 45 of 56 choice tests, ants preferred their own nest material ($p < 0.001$), while in three cases the foreign nest material was preferred ($p < 0.001$). A similar result was obtained by Hubbard (1974) who tested individual *Solenopsis invicta* workers for preference to nest materials. They were given a choice among nest material from their own, another colony, or unnested soil. Significantly more ants chose to dig in soil from their own nest ($p < 0.001$, $n = 40$; 60 ants), indicating an ability to recognize this substrate. No preference was shown between the soil from a foreign nest and that which had never been used as nest material. As these colonies were fed identically and were provided with sterilized sand for nesting, it is highly unlikely that colony-specific labels moved from nest materials to the ants. Rather recognition cues from the ants probably marked their nesting materials.

Jaisson (1980b) has documented the importance of early exposure to environmental odors. *Camponotus vagus* and *Formica polyctena* exposed to thyme, *Thymus vulgaris*, for 30 days and three weeks respectively, subsequently preferred nest tubes containing the herb to those without it. Although these results were interpreted to indicate a mechanism for the evolution of plant–ant interations, it is also reasonable to infer that environmentally acquired components of a colony-specific odor may be learned as readily as endogenous cues.

Winterbottom (1981) cross-fostered brood of *Myrmica rubra* between laboratory colonies. Workers from fostering colonies which were removed at the time foreign brood was added were returned after the eclosion of the unrelated adults. The latter individuals were singled out and attacked by the unfamiliar workers when they were presented with both fostered workers and unfamiliar siblings. Further, when returned to their natal colony, the fostered ants were attacked. These findings suggest that a genotypically specified cue is overlaid by an acquired odor and that workers are capable of distinguishing between the two.

Tissues and parts of the body

The chemistry and source(s) of recognition labels are relatively unknown. The application of analytical techniques such as gas chromatography and mass spectroscopy to this question may determine the chemical structure of the cues labeling nestmates. Here, we review experimental evidence regarding the anatomical source(s) of recognition cues as well as data on the occurrence of endogenous and acquired labels. A variety of early experiments involved rubbing workers with body fluids from foreign individuals or immersing them in solvents containing alien ants. Subsequently, these treated ants were returned to their natal nests and the response of their nestmates observed (Fielde, 1904; Goetsch, 1957). More sophisticated approaches involve dissection, bioassay, and gas chromatography of body parts (Jaffé, 1983; Jaffé and Marcuse, 1983; Winterbottom, 1981).

Individual workers of *Neoponera apicalis* are attacked when introduced to strange colonies (Fresneau, 1980). When treated with an extract from the bodies of workers to which they were introduced, such treated ants were accepted. When returned to their natal colonies, unless rinsed again with washes from nestmates, they were attacked by their sisters. These results are similar to those obtained by Goetsch (1957) and Fielde (1904). Goetsch coated ants of unspecified *Messor* species with the body fluids from non-nestmate conspecifics. Upon introduction to the non-nestmate colony, these workers were accepted, although as the odor presumably wore off, they were subjected to increasing hostility. The source of 'body fluids' was not reported, nor were any species differences in this response. Fielde (1904) submerged workers of an unspecified species in distilled water with individuals from another colony; when returned to their own colony, they were attacked. If submerged singly, the same workers were not attacked upon their return. Her results suggest that nestmates are labeled by a surface chemical of low volatility that may be transferred between individuals.

Jaffé and his co-workers (Jaffé, Bazire-Bénazet and Howse, 1979; Jaffé, 1983) experimented with *Atta cephalotes*, *A. sexdens*, and *Acromyrmex octospinosus* to attempt to localize the source of cues utilized in nestmate recognition. Laboratory colonies of these species defend both nests and foraging areas against conspecific and heterospecific ants but do not act aggressively when away from these home range areas. *Atta cephalotes* workers are recognized as non-nestmates only if they are alive and unanesthetized. Dead or carbon dioxide anesthetized foreign conspecifics are ignored, as are nestmates treated in the same manner. Dead or anaesthetized *A. sexdens* and *Acromyrmex octospinosus* workers are rapidly removed from the foraging area of conspecific colonies. Different parts of the bodies of foreign *A. cephalotes* workers elicit variable aggressive responses from workers to which they are introduced. Headless ants are ignored longer than either thoraces with the heads attached, or entire bodies, suggesting that a recognition cue originates in the head. Gas chromatograms of worker heads revealed that the relative proportions of 4-methyl-3-heptanone and 2-heptanone remain fairly constant within a colony but vary between colonies when environmental variation is controlled. Jaffé (1983) suggested that these volatile compounds probably originate in the mandibular glands and are spread over the cuticle, resulting in colony-specific recognition factors. The relative constancy of intracolonial volatile profiles supports a mean template model with dosage of the two main compounds determining colony specificity.

Nestmate discrimination in *Odontomachus bauri* may be based, in part, on volatile chemical cues originating from workers (Jaffé and Marcuse, 1983). Individuals from one colony are ordinarily attacked by those from another. Agonistic responses are initiated more frequently by resident ants when foreign workers are introduced into the nest area of *O. bauri* ($p < 0.01$, $n = 21$) even if the petri dish containing the introduced ant had previously been placed in that individual's colony ($p < 0.05$, $n = 12$). When the same introductions were performed away from the nest entrance, attacks were initiated with equal frequency by both introduced and resident ants ($n = 14$). All body parts of ants dissected alive under anesthetic as well as after death released aggression although gas chromatographic analysis of body parts revealed volatiles in only heads and gasters. Washing with methylene chloride removed the chemical cues mediating nestmate recognition: all washed ants were accepted, regardless of their natal colony, indicating a surface chemical label mediating recognition. Gas chromatography on samples of 50 workers per colony revealed that colony specificity in the volatile substances is achieved by variations in the relative proportions of components. Analysis of variance of relative quantities of volatiles from heads and gasters from five colonies showed significant intercolonial differences among their chromatograms ($p < 0.001$).

Workers of *Camponotus rufipes* are able to discriminate between nestmates and foreign conspecifics; the latter elicit alarm behavior ($p < 0.05$) and biting ($p < 0.05$) significantly more often than do nestmates (Jaffé and Sanchez, 1984). Presentation of dissected body parts indicated that cues resulting in attack of non-nestmates resided in the head. Two plastic dummies were coated with mandibular gland secretions: one from a nestmate, the other from a worker from a different colony. Both elicited alarm behavior (due to alarm pheromones in the mandibular gland secretions), whereas the dummy treated with non-nestmates extract was attacked significantly more often ($p < 0.01$, $n = 15$). Jaffé and Sanchez (1984) suggest a colony specificity in the alarm pheromone which allows for nestmate discrimination, although it is possible that another set of compounds also in the mandibular gland secretion might mediate nestmate recognition.

Winterbottom (1981) found that contents of isolated glands of the poison apparatus presented on filter paper elicited increased activity levels in foreign nests of *Myrmica scabrinodis*. This treatment did not result in aggression directed toward the filter paper. When workers of this species were smeared with Dufour's gland material from *M. rubra* and returned to their colony, significantly more aggression occurred ($p < 0.001$) than if they were so treated with glandular contents from nestmates. Similar results occurred when the poison gland secretion was used; in both treatments the level of aggression (assessed by the number of threat gestures, seizing, and sting attempts) was not significantly different from that directed toward heterospecific ants.

Gas chromatographic analysis of Dufour's glands showed that the product of *M. rubra* was strikingly different from that of *M. scabrinodis* and *M. sabuleti*, which had very similar profiles (Winterbottom, 1981). The most abundant compounds in the Dufour's gland of the latter two species were sesquiterpenes. *M. rubra* workers were treated with these compounds (obtained through preparative chromatography) and returned to their colonies. There was an initially significant ($p < 0.02$) increase in aggression toward these workers, which decreased after two days to non-significant differences. No attempt was made to characterize the source of colony level recognition, although mandibular gland extracts of all three species produced similar chromatograms, making it seem unlikely that mandibular gland products contribute to nestmate recognition.

Effects of polygyny and polydomy

Species which form large, polygynous colonies have been described by Wilson (1971) as unicolonial. This term refers to the domination by a single colony (often polygynous) of a given area, as opposed to the more typical sort of formicid population structure (multicolonial) in which many discrete (typically monogynous) colonies partition a habitat. Unicolonial species may occupy a single large nest, or many distinct adjacent nests, in which case they are said to be polydomous. Polydomous colonies may be either monogynous or polygynous. Nestmate recognition systems appear poorly developed or lacking in polygynous polydomous species, perhaps due to an increase in within-colony genetic variation.

The Pharoah's ant, *Monomorium pharaonis*, exemplifies a highly polygynous, polydomous lifestyle. Boundaries are indistinct; nests connected by odor trails readily exchange workers, and colony size is enormous; millions of workers and thousands of queens are not unusual (Wilson, 1971). Conspecific workers are readily accepted between nests in the laboratory (Petersen-Braun, 1982); however, he does not indicate whether these workers were derived from neighboring nests in the field.

Traniello (1982) studied a population of *Amblyopone pallipes*, a primitive North American ponerine. Small daughter nests (range one to 32 workers) which are typically polygynous (one to 6 queens) comprise a large polygynous polydomous population. Laboratory transfers between nests showed that within a population, which may cover a large geographic area and include many polydomous arrays, workers from a given nest do not react agonistically to workers from another nest. In transfers between populations, agonism between workers resulted.

The subterranean habit of *Stigmatomma pallipes*, also a member of the Amblyoponini, renders its behavior difficult to study under natural conditions. Creighton (1950) suggested that *S. pallipes* may form both large, unicolonial populations and small multicolonial populations. This hypothesis was supported by Haskins (1928) who studied this species in artificial nests. He found that workers from extensive colonies were less likely to attack foreign conspecifics than individuals from small isolated nests.

A population structure similar to that of *Amblyopone pallipes* is found in the Australian ponerine, *Rhytidoponera metallica*. This species forms diffuse nests in leaf litter and the upper soil layers (Haskins and Haskins, 1979). Another feature of the biology of *R. metallica* that makes it an interesting subject for study of nestmate recognition is that eggs are laid mainly by fertilized workers, as true queens are rare (Haskins and Whelden, 1965). Workers of *R. metallica* from the same population typically accept one another, workers from nearby sites may experience brief hostility, and workers from distant populations are aggressively rejected (Haskins and Haskins, 1979).

Both monogynous (*Pseudomyrmex ferruginea*) and polygynous (*P. venefica*) species of *Acacia* ant have been observed by Janzen (1973). Polygynous colonies monopolize entire stands of suitable habitat, while monogynous colonies are typically limited to single trees. Populations of polygynous colonies may number in the millions, while monogynous colonies have a maximum of 20,000–30,000 workers. Workers from monogynous colonies are very aggressive toward workers from other conspecific colonies (Mintzer, 1982); intercolonial aggression is never seen in the polygynous species, even when workers come from distant colonies.

Halliday (1983) analyzed the genetic structure of colonies of the polydomous blue morph of the Australian meat ant, *Iridomyrmex purpureus*. Trails connect the polygynous, polydomous colonies to form a supercolony so that there is worker exchange among the constituent nests. Aggressive displays at supercolony boundaries appear to limit worker traffic between supercolonies, suggesting that recognition of workers at the supercolony level occurs. Worker introductions within the same supercolony result in acceptances while between supercolonies aggressive responses occur. Within a single polydomous colony workers tend to return to the same nest entrance at night, suggesting that the component colonies retain their identities. Electrophoretic analyses of workers at specific locations within nests showed significant genetic heterogeneity indicating both that functional polygyny existed and that within the polydomous colonies workers did not assort by genotype.

European red wood ants, *Formica polyctena*, form large, polydomous colonies comprised of many daughter nests; worker exchange between nests within colonies is common and amicable (Chauvin, 1971). Mabelis (1979) observed wars between neighboring polydomous colonies indicating that colony level recognition occurred. He hypothesized that the purpose of these wars was cannibalism. This hypothesis was confirmed by Driessen, Van Raalte and DeBruyn (1984) who saw over 12,000 ants killed in a single raid!

B. Bennett (in preparation) has studied a group of sibling species in the North American *Formica fusca* complex. Nestmates are reliably discriminated and aliens attacked in a monogynous sibling species pair (*F. subaenescens, F. marcida*). In the polygynous *F. podzolica*, non-nestmate conspecifics are accepted more readily if the distance between nests is small. The probability of attack is significantly greater for workers from different supercolonies ($p < 0.001$, $n = 65$) than within supercolonies where a supercolony was defined as a habitat island, averaging 10 m in diameter.

In field transfers of workers between colonies of *Acromyrmex octospinosus*, a significantly higher number of workers from widely separated nests (>30 km) was killed than of non-nestmates from nearby nests ($p < 0.001$, Jutsum, Saunders and Cherrett, 1979). Although the population structure was not described, this result suggests a unicolonial situation with polygynous colonies. The Texas leaf cutting ant, *Atta texana*, forms extensive polydomous colonies (Echols, 1966). Exchange of workers, queens and brood within a single field collected colony was noted when laboratory daughter colonies were connected with plastic tubing. Unfortunately, between-colony transfers were not done, so it is not possible to determine whether this species fits the unicolonial pattern of inter-population discrimination.

Territorial markings

Colony territories may be marked chemically; this behavior is more typical of monogynous than polygynous colonies (Bradshaw and Howse, 1984; but see Halliday, 1983). Some ant species are capable of generating colony specific territorial markings. These pheromonal marks may be spaced throughout the home range or foraging radius of a colony or may be localized on trails. They may function to deter foragers of neighboring conspecific or even heterospecific colonies (Hölldobler and Lumsden, 1980). In some genera (e.g. *Myrmica* and *Atta*), there is evidence for interspecific recognition of pheromonal components of trail pheromones.

Haskins and Haskins (1983) split a single colony of *Rhytidoponera metallica* and maintained it in the laboratory for almost a year on different diets (A and B). When workers (which were formerly nestmates) were then introduced to one another in fingerbowls there was virtually no aggression. After separation for an additional year and a half, the nest boxes of diet group B were placed into the foraging arena of diet group A. The latter then attacked their former nestmates. When groups of ten workers from one dietary treatment were placed at the entrance of a colony on either the same or different diets, the same results were obtained as in the fingerbowl introductions (Haskins and Haskins, 1983). These results suggest the presence of an environmentally correlated colony-specific mark releasing aggression, which is deposited in the area of the nest entrance, and further emphasize the significance of the nest entrance context in releasing aggression.

Colonies of the African weaver ant, *Oecophylla longinoda*, defend territories against both foreign conspecifics and other species. This species forms large, monogynous colonies; a single colony can contain over 500,000 workers (Hölldobler and Lumsden, 1980). *O. longinoda* employs a large repertoire of pheromonal signals for recruitment to food and territorial marking (Hölldobler and Wilson, 1978). Colony-specific pheromones produced by the rectal vesicle are used to mark the territory belonging to a single colony, which may encompass several trees and up to 1,600 m^2. Intruders are rapidly recognized and repelled by workers responding to an efficient, pheromonally mediated, recruitment system. Bradshaw *et al.* (1979) found intercolonial variation in volatile cephalic compounds which may be utilized in colony or territorial recognition.

Jaffé, Bazire-Bénazet and Howse (1979) reported that workers of *Atta cephalotes* mark their foraging territories with a pheromone derived from the valves gland, located near the sting. Workers of both *A. cephalotes* and *A. sexdens* raised their abdomens significantly more often ($p < 0.001$, $n = 8$ colonies) on unmarked control paper or paper marked by *Acromyrmex octospinosus* than on paper marked by nestmates. Abdomen raising is associated with exploration of unfamiliar territory; when placed in the foraging area of another conspecific colony or one of *A. sexdens*, *A. cephalotes* workers display this motor pattern significantly less often ($p < 0.001$, $n = 8$ colonies). *Atta cephalotes* workers placed on papers marked by nestmates show a significant decrease in abdomen raising when compared with foreign conspecifics ($p < 0.05$, $n = 8$ colonies). Jaffé, Bazire-Bénazet and Howse (1979) concluded that the valves gland secretion of *A. cephalotes* contains both colony- and species- (or possibly genus-) specific components. Gas chromatography of gasters and isolated valves glands of all three species revealed a component specific to both *Atta* species which did not occur in *Acromyrmex octospinosus*.

Cammaerts and her co-workers (Cammaerts-Tricot, 1974; Cammaerts-Tricot *et al.*, 1976, Cammaerts, Evershed and Morgan, 1981) have studied the pheromones used by several species of *Myrmica*. Quantitative and qualitative differences between species in the chemical composition of secretions from the poison gland, mandibular glands, and Dufour's gland were found (Cammaerts, Evershed and Morgan, 1982), although considerable cross-species reactivity to pheromones did occur (Cammaerts, Evershed and Morgan,

1981). The Dufour's gland secretion contains volatile compounds which act as attractants and less volatile components used by foragers to mark newly explored territory. *Myrmica rubra* workers reacted to the territorial markings of *M. scabrinodis* and *M. sabuleti* but not to those of *M. ruginodis*. Chemical similarity of the compounds between *M. rubra* and *M. ruginodis* apparently explains these results (Cammaerts, Evershed and Morgan, 1981) although colony-specific territorial pheromones might be expected, as inter-colonial hostility does occur (Winterbottom, 1981). Colony-specificity in the territorial markings of *M. rubra* appears several days after the marks are deposited, although it is not certain whether these chemicals originate in the Dufour's gland; chemical analysis of glands from different colonies showed no reliable pattern of differentiation (Winterbottom, 1981).

Camponotus rufipes workers use a long-lasting pheromone from an as yet undetermined gland to mark territories (Jaffé and Sanchez, 1984). When exposed to papers marked by ants of a different colony, workers' responses depend on their location. If close to their own nest, they exhibit an increase in exploratory behavior relative to paper formerly marked by nestmates ($p < 0.05$, $n = 11$ colonies). If workers are far from their nest when given a paper marked by another colony, they flee. These results further indicate the ability of worker ants to discriminate by olfactory means the territory of their own and other colonies.

Trunk trails

Two sympatric species of harvester ant in the southwestern United States, *Pogonomyrmex barbatus* and *P. rugosus*, utilize long-lasting trunk trails for foraging and orientation. These odor trails, laid by foragers, provide long-lasting olfactory information concerning the direction and quality of food resources (Hölldobler, 1974, 1976, 1977). Hölldobler (1974, 1976) found that trunk trails of neighboring conspecific colonies never crossed, rather, they tended to diverge. When trails from different colonies were experimentally merged by providing attractive baits, intense aggression between foragers resulted. A third sympatric species, *P. maricopa*, does not use trunk trails, and was characterized by a much greater distance between colonies (Hölldobler, 1974). Trunk trails in this genus may function as colony-specific territorial markings, allowing increased nest densities by minimizing encounters between non-nestmate conspecifics.

Megaponera foetens, a termitophagous ponerine native to the central African savanna, appears to follow odor trails to nearby termite colonies (Longhurst and Howse, 1979a). Foraging strategies appear to vary with ant colony population density such that colonies in more densely populated sites utilize trunk trails, while ants in less populated areas do not (Levieux, 1966). When encounters between columns from different colonies occur, aggressive interactions result in the deaths of foragers and disruption of the column. Trunk trails in this species, as in *Pogonomyrmex* discussed above, may minimize aggressive interactions between workers from different colonies.

Paraponera clavata, the giant tropical ant, is an advanced ponerine. Despite the large size (>2 cm body length) and ferocious sting, intraspecific aggression is rare (Hermann and Young, 1980; personal observation). Colony-specific territorial pheromones may serve to reduce encounters between non-nestmates, as foragers utilize odor trails (Breed and Bennett, 1986). Hermann and Young (1980) reported that foreign foragers introduced at or near nest entrances were attacked; we saw no overt hostility under these conditions (M. D. Breed and B. Bennett unpublished).

Traniello (1980) reported that workers of *Lasius neoniger*, a North American formicine, lay colony-specific trunk trails. The active agent originates in hindgut material. These trails are not truly territorial in function as *L. neoniger* workers do not respond agonistically to trails or hindgut material from other colonies. The trails may function to reduce agonistic encounters between workers from different colonies, however, as colony density may be as high as $5 \, m^{-2}$.

Gyne and queen recognition

Foreign conspecific queens are more reliably discriminated and rejected than are workers. This tendency makes adaptive sense, as fertile foreign queens pose more of a threat to the inclusive fitness of a worker than does an unrelated worker intent on food or brood robbing. In some rare cases, related queens are rejected, or unrelated queens accepted; these unusual cases are reviewed here.

Like other ponerines, *Proceratium croceum*, a rare subterranean species, is characterized by a well-developed capacity to discriminate foreign conspecifics (Haskins, 1930). Within several weeks of leaving their natal nests, however, young gynes are no longer accepted by their worker siblings. As this result implies either a change in recognition cues over time, or the need for continual reinforcement of these cues, it would be illuminating to remove queens from their colonies for several weeks and observe the reactions of the workers upon reintroduction of the queens.

The leaf-cutter ants, all members of the tribe Attini, comprise eleven genera, all restricted to the New World. Spencer (1984) noted that queenless workers (of undetermined species) accepted foreign queens of the same species, although heterospecific queens were consistently attacked. Echols (1966) joined field collected colonies of *Atta texana* in the laboratory by means of plastic tubing. Queens were readily accepted between colonies, although as the colonies so joined were adjacent to one another in the field, these queens may have been related. Queens and alate females from *Acromyrmex octospinosus* (Jutsum, Saunders and Cherrett, 1979); *Atta cephalotes* and *A. columbica* (Rockwood, 1973) were immediately attacked when placed at nest entrances of foreign colonies.

Evesham and Cammaerts (1984) showed that workers of *Myrmica rubra* accept and cluster around foreign queens from neighboring nests, while exhibiting mild aggression toward workers from these same nests. Queens from more distant colonies (>3 km) were initially attacked but slowly accepted; workers from these colonies were always attacked. Workers of this species appear to discriminate more reliably between nestmates and non-nestmates (Winterbottom, 1981) than between alien and resident queens. This difference may be due to the variation in intranest queen relatedness in this species, which ranges from zero to 0.67 between individual queens in a single colony (Pearson, 1982). Using a series of linked laboratory nests which allowed free movement to workers, but not to queens, Evesham (1984) found that while all queens are attractive to workers, certain queens within a colony are more attractive than others. Workers preferred the queen to which they were first exposed.

Worker response to queens was studied in laboratory nests of *Monomorium pharaonis* consisting of queens of specific age classes and workers from either queenless or queenright colonies (Petersen-Braun, 1982). The virtual lack of nestmate recognition in this species allowed the creation of artificial colonies with both queens and workers from different source colonies. When groups of workers from queenless colonies were given new queens, the brood-rearing cycle of the colony was disrupted if the new queens differed in age from those with which the workers were initially paired. In the case of extreme discrepancies (e.g. workers from colonies with immature queens given older queens), aggression toward the new queens resulted. In a similar experiment, queens of four age classes (four weeks, two, three and four months) were introduced into different queenright colonies. Again, disruption of brood rearing and aggression toward the introduced queens occurred if the new queens' ages were different from that of the original queens. It appears that in *M. pharaonis*, cues from queens correspond to specific stages in their life cycles and serve to maintain group integrity. This finding is not surprising considering that reproduction in this species occurs by colony fission. Groups which bud off probably contain reproductive females in the same stage of development; non-synchronous females would result in less efficient colony growth.

Male recognition

Very little is known from any of the eusocial Hymenoptera about discrimination between nestmate and foreign males. Although males consume resources, they do not represent a threat to either worker inclusive fitness or colony fitness through theft of brood or food. Consequently male discrimination is not expected to be as reliable as foreign worker and queen recognition.

Queens of *Megaponera foetens*, a termitophagous ponerine, are fertilized by males from other colonies which follow worker-laid scent trails into the worker colonies (Longhurst and Howse, 1979b). These alien males are accepted readily by the workers, as is the case in other ponerines. Wheeler (1900) described similar responses to foreign males in *Pachycondyla harpax* and *Leptogenys elongata*.

B. Bennett (in preparation) found that workers of a polygynous species, *Formica podzolica*, accepted foreign males more readily than did workers of the closely related monogynous species, *F. subaenescens*. As with worker transfers, male rejection rates rose as distance between their natal colonies increased.

In both laboratory and field introductions, conspecific but unrelated males placed near the nest entrances of *Acromyrmex octospinosus* were consistently attacked (Jutsum, Saunders and Cherrett, 1979). The same result was obtained if the males and the colonies to which they were introduced were maintained on identical diets.

Mixed colonies

Brood (i.e. larvae and pupae) are easily transferred between conspecific and even heterospecific colonies of many ant species. Hölldobler (1977) suggests that pheromones which elicit nursing behavior are present in the exuvial fluids produced by the immatures. These chemicals may rank high in a hierarchical pheromonal system, overpowering any other colony-specific pheromones associated with the brood. The lack of colony-specific responses to brood has allowed the evolution of social parasitism or slavery, as well as the formation of mixed nests by investigators interested in the ontogeny of recognition behavior. It appears that in the majority of ant species studied, in which foreign brood is adopted, the emergent adults learn ('imprint') the recognition cues of the associated adults. Even in artificial nests comprised of different species, the rearing adults and the cross-fostered young often treat one another as nestmates (e.g. Carlin and Hölldobler, 1983). The generality of this result may be questionable, as more thorough behavioral analyses indicate quantitative differences in behavioral acts directed toward cross-fostered and true siblings (Lenoir, Isingrini and Nowbahari, 1982; Lenoir, 1984; Isingrini, Lenoir and Jaisson, 1985). As phylogenetic distance between the brood and rearing workers increases, the probability of rejection of emergent adults increases (Errard, 1984). In other words, as taxonomic distance increases it becomes less likely that the cues produced by the fostered workers will correspond to those used by the fostering species.

While foreign brood and the emergent adults are typically accepted after cross-fostering in many advanced ant species (e.g. Wilson, 1971; Jaisson, 1980a; Carlin and Hölldobler, 1983), some primitive ants are less likely to accept cross-fostered workers. In *Promyrmecia* and *Myrmecia*, unrelated brood is accepted and cared for until eclosion, when the adults are rejected (Haskins and Haskins, 1950). Schneirla (1958, 1971) found that doryline brood, unlike adult workers, is not accepted by foreign colonies and is cannibalized.

Artificial mixed colonies have been utilized by Jaisson and his co-workers to investigate the ontogeny of recognition behavior in a variety of ant species from all subfamilies. Errard (1984) created mixed colonies of pairs of *Camponotus* species; *senex, abdominalis* and unidentified species in this genus, as well as *Pseudomyrmex ferruginea*. Heterospecifics of the same genus introduced after adult eclosion were accepted until the sixth day of adult age, although amicable behavior (assessed by trophallaxis) decreased steadily with age. Carlin and Hölldobler (1986) found a similar ontogeny of nestmate recognition in six *Camponotus* species. Callows from queenless nests were transferred at various ages, acceptance was assessed by survivorship. For two- to six-day old ants a longer period of exposure was required to attain the same amount of tolerance (Errard, 1984). One of the species in each artificial nest always assumed the nurturing role, feeding and grooming both conspecifics and heterospecifics. Carlin and Hölldobler (1983) found a similar division of labor in mixed nests of *Camponotus*. In mixed nests of *C. senex* and *P. ferruginea*, if the workers of the latter were over three hours post-eclosion, the *C. senex* workers were killed. Conversely, if the *C. senex* workers were older than one day, the *P. ferruginea* workers were killed. These results corroborate earlier studies of the ontogeny of recognition in mixed species nests (C. Errard and P. Jaisson, personal communication; Jaisson, 1980a) although some species [e.g. *Trachymesopus stigmus* (Ponerinae), *Solenopsis geminata*, *Monomorium ebeninum* (Myrmicinae), and various *Azteca* species (Dolichoderinae)] never integrated with *Camponotus* (Errard, 1984). Thus, it appears that while callows may lack certain recognition cues allowing them to be accepted by different species and even genera, as phylogenetic distance between taxa increases, other cues of callows result in their rejection by adult workers.

Jaisson (1973, 1975) investigated the ontogeny of recognition behavior in formicines. Pupae of *Formica polyctena* are killed by *F. fusca* workers unless these workers were exposed to the heterospecific cocoons after eclosion. The presence of a queen in the rearing nest facilitates the 'imprinting' process (Jaisson, 1973). If the foreign cocoons are left with the *F. fusca* workers for longer than 14 days, the *F. polyctena* cocoons will always be preferred, even after six months without exposure to brood. In similar experiments, callow workers of *Formica rufa* (LeMoli and Passetti, 1978) and *F. lugubris* (LeMoli

Saunders and Cherrett, 1979). Queen recognition is somewhat more variable. Myrmeciine species may 'forget' their queens after separations of several weeks or more (Haskins and Haskins, 1950) although it is not known whether these queenless workers would then accept an unrelated queen. Workers of many species, if maintained queenless for extended periods, will accept unrelated queens (e.g. *Neivamyrmex* and other dorylines, Schneirla, 1971; *Formica fusca* group species, (B. Bennett, unpublished; *Nothomyrmecia macrops*, Hölldobler and Taylor, 1983). Queen produced cues may be more labile or require more continuous reinforcement, than those which originate from the workers themselves.

Brood recognition is less colony-specific than adult nestmate discrimination (Jaisson, 1984) and often depends upon early learning of species-specific cues. Certain species (e.g. *Lasius niger*, LeMoli, 1980) are capable of distinguishing heterospecific brood while many other species accept brood from distantly related species (Fielde, 1903; Errard, 1984).

The actual anatomical source of the recognition cues is largely unknown in ants. Mandibular gland secretions may be involved in *Atta cephalotes* (Jaffé, 1983) although in *Odontomachus bauri* both head and abdominal extracts contain colony-specific components (Jaffé and Marcuse, 1983). Increased resolution of individual variation in these secretions might allow a distinction between the applicability of mean and multiple-mean template models of nestmate recognition systems.

Colony-specific territorial markings may also be involved in nestmate recognition. These markings have been demonstrated in diverse groups of ants (e.g. Hölldobler and Wilson, 1978; Jaffé, Bazire-Bénazet and Howse, 1979; Hölldobler and Lumsden, 1980; Jaffé and Marcuse, 1983; Jaffé and Sanchez, 1984). The presence of territorial markings combined with the observation that aggression toward non-nestmates is more intense closer to the nest suggest that these substances may act as releasers of aggression and perhaps facilitate discrimination. Jaffé and Sanchez (1984) suggest that those species employing a territorial marking strategy need less efficient nestmate discrimination systems than species which do not mark and defend territories, as the probability of encountering non-nestmates is lower in the former case. These authors correlate territoriality with a trend toward use of alarm pheromones as nestmate labels (Jaffé and Sanchez, 1984) although little evidence for this conclusion exists at present.

The role of exposure early in adult life is clearly important as the cues labelling nestmates appear to be learned at this time (Jaisson, 1973, 1975, 1984; LeMoli and Passetti, 1978; LeMoli and Mori, 1982). Of course, under natural conditions, most emerging ants will be surrounded by nestmates and the rule that familiar equals kin is quite acceptable (viz. Bekoff, 1981). Slave-making species have taken advantage of this early 'imprinting' effect, in essence creating mixed colonies.

OVERVIEW AND SUMMARY

Recognition systems in highly eusocial insects provide mechanisms of defense of colony resources, of intercolonial competition for resources, and of intracolonial competition for access to reproduction. Colony defense and intercolonial competition are relatively well studied, while intracolonial competition remain less well understood. Future studies will focus on the heritability of recognition cues, the learning process involved in template formation, the possibility of the use of different templates in different contexts, and the behavioral ecology of recognition.

Perhaps the most striking feature of recognition systems in highly eusocial insects is that one recognition context, nest defense, does not require genotypically correlated recognition cues. In fact, the genetic variation within colonies of many species argues for unifying cues, such as those obtained from the environment or the queen, in contexts in which a colony characteristic is needed.

Honey bees can and do use environmentally derived intercolonial odor differences in making nestmate/non-nestmate discriminations while many ants appear to use queen-derived cues. Individual bee and ant workers also produce cues themselves. At present the understanding of the integration of different types of cues is poor.

If our understanding of nestmate recognition is incomplete, our understanding of recognition in intracolonial processes such as brood rearing and swarming is nil. The data that are available suggest the possibility that workers can behave in ways that influence the genetic nature of a colony's contribution to the nest generation. This possibility will certainly provide fruitful ground for further research.

ACKNOWLEDGMENTS

We thank Charles Michener, David Fletcher, Peter Frumhoff, Ross Crozier, Rob Page, Allen Moore, Tammy Stiller and Marc Bekoff for their comments on an earlier draft. The preparation of this review was supported in part by NSF grant BNS-82-16787.

LITERATURE CITED

Ambrose, J. T., R. A. Morse and R. Boch. 1979. Queen discrimination by honey bee swarms. *Annals of the Entomological Society of America*, **72**, 673–675.

Andrews, E. A. 1911. Observations on termites in Jamaica. *Journal of Animal Behaviour*, **1**, 193–228.

Bekoff, M. 1981. Mammalian sibling interactions, in Gubernick, D. J. and P. H. Klopfer, eds., *Parental care in mammals*, pp. 307–346. Plenum Publishing Co., New York.

Bergström, G. and J. Löfqvist, 1968. Odour similarities between the slave-keeping ants *Formica sanguinea* and *Polyergus rufescens* and their slaves *Formica fusca* and *Formica rufibarbis*. *Journal of Insect Physiology*, 14, 995–1011.

Berkelhamer, R. C. 1984. An electrophoretic analysis of queen number in three species of Dolichoderine ants. *Insectes Sociaux*, 31, 132–141.

Boch, R. and R. A. Morse. 1974. Discrimination of familiar and foreign queens by honey bee swarms. *Annals of the Entomological Society of America*, 67, 709–711.

Boch, R. and R. A. Morse. 1979. Individual recognition of queens by honey bee swarms. *Annals of the Entomological Society of America*, 72, 51–53.

Boch, R. and R. A. Morse. 1981. Effects of artificial odors and pheromones on queen discrimination by honey bees, *Annals of the Entomological Society of America*, 74, 66–67.

Boch, R. and R. A. Morse. 1982. Genetic factor in queen recognition odors of honey bees. *Annals of the Entomological Society of America*, 75, 654–656.

Bradshaw, J. W. S., R. Baker, P. E. Howse and M. D. Higgs. 1979. Caste and colony variations in the chemical composition of the cephalic secretions of the African weaver ant, *Oecophylla longinoda*. *Physiological Entomology*, 4, 27–38.

Bradshaw, J. W. S. and P. E. Howse. 1984. Sociochemicals of ants, in Bell, W. J. and R. T. Cardé, eds., *Chemical ecology of insects*, Sinauer Associates, Sunderland, Massachusetts, pp. 429–473.

Breed, M. D. 1981. Individual recognition and learning of queen odors by worker honey bees. *Proceedings of the National Academy of Science, USA*, 78, 2635–2637.

Breed, M. D. 1983. Nestmate recognition in honey bees. *Animal Behaviour*, 31, 86–91.

Breed, M. D. and B. Bennett. 1986. Mass recruitment to nectar sources in *Paraponera clavata*: a field study. *Insectes Sociaux*, in press.

Breed, M. D., B. Bennett and T. M. Stiller. 1986. Learning and genetic influences in the discrimination of colony members by social Hymenoptera, in press.

Breed, M. D., L. Butler and T. M. Stiller. 1985. Kin discrimination by worker honey bees in genetically mixed groups. *Proceedings of the National Academy of Science, USA*, 82, 3058–3061.

Breed, M. D., H. H. W. Velthuis and G. Robinson. 1984. Do worker honeybees discriminate among related and unrelated larval phenotypes? *Annals of the Entomological Society of America*, 77, 737–739.

Brian, M. V. 1983. *Social insects*. Chapman & Hall, New York.

Cammaerts, M.-C., R. P. Evershed and E. D. Morgan. 1981. Comparative study of pheromones emitted by different species of *Myrmica*, in Howse, P. E. and J.-L. Clément, eds., *Biosystematics of social insects*, Academic Press, New York, pp. 185–192.

Cammaerts, M.-C., R. P. Evershed and E. D. Morgan. 1982. Mandibular gland secretions of workers of *Myrmica rugulosa* and *M. schencki*: comparison with four other *Myrmica* species. *Physiological Entomology*, 7, 119–125.

Cammaerts-Tricot, M.-C. 1974. Production and perception of attractive pheromones by differently-aged workers of *Myrmica rubra* (Hymenoptera: Formicidae). *Insectes Sociaux*, 21, 235–248.

Cammaerts-Tricot, M.-C., E. D. Morgan, R. C. Tyler and J. C. Brackman. 1976. Dufour's gland secretion of *Myrmica rubra*: chemical, electrophysiological and ethological studies. *Journal of Insect Physiology*, 22, 927–932.

Carlin, N. F. and B. Hölldobler. 1983. Nestmate and kin recognition in interspecific mixed colonies of ants. *Science*, 222, 1027–1029.

Carlin, N. F. and B. Hölldobler. 1986. The kin recognition system of carpenter ants (*Camponotus* spp.) I: Hierarchical cues in small colonies. *Behavioral Ecology and Sociobiology*, in press.

Chauvin, R. 1971. *The world of ants*. Hill and Wang, New York.
Clément, J.-L. 1978. L'aggression interspecifique et intraspecifique des especes francaises du genre *Reticulitermes*. *Comptes Rendus des Séances de L'Académie des Sciences, Paris*, **286**, 351–354.
Clément, J.-L. 1982. Signaux de contact responsables de l'aggression interspecifique des termites du genre *Reticulitermes*. *Comptes Rendus des Séances de L'Académie des Sciences, Paris*, **294**, 635–638.
Craig, R. and R. H. Crozier. 1979. Relatedness in the polygynous ant *Myrmecia pilosula*. *Evolution*, **33**, 335–341.
Creighton, W. S. 1950. The ants of North America. *Bulletin of the Museum of Comparative Zoology, Harvard*, **104**, 1–585.
Crewe, R. M. 1982. Compositional variability, the key to the social signals produced by honey bee mandibular glands, in Breed, M. D., C. D. Michener and H. E. Evans, eds., *The biology of the social insects*, Westview Press, Boulder, Colorado, pp. 318–322.
Crozier, R. H. and M. W. Dix. 1979. Analysis of two genetic models for the innate components of colony odor in social hymenoptera. *Behavioral Ecology and Sociobiology*, **4**, 217–224.
Crozier, R., P. Pamilo and Y. C. Crozier. 1984. Relatedness and microgeographic genetic variation in *Rhytidoponera mayri*, an Australian arid-zone ant. *Behavioral Ecology and Sociobiology*, **15**, 143–150.
Dawkins, R. 1976. *The selfish gene*. Oxford University Press, Oxford, England and New York.
Del Rio Pasado, M. G. and T. M. Alloway. 1983. Polydomy in the slave-making ant, *Harpogoxenus americanus* (Emery) (Hymenoptera: Formicidae). *Psyche*, **90**, 151–162.
DeVroey, S. C. and J. M. Pasteels. 1978. Agonistic behaviour of *Myrmica rubra* L. *Insectes Sociaux*, **25**, 247–265.
Driessen, G. J. J., A. T. Van Raalte and G. J. DeBruyn. 1984. Cannibalism in the red wood ant, *Formica polyctena* (Hymenoptera: Formicidae). *Oecologia*, **63**, 13–22.
Dumpert, K. 1978. *The social biology of ants*. Pitman Publishing, Ltd., Marshfield, Massachusetts.
Echols, H. W. 1966. Compatibility of separate nests of Texas leaf-cutting ants. *Journal of Economic Entomology*, **50**, 1299–1300.
Elmes, A. W. 1984. Queen numbers in colonies of ants of the genus *Myrmica*. *Insectes Sociaux*, **27**, 43–60.
Emery, C. 1909. Uber den Urspring der dulotischen, parasitischen und myrmekophilen Ameisen. *Biologische Zentralblatt*. **29**, 352–362.
Errard, C. 1984. Evolution en fonction de l'âge, des relations sociales dans les colonies mixtes hétérospécifiques chez les fourmis des genres *Camponotus* et *Pseudomyrmex*. *Insectes Sociaux*, **31**, 185–194.
Evesham, E. J. M. 1984. The attractiveness of workers towards individual queens of the polygynous ant *Myrmica rubra* L. *Biology of Behavior*, **9**, 144–156.
Evesham, E. J. M. and M.-C. Cammaerts. 1984. Preliminary studies of worker behaviour towards alien queens, in the ant *Myrmica rubra* L. *Biology of Behavior*, **9**, 131–143.
Fielde, A. M. 1903. Artificial mixed nests of ants. *Biological Bulletin, Marine Biological Laboratory, Woods Hole*, **6**, 320–325.
Fielde, A. M. 1904. Power of recognition among ants. *Biological Bulletin, Marine Biological Laboratory, Woods Hole*, **7**, 227–250.
Fielde, A. M. 1905. The progressive odours of ants. *Biological Bulletin, Marine Biological Laboratory, Woods Hole*, **10**, 1–16.

Fletcher, D. J. C. and M. S. Blum. 1983. Regulation of queen number by workers in colonies of social insects. *Science*, **219**, 312–314.
Fletcher, D. J. C., M. S. Blum, T. V. Whett and N. Temple. 1980. Monogyny and polygyny in the fire ant, *Solenopsis invicta*. *Annals of the Entomological Society of America*, **73**, 658–661.
Free, J. B. 1954. The behaviour of robber bees. *Behaviour*, **7**, 233–240.
Fresneau, D. 1980. Fermeture des sociétés et marquage territorial chez les fourmis Ponerines du genre *Neoponera*. Biologie-Ecologie Meditérranéenne, **7**, 205–206.
Getz, W. M. 1982. An analysis of learned kin recognition in Hymenoptera. *Journal of Theoretical Biology*, **99**, 585–597.
Getz, W. M., D. Brückner, and T. R. Parisian. 1982. Kin structure and the swarming behavior of the honey bee, *Apis mellifera*. *Behavioral Ecology and Sociobiology*, **10**, 265–270.
Getz, W. M. and K. B. Smith. 1983. Genetic kin recognition: honey bees discriminate between full- and half-sisters. *Nature*, **302**, 147–148.
Goetsch, W. 1957. *The ants*. University of Michigan Press, Ann Arbor, Michigan.
Greenberg, L. 1979. Genetic component of bee odor in kin recognition. *Science*, **206**, 1095–1097.
Halliday, R. B. 1983. Social organization of meat ants *Iridomyrmex purpureus* analysed by gel electrophoresis of enzymes. *Insectes Sociaux*, **30**, 45–56.
Hamilton, W. D. 1964. The genetical evolution of social behavior, I and II. *Journal of Theoretical Biology*, **7**, 1–52.
Hangartner, W., J. M. Reichson and E. O. Wilson. 1970. Orientation to nest material by the ant, *Pogonomyrmex badius*. *Animal Behaviour*, **18**, 331–334.
Haskins, C. P. 1928. Notes on the behavior and habits of *Stigmatomma pallipes* Haldeman. *Journal of the New York Entomological Society*, **36**, 179–184.
Haskins, C. P. 1930. Preliminary notes on certain phases of the behavior and habits of *Proceratium croceum* Roger. *Journal of the New York Entomological Society*, **38**, 121–126.
Haskins, C. P. and E. F. Haskins. 1950. Notes on the biology and social behavior of the archaic ponerine ants of the genera *Myrmecia* and *Promyrmecia*. *Annals of the Entomological Society of America*, **43**, 461–491.
Haskins, C. P. and E. F. Haskins. 1979. Worker compatibilities within and between populations of *Rhytidoponera metallica*. *Psyche*, **86**, 299–313.
Haskins, C. P. and E. F. Haskins. 1983. Situation and location-specific factors in the compatibility response in *Rhytidoponera metallica* (Hymenoptera: Formicidae: Ponerinae). *Psyche*, **90**, 163–174.
Haskins, C. P. and R. M. Whelden. 1965. 'Queenlessness', worker sibship, and colony versus population structure in the formicid genus *Rhytidoponera*. *Psyche*, **72**, 87–111.
Hermann, H. R. and A. M. Young. 1980. Artificially elicited defensive behavior and reciprocal aggression in *Paraponera clavata* (Hymenoptera: Formicidae: Ponerinae). *Journal of the Georgia Entomological Society*, **15**, 8–10.
Hölldobler, B. 1974. Home range orientation and territoriality in Harvesting Ants. *Proceedings of the National Academy of Science, USA*, **71**, 3274–3277.
Hölldobler, B. 1976. Recruitment behavior, home range orientation and territoriality in harvester ants, *Pogonomyrmex*. *Behavioral Ecology and Sociobiology*, **1**, 3–44.
Hölldobler, B. 1977. Communication in social Hymenoptera, in Sebeok, T. A. ed., *How animals communicate*, Indiana University Press, Bloomington, Indiana, pp. 418–471.
Hölldobler, B. and C. J. Lumsden. 1980. Territorial strategies in ants. *Science*, **210**, 732–739.

Hölldobler, B. and C. D. Michener. 1980. Mechanisms of identification and discrimination in social Hymenoptera, in Markl, H. ed., *Evolution of social behavior: hypotheses and empirical tests*, Verlag Chemie, Weinheim, pp. 35–58.
Hölldobler, B. and R. W. Taylor. 1983. A behavioral study of the primitive ant *Nothomyrmecia macrops* Clark. *Insectes Sociaux*, **30**, 384–401.
Hölldobler, B. and E. O. Wilson. 1977. The number of queens: An important trait in ant evolution. *Naturwissenschaften*, **64**, 8–15.
Hölldobler, B. and E. O. Wilson. 1978. The multiple recruitment system of the African weaver ant *Oecophylla longinoda* (Latreille) (Hymenoptera: Formicidae). *Behavioral Ecology and Sociobiology*, **3**, 19–60.
Holmes, W. G. and P. W. Sherman. 1983. Kin recognition in animals. *American Scientist*, **71**, 46–55.
Howick, L. D. and J. W. Creffield. 1980. Intraspecific agonism in *Coptotermes acinaciformis*. *Bulletin of Entomological Research*, **70**, 17–23.
Hubbard, M. D. 1974. Influence of nest material and colony odor on digging in the ant *Solenopsis invicta* (Hymenoptera: Formicidae). *Journal of the Georgia Entomological Society*, **9**, 127–135.
Isingrini, M., A. Lenoir and P. Jaisson. 1985. Preimaginal learning as a basis of colony-brood recognition in the ant *Cataglyphis cursor*. *Proceedings of the National Academy of Science, USA*, **82**, 8545–8547.
Jaffé, K. 1983. Chemical communications systems in the ant *Atta cephalotes*, in Jaisson, P. ed., *Social insects in the tropics*, Vol. 2, Université de Paris-Nord, Paris, France, pp. 165–186.
Jaffé, K., H. Bazire-Bénazet and P. E. Howse. 1979. An integumentary pheromone secreting gland in *Atta* sp.: Territorial marking with a colony-specific pheromone in *Atta cephalotes*. *Journal of Insect Physiology*, **25**, 833–839.
Jaffé, K. and Marcuse, M. 1983. Nestmate recognition and territorial behavior in the ant *Odontomachus bauri* Emery (Formicedae: Ponerinae). *Insectes Sociaux*, **30**, 466–481.
Jaffé, K. and C. Sanchez. 1984. On the nestmate recognition system and territorial marking behaviour in the ant. *Camponotus rufipes*. *Insectes Sociaux*, **31**, 302–315.
Jaisson, P. 1973. L'imprégnation dans l'ontogenèse du comportement de soin aux cocons chez les Formicines. *Proceedings VIIth Congress, International Union for the Study of Social Insects, London*, **1973** 176–181.
Jaisson, P. 1975. L'imprégnation dans l'ontogenèse du comportements de soins aux cocons chez la jeune fourmi rousse (*Formica polyctena* Forst). *Behaviour*, **52**, 1–37.
Jaisson, P. 1980a. Les colonies mixtes plurispécifiques: une modèle pour l'étude des fourmis. *Biologie Ecologie Méditeranéenne*, **7**, 163–166.
Jaisson, P. 1980b. Environmental preference induced experimentally in ants (Hymenoptera: Formicidae) *Nature*, **286**, 388–389.
Jaisson, P. 1984. Social behaviour, in Gilbert, L. I. and G. A. Kerkut, eds., *Insect comprehensive biochemistry, physiology, and pharmacology*, Vol. 9, Pergamon Press, New York, pp. 673–694.
Jaisson, P. and D. Fresneau. 1978. The sensitivity and responsiveness of ants to their cocoons in relation to age and methods of measurement. *Animal Behaviour*, **26**, 1064–1071.
Janzen, D. H. 1973. Evolution of polygynous obligate acacia-ants in Western Mexico. *Journal of Animal Ecology*, **42**, 727–758.
Jutsum, A. R., T. S. Saunders and J. M. Cherrett. 1979. Intraspecific aggression in the leaf-cutting ant *Acromyrmex octospinosus*. *Animal Behaviour*, **27**, 839–844.

Kalmus, H. and C. R. Ribbands. 1952. The origin of odors by which honey bees distinguish their companions. *Proceedings of the Royal Society, (B)*, **140**, 50–59.
Laidlaw, H. H. and R. E. Page. 1984. Polyandry in honeybees: sperm utilization and intra-colony genetic relationships. *Genetics*, **108**, 985–997.
LeMoli, F. 1980. On the origin of slaves in dulotic ant societies. *Bollettino di Zoologia, Pubbliccato dall'Unione Zoologica Italiana*, **47**, 207–212.
LeMoli, F. and A. Mori. 1982. Early learning and cocoon nursing behaviour in the red wood ant *Formica lugubris* Zett. (Hymenoptera: Formicidae). *Bollettino di Zoologia, Pubbliccato dall'Unione Zoologica Italiana*, **49**, 93–97.
LeMoli, F. and S. Parmigiani. 1982. Intraspecific combat in the redwood ant (*Formica lugubris* Zett.). *Aggressive Behavior*, **8**, 145–148.
LeMoli, F. and M. Passetti. 1978. Olfactory learning phenomena and cocoon nursing behavior in the ant *Formica rufa*. *Bollettino di Zoologia, Pubbliccato dall'Unione Zoologica Italiana*, **45**, 389–397.
Lenoir, A. 1984. Brood-colony recognition in *Cataglyphis cursor* worker ants. *Animal Behaviour*, **32**, 942–944.
Lenoir, A., M. Isingrini and M. Nowbahari. 1982. Le comportement d'ouvrières de *Cataglyphis cursor* introduites dans une colonie étrangère de la même espèce (Hyménoptères Formicidae), in de Haro, A. and X. Espadaler, eds., *La communication chez les sociétés d'insectes. Colloque International UIEIS*, Ballaterra, Barcelona, Spain, pp. 107–114.
Levieux, J. 1966. Note preliminaire sur les colonnes de chasse de *Megaponera foetens* F. (Hymenoptera: Formicidae). *Insectes Sociaux*, **13**, 117–126.
Longhurst, C. and P. E. Howse. 1979a. Foraging, recruitment and emigration in *Megaponera foetens* (Fab.) (Hymenoptera: Formicidae) from the Algerian Guinea Savanna. *Insectes Sociaux*, **26**, 204–215.
Longhurst, C. and P. E. Howse. 1979b. Some aspects of the biology of the males of *Megaponera foetens* (Fab.) (Hymenoptera: Formicidae). *Insectes Sociaux*, **26**, 85–91.
Lubbock, J. 1878. *Ants, bees and wasps*. Appleton and Co., New York.
Mabelis, A. A. 1979. Wood ant wars: the relationship between aggression and predation in the red wood ant (*Fomica polyctena* Forst.) *Netherlands Journal of Zoology*, **29**, 451–620.
Michener, C. D. 1974. *The social behavior of the bees*. Harvard University Press, Cambridge, Massachusetts.
Mintzer, A. 1982. Nestmate recognition and incompatibility between colonies of the acacia-ant *Pseudomyrmex ferruginea*. *Behavioral Ecology and Sociobiology*, **10**, 165–168.
Mintzer, A. and S. B. Vinson. 1985. Kinship and incompatibility between colonies of the Acacia ant. *Behavioral Ecology and Sociobiology*, **17**, 75–78.
Morel, L. 1982. Mise en place des processue de regulation du comportement agressif et de la reconnaissance entre ouvrières d'une société de *Camponotus vagus* (Scop.) (Hymenoptera: Formicidae), in de Haro, A. and X. Espadaler, eds., *La communication chez les sociétés d'insectes. Colloque International UIEIS*, Ballaterra, Barcelona, Spain, pp. 127–136.
Noonan, K. C. 1986a. Kin recognition of worker brood by worker honey bees, in press.
Noonan, K. C. 1986b. Recognition of queen larvae by worker honey bees, in press.
Page, R. E. and E. H. Erickson. 1984. Selective rearing of queens by worker honey bees: kin or nestmate recognition. *Annals of the Entomological Society of America*, **77**, 578–580.

Page, R. E., R. B. Kimsey, and H. H. Laidlaw. 1984. Migration and dispersal of spermatozoa in spermathecae of queen honeybees. *Experientia*, **40**, 182–184.

Page, R. E. and E. H. Erickson. 1986a. Kin recognition and virgin queen acceptance by worker honey bees, in press.

Page, R. E. and E. H. Erickson. 1986b. Kin recognition during emergency queen rearing by worker honey bees, in press.

Pamilo, P. and R. H. Crozier. 1982. Measuring genetic relatedness in natural populations: Methodology. *Theoretical Population Biology*, **21**, 171–193.

Pamilo, P. and S-L. Varvio-Aho. 1979. Genetic structure of nests in the ant *Formica sanguinea*. *Behavioral Ecology and Sociobiology*, **6**, 91–98.

Pearson, B. 1982. Relatedness of normal queens (macrogynes) in nests of the polygynous ant *Myrmica rubra* Latreille. *Evolution*, **36**, 107–112.

Pearson, B. 1983. Intra-colonial relatedness amongst workers in a population of nests of the polygynous ant, *Myrmica rubra* Latreille. *Behavioral Ecology and Sociobiology*, **12**, 1–14.

Petersen-Braun, M. 1982. Intraspezifisches Aggressionsverhalten bei der Pharaaomeise *Monomorium pharaonis* L. (Hymenoptera, Formicidae). *Insectes Sociaux*, **29**, 25–33.

Pickens, A. L. 1934. The biology and economic significance of the western subterranean termite, *Reticultermes hesperus*, in Kofoid, C. A. ed., *Termites and termite control*, University of California Press, Berkeley, California, pp. 157–183.

Ribbands, C. R. 1954. The defense of the honeybee community. *Proceedings of the Royal Society, London, (B)* **142**, 514–524.

Rockwood, L. L. 1973. Distribution, density, and dispersion of two species of *Atta* (Hymenoptera: Formicidae) in Guanacaste province, Costa Rica. *Journal of Animal Ecology*, **42**, 803–817.

Ross, K. G. and Fletcher, D. J. C. 1985. Comparative study of genetic and social structure in two forms of the fire ant, *Solenopsis invicta* (Hymenoptera: Formicidae). *Behavioral Ecology and Sociobiology*, **17**, 349–356.

Schneirla, T. C. 1958. The behavior and biology of certain nearctic army ants. Last part of the functional season, southeastern Arizona. *Insectes Sociaux*, **5**, 215–255.

Schneirla, T. C. 1971. *Army ants*. W. H. Freeman and Company, San Francisco.

Spencer, H. 1984. Origin of classes among the 'parasol' ants. *Nature*, **51**, 125–126.

Stuart, R. J. 1985. *Nestmate recognition in leptothoracine ants: Exploring the dynamics of a complex phenomenon*. Ph.D. thesis, University of Toronto.

Sudd, J. H. 1967. *An introduction to the behavior of ants*. Arnold, London.

Taylor, R. W. 1978. *Nothomyrmecia macrops*: A living-fossil ant rediscovered. *Science*, **201**, 979–985.

Thorne, B. L. 1982. Termite-termite interactions: workers as an agonistic caste. *Psyche*, **89**, 133–150.

Traniello, J. F. A. 1980. Colony specificity in the trail pheromone of an ant. *Naturwissenschaften*, **67**, 361–362.

Traniello, J. F. A. 1982. Population structure and social organization in the primitive ant *Amblyopone pallipes* (Hymenoptera: Formicidae). *Psyche*, **89**, 65–80.

Tschinkel, W. R. and D. F. Howard. 1978. Queen replacement in orphaned colonies of the fire ant, *Solenopsis invicta*. *Behavioral Ecology and Sociobiology*, **3**, 297–310.

Van der Meer, R. K. and D. P. Wojcik. 1982. Chemical mimicry in the myrmecophilous beetle *Myrmecaphodius excavaticollis*. *Science*, **218**, 806–808.

Wallis, D. I. 1962. Aggressive behavior in the ant, *Formica fusca*. *Animal Behaviour*, **10**, 267–274.

Ward, P. S. 1980. Genetic variation and population differentiation in the *Rhytidoponera impressa* group, a species complex of ponerine ants. *Evolution*, **34**, 1060–1076.
Ward, P. S. 1983. Genetic relatedness and colony organization in a species complex of Ponerine ants. I. Phenotypic and genotypic composition of colonies. *Behavioral Ecology and Sociobiology*, **12**, 205–299.
Wheeler, W. M. 1900. *Ants*. Columbia University Press, New York.
Wilson, E. O. 1971. *The insect societies*. Harvard University Press, Cambridge, Massachusetts.
Wilson, E. O. 1974. The population consequences of polygyny in the ant *Leptothorax curvispinosus*. *Annals of the Entomological Society of America*, **67**, 781–786.
Winterbottom, S. 1981. *The chemical basis for species and colony recognition in three species of myrmicine ants (Hymenoptera: Formicidae)*. Doctoral Thesis, University of Southampton.
Yadava, R. P. S. 1970. Analysis of the components of aggressive behavior of *Apis mellifera* L. workers towards introduced queens. *American Bee Journal*, **1970**, 393.
Yadava, R. P. S. and M. V. Smith. 1971a. Aggressive behavior of *Apis mellifera* L. workers towards introduced queens. I. Behavioral mechanisms involved in the release of worker aggression. *Behaviour*, **39**, 213–236.
Yadava, R. P. S. and M. V. Smith. 1971b. Aggressive behavior of *Apis mellifera* L. workers towards introduced queens. II. Role of mandibular gland contents of the queen in releasing aggressive behavior. *Canadian Journal of Zoology*, **49**, 1179–1183.
Yadava, R. P. S. and M. V. Smith. 1971c. Aggressive behavior of *Apis mellifera* L. workers towards introduced queens. III. Relationship between the attractiveness of the queen and worker aggression. *Canadian Journal of Zoology*, **49**, 1359–1362.

CHAPTER 9

Kin Recognition in Vertebrates (Excluding Primates): Empirical Evidence

ANDREW R. BLAUSTEIN
Department of Zoology, Oregon State University, Corvallis, Oregon 97331, USA

MARC BEKOFF and THOMAS J. DANIELS
Department of Environmental, Population and Organismic Biology, University of Colorado, Boulder, Colorado 80309, USA

INTRODUCTION

An individual's ability to discriminate between, and to recognize, other animals, is important in a variety of circumstances. For example, 'species recognition' is necessary so that individuals make correct choices when they mate; 'individual recognition' is vital so that, amongst other things, individuals can discriminate between dominant and subordinate members of their social group; it may also be important in animal thinking (Griffin, 1984); and 'parent-offspring recognition' is important so that offspring may efficiently seek food and shelter from adults, who, in turn, would be under intense selective pressure not to feed and shelter distantly related kin or unrelated individuals (see Colgan, 1983 for review).

A growing body of evidence suggests that individuals of numerous species can discriminate not only between related and unrelated individuals but between close relatives and more distantly related kin. Kin recognition is inferred when the statistically significant differential treatment of individuals is based on kinship relationships (Holmes and Sherman, 1983). For example, individuals may use recognition abilities to avoid mating with close kin (Bateson, 1980, 1982, 1983) or to achieve breeding with kin when it may be beneficial (Shields, 1982). The concept of 'optimal outbreeding' (Bateson, 1978, 1983) suggests that both excessive inbreeding and excessive outbreeding may be costly. Therefore, a balance between too much inbreeding and too much outbreeding may be important and recognition of kin may facilitate this balance (see Bateson, 1983 for discussion). Kin recognition may also function in predator–prey and competitive situations by enabling cannibalistic individuals to avoid eating close relatives (Blaustein and O'Hara, 1982a) and to avoid competing for resources with kin.

Kin recognition is especially germane to kin selection theory and the concept of inclusive fitness (Hamilton, 1964; Wilson, 1975; see also Grafen, 1982). In this context, the ability to recognize kin is important because it can enhance aid-giving behaviors and preferential treatment toward kin (nepotism) and preclude the misdirection of these behaviors toward unrelated animals (Holmes and Sherman, 1983). Nonetheless, there is a paucity of information available on how kin recognition relates to the natural histories of the species being investigated and how it may be adaptive.

This chapter and the next will critically review most of the major experimental studies of kin recognition (other than parent–offspring relationships) that have been conducted with vertebrates (excluding primates). We will analyze the approaches the investigators have taken, the methods used and the conclusions drawn. We will also examine the studies from a proximate perspective that entails an investigation of the ontogeny and sensory bases of kin recognition. Finally we will analyze the studies from an ultimate perspective that considers the adaptive value of the behavior. Where possible, a species' kin recognition abilities will be examined with respect to its social ecology. Our synthesis ultimately allows predictions to be made concerning kin recognition in species that have not been studied, based upon what we know about kin recognition and the ecology of closely related, ecologically similar species.

Several mechanisms have been proposed to explain how individuals may discriminate between kin and non-kin (reviewed by Alexander, 1979; Bekoff, 1981a; Dawkins, 1982; Holmes and Sherman, 1982, 1983; Blaustein, 1983; Porter, 1986; Sherman and Holmes, 1985). These mechanisms may be used alone or in conjunction with one another and for illustrative purposes will be discussed in this section in the context of nepotism. We consider these mechanisms further in the following chapter (see also Chapters 2–4). It is necessary to keep in mind that the essence of Hamilton's (1964) inclusive fitness model is that, other things being equal, individuals should behave differently toward one another based on genetic relatedness, regardless of the mechanism(s) by which individuals determine the degree of genetic relatedness among themselves.

1. *Recognition based on spatial distribution.* If relatives are distributed predictably in space, altruistic acts might be selected for if the acts are directed preferentially toward those individuals in a particular location (e.g. Frase and Armitage, 1984). Such a location may be a home site or territory.

2. *Recognition based on familiarity and prior association.* If relatives are predictably present in specific social circumstances, recognition could occur through social learning (Alexander, 1979). Thus, individuals of the same litter, within the same nest, or those from one clutch, may learn subsequently to recognize 'familiar' individuals (see Bekoff, 1981a and Rheingold, 1985 for detailed discussions of familiarity). Relatives might also recognize one another if they predictably meet in the presence of a third individual who is familiar to each of them. One example of this might be two maternally related half-siblings from different litters that interact with their common mother (Holmes and Sherman, 1982).

3. *Recognition based on phenotype matching.* When using phenotype matching an individual learns and recalls the phenotypes of relatives or of itself if phenotypic similarity is correlated with genotypic similarity. The individual then is able to assess similarities and differences between the learned phenotype and unfamiliar conspecifics (also see Dawkins, 1982, for discussion of the 'armpit effect'). For example, if chemicals or odors are involved in kin recognition, they may have a genetic component, but must be learned for kin recognition to occur. Importantly, phenotype matching is fundamentally different from recognition based on familiarity because it allows an individual to recognize a totally unfamiliar individual. The 'genetic similarity theory' postulated by Rushton, Russell and Wells (1984) seems to be a reworking of the phenotype matching hypothesis.

4. *Recognition based on the action of recognition alleles.* Phenotypes could be used in recognition, independent of learning, if recognition alleles existed (see Blaustein, 1983). In this system, the phenotypic marker (e.g. a particular chemical or odor) and the recognition of that marker both have genetic bases.

Mechanisms 1 and 2 are actually indirect means by which kin recognition could develop. These explanations do not require that kin be recognized. Rather, it is assumed that those individuals most likely to be kin are the ones likely to be aided because of spatial distribution or prior association. Recognition errors in which non-kin are treated as related individuals, may be more likely if spatial distribution, familiarity, and prior association are the primary means of recognition. Recognition errors can also occur if mechanism three (phenotype matching) were utilized. If individuals have a similar phenotypic marker but such a marker is coded by different genes, or if the same genes coded for similar markers but the individuals were unrelated, recognition errors would be likely. The selective basis for mechanisms one and two is overall genetic relatedness, whereas for mechanism four it is the sharing of a particular gene (or small set of genes) that determines who will be aided. Mechanism three depends on either overall genetic relatedness or the particular gene(s) involved in producing the marker.

There is much controversy concerning the existence of recognition alleles. For example, Alexander and Borgia (1978) suggest that the existence of recognition alleles is unlikely for two basic reasons. First, as pointed out originally by Hamilton (1964), the actions of recognition alleles would be complex. The alleles would have to:
1. Be expressed phenotypically;
2. Cause the recognition of the phenotypic marker; and
3. Enable those individuals carrying copies of these alleles to favor other individuals also carrying the alleles (for detailed discussions see Alexander and Borgia, 1978; Alexander, 1979; Holmes and Sherman, 1982). An example is the 'green-beard' system (Dawkins, 1976) in which one such phenotypic marker might be a green-beard. Secondly, recognition alleles may be 'outlaw' alleles (Alexander and Borgia, 1978). Outlaw alleles favor themselves at the expense of all other alleles in the genome (including those at other loci).

The theoretical arguments concerning the improbability of recognition alleles have been generally accepted (examples are Kurland, 1980; Sherman, 1980a; Holmes and Sherman, 1982, 1983), although Hamilton (1964) suggested that the same *a priori* objections may be argued against the existence of assortative mating which has evolved 'despite its obscure advantages.' It is questionable that recognition alleles would be outlaws (see Ridley and Grafen, 1981; Rothstein and Barash, 1983). Acceptance of the theoretical arguments in the absence of empirical data may lead to premature rejection of the possible existence of recognition alleles.

Phenotype matching seems to be the most frequently invoked mechanism for kin recognition (see for examples Dawkins, 1982; Holmes and Sherman, 1982; Lacy and Sherman, 1983) in situations in which mechanisms one (spatial distribution) and two (familiarity) seem inapplicable. However, as discussed below, there are empirical data consistent with both phenotype matching and the existence of recognition alleles.

FISHES

An investigation of kin recognition in juvenile coho salmon (*Oncorhynchus kisutch*) was made by Quinn and Busack (1985). Salmon eggs were fertilized, incubated, and reared in the laboratory. Several full-sibling families were produced and test individuals were reared with their full-siblings. Kin recognition tests consisted of introducing a juvenile salmon into a tank and allowing it to enter one of two compartments, each containing a different stimulus. One chamber always held water conditioned with conspecific, unfamiliar, unrelated salmon of one particular family (sibship) as a standard; the other chamber contained blank water or water conditioned with either familiar siblings (individuals with which the test animals were reared) or unfamiliar siblings.

Generally, the data suggest that when given a choice of entering a test chamber containing water treated with siblings versus one treated with non-siblings, or containing sibling-treated water versus blank water, the salmon will choose the chamber with sibling-treated water. However, the data are open to question because of variable results or weak sibling preferences. For example, in the first series of tests in one experiment using fish from five families and allowing them to choose between the compartment treated with unfamiliar, unrelated conspecifics, or with familiar siblings, test individuals from only one of five families (20%) made a statistically significant choice for the familiar sibling compartment, although the pooled data were significantly different from random and in the direction of familiar siblings. Furthermore, in the first series of tests in another experiment, when individuals were given a choice of entering a chamber treated with unfamiliar, unrelated conspecifics (of the same family used in the second experiment) or a chamber treated with unfamiliar siblings, salmon from two families preferred the latter, individuals from two other families showed a random distribution of chamber choices, and individuals from the fifth family spent a statistically significant amount of time with unfamiliar non-siblings.

In a second series of tests that repeated the experiments described above, a different family of unfamiliar, unrelated fish, was used to treat the water in the standard portion of the chamber. Results of replicating the experiments were similar to one another and different from results obtained in the first series of tests. All families tested this time displayed a significant tendency to enter the sibling-treated water instead of water treated by unfamiliar, unrelated fish. The overall results of these tests are intriguing and it may be that fish from different families respond differently to related and unrelated conspecifics as Quinn and Busack (1985) suggest (see discussion of Stabell, 1982 below). Importantly, individuals from different populations, living under potentially different selective pressures, may respond differently.

Because individual salmon chose to associate with unfamiliar kin rather than unfamiliar non-kin, either phenotype matching or recognition alleles may be invoked to explain recognition. Furthermore, water-borne chemicals are implicated as the recognition cues (see also discussions in Hasler, Scholz and Horvall, 1978; Colgan, 1983, on salmon homing). Quinn and Busack's (1985) results showing that kin recognition in salmon is based on chemical cues are consistent with the results of studies of parent-offspring recognition in fish (reviewed by Barnett, 1977a). For example, in some cichlid fishes, parents will enter water containing their offspring significantly more often than water containing unrelated conspecifics (McKaye and Barlow, 1976). Furthermore, young cichlids can distinguish chemically between water in which mother fish are present and water in which no fish are present, although they do not distinguish between water in which their father is present and plain water (Barnett, 1977b).

We can only speculate as to the function of kin recognition in coho salmon. Juvenile coho salmon school under certain situations and it is possible that individuals schooling in kin groups could accrue advantages in terms of inclusive fitness compared to individuals not schooling in kin groups (discussed below). It is also possible that kin recognition is used to balance inbreeding and outbreeding because this species of salmon returns to its natal streams to spawn (Quinn and Busack, 1985).

In further support of the work by Quinn and Busack (1985), Stabell (1982) has shown that discrimination of sibling groups occurs in Atlantic salmon (*Salmo salar*). In a laboratory apparatus in which fish were given choices of associating with (a) the odors from a sibling group or tap water, and (b) a sibling group or a non-sibling group, salmon either avoided odors from the sibling group or preferred them, depending upon the population tested. Thus, this study also suggests that different populations respond to siblings in different ways. Moreover, odors from different populations may elicit differential responses in the olfactory bulbs of test fish (see Døving, Nordeng and Oakley, 1974).

AMPHIBIANS

Kin recognition has been investigated in the larvae of five species of anuran amphibians. Results of these studies and aspects of the larval ecology are summarized in Table 1 (see also Blaustein and O'Hara, 1986a). From the results of kin recognition tests in anuran larvae it is apparent that there are key differences in kin recognition behavior among species. These differences may reflect their different life histories and larval ecologies.

Tests of the tadpoles and froglets of the Cascades frog (*Rana cascadae*) and of tadpoles of the red-legged frog (*Rana aurora*) and western toad (*Bufo boreas*) were conducted by Blaustein and O'Hara (Table 1). Because of the ease in which different rearing regimes can be employed for amphibians, these studies placed particular emphasis on the ontogeny and sensory basis of kin recognition. Test individuals were reared:
1. With kin only.
2. With kin and non-kin (mixed rearing regime) by placing members of two different clutches on opposite sides of a screen within an aerated aquarium.
3. In isolation from an early embryonic stage.
4. With non-kin only by placing a test individual inside a screen surrounded by non-kin.

A rectangular tank was used to test tadpoles for kin preferences. Stimulus compartments separated from the main portion of the test tank by a screen, were placed on opposite ends of the tank and groups of tadpoles of varying numbers and composition (i.e. proportion of kin and non-kin) were placed in each compartment (for an illustration see O'Hara and Blaustein, 1981). The main portion of the tank was divided in half by a line in some tests and in three equal portions in other tests. The time that test tadpoles that were allowed to swim in the entire tank spent in the half containing a sibling group or non-sibling group was recorded in some tests. In other experiments, the time that test tadpoles spent in the third of the tank closest to siblings, the third closest to non-siblings or middle third, was recorded.

Cascades frog tadpoles

Rana cascadae tadpoles reared with kin or in mixed groups of kin and non-kin generally associate with kin in preference to non-kin in laboratory tests (O'Hara and Blaustein, 1981). Results of testing mixed-reared tadpoles show that experimental individuals preferentially chose to associate with unfamiliar siblings over unfamiliar non-siblings. They also associated preferentially with an unfamiliar stimulus group composed of siblings only over a familiar group composed of 50% siblings and 50% non-siblings. These results suggest that familiarity with other individuals is not necessary in *R. cascadae* kin recognition. However, individuals reared in mixed regimes failed to discriminate between familiar siblings and familiar non-siblings. This finding suggests that a composite 'odor' comprised of chemical cues from different sibling groups may form when tadpoles from different kin groups are confined together (see discussion in Waldman, 1985a).

These results, however, may reflect an insensitivity in the laboratory assay for kin recognition, because experimental animals from mixed-reared groups in experiments conducted under natural conditions preferred to aggregate with familiar siblings over familiar non-siblings (O'Hara and Blaustein, 1985). Two series of experiments illustrated that test individuals reared in isolation since an early embryonic stage subsequently aggregated preferentially with unfamiliar kin over unfamiliar non-kin (Blaustein and O'Hara, 1981, 1982a). Furthermore, tadpoles reared with siblings could assess the relative composition of sibling groups to some extent since test individuals chose to associate in the third of a test tank nearest the stimulus group containing 50% siblings and 50% non-siblings rather than the third containing 100% non-siblings (Blaustein and O'Hara, 1983). However, tadpoles failed to discriminate between a stimulus group composed of 75% non-siblings and 25% siblings and one containing 100% non-siblings. These data suggest the possibility of a threshold effect or that the water-borne chemical cues that *R. cascadae* tadpoles use to recognize kin (Blaustein and O'Hara, 1982b) may have become too diluted for test animals to distinguish the mixed kin/non-kin group from the pure kin group.

Because anuran eggs are laid within a jelly matrix produced by the female, it is possible that developing individual embryos may learn (become familiar with) cues that emanate from the jelly. To test this hypothesis as well as to

Table 1. Summary of the results of kin recognition tests conducted in the laboratory on the larvae of five species of anuran amphibians and some key ecological and behavioral characteristics of the species studied.

Results of kin recognition (*Rana cascadae*)	Cascades frog tadpoles (*Rana cascadae*)	Red-legged frog tadpoles (*Rana tests*)	Wood frog tadpoles (*Rana cascadae*)	Western toad tadpoles (*Rana aurora*)	American toad tadpoles (*Bufo americanus*)
	Single egg masses. Communal egg masses. 400–800 eggs/clutch. Low dispersal rates. Small cohesive aggregations usually less than 100 individuals[1] (Results below from Blaustein and O'Hara, 1981, 1982a, 1982b, 1983; O'Hara and Blaustein 1981).	Egg masses laid separately. 500–1100 per clutch. Aggregations composed of many non-kin[2]. (Results below from Blaustein and O'Hara, 1986b).	Communal egg laying. Thousands of eggs per clutch (up to 3,000). Generally do not aggregate in nature[3] (Results below from Waldman, 1984).	Communal egg laying. Thousands of eggs per clutch ($\bar{x}$ = 12,000). High dispersal rates. Large aggregations with thousands of individuals (many non-kin)[4]. (Results below from O'Hara and Blaustein 1982).	Communal egg laying. 2,000–13,000 eggs/clutch. High dispersal rates. Large aggregations with thousands of individuals (many non-kin)[5]. (Results below from Waldman, 1981, 1982).
Recognition after being reared only with kin	Yes[6]	Partial; recognition disappeared as tadpoles developed.	Yes	Yes	Yes[6]
Recognition after being reared with kin and non-kin	Yes[6]	No	Yes[7]	No	No[8]
Recognition after being reared in isolation	Yes	No	Experiments not conducted.	Experiments not conducted.	Yes
Recognition after being reared only with non-kin	Yes	No	Experiments not conducted.	No	Experiments not conducted.

| Recognition of individuals of varying degrees of relatedness | Yes; discrimination between full-siblings and both paternal and maternal half-siblings. | Experiments not conducted. | Experiments not conducted. | Experiments not conducted. | Partial; full siblings are distinguished from only paternal half-siblings, not maternal half-siblings. |
| Possible recognition mechanism(s) | Phenotype matching and/or recognition alleles. | Familiarity. | Phenotype matching, recognition alleles, familiarity. | Familiarity. | Phenotype matching, recognition alleles, familiarity. |

[1] Ecological and behavioral data are from Blaustein and O'Hara, 1981, 1982a,b; O'Hara, 1981; Nussbaum, Brodie and Storm, 1983.
[2] Nussbaum, Brodie and Storm, 1983 and references therein.
[3] Nussbaum, Brodie and Storm, 1983; Waldman, 1984; Wright and Wright, 1949 and references therein.
[4] O'Hara, 1981; O'Hara and Blaustein, 1982.
[5] O'Hara and Blaustein, 1982; Waldman, 1982 and references therein.
[6] Results corroborated in field experiments.
[7] These results are equivocal because in the mixed rearing regime tadpoles were probably not equally exposed to siblings and non-siblings (Waldman, 1983; p. 271).
[8] Tadpoles can identify kin in field experiments.

Errata

On page 294, the Column headings for Table 1 should read as follows:

Results of kin recognition tests	Cascades frog tadpoles (*Rana cascadae*)	Red-legged frog tadpoles (*Rana aurora*)	Wood frog tadpoles (*Rana sylvatica*)	Western toad tadpoles (*Bufo boreas*)	American toad tadpoles (*Bufo americanus*)

test whether tadpoles can distinguish individuals of varying degrees of relatedness, a series of jelly mass manipulation experiments and experiments using half-siblings was conducted (Blaustein and O'Hara, 1982a). Several rearing regimes were employed before animals were tested. Test animals were reared in isolation or with kin. Certain test animals had their surrounding jelly matrix removed entirely, presumably before exposure to jelly could have had an effect. Other animals had their jelly removed and were reared with a massive quantity of jelly from a non-kin egg mass.

Tadpoles preferred to associate with full-siblings over either maternal or paternal half-siblings, and preferred to associate with half-siblings (either paternal or maternal) over non-siblings. Because full-siblings were chosen over maternal half-siblings and paternal half-siblings were chosen over non-siblings, it is clear that maternal cues are not necessary for kin recognition in *R. cascadae*. Furthermore, the results of the jelly manipulation tests showed that test animals were unaffected by either the removal of jelly or addition of foreign (non-kin) jelly during development but that maternal cues may exert a stronger influence than paternal cues since maternal half-siblings were preferred over paternal half-siblings. Association tests with half-siblings showed that *R. cascadae* tadpoles can use either maternal or paternal cues in sibling recognition and that there may be a hierarchy of cue importance.

The *R. cascadae* tests in the laboratory have been corroborated in field experiments using test individuals that were reared in the laboratory and released in natural ponds in the Oregon Cascade mountains (about 200 km east of Corvallis and 64 km north of Eugene) (O'Hara and Blaustein, 1985). Thirty-one *R. cascadae* egg clutches from six populations were obtained; the larvae from each sibship were reared in separate tanks (isolated from other sibships) or with one other sibship in the same tank (mixed-rearing regime) for three to six weeks. Equal numbers (100–300 individuals) of tadpoles from two unrelated sibships were dyed with neutral red or methylene blue and released together in the ponds. For several days thereafter, single aggregations (groups with at least six tadpoles clustered within approximately 4 cm of one another) were captured repeatedly, censused for sibship composition, and again released. Control groups consisted of tadpoles from one sibship that were dyed two different colors to determine if dye color influenced tadpole aggregation tendencies. A total of 25 tests was conducted which sampled over 10,000 tadpoles from 353 experimental and 103 control aggregations.

The distribution of color compositions of aggregations in control tests was random. Color compositions of aggregations in experimental tests differed from controls and from a random distribution (Fig. 1); such aggregations were dominated by one of the two colors (sibships). Therefore, *R. cascadae* tadpoles recognized and preferred to associate with siblings in natural field conditions. Circumstances of early rearing (i.e. whether tadpoles were reared with siblings or in mixed sibling/non-sibling groups) had no influence on the preference to associate with siblings.

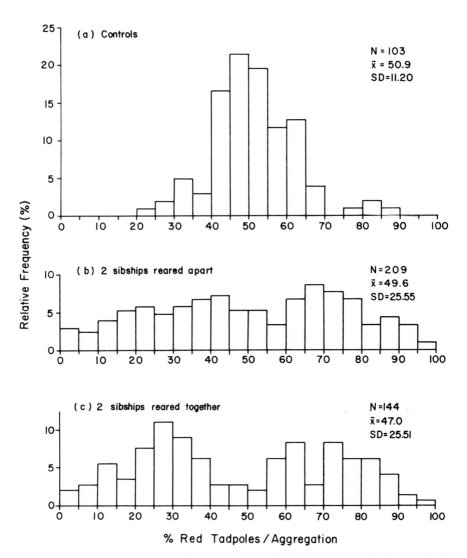

Fig. 1. Frequency distributions of the percentage of *Rana cascadae* tadpoles dyed red comprising each aggregation sampled in three experimental treatments under field conditions in the Cascade Mountain range of Oregon, USA. *Rana cascadae* tadpoles aggregated preferentially with siblings in these field experiments. *Reproduced from O'Hara and Blaustein, 1985,* Oecologia, *67, 44–51, with permission from Springer–Verlag, Heidelberg*

By blocking visual cues of test tadpoles, by impeding chemical flow from the stimulus individuals to test animals and by assessing the sound production of individuals, it was found that tadpoles distinguished between kin and non-kin using water-borne chemical cues (Blaustein and O'Hara, 1982b). It is probable that other species tested also use water-borne chemical cues (see Waldman, 1985b).

Finally, *R. cascadae* frogs retain their kin preference behavior after metamorphosis (Blaustein, O'Hara and Olson, 1984). These results pose interesting questions concerning the ontogeny of behavior and the function of kin recognition in amphibians (discussed in detail below).

Red-legged frog tadpoles

Of 140 *Rana aurora* frog tadpoles tested by Blaustein and O'Hara (1986b), only those reared with siblings and tested in early developmental stages showed a tendency to associate with kin. This tendency disappeared as tadpoles developed. Animals reared in mixed-rearing regimes, in isolation, and with non-kin only showed no preference to associate with siblings. Thus, although *R. aurora* and *R. cascadae* are closely related, they do not appear to be similar in their kin recognition behavior.

Western toad tadpoles

Bufo boreas tadpoles reared with kin preferentially associated with kin over non-kin in laboratory experiments (O'Hara and Blaustein, 1982). However, individuals reared with both kin and non-kin, and others reared only with non-kin, displayed random association with the kin and non-kin groups. The ontogeny of sibling recognition in *B. boreas* tadpoles is influenced by the rearing regime. Tadpoles reared with siblings preferred to associate with siblings over non-siblings (five of six tests; 120 tadpoles), but tadpoles exposed to siblings and non-siblings before testing showed random association (nine of nine tests; 180 tadpoles). Even when preferences for siblings were fully established following prolonged rearing with siblings, short-term exposure to non-siblings nullified these preferences and test tadpoles treated in this way displayed a random distribution with regard to kin and non-kin stimuli. However, results of rearing animals with only non-kin suggest that social preferences are neither (a) entirely labile nor (b) based solely on familiarity, because in subsequent tests tadpoles did not associate with the non-kin group members with which they were reared.

American toad tadpoles

Waldman and Adler (1979) and Waldman (1981) conducted open-field laboratory experiments on kin recognition in American toad (*Bufo americanus*) tadpoles (see also Waldman, 1986). In these tests the two sibships were dyed different colors. Equal numbers of the two sibling groups were placed in a laboratory pool simultaneously and the positions of individuals were recorded at several times for several days. For each trial, the distances from each tadpole to its nearest sibling and nearest non-sibling were recorded.

These tests suggest that *B. americanus* tadpoles reared with only siblings or in isolation are found significantly nearer their siblings than non-siblings (Waldman, 1981). Furthermore, tadpoles reared with siblings in early development only and then exposed to siblings and non-siblings in later developmental stages, retain their distance affiliation toward siblings (Waldman, 1981). However, tadpoles that were reared with siblings and non-siblings in early development and exposed to siblings only in later development did not preferentially associate with familiar siblings over non-siblings (Waldman, 1981).

Although jelly mass manipulation experiments were not conducted for *B. americanus*, tests with half-siblings reveal that a significant maternal component may be involved in recognition. Maternal half-siblings were not differentiated from full-siblings, although paternal half-siblings were distinguished from full-siblings by nearest neighbor distances (Waldman, 1981).

Field experiments generally corroborate these laboratory results (Waldman, 1982). However, there was one discrepancy; tadpoles reared continuously in mixed sibling/non-sibling groups in the laboratory failed to discriminate between siblings and non-siblings. But in field experiments, tadpoles could make this discrimination. However, there were differences in the rearing techniques and early experience of tadpoles preceding laboratory tests when compared with the animals used in the field. During early development, test individuals used in field experiments may have been exposed to siblings to a much greater extent than to non-siblings, unlike individuals reared for laboratory tests (Waldman, 1982).

Wood frog tadpoles

Tests using *Rana sylvatica* tadpoles were conducted in the laboratory by Waldman (1984) using methods described previously for *B. americanus*. In this study, Waldman tested individuals reared with only siblings and in different, mixed sibship environments.

Tadpoles reared with only siblings displayed a mean nearest-neighbor distance to siblings that was smaller than that to non-siblings. However, within successive tests, there was a trend of increasing nearest neighbor distances. These changes in distance may have been due to unexplained mortality or ontogenetic changes during the tests, either of which could have influenced the results (Waldman, 1984). However, it is unlikely that ontogenetic changes in only four days of testing (see Waldman, 1984) could have influenced the results in this way.

Interesting data emerge from tests of individuals from mixed-rearing regimes. For example, results using mixed-reared tadpoles suggest that they may be able to discriminate between familiar siblings and familiar non-siblings; the mean nearest neighbor distance between familiar siblings was smaller than for familiar non-siblings. Moreover, there was a tendency for tadpoles to associate more closely with familiar siblings than with unfamiliar siblings, suggesting an effect of familiarity. These findings seem contradictory because non-siblings reared together (familiar) were distinguishable from one another (based on nearest neighbor distances) and siblings reared separately (unfamiliar) were also distinguishable from one another. Thus, the importance of familiarity is perplexing. These results may reflect inconsistent rearing procedures since members of one sibship were reared in a single basket within a common tank with many baskets and under these circumstances tadpoles in different baskets were probably not exposed equally to siblings and non-siblings (Waldman, 1983). Thus, test animals probably responded to familiar individuals with whom they were reared.

A comparative discussion of amphibian kin recognition tests

Larvae of four of the five species of amphibians tested display a well-formed ability to recognize kin. Only *R. aurora* tadpoles failed to display this behavior throughout development. It is possible that the methods used were not sensitive enough to determine if *R. aurora* tadpoles actually can discriminate between kin and non-kin. However, results using similar techniques with *R. cascadae* and *B. boreas* and natural history comparisons, suggest that it is likely that *R. aurora* does not have the same ability to recognize kin.

The rearing regime seems to influence greatly the ontogeny of kin recognition in toad tadpoles. Both *B. boreas* and *B. americanus* larvae fail to recognize kin after being reared with kin and non-kin. However, there are differences in the ontogeny of kin recognition behavior in these two species. *B. boreas* tadpoles reared with siblings only during early development and then exposed even briefly to non-siblings, lose their tendency to associate with siblings, whereas *B. americanus* tadpoles retain their ability to associate with siblings when reared similarly. However, *B. americanus* tadpoles exposed to both non-siblings and siblings in early development do not spend significantly more time nearer to siblings. These results suggest the possibility of a sensitive period in early development during which *B. americanus* tadpoles familiarize themselves with individuals with which they are reared. These behavior patterns are ummodifiable in later ontogenetic stages (see discussion in Waldman, 1983).

Although kin recognition by *B. boreas* tadpoles seems more labile than in

B. americanus because exposure to non-siblings causes tadpoles to lose kin affiliative tendencies, the behavior patterns are not entirely modifiable and cannot be fully explained by the mechanism of familiarity. For example, individual *B. boreas* tadpoles reared with only non-kin until tested failed to associate with members of the non-kin group with which they were reared. Instead, tadpoles reared in this manner showed a random distribution within the test tank. It is possible that *B. boreas* reared with only non-kin cues were underexposed to these cues, or that the non-kin stimulus was not strong enough to counter the 'self-exposure' stimulus which may have been the most pervasive one. Similarly, the self-exposure stimulus and non-sibling stimulus could have blended and formed a composite stimulus with which tadpoles became familiar. Thus, the two stimulus choices given to *B. boreas* test individuals could have been perceived as equal.

Based on results using *B. boreas*, it is likely that familiarity (Bekoff, 1981a) is the key mechanism of recognition in larvae of this species. Familiarity also plays a role in *B. americanus*. However, since *B. americanus* tadpoles reared in isolation retain their kin affiliative tendencies, these larvae may also use phenotype matching by becoming familiar with themselves during rearing. Furthermore, maternal cues seem to be necessary in *B. americanus* recognition, since tadpoles fail to discriminate between maternal half-siblings and full-siblings.

Familiarity with other individuals seems less important in the development of kin recognition in *R. cascadae*. Tadpoles reared with only siblings, in mixed groups of siblings and non-siblings, in isolation, or with only non-siblings, still retain their kin affiliative tendencies. Moreover, *R. cascadae* tadpoles reared in a mixed group can discriminate between an unfamiliar stimulus group composed only of kin versus a totally familiar group composed of 50% kin and 50% non-kin (O'Hara and Blaustein, 1981). *R. cascadae* tadpoles do not need maternal cues to discriminate between kin and non-kin; they can distinguish paternal half-siblings from non-siblings and maternal half-siblings from full-siblings. Results of rearing individuals without their egg mass jelly or with jelly from a non-kin egg mass further suggest that *R. cascadae* do not need maternal cues for kin recognition. The results of tests with *R. cascadae* suggest that the tadpoles use phenotype matching and/or recognition alleles in discrimination.

In contrast, familiarity seems to play some role in *R. sylvatica* kin recognition because individuals associate more closely with familiar than with unfamiliar siblings in laboratory tests. However, results pertaining to mixed-reared individuals suggest that *R. sylvatica* tadpoles reared with siblings and non-siblings may also use phenotype matching to discriminate between familiar siblings and familiar non-siblings.

The results of kin recognition tests using larval anurans must be considered with regard to both the larval and adult ecology and behavior of the species under natural conditions. When considered from an evolutionary perspective it is somewhat surprising that tadpoles of several species associate preferentially with kin.

Relationship of larval anuran ecology to kin recognition

Unfortunately, little is known about the larval ecology of *R. aurora*. Two pertinent facts are:
1. There is no evidence that larval *R. aurora* aggregate with conspecifics in nature.
2. *R. aurora* is closely related to *R. cascadae* (Wright and Wright, 1949; Stebbins, 1954).

Results showing that only *R. aurora* tadpoles reared with siblings in the early developmental stages preferred to associate with kin, along with what is known about *R. aurora* larval ecology, make it difficult to interpret the importance of kin recognition in this species. However, we now know that the larvae of two closely related species (*R. aurora* and *R. cascadae*) display significantly different patterns of kin recognition behavior. This is most interesting because until recently these two species were classified as one species (Stebbins, 1954; Wright and Wright, 1949). Thus, local environmental (biotic and abiotic) conditions in which different species, or in which different populations of the same species, exist, may be important factors influencing the evolution of kin recognition behavior.

The larvae of both *B. boreas* and *B. americanus* form large aggregations (Fig. 2) consisting of many non-kin. Because *Bufo* tadpoles may be unpalatable to some predators (Wassersug, 1973; Kruse and Stone, 1984; Hews and Blaustein, 1985) and are highly conspicuous when aggregated, Wassersug (1973) suggested that their social behavior could result from kin selection. Thus, a predator that eats a tadpole and learns it is distasteful, will tend to avoid tadpoles of the same species nearby (see also Fisher's, 1930, discussion on the evolution of distastefulness in gregarious insects). The loss of one or a few tadpoles could benefit other members of the group and could favor the maintenance of sibling associations through enhancement of inclusive fitness.

For kin selection to operate, however, it is essential that siblings have an opportunity to associate sometime during development. If, for example, members of a brood randomly disperse at hatching so that future contact with siblings is unlikely, behaviors that facilitate cooperation with siblings would have little chance of evolving. Observations of the embryonic environment, larval aggregation behavior, and larval dispersal patterns of *B. boreas* in the field strongly suggest that the opportunity for tadpoles of this species to interact with siblings is low. Tadpole activity leads to mixing among sibships from early larval stages through development (see also Samollow 1980 and discussion below).

Fig. 2. The large extended black mass close to the shore-line is a typical aggregation of *Bufo boreas* tadpoles. These aggregations are generally composed of hundreds of thousands or millions of individuals from numerous kin groups

These observations do not rule out the possibility that large aggregations could be composed of several sibling cohorts and, although mixing of aggregations is frequent, limited sibling association might occur. However, field observations must be judged in conjunction with laboratory findings. While we would expect some mixing of siblings and non-siblings to occur in the field even if tadpoles preferentially associate with kin, experiments with *B. boreas* show that limited, short-term exposure to non-siblings results in a loss of sibling preference. In nature, *B. boreas* tadpoles probably encounter groups comprised of individuals from many sibships of which only a small proportion may include siblings. Thus, *B. boreas* larvae probably have little opportunity to learn and to respond to sibling-specific cues particularly during active stages, and even if sibling preferences develop during embryonic stages or immediately following hatching, subsequent encounters with non-siblings will probably alter these preferences. Therefore, there seems to be little justification to invoke kin selection as the main mechanism underlying the observed social behavior of *B. boreas* tadpoles.

A similar conclusion can be drawn concerning *B. americanus* larvae. Observations in the field (Beiswenger 1972, 1975, 1977; R. K. O'Hara, personal communication) indicate behavioral differences from and some similarities to *B. boreas*. Aggregation sizes may be smaller in *B. americanus* and the polarized, swimming schools, commonly observed in *B. boreas*, are infrequent. Otherwise, aggregation behavior is similar in these species. Aggregations form, split and combine with others, frequently within a given day. When *B. americanus* larvae become active and begin aggregating they disperse from oviposition sites and 'members of different broods freely intermingle' (Beiswenger, 1972). Also in four field experiments by Beiswenger (1972), larvae that were transferred from single aggregations in one pond to another pond 'intermingled freely' and 'randomly mixed' with resident tadpoles. If sibling preferences are manifested during some critical period prior to active, dispersing stages (Waldman, 1981), limited segregation of kin may occur, particularly if egg masses of different broods do not closely overlap and dispersal in later stages is non-random. A critical test that has not yet been conducted would be to expose *B. americanus* tadpoles to non-siblings at the time active swimming commences before testing their affiliative preferences.

The evidence suggests that larval *R. sylvatica* generally do not aggregate but commonly avoid each other in nature (Hassinger, 1972; DeBenedictis, 1974; Waldman, 1983, 1984). Therefore, there is no *a priori* reason to predict that *R. sylvatica* would display kin recognition.

In contrast, field observations of *R. cascadae* and controlled field and laboratory experiments using *R. cascadae* larvae seem consistent with a kin selection model. Observations by R. M. Storm (personal communication) for four decades, and by R. K. O'Hara (personal communication) since 1975 reveal that *R. cascadae* larvae are almost never found alone in nature.

Rather, they live in close, social aggregations composed generally of fewer than 100 individuals (Fig. 3). Larval development in the egg mass may occur in proximity to both non-kin and kin (see Blaustein and O'Hara, 1982a; Sype, 1975). Thus, there may have been strong selection for a precise kin recognition mechanism. Furthermore, field and laboratory experiments (O'Hara, 1981) suggest that *R. cascadae* larvae prefer to associate with conspecifics whether related or not over being alone (corroborated in Blaustein and O'Hara, 1983). Moreover, dispersal from sites of oviposition is low (O'Hara, 1981). Therefore, it is possible that natural aggregations of *R. cascadae are composed primarily of kin.*

Fig. 3. *Rana cascadae* tadpoles form relatively small aggregations of generally fewer than 100 individuals. Individuals are gregarious and are almost never found alone

The ability of *R. cascadae* to distinguish between individuals of various degrees of relatedness by using either paternal or maternal cues suggests that sibling recognition in this species is finely tuned. This ability may be especially important in *R. cascadae* because adults are philopatric (D. H. Olson, personal communication; personal observations) and inbreeding effects may produce individuals of varying degrees of relatedness. Recognition of half-siblings could further reduce recognition errors (see below) if altruistic or cooperative acts are important in increasing an individual's inclusive fitness. Moreover, the tadpoles are subjected to intense predation pressure (personal observation; O'Hara, 1981) and they exhibit an alarm reaction in response to injured conspecifics (Hews and Blaustein, 1985).

The alarm response could be important in warning relatives within the aggregation. Kin recognition could enhance the ability of related individuals to regroup after disturbances, such as predatory attempts, which cause tadpoles to disperse. Altruists in groups with many relatives could increase their inclusive fitness by aiding members of the group, compared to those individuals giving similar aid in groups with fewer or no siblings. Depending upon the costs and benefits involved, increases in inclusive fitness could be further enhanced if individuals aided distant as well as closely related kin. Life history characteristics could have responded to kin selection in the past, and kin recognition may have been an important mechansim facilitating larval nepotism in *R. cascadae*. Conversely, kin recognition may have evolved in association with, or as an adaptive response to, certain life history characteristics. It is also possible that kin recognition allows adult frogs to balance inbreeding and outbreeding (Blaustein, O'Hara and Olson, 1984; see also discussions in Bateson, 1980, 1982, 1983; Shields, 1982).

In the other species tested, kin selection could have been instrumental in the evolution of social behavior patterns, possibly under different environmental and social conditions. A remnant of this recognition ability may enable larvae to display nepotism under certain conditions in present environmental regimes (e.g. in *B. boreas* and *B. americanus* if a single clutch is deposited in one area of a pond or a few clutches are deposited in distinct habitats such that some segregation between progenies is maintained). Alternatively, preferences based on familiarity could be useful in habitat selection, locating food, or in responding to conspecifics (i.e. in aggregation behavior). For example, *Bufo* or *R. sylvatica* tadpoles may be using the most familiar cues—other individuals of the same species—to seek specific microhabitats. *Bufo* or *R. sylvatica* tadpoles that aggregate (with kin or non-kin) could benefit by an enhanced ability to locate and procure food (Beiswenger, 1975; Wilbur, 1977) or by decreasing the risk of predation (Hamilton, 1971; Bertram, 1978).

BIRDS

Kin recognition has been studied in various birds, including Japanese quail (*Coturnix coturnix*; Bateson, 1978, 1980, 1982, 1983), bank swallows (*Riparia riparia*; Beecher and Beecher, 1983), and Canada geese (*Branta canadensis*; Radesäter, 1976) (also see reviews in Gottlieb, 1968, 1971; Evans, 1970). These studies and those investigating other aspects of recognition reveal species differences in the ontogeny of recognition and the probable proximate mechanisms involved (Table 2).

Table 2. Results of kin recognition tests in birds and some key ecological and behavioral characteristics of the species studied.

	Japanese Quail (*Coturnix coturnix*)	Lesser Snow Geese (*Anser caerulescens*)	Canada Geese (*Branta canadensis*)	Bank Swallows (*Riparia riparia*)
Results of kin recognition tests	Natural history poorly understood[1], may be solitary at times and gregarious at times[2]. Variable pair-bond ranging from monogamy to polygamy depending on local sex ratio[2]. Female generally incubates eggs alone but male may assist[2]. (Results below from Bateson, 1982.)	Pair-bonds are usually life long and family groups stay together for almost a year[3]. Assortative mating occurs[3]. Pairing takes place when birds from many breeding colonies intermingle[3]. Birds that pair are generally from different natal colonies[3]. (Results below from Cooke, Finney and Rockwell, 1976 and Cooke, 1978.)	Pair-bonds form on wintering grounds during the early stages of migration[4]. Pair-bonds are usually life long[4]. Family groups, consisting of established pairs and young of previous year, arrive at breeding site together[4]. (Results below from Radesäter, 1976.)	Colonial; hundreds of nests in colony with synchronous reproduction[5]. Young often enter burrows other than their own[5]. Exist in family groups[5]. (Results below from Beecher and Beecher, 1983.)
Basis for kin recognition	Individuals spend more time near unfamiliar first cousins or third cousins than near familiar siblings, unfamiliar non-siblings or unfamiliar non-kin.	Individuals mate with others having the same plumage color as their parents or siblings in laboratory tests and in field observations.	In laboratory tests goslings choose to associate more often with individuals from their own rearing group over individuals they were not reared with.	Birds respond more strongly to calls of their own sibling group than to calls of unrelated birds.
Recognition of relatives of varying degrees of relatedness	Yes	Experiments not conducted.	Experiments not conducted.	Experiments not conducted.
Possible recognition mechanism(s)	Phenotype matching, recognition alleles.	Familiarity.	Unknown.	Familiarity.

[1] Bateson, 1980. [2] Cramp, 1980. [3] Cooke, 1978 (and references therein). [4] Van Wormer, 1968. [5] Beecher, Beecher and Hahn, 1981; Beecher, Beecher and Lumpkin, 1981.

Japanese quail

Bateson (1978, 1980, 1982, 1983) has used a variety of procedures to test mating preferences in Japanese Quail. In one important study (Bateson, 1982; see also discussion in Bateson, 1983), quail were reared with siblings for 30 days after they hatched, and were then isolated for an additional 30 days. When birds were sexually mature (60 days after hatching), they were tested for association preferences. One test individual was placed within a circular apparatus containing separate compartments which housed a familiar sibling, unfamiliar sibling, unfamiliar first cousin, unfamiliar third cousin or unfamiliar non-sibling. The time spent in front of each of these stimulus compartments was recorded for 30 min for each test bird. The results showed that these birds associated most often with unfamiliar first cousins. Thus, Japanese quail were able to discriminate among individuals of varying degrees of relatedness, and familiarity with the particular stimulus bird was not necessary. Japanese quail may have become familiar with themselves before tests were conducted or with the quail with which they were reared, and thus could have used phenotype matching to recognize unfamiliar individuals. They may have also used recognition alleles.

The ability to recognize individuals of varying degrees of relatedness seems to play an important role in balancing inbreeding and outbreeding in Japanese quail (see discussion in Bateson, 1983 and a similar study by Slater and Clements, 1981 on Zebra finches). Unfortunately, so little is known about the natural history of this species that it is unclear how the recognition behavior in laboratory tests relates to behavior in natural situations.

Bank swallows

In a follow-up study to their investigation of parent-offspring recognition in bank swallows (Beecher, Beecher and Hahn, 1981; Beecher, Beecher and Lumpkin, 1981), an investigation of sibling recognition was undertaken by Beecher and Beecher (1983). Two experiments were conducted to test whether swallows could recognize their siblings and to investigate the proximate mechanism of recognition.

In the first experiment, a sibling group ($n = 3-5$ birds) was removed from its burrow and confined to a small box. Tape recorded calls from their own burrow and from an unrelated sibling group were played to the test birds. Chicks gave significantly more calls in response to siblings' calls than to calls from an unrelated group. In the second experiment, test chicks were removed from their nests at eight to ten days of age and raised in isolation in the laboratory, These chicks were exposed to calls from unrelated birds for up to 20 days, at which time their responses to familiar calls from unrelated birds and unfamiliar calls from unrelated birds were assessed. Results showed that chicks approached the speaker playing familiar calls significantly more often than they approached a speaker playing unfamiliar calls.

These data suggest that bank swallows can recognize siblings and that the recognition is probably based on familiarity, the mechanism probably used in parent-offspring recognition (Beecher, Beecher and Hahn, 1981). In the second experiment it is possible that test birds could have learned components of their own calls or calls of their siblings prior to their separation and isolation at eight to ten days of age. However, this is unlikely because mature calls are apparently not developed in this species until birds are about 16 days old. To investigate more directly the importance of genetic relatedness in bank swallow sibling recognition, it would be important to conduct a test giving individuals a choice of associating near speakers playing calls of unfamiliar unrelated birds and unfamiliar related birds. To the best of our knowledge, this experiment has not been done.

Learning familiar cues is apparently efficient in bank swallow parent-offspring recognition but it is not error-free (Beecher, 1982; see below). Bank swallows are highly social and live in large colonies containing hundreds of nests (Hoogland and Sherman, 1976; Beecher, Beecher and Lumpkin, 1981). Young swallows often enter burrows other than their own and are sometimes fed by unrelated adults. Therefore, recognition errors on the part of both the parent and offspring may occur. Of course these young birds may benefit in a variety of ways from being fed by unrelated birds, and therefore they may not be entering these foreign burrows because of recognition errors. However, alien chicks are usually eventually evicted from nests (Beecher, Beecher and Lumpkin, 1981). Due to their highly social nature, numerous but as yet undetermined benefits could accrue as a result of recognizing siblings.

Canada geese

Radesäter (1976) conducted laboratory experiments in which goslings were hatched in incubators and reared in groups of five (unfortunately we do not have information on the genetic relatedness among members of these groups). Test goslings were given a choice of associating with:
1. Their own group or an unfamiliar group.
2. A single individual from their own group or a single individual from an unfamiliar group.
3. A single individual from their own group or a strange group.

The results of these experiments were ambiguous. Young (below nine days old) test goslings showed no preference for their own group versus an unfamiliar group, but older goslings chose to associate with the familiar group over the unfamiliar group. However, young goslings chose to associate with an individual member of the familiar group rather than an individual member of the unfamiliar group, but these preferences were not exhibited in older birds. Finally, young birds chose to associate with a strange group over a single familiar bird, but there was a tendency for older birds to prefer to associate with a familiar individual instead of an unfamiliar group. Because the exact rearing regime was not reported in this study the mechanism of kin recognition is unknown. Kin recognition may be important in maintaining family groups that arrive at breeding sites together (Van Wormer, 1968) and recognition abilities may play a role in mate choice as it appears to do in lesser snow geese.

Lesser snow geese

Recognition in lesser snow geese (*Anser caerulescens*) has been investigated in a variety of laboratory and field experiments. These studies have revealed positive assortative mating based on matching a single phenotypic character, plumage color (blue or white), which is controlled by a single gene (Cooke and Cooch, 1968; Cooke, Mirsky and Seiger, 1972). Cooke, Mirsky and Seiger (1972) and Cooke and McNally (1975) showed that goslings placed in a free choice situation with unfamiliar test birds significantly preferred to approach birds of parental color. The color of a gosling's siblings also influenced its preference; a gosling with siblings whose color was opposite to that of its parents chose parentally colored birds less frequently than did goslings from families in which all birds were the same color. Young birds tested for color preferences associated with birds of the same color as their parents or foster parents. When the original foster parents were removed and replaced by birds of an opposite color, the goslings' preference was for the color of the most recent foster parent. Offspring of parents of mixed colors do not show a color preference (Cooke, Finney and Rockwell, 1976).

In open-field situations in the absence of parents, birds generally associated with siblings. When birds associated with non-siblings, they displayed a preference for birds of the same color as their siblings. An individual's own color does not seem to be of general importance in the mate selection processes (Cooke, Finney and Rockwell, 1976; Cooke, 1978). In cases where a particular bird is the only representative of its color in the family, these different colored individuals chose mates of the family color (Cooke, 1978).

The overall results of studies of lesser snow geese suggest a positive assortative mating system based on familiarity with other individuals. The data indicate that recognition tendencies are not fixed during development and that plumage color is the key phenotypic marker. Although this is generally a straightforward positive assortative system, about 11% of the birds with both parents of the same color choose a mate of opposite color. Recognition among siblings may be important in retaining the integrity of families; siblings may remain together even after they leave their parents (Cooke, 1978).

MAMMALS (except primates)

Numerous studies of small mammals have provided a wealth of information on the ontogeny of kin recognition. They have also revealed important contrasts in how the behavior is manifested in different species.

Spiny mice

The pioneering and continuing research of Porter and his colleagues on kin recognition in captive spiny mice (*Acomys caharinus*) probably represents the most comprehensive study of kin recognition in a single mammalian species (Table 3). Kin recognition in spiny mice is probably mediated by more than one mechanism (Porter, Matochik and Makin, 1984; Porter, 1986). The research protocol involved introducing pairs having various combinations of relatedness and familiarity into a test cage, and quantifying the dyadic associations (huddling) (e.g. Porter, Wyrick and Pankey, 1978; Porter and Wyrick, 1979).

When littermate siblings and non-sibling cagemates were placed in the test terrarium, there was a preference for mice to huddle with their familiar siblings (Porter, Wyrick and Pankey, 1978). However, preferential huddling with siblings did not occur when test individuals were housed with non-siblings for five days prior to being tested, suggesting that exposure to non-siblings nullifies sibling preferences. Test individuals that were isolated for three to five days from the day of weaning retained their huddling preference for siblings but those isolated for eight days failed to show this preference (Porter and Wyrick, 1979). If, however, test individuals were exposed to siblings for as few as 4 h after their isolation period, their preference for huddling with siblings was reinstated.

The role of experience in sibling recognition was investigated further when Porter, Tepper and White (1981) cross-fostered pups on the day of birth, ten days after birth or 20 days after birth. Three classes of dyadic pairings were possible in tests of huddling preferences:
1. Biological sibling pairings
2. Foster littermate pairings.
3. Pairings of unrelated, unfamiliar weanlings.

The results showed that foster littermates are preferred huddling partners over unfamiliar biological siblings and over unrelated, unfamiliar animals.

Moreover, the data suggest that maternal influence is important in spiny mouse kin recognition; siblings that were reared apart but nursed by the same lactating female showed significantly more pairings than non-sibling pairings (Porter, Tepper and White, 1981). Siblings reared apart and nursed by different lactating females displayed huddling preferences that were not significantly different from pairings of non-siblings nursed by different females. Finally, non-siblings reared apart but nursed by the same female displayed significantly more dyadic pairings than did unfamiliar non-siblings

Table 3. Summary of the results of representative kin recognition experiments in rodents and some key ecological and behavioral characteristics of the species studied.

	Spiny mouse (*Acomys caharinus*)	House mouse (*Mus musculus*)	White-footed mouse (*Peromyscus leucopus*)	Deer mouse (*Peromyscus maniculatus*)
Summary of kin recognition tests	Highly social. 1–4 young/litter. Adult females in social group assist with birth and tend young[1]. (Results below from Porter and Wyrick, 1979; Porter, Wyrick and Pankey, 1978; Porter, Tepper and White, 1981; Porter, Matochik and Makin, 1983.)	Highly social[2]. Group territorial[2]. Dispersal is variable with some individuals dispersing great distances; others do not disperse far from natal territory[3].	Highly social[4]. Relatively stable home ranges[5]. (Results below from Grau, 1982.)	Can be highly social with cooperative family units nesting together[6,7,8]. Dispersal may be related to social factors[9]. (Results below from Dewsbury, 1982.)
Basis of kin recognition	Siblings huddle more often with each other than with non-siblings in laboratory tests	(Results from Kareem and Barnard, 1982; Kareem, 1983.) Kin display fewer aggressive behaviors than non-kin in laboratory tests.	Individuals investigated related strangers more than unrelated strangers in the laboratory.	Reduced breeding between siblings as compared to non-siblings in laboratory tests.
Importance of *in utero* environment	Important. Pups suckling from same female interact as if kin even if not previously exposed to one another. Odor cues transferred in milk, saliva or urine.	Not important (see below).	(Results from Yamazaki et al., 1976, 1980.) In mate choice tests congenic males of various strains generally choose females that differ at a single locus in the major histocompatibility complex (*H-2*). Experiments not conducted but recognition of non-littermate siblings could be due to common *in utero* environment.	Experiments not conducted.

	Cactus mouse (*Peromyscus eremicus*)	Prairie vole (*Microtus ochrogaster*)	Gray-tailed vole (*Microtus canicaudus*)	Belding's ground squirrel (*Spermophilus beldingi*)
Summary of kin recognition tests	Loose social structure with transient pairing[10] (Results below are from Dewsbury, 1982.)	Highly social family units with pair bonding[11,12]. Populations undergo periodic fluctuations (cycles)[13] (Results below are from Gavish, Hoffman and Getz, 1984.)	Highly social with male parental care[14]. Populations undergo periodic fluctuations (cycles)[15] (Results below are from Boyd and Blaustein, 1985.)	Social; single family female kin clusters[17,18]. Females are sedentary between years and males permanently disperse from birthplace[16,17]. (Results are from Holmes and Sherman, 1982, 1983.)
Basis of kin recognition	Reduced breeding between siblings as compared to non-siblings in laboratory tests.	Reduced breeding in pairs reared together compared to pairs that had been reared apart.	Reduced breeding in pairs reared together compared to pairs that had been reared apart.	Pairs reared together, regardless of relatedness, are less aggressive than pairs reared apart in laboratory tests.
Importance of *in utero* environment	Experiments not conducted.	Experiments not conducted.	Experiments not conducted.	Important? (see below)
Recognition of individuals of varying degrees of relatedness	Experiments not conducted.	Experiments not conducted.	Experiments not conducted.	Yes; paternal half-siblings differed from unfamiliar non-siblings in passive and aggressive behavior.
Possible recognition mechanism(s)	Familiarity, phenotype matching.	Recognition alleles, phenotype matching.	Phenotype matching, recognition alleles, familiarity.	Familiarity.
Other relevant Studies		Gilder and Slater, 1978; D'Udine and Partridge, 1982.	Vestal and Hellack, 1978	Hill, 1974.

Table 3. (continued)

Summary of kin recognition tests	Cactus mouse (*Peromyscus eremicus*)	Prairie vole (*Microtus ochrogaster*)	Gray-tailed vole (*Microtus canicaudus*)	Belding's ground squirrel (*Spermophilus beldingi*)
Recognition of individuals of varying degrees of relatedness	Experiments not conducted.	Experiments not conducted.	Experiments not conducted.	Yes; full-sisters fight significantly less often than half-sisters as yearlings. Paternal half-sibling tests not conducted.
Possible recognition mechanism(s)	Familiarity	Familiarity.	Familiarity.	Familiarity, phenotype matching, recognition alleles.
Other relevant studies		Wilson, 1982; Huck and Banks, 1979		Holmes, 1986a,b

Results of kin recognition tests	Arctic ground squirrel (*Spermophilus parryii*)	13-Lined ground squirrel (*Spermophilus tridecemlineatus*)	Richardson's ground squirrel (*Spermophilus richardsonii*)
	Highly social[17,18]. Females remain in natal area and males may maintain territories beyond breeding season[18,19]. In some populations males permanently disperse from birth place[18]. Communal nesting by related females[18,19]. (Results below from Holmes and Sherman, 1982.)	May be asocial or somewhat social[17,20,21]. Much aggression between kin in field[17]. Juvenile females tend to remain closer to natal area than juvenile males, so groups of closely related females may occur[17]. (Results below from Holmes, 1984b.)	Social; single family female kin clusters[17]. Sons disperse from natal area, whereas daughters may remain at natal site throughout life[17].

Basis of kin recognition	Pairs reared together less aggressive than pairs reared apart in laboratory tests.	Pups reared together, regardless of relatedness, displayed less exploratory behavior toward one another than pups reared apart.	(Results from Davis, 1982.) Siblings displayed different frequencies of aggressive and amicable behavior toward one another compared to non-siblings in laboratory tests.	(Results from Sheppard and Yoshida, 1971.) Siblings displayed different frequencies of aggressive and amicable behavior toward one another than non-siblings in laboratory tests.
Importance of *in utero* environment	Important? Effect of relatedness only important in female pairings not male-male or male-female pairings. Sisters reared apart less aggressive toward each other than unrelated females reared apart.	Not important.	Questionable. Individuals may recognize sibs in the absence of postpartum familiarity but there were greater frequencies of certain behaviors between uterine sibs reared together compared to non-siblings reared together.	Experiments not conducted.
Recognition of individuals of varying degrees of relatedness	Experiments not conducted.	Experiments not conducted.	Experiments not conducted.	Experiments not conducted.
Possible recognition mechanism(s)	Familiarity, phenotype matching, recognition alleles.	Familiarity.	Phenotype matching, recognition alleles, familiarity.	?
Other relevant studies	McLean, 1982	Schwagmeyer, 1980 Vestal and McCarley, 1984	Davis, 1984a,b,c	

[1] Delany and Happold, 1979. [2] Anderson, 1970. [3] examples are DeLong, 1967; Crowcroft and Rowe, 1957; Selander, 1970. [4] Myton, 1974. [5] Stickel, 1968. [6] Eisenberg, 1962. [7] Eisenberg, 1968. [8] Hansen, 1957. [9] Fairbairn, 1978. [10] Veal and Caire, 1979. [11] Getz and Carter, 1980. [12] Getz, Carter and Gavish, 1981. [13] Krebs and Myers, 1974. [14] S. K. Boyd, personal communication. [15] Boyd and Blaustein, 1985. [16] Sherman, 1977. [17] Michener, 1983. [18] Holmes and Sherman, 1982. [19] McLean, 1982. [20] Schwagmeyer, 1980. [21] Holmes, 1984b.

that nursed from different females. Interestingly, exposure to a common mother did not result in the same degree of preferential huddling as did common maternal exposure coupled with early exposure of test pups to one another. These results reveal that relatively brief exposure in the home cage is not, by itself, sufficient for recognition.

Porter, Matochik and Makin (1983) suggest that phenotype matching may play a role along with familiarity in the development of spiny mouse kin recognition. They artificially odorized pups for several days and either housed them with littermate pups given the same odor, or in isolation. Pups reared in groups or in isolation preferred to associate with unfamiliar pups of the same odor. Furthermore, individuals that were separated from most members of their litter for nine days but were housed with one pup from their litter, subsequently huddled preferentially with one another.

The results of these latter experiments contrasted somewhat with the earlier work which showed that pups reared in isolation for eight days (with no artificial odor) prior to testing, displayed no reliable preferences to huddle with siblings (Porter and Wyrick, 1979). That pups reared in isolation, but scented artificially, could recognize unfamiliar pups scented with the same odor may reflect differences in the cues used in recognition (Porter, Matochik and Makin, 1983, 1984). Porter, Matochik and Makin (1983) believe that the artificial odors may have been more intense than natural odor cues. Moreover, although animals scented with the same artificial odor shared an identical recognition cue, natural littermate odors are probably similar, but not identical. The results of these experiments suggest that phenotype matching with littermates may be important in spiny mouse kin recognition. These results, together with previous tests of anosmic individuals (Porter, Wyrick and Pankey, 1978), provide evidence that the cue used in spiny mouse kin recognition is odor.

Although there is a paucity of information concerning the social behavior of spiny mice in nature, laboratory studies have shown that this species is highly social. Familiar siblings share food more often than unfamiliar non-siblings in laboratory tests (Porter, Moore and White, 1981). Adult females within a social group assist other females with birth and help tend the young (Delany and Happold, 1979). Males also seem to behave in this way (R. H. Porter, personal communication). The level of relatedness between the helpers and female recipient has not been accurately quantified (R. H. Porter, personal communication). However, kin recognition may be important in maintaining the family group. Furthermore, young *Acomys* probably become familiar with several individuals other than their littermate siblings and parents. If the individuals aiding pregnant females and tending offspring are likely to be kin, then familiarity (Porter, Matochik and Makin, 1984) is probably one efficient mechanism for discriminating between kin and non-kin in this species.

House mice

Several studies have investigated various aspects of recognition in house mice (*Mus musculus*) (Table 3). The comprehensive research of Kareem and Barnard (1982) and Kareem (1983) utilized affiliative, aggressive, and non-social, neutral behaviors (exploration) to analyze the effects of familiarity and relatedness in kin recognition. Pairs of mice were observed in a plastic cage and behavioral interactions of the mice were quantified for 5 min after a partition separating the two test animals was lifted.

In the first series of experiments, Kareem and Barnard (1982) reared mice with their siblings and obtained three sets of pairwise treatments for testings. These were:
1. Full-sibling cagemates.
2. Paternal half-sibling non-cagemates.
3. Non-sibling, non-cagemates.

The results showed that adult male/full-sibling pairs, adult female/full-sibling pairs, and half-sibling pairs behaved differently from non-sibling pairs, although behavior varied with sex. There were also behavioral differences between half-siblings and full-siblings in that half-sibling pairs performed certain behaviors with different frequencies than did full-sibling pairs.

Another set of experiments was performed using only adult male mice to investigate in more detail the importance of familiarity versus genetic relatedness in kin recognition. Litters containing mice of known relatedness were arranged to produce rearing regimes composed of individuals of different degrees of relatedness and familiarity. Pairs of familiar half-siblings, familiar non-siblings, unfamiliar half-siblings, and unfamiliar non-siblings were tested. Differences in behavior between half-siblings and non-siblings were found, but kinship effects disappeared completely when animals that were familiar with one another were tested.

A more recent study by Kareem (1983) analyzed the effects of prolonged familiarity on kin recognition in adult house mice. Mice were given increasing periods together in 'families' composed of litters from two different sibships produced through cross-fostering. Thus, all individuals were associated with both siblings and non-siblings. Cross-fostering was conducted at birth, and at seven, 14, 21, 30 and 50 days after birth. In an additional treatment, pups were reared by their own parents or foster parents until weaning and then placed with sibling littermates until testing. Behavioral interactions among pairs were quantified as in Kareem and Barnard (1982).

When tested as adults, previously separated siblings displayed differences in behaviors as the period of familiarity increased, with the most pronounced differences in animals having had 21 or more days of contact. Unfamiliar siblings that were separated at birth displayed differences in behavior when compared with unfamiliar non-siblings. Unfamiliar siblings spent less time investigating each other and showed more passive body contact and total activity compared with unfamiliar non-siblings. Therefore, familiarity plays a role in kin recognition but as unfamiliar siblings were reared by different parents before weaning and were separated from one another until testing, post-natal maternal effects may not be necessary. Results using paternal half-siblings further support this contention. Although familiarity *per se* seems to have played the major role in these tests (Kareem and Barnard, 1982; Kareem, 1983), phenotype matching by individuals reared together may also be important, especially because each animal in the latter tests (Kareem, 1983) was housed with a sibling.

Recognition in house mice has also been studied by analyzing mating preferences. The results of these analyses have revealed a recognition system, the *H-2* major histocompatibility complex (MHC) (Klein, 1975, 1979), that appears to be coded by alleles at one locus that are also concerned with odor production (reviews in Yamazaki *et al.*, 1980; Boyse *et al.*, 1982; Chapters 2, 3 and 13). Moreover, different allelic forms influence the quantity and/or chemistry of the odors and house mice are able to detect these differences.

Experiments using highly inbred strains of mice that differed from one another only at the *H-2* locus showed that males from different congenic strains that were given a choice of females with the same or different *H-2* types, generally chose the dissimilar female (Yamazaki *et al.*, 1976). By using a Y-maze whose two arms were differentially scented with currents of air passing through cages occupied by mice (Yamazaki *et al.*, 1979) or by using urine that was produced by mice differing at the *H-2* locus (Yamaguchi *et al.*, 1981), it was also shown that differences were detected using olfactory cues. The olfactory-based mating preference is probably controlled by two linked genes in the region of *H-2*, one for the female signal and one for the male receptor (Yamazaki *et al.*, 1976). The results from a recent study indicate that pregnant females showed a greater incidence of pregnancy blocking (i.e. Bruce Effect; Bruce, 1959, 1960) when exposed to males or females that differed in *H-2* type from the stud male, compared with females exposed to individuals having the same *H-2* type as the stud male (Yamazaki *et al.*, 1983). Chemosensory recognition of *H-2* types must affect the hormonal status of pregnant females (Yamazaki *et al.*, 1983). These results suggest that a female's recognition of a strain difference between stud and unfamiliar males may be under intense selective pressure (see Labov, 1979).

Studies of the ability of house mice to discriminate between genotypes differing at the *T*-locus closely parallel those discussed above concerning mating preferences and the *H-2* major histocompatibility complex. Lenington (1983), Lenington and Egid (1985) and Egid and Lenington (1985) have shown, in laboratory choice experiments, that both male and female house mice can discriminate between members of the opposite sex based on genotypic differences in odor production. Both males and females preferred to associate with members of the opposite sex that were homozygous-wild $(+/+)$ over those that carried a wild and a t-allele $(+/t)$. A female's preference appears to be related to the genotypes of her parents. Females with one heterozygous parent carrying a wild and a t-allele were more likely to prefer a heterozygous male than a female whose parents were both homozygous-wild (Lenington and Egid, 1985). Moreover, the preferences of females for homozygous-wild male odor is a response to differences in male odor production associated with the T-locus and not the closely linked H-2 locus, because males used as odor donors were similar at the H-2 locus. Importantly, in studies investigating H-2 recognition (e.g. Yamazaki *et al.*, 1979; Yamaguchi *et al.*, 1981) mice were similar in T-locus genotype but differed at the H-2 locus (Lenington and Egid, 1985).

Thus, the ability to discriminate between genotypes at the H-2 and T-loci probably evolved independently (Lenington and Egid, 1985). The source of the odor cue produced by males involves the presence of a t-allele. The genetic basis for the female odor cue is unknown. The T-locus produces cells that function in cell–cell recognition during early embryonic life but after about ten days of development, this immunological function is apparently taken over by the H-2 locus.

More than 30 different t-alleles have been identified from wild house mouse populations (Bennett, 1975; Lenington and Egid, 1985; see also Levine, Rockwell and Grossfield, 1980), and these are divided into two general classes, lethal and semi-lethal, depending upon how they are expressed phenotypically. Possession of homozygous lethal alleles results in death during fetal development. Possession of semi-lethal alleles causes sterility in homozygous males (Lenington and Egid, 1985). The frequency of t-alleles in wild populations of house mice is considerably lower than is predicted on the basis of selection against homozygotes (Bennett, 1975; Lenington and Egid, 1985). Because of the potential deleterious effects on fitness, it would be advantageous to be able to distinguish between individuals having a lethal or semi-lethal genotype from those that do not.

Studies of the MHC and mating preferences in house mice are intriguing because they provide evidence for the possible existence of recognition alleles (discussed in detail in Chapters 2, 3, 10, etc.). Studies of the house mouse MHC and T-locus along with results obtained by Kareem and Barnard (1982) and Kareem (1983) (see also studies by Bowers and Alexander, 1967; Gilder and Slater, 1978; and D'Udine and Partridge, 1982) suggest that in house mice, more than one mechanism of recognition may be operating together. It is also possible that different mechanisms may operate most efficiently under different environmental and social conditions. As in spiny mice, the phenotypic basis of recognition is odor. Olfactory cues also play important roles in recognition in numerous contexts in many other species of small mammals (Halpin, 1980; Blaustein, 1981).

Although kin recognition in house mice has been studied only in highly inbred laboratory strains, the results are not surprising. Feral house mice are highly social and tend to live in small, close-knit family groups (Crowcroft and Rowe, 1957; Anderson, 1970; Selander, 1970; Baker, 1981). Populations of house mice inhabiting barns, sheds, corn ricks or other buildings where food and shelter are available for relatively long periods, may become divided into tribes each of which consists of a dominant male, several breeding females (including the daughters of the dominant male), and several subordinate males that probably do not breed often (Selander, 1970). These groups may contain five to 80 individuals but generally have ten or fewer mice. Both male and female territoriality contribute to maintaining group integrity. Intertribal migration is rare and generally involves only females.

Kin recognition may function in house mice to maintain cohesion within family units, especially after disturbances that cause a temporary breakdown of the social structure. For example, when a dominant male of a particular tribe dies, males from outside the group may enter, and social bonds within the tribe may be disrupted (A. R. Blaustein, personal observations from laboratory colonies). Tribes may represent highly inbred units in which optimal outbreeding may be important. Discriminating between related and unrelated individuals may be essential in this context. Results of the MHC mating studies are generally consistent with Bateson's (1983) optimal outbreeding hypothesis. Furthermore, the social structure of house mouse tribes makes it likely that most individuals within the group are related. Familiar individuals, therefore, are probably related and familiarity would, in most cases, be an efficient cue for kin recognition in this rodent.

Norway rats

Sibling recognition was investigated in captive Norway rats (*Rattus norvegicus*) by Hepper (1983). He used individuals that were reared only with their siblings, or with non-kin from several different sibships. Test animals were placed in a Y-maze-like apparatus and allowed to crawl to either of two sides of the apparatus. Once an animal touched one of two particular 'goal boxes' that were situated at opposite ends of the maze, and which contained rats of a particular kin group, Hepper concluded that the animal preferred the particular stimulus group housed within the goal box. Various combinations of stimulus rats were used in these experiments.

The results suggested that Norway rats may be able to discriminate between siblings and non-siblings. Because the infant rats used in these tests had not yet opened their eyes, olfactory cues were implicated as the cue used in discrimination. However, the results are confounded by the effects of familiarity (see Coppersmith and Leon, 1984). Young rats were not immediately separated from one another after they were born. Thus, tests in which rats were supposedly 'unfamiliar' with one another were actually conducted with individuals that had had at least two days of post-natal contact before being separated. Furthermore, although many combinations of stimulus choices were given to test rats, several important choices were not provided. For example, one necessary test would have rats choose between familiar, unrelated animals and unfamiliar, related individuals. This test would provide some clue as to the importance of familiarity in rat sibling recognition. A more powerful series of experiments using half-siblings would also provide important information on the role of familiarity in sibling recognition.

Norway rats have a complex social system (Calhoun, 1962). There is evidence that they assemble in social groups and that individuals may be associated with a particular group for months, although the groups may move occasionally (Calhoun, 1962). It is possible that kin recognition could help to maintain group structure.

White-footed mice, deer mice and cactus mice

Grau (1982) studied kin recognition in white-footed mice (*Peromyscus leucopus*) that were reared with siblings for several weeks and then kept in groups of four with each group containing at least two siblings and one non-sibling. Recognition was tested by observing the behavioral interactions of a pair of mice in a circular arena and quantifying the behaviors in a manner similar to that described above for house mice (Kareem and Barnard, 1982; Kareem, 1983). Five pairing categories were tested:
1. Littermate sibling cagemates (born in the same litter and housed together since birth).
2. Littermate sibling non-cagemates.
3. Non-littermate sibling, non-cagemates (born to the same parents but in different litters and having no previous contact).
4. Non-sibling, cagemates.
5. Non-sibling, non-cagemates.

Results (Table 3) indicated that familiarity plays a role in recognition (e.g. both males and females huddled more often with familiar mice of the same sex rather than with unfamiliar mice) but that prior association was not necessary, because differences in behavior were observed between siblings born in different litters (related but unfamiliar) and unrelated unfamiliar individuals. These results suggested that more than one mechanism of recognition may be operating in white-footed mice. Recognition of unfamiliar individuals implies that phenotype matching or recognition alleles may be important. As Grau (1982) suggested, the ecology of *P. leucopus* provides the possibility of related individuals interacting in nature. They are highly social (Myton, 1974), and they may have relatively stable but gradually shifting home ranges (Stickel, 1968). It is possible that kin recognition is used so that kin can efficiently cooperate with one another (Grau, 1982). For example, kin group cohesion and territorial defense may be more efficiently maintained by being able to recognize kin, but the actual functions of kin recognition in *P. leucopus* are unknown.

Dewsbury's (1982) study of mating behavior in deer mice (*P. maniculatus*) and cactus mice (*P. eremicus*) suggested that familiarity is important in mate choice, an assay of recognition. Offspring production and rearing were investigated using several categories: *P. maniculatus* siblings and non-siblings; *P. eremicus* non-siblings reared apart or together through cross-fostering; *P. eremicus* siblings reared together for four, eight or 24 days before pairing.

These experiments showed a non-significant trend for reduced production of litters by sibling pairs of *P. maniculatus* when compared with non-sibling pairs. However, there was a significant difference among the six groups of *P. eremicus* in the proportions of pairs producing litters. Significantly more non-siblings than siblings produced litters and more non-siblings than cross-fostered pairs produced litters. The probability of breeding was the primary measure affected, rather than latency to breed, litter size, number of litters or success of rearing young. Furthermore, avoidance of incestuous breeding declined as siblings were isolated from each other for longer periods prior to pairing. The results suggest that individuals familiar with one another are less likely to breed.

P. maniculatus, like *P. leucopus*, is highly social and may be found in family units nesting together (Hansen, 1957; Eisenberg, 1962, 1968). In light of Dewsbury's (1982) results, it is possible that kin recognition in *P. maniculatus* may function to balance inbreeding and outbreeding. Furthermore, live-trapping data indicate that members of this species travel in pairs to some extent (Jenkins and Llewellyn, 1981). Although it is not known if these are sibling pairs, numerous benefits could accrue to mice that travel in sibling pairs (see Blaustein and Rothstein, 1978 for discussion).

Little is known of the social behavior of *P. eremicus* (see Veal and Caire, 1979; Dewsbury, 1982). As in other rodent species, kin recognition may facilitate optimal outbreeding (see discussion in Dewsbury, 1982 and references therein).

The *Peromyscus* studies reveal that familiarity plays a significant role in recognition in all three species. The social systems of *P. leucopus* and *P. maniculatus* make it likely that familiarity is associated with relatedness, and thus an efficient means of recognition.

Voles

Because voles undergo periodic population fluctuations that may be influenced by intraspecific behavioral interactions (Krebs and Myers, 1974; Charnov and Finerty, 1980; Charnov, 1981; Bekoff, 1981b), they are good species to use in studies of kin recognition. Furthermore, voles display different social systems, and elucidation of kin recognition abilities as they relate to social systems may provide answers to questions about the function of kin recognition in these rodents.

Wilson (1982) conducted a comparative study of the social behavior and recognition abilities of the meadow vole (*Microtus pennsylvanicus*) and the prairie vole (*M. ochrogaster*). These species were ideal for comparison because *M. ochrogaster* is highly social, whereas *M. pennsylvanicus* is virtually asocial.

Voles used in tests were generally obtained from a breeding stock (although some wild caught *M. pennsylvanicus* were used) and reared in sibling groups. Wild caught voles were housed together as if they were siblings. Procedures used to quantify behavior were similar to those used in the mouse studies described above. Various combinations of age, sex, and sibling/non-sibling pairings of juvenile voles were observed.

Although there were some differences in behavior among age and sex classes within a species, the overall results showed that prairie voles engaged in more huddling and other contactual behaviors than did meadow voles. Prairie voles also differentiated more clearly than meadow voles between siblings and non-siblings in their social behavior. Meadow voles did, however, display some differences in encounters with siblings and with non-siblings. Significant variations in social behavior between these two species were apparent after about three weeks of age. Furthermore, the behaviors of wild caught, and laboratory reared *M. pennsylvanicus* were similar. Because siblings tested were familiar with one another, the role of genetic relatedness and familiarity were confounded. However, at least two important results emerged:

1. Familiar animals treated one another differently than did unfamiliar animals, indicating that familiarity is of prime importance in the social behavior of these voles.
2. Significant differences in behavior between species with differing social systems were observed.

Studies of mating behavior of *M. ochrogaster* by Gavish, Hoffman and Getz (1984; see also McGuire and Getz, 1981) and of gray-tailed voles (*M. canicaudus*) by Boyd and Blaustein (1985) further illustrate the importance of familiarity in vole recognition (see also similar studies showing the importance of familiarity in mating behavior of Mongolian gerbils, *Meriones unguiculatus*, by Ågren, 1981, 1984). These studies used methods similar to those of Dewsbury (1982) in his study of *Peromyscus* mating behavior.

Gavish, Hoffman and Getz (1984) showed that of ten pairs of unrelated *M. ochrogaster* in which pair formation occurred shortly after birth through cross-fostering, only one pair (10%) produced a litter within 60 days. In contrast, eleven of twelve pairs (92%) (similarly formed) of unfamiliar siblings (siblings reared apart) produced litters. Unrelated voles that were paired before they matured (at 14 days of age) produced significantly fewer litters than did unrelated voles that were paired after they were sexually mature (at 21 days of age), suggesting that familiarity between non-siblings, before reproductive maturity, decreases the likelihood that they will mate. Interestingly, separating 21-day old siblings for eight days before pairing overcame incest avoidance, whereas a 15-day separation period was required for successful breeding in siblings that were housed together for 50 days. These results suggest that sibling recognition and breeding among sibling (or familiar) pairs depend on association before maturity.

Similarly, Boyd and Blaustein (1985) showed that familiar pairs of *M. canicaudus* produced fewer litters than unfamiliar pairs, regardless of relatedness. There were no apparent differences in litter size or pup viability between reproductive siblings and reproductive non-siblings. Individuals that were separated for five or twelve days from potential partners with which they were originally reared continued to demonstrate mating avoidance.

Prairie voles are extremely social, and may be one of the few species that form pair bonds and exist in family units (Getz and Carter, 1980; Getz, Carter and Gavish 1981; see discussion in Wilson, 1982; L. L. Getz, personal communication). Prairie vole young probably do not disperse when weaned, but may remain with their parents through the birth and weaning of a subsequent litter (Thomas and Birney, 1979; Wilson, 1982). Thus, Wilson (1982) believes that young prairie voles may interact almost exclusively with parents and siblings for four to five weeks after they first emerge from the nest. This system provides highly favorable conditions for promoting sibling interactions and the degree of familiarity among individuals appears to be closely associated with genetic relatedness (Bekoff, 1981a).

Conversely, young meadow voles may rapidly disperse from natal sites (Getz, 1972). Social bonds among members of a litter probably do not form as in the prairie vole, and family units probably do not exist. Conditions promoting sibling interactions *per se* are not as favorable as in *M. ochrogaster* but individuals may interact with familiar neighbors with which they may form social bonds (Wilson, 1982).

Unfortunately, there is very little information about the social behavior of *M. canicaudus*. The species is highly social and relatively non-aggressive in laboratory situations (A. R. Blaustein, personal observations) and both maternal and paternal care have been observed (S. K. Boyd, personal communication). Information on a closely related species (*M. montanus*) suggests that pregnant females do not move to new nests and travel short distances only (Jannett, 1980). If these movement patterns are similar in *M. canciaudus*, it is unlikely that nestlings of varying degrees of relatedness will associate with one another. Because there is a high probability that members of different *M. canicaudus* litters do not mix, familiarity can serve as a general kin recognition cue and may also be important in mating behavior to facilitate incest avoidance.

The relative importance of familiarity and genetic relatedness in the population fluctuations of voles are also of interest (Charnov and Finerty, 1980; Bekoff, 1981b; Charnov, 1981). This relationship will be discussed in the following chapter.

Ground squirrels

Studies of ground squirrel kin recognition are important because they have provided comparative information on ontogeny, and also because they show that there are marked differences among species in their kin recognition abilities. Moreover, there is a good deal of information on the behavior of wild individuals of most species that have been tested in the laboratory. Possible functions of kin recognition can be discussed, therefore, with respect to the behavior of natural populations.

Belding's and Arctic ground squirrels

A comparative study of kin recognition in Belding's (*Spermophilus beldingi*) and Arctic ground squirrels (*S. parryii*) was reported by Holmes and Sherman (1982). This study is especially important because both species were tested using similar laboratory techniques, and both species are highly social.

Belding's ground squirrels have been studied in marked populations for several years (see Holmes and Sherman, 1982, for details). They are active above ground for about five months of the year. Females are sexually receptive for only a few hours on a single afternoon each year, during which time each female mates with several different males. Females rear one litter of three to six young per year, and because females nest alone, young animals interact only with each other and their mother until they leave the burrow. In captivity, rearing familiarity, rather than genetic relatedness, appears to be important in dam–young discrimination (Holmes, 1984a). After young squirrels leave the burrow, they may interact with neighboring conspecifics. Nestmates will share their burrow for a few weeks after emergence.

Males typically disperse from natal sites, whereas females generally remain where they were born. Thus, males seldom interact with their matrilineal relatives (Holmes and Sherman, 1982) but females are surrounded by relatives of varying degrees of relatedness. Furthermore, females give alarm calls in the presence of potential mammalian predators more often than expected by chance, while males give alarm calls less often than expected by chance (Sherman, 1977). Females also seem to cooperate (for example, against infanticidal conspecifics) more with close kin than with more distantly related squirrels (Holmes and Sherman, 1982). These observations led Sherman

(1977) to argue that alarm calling in Belding's ground squirrels functions to warn relatives, and that this nepotistic behavior could be facilitated by kin recognition (Holmes and Sherman, 1982). However, there is some controversy concerning the overall importance of some forms of cooperation in Belding's ground squirrels (e.g. see Michener, 1982) and whether or not alarm calling is nepotistic toward collateral kin *per se* (Sherman, 1980b, Shields, 1980).

Arctic ground squirrels also are highly social (Michener, 1983). Females produce one litter per year (five to six young) and nurse their young alone (Holmes and Sherman, 1982). Infants interact only with each other and their mother until they are about 20–25 days old. Just before juveniles emerge, females may transfer them to new burrow systems and two or three females sometimes place pups in the same burrows (Holmes and Sherman, 1982). These females are usually mothers and daughters or sister pairs, and clumping seems to occur selectively with close relatives rather than randomly with nearest neighbors that are not always close kin (Holmes and Sherman, 1982). In this species, adult males may maintain territories beyond the breeding season, defending an area that overlaps the smaller ranges of several adult females and their offspring (Michener, 1983).

In one *S. parryii* population, males dispersed and only females remained (McLean, 1982), thereby presenting a situation similar to that seen in *S. beldingi* (Sherman, 1977). McLean (1982) showed that close female kin (sisters and mothers/daughters) had greater home range overlap and interacted more amicably and less aggressively than did more distantly related females. Furthermore, closely related females clumped their young at emergence, whereas more distantly related females did not. These results indicated that closely related females share resources more frequently than do more distantly related females. Thus, nepotism may be an important aspect of *S. parryii* social behavior as in *S. beldingi*.

Animals tested for recognition abilities by Holmes and Sherman (1982) were obtained by live-trapping pregnant females in the field. Intraspecific cross-fostering was conducted when two females produced litters within three hours of each other. Four categories of infant pairs were obtained: (a) siblings reared together, (b) siblings reared apart, (c) non-siblings reared together, and (d) non-siblings reared apart. In all rearing regimes, almost all pups in each group were reared with at least one sibling and one non-sibling. Because the frequency of multiple mating and multiple paternity is unknown for *S. parryii* (Holmes and Sherman, 1982) and is high in *S. beldingi* (Hanken and Sherman, 1981), it was not known for certain whether 'siblings' were full-siblings or half-siblings.

Laboratory tests for recognition were conducted by observing the behavior of pairs in an arena for 5 min. Behavioral encounters were quantified in a manner similar to the studies of mice described above. In addition to these laboratory tests, cross-fostering of *S. beldingi* was conducted in field experiments (Holmes and Sherman, 1982). Pregnant females were obtained and soon after parturition, young animals were individually marked. At various ages (one to 50 days), young squirrels were cross-fostered into the nests of free-living lactating females. An infant was recorded as 'accepted' if it remained with its nestmates for at least one week after the litter into which it was placed emerged from its burrow. Young introduced after a specific litter emerged were considered accepted if they continued to use the specific burrow for one week. Sibling recognition was studied only in yearling females.

Laboratory tests of kin recognition gave similar results for Arctic and Belding's ground squirrels. Juveniles reared together, regardless of their genetic relatedness, were significantly less aggressive toward one another than pairs reared apart. Siblings and non-siblings reared together did not show differences in aggressive behavior. Only sisters reared apart were significantly less aggressive than unrelated females reared apart, whereas relatedness did not seem to influence the behavior of male–male or female–male pairs that were reared apart. There were no differences in behavior of siblings reared in pure sibling litters when compared to siblings reared in mixed sibling litters.

Acceptance of young into a foster mother's nest was most successful up to 23–25 days after birth, corresponding to the time that young animals usually emerge from their burrows. To investigate the ontogeny of sibling recognition in more detail, cross-fostered females that were accepted as young were observed as yearlings. Accepted foster nestmates and littermate sisters behaved similarly toward one another. Furthermore, although sample sizes were small, young transferred after their target litter emerged, were aggressive towards, and cooperated less often with, each other than did littermates or young that were transferred early. Thus, interactions between young transferred prior to weaning and foster sisters were indistinguishable from littermates, but the behavior of young transferred after weaning and foster sisters was similiar to non-kin interactions.

In addition to the experiments discussed above, analyses of previous observations of full and maternal half-sibling *S. beldingi*, that were identified through electrophoresis, were made (Holmes and Sherman, 1982). The analysis revealed that full-sisters fought significantly less often than did half-sisters and they aided each other more than they aided half-sisters. Thus, full-sisters were treated preferentially over half-sisters.

Laboratory and field experiments, combined with field observations, suggest that familiarity is an important component of kin recognition in both *S. beldingi* and *S. parryii*. However, because *S. beldingi* differentiated between full- and maternal half-siblings, phenotype matching and/or recognition alleles must also play a role (see also Holmes, 1986a,b). The results obtained from the field seem somewhat contradictory because *S. beldingi* individuals seem to utilize familiarity to learn to recognize other animals prior to weaning, yet they can also distinguish between full and maternal half-siblings—individuals that shared a common uterine and nest environment. This observation suggests that it is unlikely that recognition is based solely on familiarity.

Thirteen-lined ground squirrels

There are at least two studies suggesting that nepotism and kin recognition are important for 13-lined ground squirrels (*S. tridecemlineatus*) (Schwagmeyer, 1980; Holmes, 1984b). However, this species differs in its social behavior and ecology from *S. beldingi* and *S. parryii*.

Michener (1983) categorizes *S. tridecemlineatus* as being less social than *S. beldingi* and *S. parryii*. In some populations, males and females do not share territories and social interactions between kin may be aggressive. Males generally disperse farther from natal areas than do females (Rongstad, 1965; Michener, 1983). The female and her offspring may form a single cohesive family that is generally not sociable with other groups, and thus juveniles from different litters do not usually mix (Michener, 1983).

Schwagmeyer (1980) studied alarm calling in *S. tridecemlineatus* by experimentally assessing the alarm calls of these squirrels in the field in response to approaches by humans, canids and naturally occurring avian predators. Males never emitted calls during any approaches by humans or dogs. Audible vocalizations were rarely emitted by non-parous females approached by dogs or humans, but parous females commonly vocalized under these circumstances. The peak incidence of vocalization coincided with litter emergence. Alarm calls were rarely emitted in response to avian predators. Juveniles of both sexes called in response to both types of predators but much less frequently than did adult, parous females. These data suggest that alarm calling functions to warn relatives (see Sherman, 1977). However, because the calls were most often given by mothers and their offspring, alarm calling by 13-lined ground squirrels does not provide for a clear case of nepotism toward collateral kin; it may be more appropriate to classify this behavior as a form of parental care (see Shields, 1980).

Holmes' (1984b) methods for studying kin recognition in *S. tridecemlineatus* were similar to those described previously (Holmes and Sherman, 1982). By studying pairs of 13-lined ground squirrels in an encounter cage for 5 min, he noted that pups reared together, whether biological or foster siblings, treated each other similarly. Pups that were reared apart, regardless of their relatedness to one another, also behaved similarly. The different behavior patterns seem to be based on rearing familiarity. The lack of differences in behavior between siblings reared apart and non-siblings reared apart suggests that *in utero* association or early post-natal association is not crucial for recognition in this species. Moreover, olfactory impairment through zinc sulfate application eliminated the differential treatment of familiar individuals, suggesting that odors serve as proximate recognition cues.

Richardson's ground squirrels

Richardson's ground squirrels (*S. richardsonii*) are social. Males may defend territories prior to, and during, the mating period only (Yeaton, 1972). In some populations, female territoriality is manifested from one week after emergence until hibernation (Yeaton, 1972). Female territories may be retained from year to year by one individual, or may be transferred from a female of one generation to a related female of the next generation (Yeaton, 1972). Mothers and offspring form social family units and juveniles from different litters are not known to mix (Michener, 1983). Observations by Yeaton (1972) suggest that individuals recognize one another in natural field situations because fighting of related adult females was seen much less frequently than fighting between unrelated adult females. Furthermore, cohesive interactions occurred most frequently between litter mates.

To study kin recognition in *S. richardsonii*, Davis (1982) cross-fostered young within 24 h of birth. Litters composed of uterine siblings and fostered non-siblings were caged together for 37 days and then separated and housed individually. When squirrels were about 110 days old, they were tested for recognition during 10 min trials in a laboratory arena. Social encounters were observed for the following types of pairs: (a) siblings reared together, (b) siblings reared apart, (c) non-siblings reared together and (d) non-siblings reared apart.

Davis (1982) found that siblings reared together and siblings reared apart treated each other similarly. Furthermore, squirrels did not differentiate between non-siblings reared apart and non-siblings reared together. In trials involving test animals that were reared apart, significant differences in behavior were observed between siblings reared apart and non-siblings reared apart. The results of Davis' study suggest that *S. richardsonii* can discriminate between kin and non-kin and the ability to do so can arise without prior association. However, the frequencies of one of five quantified behavioral components (number of contacts between individuals) did differ among familiar versus unfamiliar animals, suggesting that familiarity may play some role in kin recognition.

The study of *S. richardsonii* by Sheppard and Yoshida (1971) provides additional evidence that these squirrels can recognize kin. They observed that the frequencies of aggressive and amicable behaviors differed between siblings and non-siblings in laboratory tests. Unfortunately, the possible mechanism(s) of recognition is unknown due to a lack of information on rearing conditions.

Observations of other sciurids suggest that they also may have an ability to discriminate between kin and non-kin. For example, behavioral discrimination seems to account for some inbreeding avoidances shown by black-tailed prairie dogs (*Cynomys ludovicianus*) (Hoogland, 1982). Likewise, kinship and kin recognition may be important in the reproductive and social dynamics of yellow-bellied marmots (Armitage and Johns, 1982).

Summary of kin recognition in ground squirrels

Laboratory and field experiments, in conjunction with observations of individuals under natural conditions, suggest that several species of ground squirrels have the ability to discriminate between kin and non-kin. Of the species in which kin recognition has been studied, *S. beldingi*, *S. parryii*, and *S. richardsonii* are generally highly social, whereas *S. tridecemlineatus* is less social (Michener, 1983). In all species but *S. richardsonii* (Davis, 1982) familiarity is a significant factor in recognition. In *S. tridecemlineatus*, familiarity alone seems to be the mechanism of recognition (Holmes, 1984b).

There are notable contradictory results pertaining to recognition in *S. parryii* and *S. beldingi*. In general, familiar individuals treated each other similarly in these species. Yet in both *S. parryii* and *S. beldingi*, sisters reared apart were less aggressive toward one another than unrelated females reared apart, whether the unrelated females were reared with siblings alone or with both siblings and non-siblings (Holmes and Sherman, 1982). Furthermore, in *S. beldingi*, full-sisters and maternal half-sisters treated each other differently as yearlings. Because these kin shared both a common uterine and post-natal environment, familiarity alone cannot account for this discrimination. Therefore, recognition in *S. beldingi* and *S. parryii* must be achieved either through phenotype matching or by recognition alleles. An ability to recognize unfamiliar individuals or individuals that vary in their degree of relatedness may be especially important in these species.

In *S. beldingi*, multiple paternity is prevalent (Hanken and Sherman, 1981); the result is naturally occurring maternal half-siblings and full-siblings. If nepotism is important in this species (Sherman, 1977), those individuals favoring kin (regardless of the degree to which the aided individuals are related to the altruist) may achieve inclusive fitness benefits (if benefits of the nepotistic behavior outweigh the costs) over those individuals that do not discriminate between kin of varying degrees of relatedness. To be able to discriminate between maternal half-siblings and full-siblings, an ability to recognize other individuals in the absence of previous association is necessary.

In *S. parryii*, females carry their pups to a common burrow just prior to weaning (McLean, 1982). Nest sharing is usually among related females, producing a situation in which recognition must be learned prior to weaning or must be achieved through a mechanism of recognizing unfamiliar individuals—phenotype matching or recognition alleles. Because *S. tridecemlineatus* is relatively asocial, with less spatial overlap among individuals than in the other species (Holmes, 1984b), familiarity alone seems to be an efficient mechanism of recognition in this species.

ACKNOWLEDGMENTS AND LITERATURE CITED

See the end of Chapter 10.

Kin Recognition in Animals
Edited by D. J. C. Fletcher and C. D. Michener
© 1987 John Wiley & Sons Ltd

CHAPTER 10

Kin Recognition in Vertebrates (Excluding Primates): Mechanisms, Functions, and Future Research

ANDREW R. BLAUSTEIN
Department of Zoology, Oregon State University, Corvallis, Oregon 97331, USA

MARC BEKOFF and THOMAS J. DANIELS
Department of Environmental, Population and Organismic Biology, University of Colorado, Boulder, Colorado 80309, USA

KIN RECOGNITION MECHANISMS RECONSIDERED

The recognition systems exhibited by the species discussed in the preceding chapter require that one or more phenotypic markers convey a message indicating the degree of relatedness. These cues are displayed by a sender and must be received by a target individual that then makes the appropriate motor response. Signals that are important in recognition may be genetically specified and may become functional only in certain developmental stages. They may also be influenced by environmental components and, thus, potentially convey a variable signal. Individuals may recognize other animals because of their unique traits, or they may recognize groups sharing traits (see Breed and Bekoff, 1981). Theoretically, there are several ways in which individuals could classify conspecifics as to their degess of relatedness (for theoretical considerations see discussions in Crozier and Dix, 1979; Rothstein, 1980; Breed and Bekoff, 1981; Getz, 1981; Beecher, 1982; Lacy and Sherman, 1983; see also Chapters 2, 3, 4 and 13).

Sensory bases of kin recognition

Visual, auditory and chemical cues are used in vertebrate kin recognition.

Animals should transmit their signals in a medium that is subject to minimal distortion (see discussion in Wilson, 1975). More than one signal may be involved in the recognition process—a primary signal that enables an animal to orient toward a conspecific group or individual, and a secondary signal that is used for fine-tuning the primary signal, thus allowing for discrimination between kin and non-kin or between close and distant relatives.

For example, in the aquatic forms, both fishes and anuran amphibian larvae may use visual cues or stimuli sensed by the lateral line to orient toward, and swim with, conspecifics (e.g. Wassersug and Hessler, 1971; Wassersug, 1973; Partridge and Pitcher, 1980; Katz, Potel and Wassersug, 1981; O'Hara, 1981; Partridge, 1982). However, the more fine-tuned ability to recognize kin appears to be mediated by chemical cues (Blaustein and O'Hara, 1982b; Stabell, 1982; Quinn and Busack, 1985). The cues may be fixed or variable depending upon the species.

Birds may use either fixed genetic cues such as plumage coloration (Cooke, Finney and Rockwell, 1976; Cooke, 1978) or variable cues such as vocalizations, for recognition (e.g. Beecher and Beecher, 1983; Røskaft and Espmark, 1984). Both are efficiently transmitted by birds.

Olfactory cues are important in mammalian communication and they are implicated as the sole basis for kin recognition in several species (e.g. Yamazaki *et al.*, 1976; Porter, Wyrick and Pankey, 1978; Yamaguchi *et al.*, 1981; Holmes, 1984b). These chemical signals may vary with diet or other environmental parameters (see review in Blaustein, 1981) or they may be fixed, as is implied by several studies discussed.

Recognition of signals

By far the most common mechanism used to discriminate between kin and non-kin appears to be familiarity, i.e. individual recognition. Familiarity plays some role in the recognition abilities of the larvae of both toad species and in the wood frog and red-legged frog larvae discussed in the preceding chapter. At least two of the bird species discussed in the preceding chapter use familiarity for recognition. Recognition in most of the mammalian species is influenced, at least to some degree, by familiarity. However, available evidence suggests that many species are capable of utilizing more than one kin recognition mechanism and that they may use different mechanisms alone or in conjunction with one another.

As discussed elsewhere (Bekoff, 1981a; Holmes and Sherman, 1982), familiarity is an efficient kin recognition mechanism when there is a high probability that the individuals with whom a particular animal interacts are related to it. However, there are several ecological and social situations that may favor phenotype matching or the use of recognition alleles (see discussion in Holmes and Sherman, 1982 for details of phenotype matching).

For example, if individuals disperse from their birth sites before they have sufficient exposure to littermates or other kin, or if they live in large groups, familiarity may be an inefficient mechanism of recognition. If multiple fertilization of females occurs, clutches or litters may be composed of full and maternal half-siblings. If males mate with more than one female, paternal half-siblings may be produced. In these situations a more fine-tuned mechanism of recognition, such as phenotype matching or the use of recognition alleles, may be employed. These mechanisms would also be efficient if young develop in close proximity to, or are grouped with, non-kin. Intraspecific parasitism would be a special case of this situation.

Maternal labeling

In several species, maternal labeling may be important in the recognition process. For example, individuals sharing the same uterine environment may be labeled *in utero* and may smell more similar when compared to non-kin or to paternal half-siblings. Moreover, in many species of mammals, licking the offspring after birth may label the young with a characteristic odor (see Gubernick, 1980). An analogous situation may occur in egg-laying species. For example, eggs from one clutch laid by a frog may have similar chemicals (cues) incorporated into their cytoplasm or into the acellular jelly mass that surrounds them (see Waldman, 1981; Blaustein and O'Hara, 1982a for discussions). Alternatively or additionally, environmental components may be incorporated into the jelly mass after eggs are laid, thus labeling certain specific clutches with local environmental components.

Maternal labeling may be important for recognition in some species, but it is less important in others. For example, in mammals, the recognition abilities of spiny mice and some species of ground squirrels may depend on maternal labeling (see Porter, Tepper and White, 1981, and Table 3, Chapter 9). Yet house mice have the ability to recognize paternal half-siblings (Kareem and Barnard, 1982). American toad tadpoles fail to discriminate between full-siblings and maternal half-siblings, suggesting the importance of maternal cues in kin recognition in this species (Waldman, 1981). Conversely, Cascades frog tadpoles can distinguish between full-siblings and maternal half-siblings and between paternal half-siblings and non-siblings (Blaustein and O'Hara, 1982a). Furthermore, when these tadpoles were reared experimentally without jelly or only with jelly from a non-kin group, they still preferred to associate with kin rather than with non-kin. Thus, Cascades frog tadpoles do not need maternal cues for kin recognition. Further tests investigating the influence of the uterine environment in kin recognition and post-natal behavior are warranted (Vom Saal, 1981).

Phenotype matching or recognition alleles?

The problem of how kin recognition is achieved in the absence of obvious opportunities to learn phenotypic characters of kin is the subject of some debate (see Chapter 3). There are two possible mechanisms:
1. Phenotype matching achieved by comparing one's own phenotype with the phenotype of unfamiliar individuals, or
2. Recognition alleles (see discussion in preceding chapter).

There are numerous cases discussed in Chapter 9 in which the investigators have invoked phenotype matching as the mechanism of recognition when, in fact, it is virtually impossible to distinguish between phenotype matching through self inspection and the use of recognition alleles. A few examples illustrating the difficulty of distinguishing between these two mechanisms are warranted.

In their work on ground squirrels, Holmes and Sherman (1982) reported that Belding's ground squirrels that were electrophoretically identified as full-sisters and maternal half-sisters (within the same litter) treated each other differently as yearlings when aggressive and amicable behaviors were quantified. Full-sisters fought significantly less often than did half-sisters and they chased each other significantly less often from territories than did half-sisters. Both types of siblings shared a common nest and even a common uterus. Holmes and Sherman (1982) argued that '... sibling recognition in *S. beldingi* is augmented by some mechanism in addition to simple association in the natal burrow'. They also suggested that phenotype matching may be involved in differentiating between maternal half-siblings and full-siblings (see Holmes and Sherman, 1982). Individuals could learn their own phenotypic trait, such as an odor, and then assess the similarities and differences between themselves and other individuals with whom they have contact. Thus, full-siblings may smell more similar to one another than do half-sibling ground squirrels.

Wu *et al.* (1980) showed that unfamiliar infant paternal half-sibling macaques (*Macaca nemestrina*) were preferred in association tests over unfamiliar non-relatives. The results were intriguing because maternal learning effects were absent; only paternal half-siblings were used. Therefore maternal 'labeling' or 'imprinting' is unlikely (but see Fredrickson and Sackett, 1984; see Blaustein and O'Hara, 1982a for discussion).

Phenotype matching was subtly invoked by Wu *et al.* (1980) as the mechanism for macaque kin recognition. They suggested that an individual's experience of itself could be the basis of later social preferences for unfamiliar animals. However, the monkeys were allowed to interact with non-relatives prior to testing and the later preference for relatives may be based on a model similar to the one proposed by Bateson (1978, 1980, 1982) to explain mate selection, in which individuals choose mates which are slightly different from individuals to whom they were reared. Furthermore, an additional study suggests that familiarity may play a more important role in macaque kin recognition than was originally realized and the results reported by Wu *et al.* need to be reevaluated (Fredrickson and Sackett, 1984). For example, Fredrickson and Sackett (1984) in their subsequent study of kin preferences in pig-tailed macaques found that kinship had neither independent nor additive influences.

Studies of Cascades frog (*Rana cascadae*) tadpoles are particularly relevant to this discussion because of the various rearing regimes employed (Blaustein and O'Hara 1981, 1982a, 1983; O'Hara and Blaustein 1981; Chapter 9). In the first series of experiments, animals reared with siblings and those reared in a mixed group of siblings and non-siblings preferred to associate with siblings rather than non-siblings (O'Hara and Blaustein 1981). In other experiments, individuals reared in total isolation from an early embryonic stage or those reared with only non-kin preferred to associate with unfamiliar siblings instead of unfamiliar non-siblings (Blaustein and O'Hara 1981, 1983). Additionally, tadpoles reared with siblings or in isolation preferred to associate with full-siblings over half-siblings (either maternal half-siblings or paternal half-siblings) and half-siblings over non-siblings (either maternal or paternal; Blaustein and O'Hara, 1982a). In the latter series of experiments, it was shown that the jelly mass that surrounds each clutch was unimportant in influencing kin recognition.

These experiments reveal that tadpoles can differentiate between individuals of varying degrees of relatedness, even if they are reared in mixtures of siblings and non-siblings, in isolation, with full-siblings, or with non-kin only. Holmes and Sherman (1982) interpreted the results of the first two series of experiments (Blaustein and O'Hara 1981; O'Hara and Blaustein 1981) as a possible example of phenotype matching. According to proponents of phenotype matching, tadpoles reared in isolation could learn their own cues and later match their own phenotypes with other individuals, even totally unfamiliar ones, and then make their association choice. But, this explanation does not necessarily account for cases in which tadpoles choose to associate with unfamiliar siblings rather than unfamiliar non-siblings after the test animals were reared with only non-kin. It is possible that tadpoles reared with non-siblings are still 'closer to themselves'. Thus, an individual's own odor would be the most important template against which it would match its own phenotype with other phenotypes.

Isolation experiments, or those using paternal half-siblings, obviously yield equivocal results as far as differentiating between recognition by phenotype matching or by recognition alleles. It may be impossible to falsify the phenotype matching hypothesis (see Blaustein, 1983). Unequivocal support for the recognition allele hypothesis could only be achieved by experimentally masking the ability of an individual to perceive the phenotypic marker in question before it becomes sensitive to the particular cue.

The results concerning ground squirrels (Holmes and Sherman 1982), tadpoles (Blaustein and O'Hara 1981, 1982a, 1983; O'Hara and Blaustein 1981), and possibly macaques (Wu *et al.*, 1980) are consistent with explanations of recognition using phenotype matching and/or recognition alleles. There actually may be no difference between the two mechanisms. Phenotype matching merely specifies a learned component by which genetically encoded information is manifested. For example, if the members of a population have identical genes enabling them to learn their individual phenotypic markers and prompting them to aid others with the same markers, then the difference between the two mechanisms is trivial. Proponents of phenotype matching seem to require that there be no learning before recognition alleles are considered. However, virtually no complicated behavior appears to be totally independent of learning.

The theoretical objections (discussed in Chapter 9) against the existence of recognition alleles must be reevaluated with reference to the *H-2* major histocompatibility complex in mice. It is possible that this complex is an example of a single locus, genetic recognition system. Importantly, one of the functions of this system is to produce antigens responsible for self/non-self recognition. There may be an analogy with recognition at another level, namely kin recognition.

The *H-2* locus is extremely polymorphic (Klein, 1979) and could potentially encode the numerous bits of information necessary for such a genetic recognition system. Klein (1979) identified 56 different alleles at the *H-2K* locus and 45 at the *H-2D* locus which can potentially create 2,500 combinations, most of which do occur in wild house mice. The high degree of *H-2* heterozygosity and the fact that there are many more loci in the *H-2* complex, suggest that the variability of this complex in natural populations of house mice is in Klein's (1979) words 'extraordinary'. Moreover, the genes in this region (e.g. *H-2K* and *H-2D*) display extremely high mutation rates (Steele, 1979). The estimated rate of the *H-2K* and *H-2D* gene products ranges from 1/200 to 1/1,400 mutations per haploid gene per generation (Steele, 1979). These rates are 25–1000 times higher than rates in non-MHC genes (Steele, 1979; see also Treisman, 1981). Obviously, the extreme polymorphism at the *H-2* locus could potentially encode for numerous bits of recognition information, a necessary requirement for such a system.

As Scofield *et al.* (1982) suggest, the histocompatibility complex in vertebrates may have evolved from gamete self/non-self recognition which prevented self-fertilization in hermaphrodites. Therefore, self/non-self discrimination in primitive chordates may represent an evolutionary precursor to the MHC (Scofield *et al.*, 1982; see also Grosberg and Quinn, 1986). Eventually, selection for functions other than outbreeding may have arisen in vertebrates through the MHC. Our arguments concerning the house mouse MHC certainly suggest the possibility of a single locus, genetic recognition system.

However, even the case of positive assortative mating in house mice based on *H-2* locus differences cannot rule out the possibility of phenotype matching through self inspection or by learning the cues of individuals with whom test subjects were reared (see Buckle and Greenberg, 1981; Porter, Matochik and Makin, 1983; Chapters 7 and 9).

Finally, what is the evidence for phenotype matching through self inspection? Lacy and Sherman (1983) suggest that studies showing recognition of paternal half-siblings and recognition of kin after being reared in isolation are consistent with this mechanism. Dawkins (1982) and numerous others (e.g. Alexander, 1979; Holmes and Sherman, 1982) concur. So do we. However, as we have argued, the results are also consistent with the recognition allele hypothesis. One study that is often cited as providing a convincing argument for phenotype matching through self inspection is by Salzen and Cornell (1968). In this study, chickens reared in isolation subsequently showed a tendency to associate with other chickens that were dyed the same color. However, if the chickens were unable to see their reflection in the drinking water, they failed to choose chickens of the same color. This experiment does not provide unequivocal evidence for phenotype matching through self inspection because the reflection of a bird in the drinking water may have been functionally equivalent to the presence of another individual bird. Thus, recognition may have been achieved by familiarity. With the information available, we must conclude that unequivocal evidence for phenotype matching through self inspection is meager at best.

Although we do not necessarily endorse their existence, we feel that the possibility of recognition alleles should not be ruled out, especially on theoretical grounds. Available empirical evidence appears to be consistent with both phenotype matching through self inspection and recognition allele mechanisms. Both explanations allow an individual to recognize other animals with whom they are unfamiliar. Furthermore, both mechanisms function similarly and lead to the same evolutionary predictions.

POSSIBLE FUNCTIONS OF KIN RECOGNITION

The studies described in the previous chapter depict two basic potential functions of kin recognition:
1. Optimal outbreeding as illustrated by Japanese quail, some rodent species, and possibly in one species of frog.
2. Directing of behaviors preferentially toward kin as exemplified by
   (i) the warning calls of certain ground squirrels and potential warning behaviors of anuran larvae (see also Rohwer, Fretwell and Tuckfield 1976),
   (ii) The maintenance of cohesive 'family' units as depicted in snow geese, several rodents, and Cascades frog tadpoles,

(iii) helping; and
(iv) interference in agonistic encounters.

In addition to these functions, kin recognition may also play an important role in intraspecific competition, cannibalism (including infanticide), and habitat selection. By elucidating the possible importance of kin recognition to these phenomena we may substantially alter the ways in which we view, and understand, these and other ecological processes.

For example, the phenomenon of cycling in small mammal populations has been the subject of intense research and debate for many years (see Krebs and Myers, 1974 for a review). Recently, a novel approach to this problem was formulated by Charnov and Finerty (1980) and Bekoff (1981b) using kin selection theory and recognition. Charnov and Finerty (1980) hypothesized that when vole populations are low, close relatives probably interact most frequently, and aggressive encounters are diminished because of relatively high coefficients of relatedness. Demographic fluctuations, therefore, could be influenced by changes in behavior as manifested in reproduction or aggression (see Krebs and Myers, 1974, for a discussion of the role of aggression and dispersal in cycling small mammal populations). It is also likely that voles in small (low dispersal) populations are more familiar with one another (regardless of their relatedness). Thus, increased familiarity may be the cue resulting in decreased aggression (Bekoff, 1981b).

Furthermore, increased familiarity might also result in decreased reproduction. This hypothesis is consistent with the studies of Gavish, Hoffman and Getz (1984) and Boyd and Blaustein (1985). Increased reproduction and subsequent population growth may occur only after sufficient emigration ensues. It is possible that the coefficient of relatedness (r) and the coefficient of familiarity (f; *sensu* Bekoff, 1981a) are directly correlated, and assessment of f by voles is a means of avoiding incestuous mating that could lead to inbreeding depression (see Bekoff, 1981a). Regardless of the relationship between r and f, it is important that increased familiarity in low vole populations could yield diminished breeding and this could greatly influence population cycling. Studies assessing the relationships between coefficients, r and f, in natural populations, are warranted.

Kin recognition may also be potentially important in intraspecific competition. The importance of competition in structuring communities is presently being, and probably forever will be, debated (e.g. Wiens, 1977; Diamond, 1978; Connell, 1980, 1983; Schoener, 1983), but ecologists generally agree that competition is an important ecological force at least at certain times in certain situations. Intraspecific competition for limited resources may be especially intense because members of the same species at the same life stage are most likely to be more ecologically similar than are members of different species. We suggest that if intraspecific competition is important, individuals would benefit if they competed most intensely with unrelated members of their species. The direction of intraspecific competition toward unrelated or distantly related individuals and away from closely related individuals could be maintained through kin selection. Kin recognition could facilitate this skewed competition. Experimental tests of intraspecific competition that also investigate the role of relatedness would add to our understanding of the competition process.

Kin recognition may also play a role in cannibalism. Cannibalism seems to be an important force in nature (see Fox, 1975; Polis, 1981 for reviews). An argument similar to the one expressed above regarding the benefits of reducing competitive behavior with relatives can be put forth with regard to cannibalism. Those individuals that kill and eat members of their species but avoid killing and eating relatives may achieve benefits over those individuals that do not discriminate in their cannibalistic behavior. Kin selection could maintain such selective predation.

A similar argument could be made for infanticide (Hrdy, 1979; Polis, Myers and Hess, 1984) which seems to be a relatively widespread phenomenon correlated with special demographic conditions (Eisenberg, 1981). Those individuals that practice infanticide and have an ability to recognize kin, could benefit by selectively killing young to whom they are not or only distantly related, thereby avoiding the killing of individuals with whom they share genes. A recognition mechanism that allowed an infanticidal animal to distinguish young that are unrelated from young that are related (e.g. cousins) could also be maintained through kin selection. Indeed, experiments using guppies (*Poecilia reticulata*) show that females preferentially cannibalize unrelated fry over related fry (Loekle, Madison and Christian, 1982).

Habitat selection also may be achieved through kin recognition. Individuals may settle where related individuals have already settled. Thus, being able to recognize kin may enable individuals to settle more efficiently in suitable areas. Some empirical evidence exists suggesting that siblings of certain vertebrates may disperse to, and settle in, the same area (e.g. voles, Hilborn, 1975; spruce grouse, Keppie, 1980). Furthermore, settlement by larval invertebrates may depend on recognizing kin (Keough, 1984). The larval ecologies of western toads, American toads, and especially wood frogs, suggests that these animals may be using kin recognition, as reflected in their recognizing 'familiar cues', in habitat selection. O'Hara's (1981) experimental work supports the contention that anuran larvae of several species may orient towards conspecifics when selecting a habitat.

RECOGNITION ERRORS AND BEHAVIORAL POLYMORPHISM

Kin recognition appears to be a polymorphic trait. Analyses of available data clearly reveal that certain individuals within a population or experimental group may exhibit kin recognition behavior, whereas others do not. Not all individuals that are tested for kin recognition abilities respond as theory would predict. Similarly, certain populations may be composed primarily of individuals exhibiting kin recognition behavior, whereas other populations may be composed primarily of individuals that do not. This behavioral polymorphism could be maintained by fluctuating environmental conditions and varying selective pressures. Therefore, intraspecific variability in kin recognition behavior may be as widespread as interspecific variability. Single individuals may also show plasticity in kin recognition so that it is manifested under certain environmental conditions and not others. For example, when ecological conditions are variable and at times when food is not limited, kin recognition, with respect to directing intraspecific competition toward non-kin, may not be manifested.

It is clear that benefits can accrue to individuals that have an ability to recognize kin and then behave preferentially towards them. In many cases it seems likely that individuals who cannot recognize kin would be penalized in terms of their genetic contributions to future generations. Selective pressures may be intense in favoring kin recognition mechanisms. Why then do recognition errors occur?

There are limits to the recognition abilities of organisms in the various contexts in which they exist. For example, in natural populations of Belding's ground squirrels, yearlings may cooperate with unrelated foster sisters or treat genetic sisters as if they are unrelated (Sherman, 1980a). Yet full-sibling Belding's ground squirrels are treated differently than maternal half-siblings in nature (Holmes and Sherman, 1982), suggesting that members of this species do have a capacity to recognize others with some precision. Bank swallows may enter the wrong nests and may be fed by unrelated parent birds before they are expelled (Hoogland and Sherman, 1976: Beecher, Beecher and Lumpkin, 1981), suggesting possible errors in both recognition of nests by young swallows and in parent-offspring recognition. Numerous reports show that young of many species are often reared by unrelated foster parents (reviewed in Riedman, 1982), and the young and alloparent may even be of different species, indicating the possibility of gross recognition errors.

Perhaps the most obvious examples of recognition errors are from studies of avian brood parasitism. Both intraspecific (Yom-Tov, 1980) and interspecific (Rothstein, 1982) avian brood parasitism can lead to severe losses in fitness to the host individual. Yet Rothstein's (1975, 1982) work has shown that numerous failures in recognition occur. For example, some avian brood parasites leave eggs that are quite different morphologically from those of the host species in host nests to develop and eventually to replace the host's nestlings.

Finally, in many of the experimental kin recognition studies discussed in the preceding chapter, the data indicate a statistically significant trend in which kin are distinguished from non-kin most of the time. However, it is possible that many of the 'limits' to recognition exhibited by animals in laboratory tests may actually be the results of laboratory techniques that are not sensitive enough to detect some of the more subtle behaviors that might demonstrate kin recognition. Nonetheless, the numerous 'recognition errors' that occur under natural conditions warrant an explanation. For example, there may be errors due to decreased efficiency in signal transmission, and recognition errors may occur if individuals of varying kin groups overlap in their phenotypic markers (Breed and Bekoff, 1981).

Another explanation of errors in recognition may stem from our own expectations as investigators. Although we call them 'errors' it is possible that animals are not necessarily making mistakes in the way in which they direct different behavior patterns towards other individuals. Perhaps most important of all is motivation. Under the conditions of the experiment an animal may not choose to behave in a way that reveals to us its ability to distinguish kinship levels. In his excellent discussion of optimality theory, Rothstein (1982) succinctly points out that not all behaviors are optimal. Optimal behavior is something many biologists have come to expect. Rothstein (1982) and others (e.g. Curio, 1973; Lewontin, 1978; Maynard Smith, 1978; Gould and Lewontin, 1979) have cautioned investigators in their quest for 'adaptive value' explanations. There are several good reasons that natural selection may fail to produce optimal solutions and why non-adaptive traits may exist (Rothstein, 1982; see also discussion in Gould and Lewontin, 1979).

It is interesting to note that of all the species discussed in the previous chapter, only one (*Rana aurora* tadpoles) did not show a well developed (determined statistically) ability to discriminate between kin and non-kin. The lack of similar data for other species may be because negative results are difficult to publish or to advertise otherwise. Nonetheless, it is important to know which species do not display kin recognition so that we can more fully understand the phenomenon.

Finally, if kin recognition behavior is at least partially determined genetically (see discussion in Blaustein, 1983), then the behavior may rapidly (within evolutionary time) become fixed within a species' repertoire because of its selective value (see discussion in Rothstein, 1975). Rothstein's (1975) model of evolutionary rates in host defenses against avian brood parasitism suggests that once a species obtains at least a partially genetically determined behavior to recognize and reject parasitic eggs, the behavior may spread in as few as 20–100 years. As yet, an evolutionary rate model does not exist for kin recognition behavior. Yet kin recognition and recognition of parasitic eggs are similar problems. One of the reasons why kin recognition may be much more prevalent than investigators originally thought, may lie in its great selective value and its potentially rapid spread through populations. Animals that form social units or become social at least during some time in their life cycle, may all have the potential to develop a kin recognition system.

DIRECTIONS FOR FUTURE RESEARCH AMONG VERTEBRATES

There is enormous potential for future research that can be approached from many directions. These include studying problems in methodological (sampling, analytical) procedures and ontogenetic, ecological and evolutionary aspects of kin recognition, all of which are interrelated. There is also a need to investigate the kin recognition ability of additional species. For example, fishes are one vertebrate group that has been relatively understudied, but could potentially provide important information because of their diverse life histories (Shapiro, 1983). As mentioned above, several species of fishes have a well developed, chemically mediated, parent-offspring recognition system. Furthermore, in a genetic analysis of allele frequencies and rare allele distributions in two tuna species, there is intriguing evidence for the possibility of a cohesiveness of related fishes in schools (Sharp, 1978). Sharp observed that where more than one very rare allele is encountered in a large sample, the individuals exhibiting the rare alleles are often the same length or vary by 1 cm or less. He believes that this is highly unlikely unless the fish are related.

Studies of kin recognition in reptiles are also warranted. To our knowledge, such analyses have not been made. Species, sexual and social status recognition have been documented in some reptile species (Colgan, 1983; Wilson, 1975). Therefore, the potential for kin recognition based on visual or chemical cues (Stoddart, 1980; Wilson, 1975) exists in the more social reptiles.

There is also a need to replicate some of the experimental studies of species already tested, using individuals from a number of different populations. These analyses could provide information on how widespread kin recognition is within a particular species and may give insight into the selective pressures associated with the evolution of kin recognition in different and fluctuating environments.

Further work concerning the ontogeny of kin recognition throughout the overlapping stages that categorize development from the prenatal period through senescence (Bekoff, 1985) is especially important. Vom Saal (1981) stressed that the intrauterine environment could influence future recognition. He also suggested that a wide range of behaviors may be influenced by the intrauterine environment and that these behaviors can have potentially profound effects on population structure and reproductive success (see also Vom Saal, 1986). Developmental analyses of the relationship between r and f (the coefficient of familiarity; Bekoff, 1981a) also are warranted in natural populations.

The mechanism of phenotype matching should also be examined in more detail. In the particular case of phenotype matching through self-inspection, it would be important to determine the precise time during development when 'awareness' of the phenotypic characteristics used in kin recognition first appears (Bekoff, 1981a). For example, we can ask if phenotype matching occurs before an animal begins grooming itself. If not, auto-grooming, which occurs early in life in various species, could provide a good opportunity for an animal to learn aspects of its own phenotype. Therefore the time at which auto-grooming emerges is a logical point in the development of an individual to look for the first occurrence of kin recognition via phenotype matching.

More tests of animals at various ages and stages of development are needed to understand more fully the development of kin recognition. Especially important are studies that investigate kin recognition in animals that undergo metamorphosis, such as insects and amphibians. Is kin recognition manifested in both the larval and adult stages of life? If so, are there different selective pressures at these stages requiring different mechanisms of kin recognition? With regard to amphibians, there is good evidence that chemical cues are important in individual recognition in salamanders (e.g. Madison, 1975; Simon and Madison, 1984). Kin recognition studies of larval and adult salamanders similar to those conducted on anuran amphibians would enhance our knowledge of kin recognition in amphibians.

There seems to be plasticity in the mechanisms animals, even conspecifics, use to distinguish between kin and non-kin under controlled conditions. Perhaps animals use the most simple mechanism of recognition, probably familiarity in most species, as a primary means to discriminate between kin and non-kin and in many species a back-up machanism perhaps phenotype matching through self inspection, is used when the primary mechanism fails. It is possible that animals have the potential to behave in a variety of ways that are not generally exhibited until there are strong (or different) environmental pressures to do so. Therefore, future research should investigate the ability of a species to utilize more than one mechanism of recognition under field conditions. These data will also give us an idea of the magnitude of natural variation in recognition abilities. Moreover, these studies could investigate the possibility of a hierarchy of mechanism usage.

Related to this would be investigations of whether certain proximate cues are used preferentially to distinguish kin from non-kin. For example, olfactory or water-borne chemical cues seem to be the most common proximate cues used in most of the species studied. A number of questions could be investigated. For example:
1. Are chemical cues the most efficient for recognition?
2. If an individual lost its ability to use the primary sense for recognition, could it use a different modality to distinguish between kin and non-kin?
3. Is there a hierarchy of cue importance?

Patterns of recognition can also influence the structure of a population and various life history characteristics. When kinship affects inclusive fitness (Hamilton, 1964; Grafen, 1982), one might expect that sociality, dispersal, spacing, mating, care-giving and competitive and foraging tactics would be influenced by these relationships (Bekoff, Daniels and Gittleman, 1984; Bekoff and Wells, 1986). For example, individuals may direct their competitive or predatory behaviors toward unrelated conspecifics and thus achieve reproductive benefits compared with animals that do not direct their behaviors in this way. Thus, we stress that future ecological studies would benefit from an analysis that includes consideration of the kin composition of the species being studied. Carefully designed field experiments will aid in our understanding of the interplay between kinship and ecological processes.

Individual variation in recognition must be considered. For example, it is obvious that individuals of many species can discriminate between kin and non-kin. However, it also seems clear that not all individuals in species in which recognition has been demonstrated display the ability to do so. It is important to stress that an individual may demonstrate recognition abilities that vary temporally, both during development and in fluctuating environments. Finally, it has also been shown in some cases, that under laboratory conditions, test animals do not behave precisely as they would in nature. Thus, we may ask, what is the fate of those individuals that cannot or do not discriminate between kin and non-kin? Do these individuals actually have lower reproductive success, assuming it can be accurately and reliably determined, than individuals that can discriminate between kin and non-kin? Where t-alleles are involved, for instance, mice that cannot discriminate between individuals having the t-allele during mate selection will probably have a lower reproductive success than those that can make this differentiation. Finally, by stressing the importance of studying identified individuals over long periods of time, it is obvious that we need to consider whether 'statistical significance' can always be simply equated with 'biological significance', especially with respect to demonstrated skills that may influence individual patterns of growth, maintenance, and reproduction. Only after large numbers of individuals of a particular species are tested in numerous replicates (ideally both in the laboratory and in the field where data on natural variation can be gathered) and only when detailed natural and life history information has been obtained, will we be able to answer reliably questions about the role of kin recognition in the evolution of various taxa.

ACKNOWLEDGMENTS (CHAPTERS 9 AND 10)

We thank Beth Bennett, Kathy Blaustein, Michael Breed, Jennifer Fewell, Alcinda Lewis, Allen Moore, Mary Jo Moore, Richard K. O'Hara, Deanna H. Olson and Tammy Stiller for critically reviewing various drafts of our chapters. A. R. Blaustein wishes especially to thank Richard O'Hara with whom he conducted all research on kin recognition in western species of anuran larvae and Janna Ellingson and Colleen Lenihan for their help with library research. Discussions with Paul K. Anderson, Michael Beecher, Bruce Coblentz, John Chance, Lowell Getz, Mark Hixon, Sarah Lenington, Joseph Levitch, Gail Michener, Richard Porter, Steve Rothstein, Paul Samollow, Robert M. Storm, John Stryker, M. Tashman and J. A. Wapner were extremely helpful. We are especially grateful to Suzi Brubaker and Linda Olds for typing the various drafts of our manuscripts. The authors gratefully acknowledge the advice and critical comments by the editors, David Fletcher and Charles Michener. Marc Bekoff was supported by a John Simon Guggenheim Memorial Foundation Fellowship when he formulated the coefficient of familiarity. Thomas Daniels was supported by a University of Colorado Fellowship during preparation of our chapters. Financial support for research and for writing our reviews was provided by a National Geographic Society Grant and Grants BNS-8120203 and BNS-8406256 from the National Science Foundation to A. R. Blaustein and R. K. O'Hara.

LITERATURE CITED (FOR CHAPTERS 9 AND 10)

Ågren, G. 1981. Two laboratory experiments on inbreeding avoidance in the Mongolian gerbil. *Behavioral Processes*, **6**, 291–297.
Ågren, G. 1984. Incest avoidance and bonding between siblings in gerbils. *Behavioral Ecology and Sociobiology*, **14**, 161–169.
Alexander, R. D. 1979. *Darwinism and human affairs*. University of Washington Press. Seattle.
Alexander, R. D. and G. Borgia. 1978. Group selection, altruism, and the levels of organization of life. *Annual Review of Ecology and Systematics*, **9**, 449–474.
Anderson, P. K. 1970. Ecological structure and gene flow in small mammals. *Symposium Zoological Society of London*, **26**, 299–325.
Armitage, K. B. and D. W. Johns. 1982. Kinship, reproductive strategies and social dynamics of yellow-bellied marmots. *Behavioral Ecology and Sociobiology*, **11**, 55–63.
Baker, A. E. M. 1981. Gene flow in house mice: Behavior in a population cage. *Behavioral Ecology and Sociobiology*, **8**, 83–90.
Barnett, C. 1977a. Aspects of chemical communication with special reference to fish. *Bioscience Communications*, **3**, 331–392.
Barnett, C. 1977b. Chemical recognition of the mother by young of the cichlid fish, *Cichlasoma citrinellum*. *Journal of Chemical Ecology*, **3**, 461–466.
Bateson, P. 1978. Sexual imprinting and optimal outbreeding. *Nature*, **273**, 659–660.
Bateson, P. 1980. Optimal outbreeding and the development of sexual preferences in Japanese quail. *Zeitschrift für Tierpsychologie*, **53**, 231–244.
Bateson, P. 1982. Preferences for cousins in Japanese quail. *Nature*, **295**, 236–237.
Bateson, P. 1983. Optimal outbreeding, in Bateson, P. ed., *Mate choice*, Cambridge University Press, Cambridge, England, pp. 257–277.
Beecher, M. D. 1982. Signature systems and kin recognition. *American Zoologist*. **22**, 477–490.

Beecher, I. M. and M. D. Beecher. 1983. Sibling recognition in bank swallows (*Riparia riparia*). *Zeitschrift für Tierpsychologie*, **62**, 145–150.
Beecher, M. D., I. M. Beecher and S. Hahn. 1981. Parent-offspring recognition in bank swallows (*Riparia riparia*): II. Development and acoustic basis. *Animal Behaviour*, **29**, 95–101.
Beecher, M. D., I. M. Beecher and S. Lumpkin. 1981. Parent-offspring recognition in Bank swallows (*Riparia riparia*): I. Natural history. *Animal Behaviour*. **29**, 86–94.
Beiswenger, R. E. 1972. *Aggregative behavior of tadpoles of the American toad,* Bufo americanus, *in Michigan*. Ph.D. Thesis. University of Michigan, Ann Arbor, Michigan.
Beiswenger, R. E. 1975. Structure and function in aggregations of tadpoles of the American toad *Bufo americanus. Herpetologica.* **31**, 222–233.
Beiswenger, R. E. 1977. Diel patterns of aggregative behavior in tadpoles of *Bufo americanus* in relation to light and temperature. *Ecology* **58**, 98–108.
Bekoff, M. 1981a. Mammalian sibling interactions: Genes, facilitative environments, and the coefficient of familiarity, in Gubernick, D. and P. N. Klopfer, eds., *Parental care in mammals*, Plenum Press, New York, pp. 307–346.
Bekoff, M. 1981b. Vole population cycles: kin selection or familiarity? *Oecologia* **48**, 131.
Bekoff, M. 1985. Evolutionary perspectives of behavioral development. *Zeitschrift für Tierpsychologie*, **69**, 166–167.
Bekoff, M., T. J. Daniels and J. L. Gittleman. 1984. Life history patterns and the comparative social ecology of carnivores. *Annual Review of Ecology and Systematics*, **15**, 191–232.
Bekoff, M. and M. C. Wells. 1986. Social ecology and behavior of coyotes *Advances in The Study of Behavior*, **16**, 251–338.
Bennett, D. 1975. The *T*-locus of the mouse. *Cell*, **6**, 441–454.
Bertram, B. C. R. 1978. Living in groups: predators and prey. In Krebs, J. R. and N. B. Davies eds., *Behavioural ecology: an evolutionary approach*, Sinauer Associates, Inc., Sunderland, Massachusetts, pp. 64–96.
Blaustein, A. R. 1981. Sexual selection and mammalian olfaction. *American Naturalist*. **117**, 1006–1010.
Blaustein, A. R. 1983. Kin recognition mechanisms: phenotypic matching or recognition alleles? *American Naturalist*, **121**, 749–754.
Blaustein, A. R. and R. K. O'Hara. 1981. Genetic control for sibling recognition? *Nature*, **290**, 246–248.
Blaustein, A. R. and R. K. O'Hara. 1982a. Kin recognition in *Rana cascadae* tadpoles: maternal and paternal effects. *Animal Behaviour*, **30**, 1151–1157.
Blaustein, A. R. and R. K. O'Hara. 1982b. Kin recognition cues in *Rana cascadae* tadpoles. *Behavioral and Neural Biology*, **36**, 77–87.
Blaustein, A. R. and R. K. O'Hara. 1983. Kin recognition in *Rana cascadae* tadpoles: Effects of rearing with non-siblings and varying the strength of the stimulus cues. *Behavioral and Neural Biology*, **39**, 259–267.
Blaustein, A. R. and R. K. O'Hara. 1986a. Kin recognition in tadpoles. *Scientific American*, **254**, 108–116.
Blaustein, A. R. and R. K. O'Hara. 1986b. An investigation of kin recognition in red-legged frog (*Rana aurora*) tadpoles. *Journal of Zoology (London)*, (A), **209**, 347–353.
Blaustein, A. R., R. K. O'Hara and D. H. Olson. 1984. Kin preference behaviour is present after metamorphosis in *Rana cascadae* frogs. *Animal Behaviour*, **32**, 445–450.
Blaustein, A. R. and S. I. Rothstein. 1978. Multiple captures of *Reithrodontomys megalotis*: social bonding in a mouse? *American Midland Naturalist*, **100**, 376–383.

Bowers, J. M. and B. K. Alexander. 1967. Mice: individual recognition by olfactory cues. *Science*, **158**, 1208–1210.
Boyd, S. K. and A. R. Blaustein. 1985. Familiarity and inbreeding avoidance in the gray-tailed vole. *Journal of Mammalogy*, **66**, 348–352.
Boyse, E. A., G. K. Beauchamp, K. Yamazaki, J. Bard and L. Thomas. 1982. A new aspect of the major histocompatibility complex and other genes in the mouse. *Oncodevelopmental Biology and Medicine*, **4**, 101–116.
Breed, M. D. and M. Bekoff. 1981. Individual recognition and social relationships. *Journal of Theoretical Biology*, **88**, 589–593.
Bruce, H. M. 1959. An exteroceptive block to pregnancy in the mouse. *Nature*, **184**, 105.
Bruce, H. M. 1960. A block to pregnancy in the mouse caused by proximity to strange males. *Journal of Reproduction and Fertility*, **1**, 96–103.
Buckle, G. R. and L. Greenberg. 1981. Nestmate recognition in sweat bees (*Lasioglossum zephyrum*): does an individual recognize its own odour or only odours of its nestmates? *Animal Behaviour*, **29**, 802–809.
Calhoun, J. B. 1962. *The ecology and sociology of the Norway rat.* U.S. Public Health Service Publication No. 1008.
Charnov, E. L. 1981. Vole population cycles: ultimate or proximate explanation? *Oecologia*, **48**, 132.
Charnov, E. L. and J. P. Finerty. 1980. Vole population cycles: A case for kin selection? *Oecologia*, **45**, 1–2.
Colgan, P. 1983. *Comparative social recognition.* John Wiley and Sons, New York.
Connell, J. H. 1980. Diversity and the coevolution of competitors, or the ghost of competition past. *Oikos*, **35**, 131–138.
Connell, J. H. 1983. On the prevalence and relative importance of interspecific competition: evidence from field experiments. *American Naturalist*, **122**, 661–696.
Cooke, F. 1978. Early learning and its effect on population structure. Studies of a wild population of snow geese. *Zeitschrift für Tierpsychologie*, **46**, 344–358.
Cooke, F. and F. G. Cooch. 1968. The genetics of polymorphism in the snow goose, *Anser caerulescens*. *Evolution*, **22**, 289–300.
Cooke, F., G. H. Finney and R. F. Rockwell. 1976. Assortative mating in lesser snow geese (*Anser caerulescens*). *Behavior Genetics*, **6**, 127–140.
Cooke, F. and C. M. McNally. 1975. Mate selection and colour preferences in lesser snow geese. *Behaviour*, **53**, 151–170.
Cooke, F., P. J. Mirsky and M. B. Seiger. 1972. Color preferences in the lesser snow goose and their possible role in mate selection. *Canadian Journal of Zoology*, **50**, 529–536.
Coppersmith, R. and M. Leon. 1984. Enhanced neural response to familiar olfactory cues. *Science*, **225**, 849–851.
Cramp, S. (ed.) 1980. *Handbook of the birds of Europe, the Middle East and North Africa: the birds of the Western Palearctic. Volume II.* Oxford University Press, Oxford, England.
Crowcroft, P. and F. P. Rowe. 1957. The growth of confined colonies of the wild house mouse (*Mus musculus* L.) *Proceedings of the Zoological Society of London*, **129**, 359–370.
Crozier, R. H. and M. W. Dix. 1979. Analysis of two genetic models for the innate components of colony odor in social hymenoptera. *Behavioral Ecology and Sociobiology*, **4**, 217–224.
Curio, E. 1973. Towards a methodology of teleonomy. *Experientia*, **29**, 1045–1058.
Davis, L. S. 1982. Sibling recognition in Richardson's ground squirrels (*Spermophilus richardsonii*). *Behavioral Ecology and Sociobiology*, **11**, 65–70.

Davis, L. S. 1984a. Behavioral interactions of Richardson's ground squirrels: asymmetries based on kinship, in Murie, J. O. and G. R. Michener eds., *The biology of ground dwelling squirrels.* University of Nebraska Press, Lincoln, Nebraska, pp. 424–443.
Davis, L. S. 1984b. Alarm calling in Richardson's ground squirrels (*Spermophilus richardsonii*). *Zeitschrift für Tierpsychologie*, **66**, 152–164.
Davis, L. S. 1984c. Kin selection and adult female Richardson's ground squirrels: a test. *Canadian Journal of Zoology*, **62**, 2344–2348.
Dawkins, R. 1976. *The selfish gene.* Oxford University Press, Oxford, England.
Dawkins, R. 1982. *The extended phenotype.* Oxford University Press, Oxford, England.
DeBenedictis, P. A. 1974. Interspecific competition between tadpoles of *Rana pipiens* and *Rana sylvatica*: an experimental field study. *Ecological Monographs*, **44**, 129–151.
Delany, M. J. and D. C. D. Happold. 1979. *Ecology of African mammals.* Longman Group Limited, London.
DeLong, K. T. 1967. Population ecology of feral house mice. *Ecology*, **48**, 611–634.
Dewsbury, D. A. 1982. Avoidance of incestuous breeding between siblings in two species of *Peromyscus* mice. *Biology of Behavior*, **7**, 157–169.
Diamond, J. M. 1978. Niche shifts and the rediscovery of interspecific competition. *American Scientist*, **66**, 322–331.
Døving, K. B., H. Nordeng and B. Oakley. 1974. Single unit discrimination of fish odours released by char (*Salmo alpinus* L.) populations. *Comparative Biochemistry and Physiology*, **47A**, 1051–1063.
D'Udine, B. and L. Partridge. 1982. Olfactory preferences of inbred mice (*Mus musculus*) for their own strain and for siblings: effects of strain, sex and cross fostering. *Behaviour*, **78**, 314–324.
Egid, K. and S. Lenington. 1985. Responses of male mice to odors of females: Effects of *T* and *H-2* locus genotype. *Behavior Genetics*, **15**, 287–295.
Eisenberg, J. F. 1962. Studies on the behavior of *Peromyscus maniculatus gambelii* and *Peromyscus californicus parasiticus*. *Behaviour*, **19**, 177–207.
Eisenberg, J. F. 1968. Behavior patterns, in King, J. A., ed., *Biology of Peromyscus.* Special Publication No. 2, American Society of Mammalogists, pp. 541–595.
Eisenberg, J. F. 1981. *The mammalian radiations: an analysis of trends in evolution, adaptation and behavior.* University of Chicago Press, Chicago, Illinois.
Evans, R. M. 1970. Imprinting and mobility in young ring-billed gulls, *Larus delawarensis*. *Animal Behaviour Monographs*, **3**, 193–248.
Fairbairn, D. J. 1978. Dispersal of deer mice (*Peromyscus maniculatus*): Proximal causes and effects on fitness. *Oecologia*, **32**, 171–193.
Fisher, R. A. 1930. *The genetical theory of natural selection.* Clarendon Press, Oxford, England.
Fox, L. R. 1975. Cannibalism in natural populations. *Annual Review of Ecology and Systematics*, **6**, 87–106.
Frase, B. A. and K. B. Armitage. 1984. Foraging patterns of yellow-bellied marmots: role of kinship and individual variability. *Behavioral Ecology and Sociobiology*, **16**, 1–10.
Fredrickson, W. T. and G. P. Sackett. 1984. Kin preferences in primates (*Macaca nemestrina*): relatedness or familiarity? *Journal of Comparative Psychology*, **98**, 29–34.
Gavish, L., J. Hoffman and L. L. Getz. 1984. Sibling recognition in the prairie vole, *Microtus ochrogaster*. *Animal Behaviour*, **32**, 362–366.
Getz, L. L. 1972. Social structure and aggressive behavior in a population of *Microtus pennsylvanicus*. *Journal of Mammalogy*, **53**, 310–317.

Getz, L. L. and C. S. Carter. 1980. Social organization in *Microtus ochrogaster* populations. *The Biologist*, **62**, 56–69.
Getz, L. L., C. S. Carter and L. Gavish. 1981. The mating system of the prairie vole, *Microtus ochrogaster*: field and laboratory evidence for pair bonding. *Behavioral Ecology and Sociobiology*, **8**, 189–194.
Getz, W. M. 1981. Genetically based kin recognition systems. *Journal of Theoretical Biology*, **92**, 209–226.
Gilder, P. M. and P. J. B. Slater. 1978. Interest of mice in conspecific male odours is influenced by degree of kinship. *Nature*, **274**, 364–365.
Gottlieb, G. 1968. Prenatal behavior of birds. *Quarterly Review of Biology*, **43**, 148–174.
Gottlieb, G. 1971. *Development of species identification in birds: an inquiry into the prenatal determinants of perception.* University of Chicago Press, Chicago, Illinois.
Gould, S. J. and P. F. Lewontin. 1979. The spandrels of San Marco and the Panglossian paradigm: A critique of the adaptationist programme. *Proceedings of the Royal Society of London B*, **205**, 581–598.
Grafen, A. 1982. How not to measure inclusive fitness. *Nature*, **298**, 425–426.
Grau, J. H. 1982. Kin recognition in white-footed deermice (*Peromyscus leucopus*). *Animal Behaviour*, **30**, 497–505.
Griffin, D. R. 1984. *Animal thinking.* Harvard University Press. Cambridge, Massachusetts.
Grosberg, R. K. and J. F. Quinn. 1986. The genetic control and consequences of kin recognition by the larvae of a colonial marine invertebrate. *Nature*, **322**, 456–459.
Gubernick, D. J. 1980. Maternal 'imprinting' or maternal 'labeling' in goats? *Animal Behaviour*, **28**, 124–129.
Halpin, Z. T. 1980. Individual odors and individual recognition: review and commentary, *Biology of Behavior*, **5**, 233–248.
Hamilton, W. D. 1964. The genetical evolution of social behavior. I. II. *Journal of Theoretical Biology*, **7**, 1–52.
Hamilton, W. D. 1971. Geometry for the selfish herd. *Journal of Theoretical Biology*, **31**, 295–311.
Hanken, J. and P. W. Sherman. 1981. Multiple paternity in Belding's ground squirrel litters. *Science*, **212**, 351–353.
Hansen, R. M. 1957. Communal litters of *Peromyscus maniculatus*. *Journal of Mammalogy*, **38**, 523.
Hasler, A. D., T. Scholz and R. M. Horvall. 1978. Olfactory imprinting and homing in salmon. *American Scientist*, **66**, 347–355.
Hassinger, D. D. 1972. *Early life history and ecology of three congeneric species of Rana in New Jersey.* Ph.D. Thesis. Rutgers University, New Brunswick, New Jersey.
Hepper, P. G. 1983. Sibling recognition in the rat. *Animal Behaviour*, **31**, 1177–1191.
Hews, D. K. and A. R. Blaustein. 1985. An investigation of the alarm response in *Bufo boreas* and *Rana cascadae* tadpoles. *Behavioral and Neural Biology*, **43**, 47–57.
Hilborn, R. 1975. Similarities in dispersal tendency among siblings in four species of voles (*Microtus*). *Ecology*, **56**, 1221–1225.
Hill, J. L. 1974. *Peromyscus*: effect of early pairing on reproduction. *Science*, **13**, 1042–1044.
Holmes, W. G. 1984a. Ontogeny of dam-young recognition in captive Belding's ground squirrels (*Spermophilus beldingi*). *Journal of Comparative Psychology*, **98**, 246–256.
Holmes, W. G. 1984b. Sibling recognition in 13-lined ground squirrels: effects of genetic relatedness, rearing association and olfaction. *Behavioral Ecology and Sociobiology*, **14**, 225–233.

Holmes, W. G. 1986a. Identification of paternal half-siblings by captive Belding's ground squirrels. *Animal Behaviour*, **34**, 321–327.
Holmes, W. G. 1986b. Kin recognition by phenotype matching in female Belding's ground squirrels. *Animal Behaviour*, **34**, 38–47.
Holmes, W. G. and P. W. Sherman. 1982. The ontogeny of kin recognition in two species of ground squirrels. *American Zoologist*, **22**, 491–517.
Holmes, W. G. and P. W. Sherman. 1983. Kin recognition in animals. *American Scientist*, **71**, 46–55.
Hoogland, J. L. 1982. Prairie dogs avoid extreme inbreeding. *Science*, **215**, 1639–1641.
Hoogland, J. L. and P. W. Sherman. 1976. Advantages and disadvantages of bank swallow (*Riparia riparia*) coloniality. *Ecological Monographs*, **46**, 33–58.
Hrdy, S. B. 1979. Infanticide among mammals: a review, classification and examination of the implications for the reproductive strategies of females. *Ethology and Sociobiology*, **1**, 13–40.
Huck, V. W. and E. M. Banks. 1979. Behavioral components of individual recognition in the collared lemming (*Dicrostonyx groenlandicus*). *Behavioral Ecology and Sociobiology*, **5**, 85–90.
Jannett, F. J., Jr. 1980. Social dynamics of the montane vole, *Microtus montanus*, as a paradigm. *The Biologist*, **62**, 13–19.
Jenkins, S. H. and J. B. Llewellyn. 1981. Multiple captures of *Peromyscus*: age, sex and species differences. *Journal of Mammalogy*, **62**, 639–641.
Kareem, A. M. 1983. Effects of increasing periods of familiarity on social interactions between male sibling mice. *Animal Behaviour*, **31**, 919–926.
Kareem, A. M. and C. J. Barnard. 1982. The importance of kinship and familiarity in social interactions between mice. *Animal Behaviour*, **30**, 594–601.
Katz, L. C., M. J. Potel and R. J. Wassersug. 1981. Structure and mechanisms of schooling in tadpoles of the clawed frog, *Xenopus laevis*. *Animal Behaviour*, **29**, 20–33.
Keppie, D. M. 1980. Similarity of dispersal among sibling male spruce grouse. *Canadian Journal of Zoology*, **58**, 2102–2104.
Keough, M. J. 1984. Kin recognition and the spatial distribution of larvae of the bryozoan *Bugula neritina* (L). *Evolution*, **38**, 142–147.
Klein, J. 1975. *Biology of the mouse histocompatibility-2 complex*. Springer-Verlag, New York.
Klein, J. 1979. The major histocompatibility complex of the mouse. *Science*, **203**, 516–521.
Krebs, C. J. and J. H. Myers. 1974. Population cycles in small mammals. *Advances in Ecological Research*, **4**, 267–399.
Kruse, K. C. and B. M. Stone. 1984. Largemouth bass (*Micropterus salmoides*) learn to avoid feeding on toad (*Bufo*) tadpoles. *Animal Behaviour*, **32**, 1035–1039.
Kurland, J. A. 1980. Kin selection theory: a review and selected bibliography. *Ethology and Sociobiology*, **1**, 255–274.
Labov, J. B. 1979. *Pregnancy blocking in house mice* (Mus musculus) *and other mammals: Sociobiological implications and adaptive advantages for females*. Ph.D. Thesis. University of Rhode Island, Kingston, Rhode Island.
Lacy, R. C. and P. W. Sherman. 1983. Kin recognition by phenotype matching. *American Naturalist*, **121**, 489–512.
Lenington, S. 1983. Social preferences for partners carrying 'good genes' in wild house mice. *Animal Behaviour*, **31**, 325–333.
Lenington, S. and K. Egid. 1985. Female discrimination of male odors correlated with male genotype at the *T*-locus: a response to *T*-locus or *H-2* locus variability? *Behavior Genetics*, **15**, 53–67.

Levine, L., R. F. Rockwell and J. Grossfield. 1980. Sexual selection in mice. V. Reproductive competition between $+/+$ and $+/t^{w5}$ males. *American Naturalist*, **116**, 150–156.

Lewontin, R. C. 1978. Fitness, survival, and optimality, in Horn, D. H., R. Mitchell and G. R. Stairs, eds., *Analysis of ecological systems*. Ohio State University Press, Columbus, Ohio, pp. 3–21.

Loekle, D. M., D. M. Madison and J. J. Christian. 1982. Time dependency and kin recognition of cannibalistic behavior among poeciliid fishes. *Behavioral and Neural Biology*, **35**, 315–318.

Madison, D. M. 1975. Intraspecific odor preferences between salamanders of the same sex: dependence on season and proximity of residence. *Canadian Journal of Zoology*, **53**, 1356–1361.

Maynard Smith, J. 1978. Optimization theory in evolution. *Annual review of ecology and systematics*, **9**, 31–56.

McGuire, M. R. and L. L. Getz. 1981. Incest taboo between sibling *Microtus ochrogaster*. *Journal of Mammalogy*, **62**, 213–215.

McKaye, K. R. and G. W. Barlow. 1976. Chemical recognition of young by the midas cichlid, *Cichlasoma citrinellum*. *Copeia*, **1965**, 276–282.

McLean, I. G. 1982. The association of female kin in the Arctic ground squirrel *Spermophilus parryii*. *Behavioral Ecology and Sociobiology*, **10**, 91–99.

Michener, G. R. 1982. Infanticide in ground squirrels. *Animal Behaviour*, **30**, 936–938.

Michener, G. R. 1983. Kin identification, matriarchies, and the evolution of sociality in ground-dwelling sciurids, in Eisenberg, J. F. and D. G. Kleiman, eds., *Advances in the Study of mammalian behavior*. Special publication No. 7, The American Society of Mammalogists, pp. 528–572.

Myton, B. 1974. Utilization of space by *Peromyscus leucopus* and other small mammals. *Ecology*, **55**, 277–290.

Nussbaum, R. A., E. D. Brodie and R. M. Storm. 1983. *Amphibians and reptiles of the Pacific Northwest*. The University Press of Idaho, Moscow, Idaho.

O'Hara, R. K. 1981. Habitat selection behavior in three species of anuran larvae: environmental cues, ontogeny and adaptive significance. Ph.D. Thesis, Oregon State University, Corvallis, Oregon.

O'Hara, R. K. and A. R. Blaustein. 1981. An investigation of sibling recognition in *Rana cascadae* tadpoles. *Animal Behaviour*, **29**, 1121–1126.

O'Hara, R. K. and A. R. Blaustein. 1982. Kin preference behavior in *Bufo boreas* tadpoles. *Behavioral Ecology and Sociobiology*, **11**, 43–49.

O'Hara, R. K. and A. R. Blaustein. 1985. *Rana cascadae* tadpoles aggregate with siblings: an experimental field study. *Oecologia*, **67**, 44–51.

Partridge, B. L. 1982. The structure and function of fish schools. *Scientific American*, **246**, 114–123.

Partridge, B. L. and T. J. Pitcher. 1980. The sensory basis of fish schools: relative roles of lateral line and vision. *Journal of Comparative Physiology*, **135**, 315–325.

Polis, G. A. 1981. The evolution and dynamics of intraspecific predation. *Annual Review of Ecology and Systematics*, **12**, 225–251.

Polis, G. A., C. A. Myers and W. R. Hess. 1984. A survey of intraspecific predation within the class Mammalia. *Mammal Review*, **14**, 187–198.

Porter, R. H. 1986. Kin recognition: functions and mediating mechanisms, in Crawford, C., M. Smith and P. Krebs, eds., *Sociobiology and psychology: ideas, issues and findings*, Lawrence Erlbaum, Hillsdale, New Jersey, in press.

Porter, R. H., J. A. Matochik and J. W. Makin. 1983. Evidence for phenotype matching in spiny mice (*Acomys caharinus*). *Animal Behaviour*, **31**, 978–984.

Porter, R. H., J. A. Matochik and J. W. Makin. 1984. The role of familiarity in the development of social preferences in spiny mice. *Behavioral Processes*, **9**, 241–254.

Porter, R. H., J. A. Matochik and J. W. Makin. 1986. Discrimination between full sibling spiny mice (*Acomys caharinus*) by olfactory signatures. *Animal Behaviour*, **34**, 1182–1188.

Porter, R. H., J. D. Moore and D. M. White. 1981. Food sharing by sibling vs. nonsibling spiny mice (*Acomys cahirinus*). *Behavioral Ecology and Sociobiology*, **8**, 207–212.

Porter, R. H., V. J. Tepper and D. M. White. 1981. Experiential influences on the development of huddling preferences and 'sibling' recognition in spiny mice. *Developmental Psychobiology*, **14**, 375–382.

Porter, R. H. and M. Wyrick. 1979. Sibling recognition in spiny mice (*Acomys cahirinus*): influence of age and isolation. *Animal Behaviour*, **27**, 761–766.

Porter, R. H., M. Wyrick and J. Pankey. 1978. Sibling recognition in spiny mice (*Acomys cahirinus*). *Behavioral Ecology and Sociobiology*, **3**, 61–68.

Quinn, T. P. and C. A. Busack. 1985. Chemosensory recognition of siblings in juvenile coho salmon (*Oncorhynchus kisutch*). *Animal Behaviour*, **33**, 51–56.

Radesäter, T. 1976. Individual sibling recognition in juvenile Canada geese (*Branta canadensis*). *Canadian Journal of Zoology*, **54**, 1069–1072.

Rheingold, H. L. 1985. Development as the acquisition of familiarity. *Annual Review of Psychology*, **36**, 1–17.

Ridley, M. and A. Grafen. 1981. Are green-beard genes outlaws? *Animal Behaviour*, **29**, 954–955.

Riedman, M. L. 1982. The evolution of alloparental care and adoption in mammals and birds. *Quarterly Review of Biology*, **57**, 405–435.

Rohwer, S., S. D. Fretwell and R. C. Tuckfield. 1976. Distress screams as a measure of kinship in birds. *American Midland Naturalist*, **96**, 418–430.

Rongstad, O. J. 1965. A life history study of 13-lined ground squirrels in southern Wisconsin. *Journal of Mammalogy*, **46**, 76–87.

Røskaft, E. and Y. Espmark. 1984. Sibling recognition in the rook (*Corvus frugilegus*). *Behavioural Processes*, **9**, 223–230.

Rothstein, S. I. 1975. Evolutionary rates and host defenses against avian brood parasitism. *American Naturalist*, **109**, 161–176.

Rothstein, S. I. 1980. Reciprocal altruism and kin selection are not clearly separable phenomena. *Journal of Theoretical Biology*, **87**, 255–261.

Rothstein, S. I. 1982. Successes and failures in avian egg and nestling recognition with comments on the utility of optimality reasoning. *American Zoologist*, **22**, 547–560.

Rothstein, S. I. and D. P. Barash. 1983. Gene conflicts and the concepts of outlaw and sheriff alleles. *Journal of Social and Biological Structures*, **6**, 367–379.

Rushton, J. P., R. J. H. Russell and P. A. Wells. 1984. Genetic similarity theory: beyond kin selection. *Behavior Genetics*, **14**, 179–193.

Salzen, E. A. and J. M. Cornell. 1968. Self-perception and species recognition in birds. *Behaviour*, **30**, 44–65.

Samollow, P. B. 1980. Selective mortality and reproduction in a natural population of *Bufo boreas*. *Evolution*, **34**, 18–39.

Schoener, T. W. 1983. Field experiments on interspecific competition. *American Naturalist*, **122**, 240–285.

Schwagmeyer, P. L. 1980. Alarm calling behavior of the 13-lined ground squirrel. *Behavioral Ecology and Sociobiology*, **7**, 195–200.

Scofield, V. L., J. M. Schlumpberger, L. A. West and I. L. Weissman. 1982. Protochordate allorecognition is controlled by a MHC-like gene system. *Nature*, **295**, 499–502.

Selander, R. K. 1970. Behavior and genetic variation in natural populations. *American Zoologist*, **10**, 53–66.

Shapiro, D. Y. 1983. On the possibility of kin groups in coral reef fishes, in Reaka, M. L., ed., *The ecology of deep and shallow coral reefs*, NOAA'S Undersea Research Program, Volume 1. No. 1, pp. 39–45.

Sharp, G. D. 1978. Behavioral and physiological properties and their effect on vulnerability to fishing gear, in Sharp, G. D. and A. E. Dixon, eds., *The physiological ecology of tunas*, Academic Press, New York, pp. 397–449.

Sheppard, D. H. and S. M. Yoshida. 1971. Social behavior in captive Richardson's ground squirrels. *Journal of Mammalogy*, 52, 793–799.

Sherman, P. W. 1977. Nepotism and the evolution of alarm calls. *Science*, **197**, 1246–1253.

Sherman, P. W. 1980a. The limits of ground squirrel nepotism, in Barlow, G. W. and J. Silverberg, eds., *Sociobiology: beyond nature/nurture?* Westview Press, Boulder, Colorado, pp. 505–544.

Sherman, P. W. 1980b. The meaning of nepotism. *American Naturalist*, 116, 604–606.

Sherman, P. W. and W. G. Holmes. 1985. Kin recognition: issues and evidence. *Fortschritte der Zoologie*, **31**, 437–460.

Shields, W. M. 1980. Ground squirrel alarm calls: nepotism or parental care? *American Naturalist*, **116**, 599–603.

Shields, W. M. 1982. *Philopatry, inbreeding, and the evolution of sex*, State University of New York Press, Albany, New York.

Simon, G. S. and D. M. Madison. 1984. Individual recognition in salamanders: cloacal odours. *Animal Behaviour*, **32**, 1017–1020.

Slater, P. J. B. and F. A. Clements. 1981. Incestuous mating in Zebra finches. *Zeitschrift für Tierpsychologie*, **57**, 201–208.

Stabell, O. B. 1982. Detection of natural odorants by Atlantic salmon parr using positive rheotaxis olfactometry, in Brannon, E. L. and E. O. Salo, eds., *Proceedings of the salmon and trout migratory behaviour symposium*. School of Fisheries, University of Washington, Seattle, Washington, pp. 71–78.

Stebbins, R. C. 1954. *Amphibians and reptiles of western North America*. McGraw-Hill Book Company, New York.

Steele, E. J. 1979. *Somatic selection and adaptive evolution: on the inheritance of acquired characters*. University of Chicago Press, Chicago, Illinois.

Stickel, L. F. 1968. Home range and travels, in J. A. King ed., *Biology of* Peromyscus. Special Publication No. 2 American Society of Mammalogists, pp. 373–411.

Stickel, L. F. 1979. Population ecology of house mice in unstable habitats. *Journal of Animal Ecology*, **48**, 871–877.

Stoddart, D. M. 1980. *The ecology of vertebrate olfaction*. Chapman and Hall, New York.

Sype, W. E. 1975. *Breeding habits, embryonic thermal requirements and embryonic and larval development of the Cascade frog*, Rana cascadae Slater. Ph.D. Thesis. Oregon State University, Corvallis, Oregon.

Thomas, J. A. and E. C. Birney. 1979. Parental care and mating system of the prairie vole, *Microtus ochrogaster*. *Behavioral Ecology and Sociobiology*, **5**, 171–186.

Treisman, M. 1981. The significance of immunity restriction by the major histocompatibility complex, and of the occurrence of high polymorphism at MHC loci: two hypotheses. *Journal of Theoretical Biology*, **89**, 409–421.

Van Wormer, J. 1968. *The world of the Canada goose*. J. B. Lippincott Company, Philadelphia.

Veal, R. and W. Caire. 1979. *Peromyscus eremicus*. *Mammalian species*, **118**, 1–6.

Vestal, B. M. and J. J. Hellack. 1978. Comparison of neighbor recognition in two species of deermice (*Peromyscus*) *Journal of Mammalogy*, **59**, 339–346.

Vestal, B. M. and H. McCarley. 1984. Spatial and social relations of kin in thirteen-lined and other ground squirrels, in Murie, J. O. and G. R. Michener, eds., *The biology of ground dwelling squirrels*. University of Nebraska Press, Lincoln, Nebraska, pp. 404–423.
Vom Saal, F. S. 1981. Variation in phenotype due to random intrauterine proximity of male and female fetuses in rodents. *Journal of Reproduction and Fertility*, **6**, 633–650.
Vom Saal, F. S. 1986. The intrauterine position phenomenon: effects on physiology, aggressive behavior and population dynamics in house mice, in Blanchard, C., K. Flannelly and R. Blanchard, eds., *Biological perspectives on aggression*, in press.
Waldman, B. 1981. Sibling recognition in toad tadpoles: the role of experience. *Zeitschrift für Tierpsychologie*, **56**, 341–358.
Waldman, B. 1982. Sibling association among schooling toad tadpoles: field evidence and implications. *Animal Behaviour*, **30**, 700–713.
Waldman, B. 1983. *Kin recognition and sibling association in anuran amphibian larvae*. Ph.D. Thesis. Cornell University, Ithaca, New York.
Waldman, B. 1984. Kin recognition and sibling association among wood frog (*Rana sylvatica*) tadpoles. *Behavioral Ecology and Sociobiology*, **14**, 171–180.
Waldman, B. 1985a. Sibling recognition in toad tadpoles: are kinship labels transferred among individuals? *Zeitschrift für Tierpsychologie*, **68**, 41–59.
Waldman, B. 1985b. Olfactory basis of kin recognition in toad tadpoles. *Journal of Comparative Physiology*, **156**, 565–577.
Waldman, B. 1986. Preference for unfamiliar siblings over familiar non-siblings in American toad (*Bufo americanus*) tadpoles. *Animal Behaviour*, **34**, 48–53.
Waldman, B. and K. Adler. 1979. Toad tadpoles associate preferentially with siblings. *Nature*, **282**, 611–613.
Wassersug, R. J. 1973. Aspects of social behavior in anuran larvae, in J. L. Vial ed., *Evolutionary biology of the anurans*. University of Missouri Press, Columbia, Missouri, pp. 273–297.
Wassersug, R. J. and C. M. Hessler. 1971. Tadpole behaviour: aggregation in larval *Xenopus laevis*. *Animal Behaviour*, **19**, 386–389.
Wilbur, H. M. 1977. Density-dependent aspects of growth and metamorphosis in *Bufo americanus*. *Ecology*, **58**, 196–200.
Wiens, J. A. 1977. On competition and variable environments. *American Scientist*, **65**, 590–592.
Wilson, E. O. 1975. *Sociobiology: the new synthesis*. Belknap Press of Harvard University Press, Cambridge, Massachusetts.
Wilson, S. C. 1982. The development of social behaviour between siblings and non-siblings of the voles *Microtus ochrogaster* and *Microtus pennsylvanicus*. *Animal Behaviour*, **30**, 426–437.
Wright, A. H. and A. A. Wright. 1949. *Handbook of frogs and toads of the United States and Canada*, Cornell University Press, Ithaca, New York.
Wu, H. M. H., W. G. Holmes, S. R. Medina and G. P. Sackett. 1980. Kin preference in infant *Macaca nemestrina*. *Nature*, **285**, 225–227.
Yamaguchi, M., K. Yamazaki, G. K. Beauchamp, J. Bard, L. Thomas and E. A. Boyse. 1981. Distinctive urinary odors governed by the major histocompatibility locus of the mouse. *Proceedings of the National Academy of Sciences, USA*, **78**, 5817–5820.
Yamazaki, K., G. K. Beauchamp, C. J. Wysocki, J. Bard, L. Thomas and E. A. Boyse. 1983. Recognition of *H-2* types in relation to the blocking of pregnancy in mice. *Science*, **221**, 186–188.

Yamazaki, K., E. A. Boyse, V. Mike, H. T. Thaler, B. J. Mathieson, J. Abbot, J. Boyse, Z. A. Zayas and L. Thomas. 1976. Control of mating preferences in mice by genes in the major histocompatibility complex. *Journal of Experimental Medicine*, **144**, 1324–1335.

Yamazaki, K., M. Yamaguchi, L. Baranoski, J. Bard, E. A. Boyse and L. Thomas. 1979. Evidence for the use of Y-maze differentially scented by congenic mice of different major histocompatibility types. *Journal of Experimental Medicine*, **150**, 755–760.

Yamazaki, K., M. Yamaguchi, E. A. Boyse and L. Thomas. 1980. The major histocompatibility complex as a source of odors imparting individuality among mice, in Muller-Schwarze, D. and R. M. Silverstein, eds., *Chemical Signals*, Plenum Press, New York, pp. 267–273.

Yeaton, R. I. 1972. Social behavior and social organization in Richardson's ground squirrel (*Spermophilus richardsonii*) in Saskatchewan. *Journal of Mammalogy*, **53**, 139–147.

Yom-Tov, Y. 1980. Intraspecific nest parasitism in birds. *Biological Reviews*, **55**, 93–108.

Kin Recognition in Animals
Edited by D. J. C. Fletcher and C. D. Michener
© 1987 John Wiley & Sons Ltd

CHAPTER 11

Kin Recognition in Non-human Primates

JEFFREY R. WALTERS
Department of Zoology, North Carolina State University, Campus Box 7617, Raleigh, North Carolina 27695, USA

INTRODUCTION

Non-human primates exhibit complex social behavior and well developed learning abilities. Given these attributes, that they recognize kin and treat them differently from non-kin is not as much a revelation as in some other animals. In this chapter I will discuss differential behaviour toward kin by non-human primates and the kin recognition mechanisms that enable such behavior. I will argue that non-human primates are unusual in the extent to which they exhibit kin biases in behavior and perhaps in the scope of their recognition abilities. Kin are apparently recognized through social interaction, by mechanisms that in many ways are less remarkable than those used by animals such as bees and tadpoles. Any statements about kin recognition must be tentative, however, because data on mechanisms are almost totally lacking for non-human primates. The intent of this review, therefore, is to define the key research questions rather than to provide answers.

There are two aspects of social behavior in which differential treatment of kin is especially important. The first is sexual interaction. Mating with close kin often has negative effects on fitness, and recently Bateson (1983) has proposed that mating with distant relatives is advantageous over mating with non-relatives. Recognizing close kin to avoid inbreeding may therefore be of great advantage, and precise recognition of distant kin may provide additional advantages in mate selection.

The second aspect of behavior in which differential treatment of kin is of special interest is performance of altruistic and harmful behaviors. By altruistic behaviors I mean those of immediate benefit to the recipient and immediate cost to the performer; harmful behaviors are those that are the reverse. The theory of kin selection outlines the positive effect on inclusive fitness of preferentially directing altruistic behavior toward relatives and harmful behavior toward non-relatives (Hamilton, 1964; Michod, 1982).

There may also be other reasons why the ability to recognize kin might be advantageous to a non-human primate or any other animal (see previous chapters). The various proposed mechanisms by which kin might be distinguished can be summarized by the four basic categories outlined by Holmes and Sherman (1983). First, relatives might be predictably distributed in space, and individuals might respond differentially toward others according to location. Second, relatives might predictably encounter one another in particular social contexts, and individuals might then respond differentially toward those others encountered in these contexts. Third, individuals might be capable of phenotype matching, that is, they might establish some template of their own phenotype, and then respond differentially toward those others that most closely resemble that template. Finally, recognition alleles may exist that indicate lineage, enabling individuals to respond differentially toward others possessing the allele. In all these cases, recognition may occur on an interaction by interaction basis, or the recognition mechanism may provide the basis for learned individual recognition that then guides subsequent interaction.

KIN BIASES IN THE PERFORMANCE OF HARMFUL AND ALTRUISTIC BEHAVIORS

Kin biases in behavior can occur at two levels in group-living species, between social units and within social units (Alexander, 1974). For primates, attention has been focused on kin biases within groups, whereby an individual treats other group members differently according to their kin relationship. This will be reviewed in detail below. First, however, I will discuss a more neglected topic, biases inherent in treating group members differently from non-group members.

Kin biases and group living

Many non-human primates live in social groups, and behave altruistically toward other members of their group. For example group members cooperate in defense against predators (Crook, 1970; Cheney and Wrangham, 1986) and give alarm calls that warn others of danger (Seyfarth, Cheney and Marler, 1980; Seyfarth, 1986). Group members may also cooperate in defending resources or territories against other groups (Hamilton, Buskirk and Buskirk, 1976; Robinson, 1979; Goodall et al., 1979; Mitani and Rodman, 1979), and are often highly aggressive toward non-group members, especially strangers (Marler, 1976). Other altruistic behaviors exhibited by non-human primates such as grooming are also directed much more often at group members, even non-relatives, than at outsiders.

Kin recognition is involved in differential behavior toward group members only to the extent that the evolution of the behavior in question is dependent on the average relatedness of group members, and thus on kin selection. Population structure is critical in this regard (Wade, 1979, 1980; Michod, 1982; Moore and Ali, 1984). Population structure may be divided into two components, mating structure and interaction structure (Wade, 1980). The best studied population structure in non-human primates is one in which groups consist of related females and their offspring, plus associated males (Wrangham, 1980; Moore, 1984; Periera and Altmann, 1985; Pusey and Packer, 1986). Males usually emigrate from their natal groups, and avoid mating with close kin within their natal groups even when they do not emigrate (see below). The interaction structure in this system, known as the female-bonded group system (Wrangham, 1980; Moore, 1984), thus results in the association of maternal kin. The polygynous nature of the mating structure (see below) also leads to high relatedness within groups, especially within age cohorts, but through paternal lines (J. Altmann, 1979). Indeed, under certain demographic conditions paternal relatedness may make as great a contribution to average relatedness within female-bonded groups as maternal relatedness (Melnick and Kidd, 1983). However, the mating structure, specifically that males are generally unrelated to the females with which they breed, prevents groups from becoming highly inbred. Wade (1979) therefore concluded that the population structure of most primate species is not conducive to the evolution of extreme altruism (see also Wade, 1980). The overall level of inbreeding nevertheless is quite high relative to random breeding.

Recognition of relatives within groups changes the interaction structure to increase the average relatedness of social partners, and thereby may permit the evolution of more extreme altruism. I am distinguishing behavior concentrated on relatives within groups, discussed below (see 'Matrilineal kin bias within social groups' below), from overall levels of cooperation characterizing social groups. The issue in the latter case is to what extent particular levels of cooperation can be attributed to average relatedness of group members resulting from the population structure.

Some of the cooperative benefits that primates derive from group living are widespread among animals, and occur even in groups of unrelated individuals. These include benefits gained from alarm calling, improved foraging efficiency, and some forms of cooperative defense against predators. Similar behavior occurs for example in mixed species flocks of birds (Morse, 1980, Chapter 12) and in temporary flocks of wintering sanderlings (*Calidris alba*) comprised of unrelated individuals (Myers, 1983). The benefits derived from these forms of cooperation are presumably sufficiently large that high relatedness (i.e. kin selection) is not necessary for their evolution. Behavior such as cooperative defense of resources, allogrooming, and other forms of cooperative defense against predators (e.g. physical attacks on predators) are not as widespread, and their evolution may well be attributable to the levels of relatedness characteristic of primate groups. On the other hand, even more extreme forms of altruism occur in other kinds of animals. Food sharing (Bertram, 1975) and cooperative hunting (Busse, 1978) in social carnivores and cooperative raising of young in certain birds (Emlen and Vehrencamp, 1983) are examples. That primate social units may be less inbred than the social units of these more cooperative species is one explanation of the absence of such levels of altruism among the former (Wade, 1979).

Predictable social context, specifically group membership, is the mechanism of kin recognition associated with this level of selection for altruism. Bekoff (1981) pointed out that where kin recognition is based on social familiarity, the power of selection is limited by the correlation between familiarity and relatedness. In the case of group-level cooperation this correlation is the average relatedness within groups, and Bekoff's statement is equivalent to the theoretical conclusion that the level of inbreeding determines the power of kin selection (Breden and Wade, 1981; Michod, 1982).

Alternatives to kin selection as explanations of group altruism exist (Moore, 1984; Moore and Ali, 1984). These are generally a delayed benefit to the altruist or some form of reciprocity, and they can be applied to even the most extreme forms of altruism (e.g. Packer and Pusey, 1982; Emlen and Vehrencamp, 1983; Brown, Sanderson and Michod, 1983). Determining the extent to which kin selection rather than these alternatives is responsible for cooperative group behavior in primates requires a precision in measuring the benefits and costs of social behavior that is far beyond current methodologies. One can, however, compare species to determine whether group-level cooperation varies with relatedness within groups in a manner consistent with the hypothesis of kin selection.

Variation among primates in group-level cooperation

Female-bonded groups

Variation in dispersal patterns and mating structure among species characterized by female-bonded groups may lead to differences in average relatedness within groups. For example, in bonnet macaques (*Macaca radiata*) male migration is less frequent than in related species, which presumably leads to higher levels of average relatedness of female group members. Females groom and form alliances with others that are not close maternal kin unusually often in this species (Defler, 1978; Wade, 1979; Moore and Ali, 1984; Gouzoules and Gouzoules, 1986), suggesting that group-level cooperation may indeed vary with average relatedness.

Species also differ in the frequency with which females emigrate (Moore, 1984; Gouzoules, 1984; Pusey and Packer, 1986); this difference could similarly affect average relatedness within groups and thus levels of cooperation. However, migrating females may form new groups rather than transfer to an existing group (Pusey and Packer, 1986), with the result that inbreeding is not reduced as much. In some species males tend to join groups containing other males that are maternal kin (Cheney, 1983), a phenomenon that should increase inbreeding above levels due to matrilineal kinship and mating structure. Whether these dispersal patterns affect group-level cooperation should be explored.

An aspect of mating structure that may have a large effect on average relatedness within groups is the number of males contributing to reproduction at any one time. In species in which several reproductively active adult males coexist in groups, there is typically great variability among males in reproductive success (Hausfater, 1975; Packer, 1979a). This increases patrilineal relatedness, especially within age cohorts (J. Altmann, 1979). More extreme are species characterized by one-male or age-graded-male social systems, in which one male fathers all or nearly all offspring in the group until he is replaced by a new male (Wrangham, 1980; Gouzoules, 1984). Relatedness within groups is thus affected by male tenure, as well as degree of polygyny, which determines, for example, the frequency of full-siblings (Gouzoules, 1984). Gouzoules (1984) has proposed that the high levels of infant-handling, low rates of aggression, lack of alliances and generally reduced orientation toward close kin that characterize colobines (McKenna, 1979a) in comparison to cercopithecines may reflect differences in levels of relatedness within groups due to the predominance of age-graded-male and one-male social systems in colobines. Such differences may consistently occur between age-graded and multi-male systems.

Other social systems

There are several social structures other than female-bonded groups among non-human primates. These include solitary species, family groups, and non-female-bonded groups (Wrangham, 1980; Moore, 1984; Pereira and Altman, 1985). In some solitary species, e.g. galagos, it appears that the interaction structure is similar to that characteristic of female-bonded groups. Dispersal is male-biased (Pusey and Packer, 1986), and female kin may associate in sleeping groups in which grooming and perhaps other altruistic behaviors are exchanged (Charles-Dominique, 1977; Clark, 1978; Bearder, 1986). Such interaction could represent group-level cooperation evolved through kin selection. Presumably the ecology of such species prohibits ranging in groups, and thus limits opportunity for the evolution of further cooperative behavior.

In species living in family groups (gibbons, callitrichids), social units are typically composed of a mated pair of adults and their offspring, although there may be additional group members in some callitrichids (Gouzoules, 1984; Leighton, 1986; Goldizen, 1986). Relatedness within groups is unusually great in these species (Moore and Ali, 1984), and many appear to exhibit unusually high levels of cooperation. In addition to defense of territory, allogrooming, and the other usual forms of cooperation, group members cooperate in caring for young. In captivity immature callitrichids carry and otherwise tend their younger siblings regularly (Box, 1977; Hoage, 1977; Ingram, 1977; Goldizen, 1986). Furthermore, field data suggest that the extra adults found in some callitrichid groups behave like helpers in cooperatively breeding birds (Goldizen, 1986). It is interesting that what appear to be extreme forms of cooperation occur in those species in which relatedness within social units is probably high.

In the non-female-bonded group social structure, females emigrate regularly, so that groups are composed primarily of unrelated females and their dependent offspring. In some species males often remain in their natal groups as adults, e.g. red colobus (*Colobus badius*) and chimpanzees (*Pan troglodytes*), whereas in others males also usually emigrate, e.g., gorillas (*Gorilla gorilla*) (Pusey and Packer, 1986). Levels of cooperation within groups in the former species might be expected to be similar to those found in female-bonded groups, because the level of inbreeding, although produced by a different social structure, should be similar. In species in which both sexes emigrate, average relatedness within groups may be lower than in other species, and there is some evidence to suggest that general levels of cooperation are lower in such groups (Harcourt, 1979; Wrangham, 1982; Walter and Seyfarth, 1986).

Conclusions

There are intriguing indications that the level of cooperation within primate groups may vary with average relatedness of group members as determined by the type of social structure. This level of kin bias has received much less attention than differential behavior directed toward relatives within groups (see below), and needs to be investigated much more fully. Specifically, data on levels of inbreeding within social units such as those collected by Melnick and Kidd (1983) are needed from a variety of species to determine the relationships between social structure and inbreeding, and the correlation of these with social behavior. Also, the level of cooperative behavior in various species needs to be documented, and the nature of group level cooperation carefully defined. Direct benefits to individuals should be separated from indirect inclusive fitness benefits in exploring alternatives to kin selection as explanations for the kinds and levels of cooperation observed. If effects of differential behavior toward recognized kin within groups are removed, are there corresponding differences in the frequency of grooming and other forms of cooperation among these three types of primates? Has territoriality evolved in species living in one-male groups (Mitani and Rodman, 1979) due to benefits to the male, groups benefits or both? These are the kinds of questions that must be addressed in exploring aspects of kin recognition.

Matrilineal kin bias within social groups

That individuals behave differently toward close maternal kin than toward non-kin within social groups is well documented among non-human primates. The pertinent data have been reviewed in detail in several recent papers (Gouzoules, 1984; Pereira and Altmann, 1985; Gouzoules and Gouzoules, 1986; Walters and Seyfarth, 1986). Among the presumably altruistic behaviors for which kin bias has been shown are allogrooming, alliance formation and food sharing (Silk, 1986). Other behaviors for which evidence of kin bias exists include tolerance during feeding (co-feeding), spatial proximity, play, carrying and otherwise caring for infants, and aggression.

Food sharing

Food sharing is rare in non-human primates, occurring only in callitrichids, titi monkeys (*Callicebus torquatus*), douc langurs (*Pygathrix nemaeus*) and chimpanzees (Brown and Mack, 1978; Starin, 1978). In most species it is restricted to mothers sharing with their offspring (e.g. chimpanzees, Silk, 1978, 1979). The callitrichids are exceptional in that food sharing between males and young also occurs, but again the behavior is restricted to parent and offspring (Goldizen, 1986).

Grooming

Allogrooming is nearly universal among primates, and there is good evidence of kin bias in its distribution from a number of species. But important questions about the exact nature of this bias remain. First, data are totally lacking, beyond the mother-infant bond, for prosimians and New World monkeys (Gouzoules and Gouzoules, 1986). Kin bias in grooming is well established only for apes and cercopithecines; evidence is still rather sparse for colobines. The generality of the kinship effect among primates therefore is unclear.

Documenting kin bias requires known genealogies, and therefore, in the case of long-lived species such as primates, years of continuous research. In fact, in many cases where kin bias in behavior has been shown, genealogies were originally inferred from behavior rather than known from birth records (Walters, 1981), a practice that is dangerously circular (Bekoff, 1981). This is particularly relevant to the topic of kin recognition because non-relatives that groom particularly often are likely to be classified as kin, resulting in overestimation of kin bias (Walters, 1981). Such errors cannot account for the entire kinship effect (Walters, 1981), but contribute to a second problem related to kin bias, namely documenting its magnitude and importance relative to other factors affecting grooming.

Correctly identifying kin is but a part, and an easily corrected one, of the difficulty of measuring kin bias. A more insidious problem is the effect of spatial proximity on behavior. Most studies in which effects of kinship on grooming have been measured involve large, expanding groups in which individuals are literally surrounded by kin (e.g. Sade, 1972; Kurland, 1977). That is, most individuals have many maternal relatives living in their social group, and because there is a strong kin bias in spatial proximity (see below) they are typically near those relatives. Any tendency to interact most with the nearest individual will produce a kin bias in behavior.

If this passive mechanism can account for kin biases in behavior, kin recognition can still be said to occur, but its potential effect on behavior will be less than if a more active means of recognition occurs. Specifically, compared to large expanding groups, kin bias would be much less in other demographic contexts such as small stable groups, where individuals would have a smaller (and more variable) number of maternal kin available as social partners. If individuals behave altruistically toward unrelated neighbors in such situations, differential behavior toward kin within groups might be of little significance in the evolution of social behavior because large expanding groups are presumably unusual in the evolutionary history of most species. Most such groups today are maintained by provisioning (Sade, 1972; Kurland, 1977).

There are several ways in which the problem of spatial distribution may be circumvented. Essentially the problem is that all individuals do not have equal opportunity to interact, and it is important to know whether kin bias reflects differential tendencies to interact in equivalent circumstances rather than differential opportunities to interact. Within large expanding groups the solution has been to use relative rates of interaction, that is, to measure frequency of interaction relative to opportunity. This is typically done by standardizing rate of interaction per time spent in proximity (Kurland, 1977; Berman, 1978, 1982). Such measures, although useful, may themselves be biased to the extent that interaction causes proximity.

A better solution is to measure effects of kinship on behavior in smaller, stable or even declining groups. Spatial structure may still contribute to kin bias in such groups, but to a lesser extent than in large expanding ones. Studies of small groups are also necessary to determine if results from large groups actually underestimate the importance of kinship. In large groups not only altruistic behaviors but also harmful ones and those that are not obviously harmful or altruistic (neutral behaviors, e.g. embracing) are directed primarily at kin, presumably due to proximity effects (Kurland, 1977; Berman, 1978, 1982). Kin biases might actually be more striking in small groups where individuals have greater opportunity to interact with non-kin, if they restrict this interaction to harmful or neutral behaviors.

A final problem in measuring kin bias in behavior is separating effects of kinship from those of other important factors. Clearly factors such as age and sex affect rates of interaction, but too often these are not controlled in estimating effects of kinship. Other factors can be extremely important in particular instances. For example, in cercopithecines lactating females are groomed unusually frequently, presumably because others are attracted to the female's infant (Fig. 1), and this can confound the measurement of kin bias in grooming (Seyfarth, 1976, 1978; Walters and Seyfarth, 1986).

Fig. 1. Many factors other than kinship have strong effects on grooming interaction. In many species, lactating females are groomed frequently by non-relatives, who apparently are attracted to the new infant. Here a female yellow baboon with an infant is groomed by an unrelated juvenile female

Given these problems, what can be said about the effect of kinship on grooming in non-human primates? There is no question that in all the species studied to date mothers preferentially groom their offspring and that maternal siblings preferentially groom one another (Gouzoules and Gouzoules,

1986). Grooming is not only more frequent, but also more reciprocal in such kin dyads compared to dyads of non-kin (Walters, 1981; Silk, 1982). There is some evidence for preferential grooming between grandparents and their grandoffspring, and between aunts and their nieces and nephews (Yamada, 1963; Kurland, 1977; Grewal, 1980a), but this is not a consistent result (Gouzoules, 1984). Because evidence of differential grooming of individuals more distantly related than siblings comes only from large expanding groups, it must be considered equivocal (Gouzoules, 1984; Moore, 1984; Gouzoules and Gouzoules, 1986).

Data from small groups reveal clear kinship effects, indicating that proximity cannot account for the entire kin bias in grooming (e.g. Cheney, 1978a; Rowell and Olson, 1983). Kin bias was particularly striking in a small group of yellow baboons (*Papio cynocephalus*) that I studied, in the sense that bias in the absolute frequency of grooming contrasted with lack of bias in frequencies of neutral behaviors such as embracing (Walters, 1981, in preparation). Finally, both my study and Dunbar's (1984) study of geladas (*Theropithecus gelada*) indicate that individuals do not form grooming relationships comparable to those among kin if they have no close relatives in their group. This suggests that chance demographic events may have important effects on the lives of individuals (Altmann and Altmann, 1979). Specifically, whether an individual happens to have several close maternal kin in its group can have a dramatic effect on the frequency and kind of its grooming interactions.

Kin bias in grooming among siblings and between mother and offspring cuts across social systems. Although most of the data come from multi-male groups, similar results were obtained from patas monkeys (*Erythrocebus patas*) (Rowell and Olson, 1983) and geladas (Dunbar, 1979, 1984), which live in one-male groups. These are all species in which female kin typically remain together in the same social unit as adults. In social systems in which females emigrate, frequent grooming still occurs between mothers and their offspring, and among immature siblings. Chimpanzees (Pusey, 1983) and gorillas (Fossey, 1979) are examples. Also, on those few occasions when mother and daughter or sisters do live together as adults, they groom one another preferentially (Harcourt, 1979; Pusey, 1983, Sigg *et al.*, 1982; Stewart and Harcourt, 1986). Perhaps close kinship may also account for the strong grooming relationships observed among males in some species characterized by female emigration (Struhsaker, 1975; Bygott, 1979; Nishida, 1979; Goodall *et al.*, 1979).

It is not known whether differential grooming according to kinship occurs in species ranging solitarily or in family groups. In at least some such species, group composition is presumably sufficiently variable to provide an opportunity for discrimination to evolve (see Sherman, 1980). This would produce additional kin bias above that resulting from group-level cooperation.

Little can be said about the magnitude of kin bias in grooming. Attempts to measure kin bias have largely consisted of correlating grooming frequency with relatedness (Kurland, 1977). I attempted to measure kin bias by entering kinship along with several other variables in a linear model analysis of grooming frequencies of juvenile yellow baboons (Walters, in preparation). Kinship was consistently the most important factor in the grooming of the juveniles with various other age-sex classes, accounting for 40–50% of the variation among dyads. This analysis included primarily sisters and mother-daughter pairs among the kin dyads.

In summary, the data available indicate that kin bias in grooming is widespread among species living in female-bonded and non-female-bonded groups, and might occur in other social systems. At least maternal siblings are distinguished, and more distant relatives may be. Pronounced kin bias in small groups implies a recognition mechanism more active than differential behavior according to spatial proximity. Finally, the effect is apparently sufficiently large to be of major importance in the social lives of individuals.

Alliance formation

In many primate species individuals sometimes enter agonistic interactions between other individuals and somehow assist one of the participants against the other. Such behavior is termed alliance formation. Alliance formation (Fig. 2) is not as widespread as grooming, and its study has been restricted to apes and cercopithecines, where it is most common. In these species individuals preferentially form alliances with kin (Gouzoules, 1984; Pereira and Altmann, 1985; Gouzoules and Gouzoules, 1986; Walters and Seyfarth, 1986). Alliances are of many types, and kin bias is more pronounced in some types than others. For example unrelated adult male baboons frequently form alliances when competing with dominant males over access to females (Packer, 1977), but when a young baboon is attacked by an adult male it is typically its close kin that defend it (Walters, 1980). The difficulties inherent in separating the various kinds of alliances and observing a sufficient sample of each, in addition to controlling for effects of age, sex, spatial proximity and other factors as discussed above, have precluded precise measurement of kin bias in alliance formation.

Some generalizations are nevertheless possible. First, substantial evidence of differential behavior toward kin exists only for mother–offspring pairs and siblings (Gouzoules, 1984; Gouzoules and Gouzoules, 1986), although there is some evidence of differential behavior toward more distant kin (Massey, 1977; Kurland, 1977; Kaplan, 1978). Massey's (1977) study is particularly compelling because she worked with a small group and analyzed different kinds of alliances separately. Another apparent trend is that kin bias is especially strong where costs of alliance formation are high or benefits are low (Massey, 1977; Walters, 1980; Silk, 1982). That is, alliances in which the opponent is particularly dangerous or in which there is little immediate gain are especially likely to involve kin.

Alliance formation involving close maternal kin has been reported in female-bonded groups among not only females and their offspring, but also among males that immigrate to the same group (Miller, Kling and Dicks, 1973; Meikle and Vessey, 1981; Colvin, 1983; Cheney and Seyfarth, 1983). Alliances also occur among related adult males in non-female-bonded species such as chimpanzees in which males remain in their natal groups (Bygott, 1979; Walters and Seyfarth, 1986). Data on alliances thus lead to similar conclusions as those for grooming, but in a more restricted set of species.

Aggression

Effects of kinship on aggressive interaction are less clear than those on grooming or alliances. Existing data come almost entirely from baboons and macaques. In these species, the typical result in large groups is that, contrary to expectation, frequency of aggression is positively correlated with kinship (Kurland, 1977). If frequency of aggression is measured relative to time spent in proximity, either kinship has no effect on aggression or aggression is still more frequent among relatives (Berman, 1978, 1982). Both Massey (1977) and I (J. R. Walters, in preparation) obtained similar results in studies of small groups; there was no kin bias in aggressive interaction.

Although there is no general tendency to avoid directing aggressive behavior at kin among non-human primates, kinship may still affect aggression in important ways. There is evidence that aggression is less severe among relatives than among non-relatives (Kurland, 1977; Silk, Samuels and Rodman, 1981; Walters and Seyfarth, 1986; Gouzoules and Gouzoules, 1986). Infanticide, a phenomenon in which males kill infant troop members that are not their offspring, is a particularly dramatic example of intense aggression among non-relatives (Hrdy, 1979; Gouzoules, 1984; Struhsaker and Leland, 1986).

It may be that kin are less predisposed to interact aggressively than non-kin in a particular context (Walters and Seyfarth, 1986), but this is difficult to document. In addition to spatial proximity, confounding factors such as age and dominance relationship make it particularly difficult to compare aggressive tendencies toward kin and non-kin. For example, immature siblings are characterized by especially high rates of aggression (Berman, 1978, 1982; Walters, in preparation). Siblings may compete in ways that non-kin do not, for example for dominance status (Walters, 1986) or for resources provided by the mother (Berman, 1978, 1982). The frequency of

Fig. 2. Individuals often intervene in ongoing agonistic interactions involving others. Such alliances are especially prominent in cercopithecines, and have been shown to be strongly related to kinship. Here an adult female threatens a juvenile yellow baboon who has retreated to the protection of an ally

aggression among, say, adult females with established dominance relationships might be a better indicator of kin bias than the frequency of aggression among immatures. Clearly much more study is needed to determine how kinship affects aggressive interaction.

In conclusion, data on kin bias in aggressive interaction, although of theoretical interest with respect to effects of kin selection on the deployment of harmful behavior, add little in regard to kin recognition beyond that deduced from data on grooming and alliance formation.

Spatial proximity

As noted above, there is a widespread kin bias in spatial proximity among non-human primates (Gouzoules and Gouzoules, 1986). This has been documented for a variety of species, in both large (e.g. Grewal, 1980a) and small (e.g. Rowell and Olson, 1983) groups. The effect has been shown to extend even to distant kin in large cercopithecine groups (Kurland, 1977; Berman, 1978, 1982). This pattern could be interpreted as a kin recognition mechanism (see above) or as a kin bias based on some other recognition mechanism.

That is, individuals may recognize their relatives and associate with them spatially based on that recognition (Kummer, 1978). If kin are more inclined to behave altruistically and less inclined to behave harmfully toward an individual than are non-relatives, such an association pattern might be expected to evolve. Mutual tolerance at feeding sites among kin (Kawai, 1965; Stein, 1984) may be viewed as an extension of this phenomenon.

Patterns in spatial association may therefore provide evidence of recognition of both close and distant kin, but this depends on whether association is produced by an active or passive mechanism, and this is not known.

Other behavior

Kin bias has been reported for several other behaviors. Maternal kin have been observed to care for orphaned infants in several species (e.g. Marsden and Vessey, 1968; Hasegawa and Hiraiwa, 1980; Hamilton, Busse and Smith, 1982). Care in such cases consists of carrying, grooming, sleeping with, and defending the infant. Actual care of infants must be distinguished from the handling of and general interest in infants that characterizes many primates (Hrdy, 1976). The latter is not clearly altruistic or harmful, and is not characterized by a strong kin bias (J. R. Walters, in preparation; see also Hrdy, 1976; Berman, 1978, 1982). True care of infants presumably is altruistic, and is rarely performed by individuals other than the mother if she is living. It will thus be interesting to determine if kin bias within groups occurs in those callitrichids that live in extended family groups in which several individuals share in infant care (Goldizen, 1986).

Many of the other kin biases in behavior reported can be attributed to spatial proximity (e.g. embraces in rhesus macaques, *Macaca mulatta*, Berman, 1978, 1982). One possible exception is play. Factors such as age and sex have important effects on play partner preferences (Walters, 1986), but when their effects are controlled, a kin bias emerges (Fedigan, 1972; Cheney, 1978b; Caine and Mitchell, 1979; Pereira and Altmann, 1985; Stewart and Harcourt, 1986). The effect appears to be small relative to those of age and sex; nevertheless, data on play provide further evidence of active recognition of kin. Other behaviors shed no additional light on kin recognition and therefore will not be discussed.

Kin recognition mechanisms and matrilineal kin bias

Evidence of kin recognition

Data on matrilineal kin bias within groups indicate an ability to recognize at least close kin, and possibly distant kin. If it can be clearly shown that distant kin are treated differently from non-kin, recognition will be implied; however, failure to behave differentially toward distant kin does not evince failure to recognize them. Not only kinship but also the costs and benefits of altruistic and harmful behaviors in theory will determine their performance. For no behavior is it clear how close kin must be for differential treatment to be favored by kin selection (Altmann and Walters, 1978; Altmann, 1979b; Weigel, 1981; Silk, 1986). Neither is it clear that frequency of behaviour should be linearly correlated with relatedness (e.g. Kurland, 1977). Instead, it may be that the benefits and costs of a particular behavior are such that it always pays to direct the behavior at those in certain kin classes, but never at those in others (S. A. Altmann, 1979). If this is the case, differential treatment of, for example, cousins may not be favoured even though they may be recognized. One may conclude, then, that close kin are recognized, and that distant kin may or may not be. Of course, for recognition of cousins to evolve there must be selective pressure to recognize them (Sherman, 1980). Therefore if it were clear they were not treated specially in any way, their recognition would be unlikely.

In a similar vein, it is not clear to what extent individuals treat their mothers and offspring differently from their maternal siblings, nor is it clear from theory how they should. Certainly the two kin classes are distinguished and there are differences during the period the youngster is dependent upon the mother. Thereafter, whether the two kin classes, for example, groom equally is unclear, correlations between frequency of grooming and relatedness that indicate otherwise being confounded by age and proximity effects (Kurland, 1977). With respect to altruistic interaction, the relationship between an orphaned juvenile female yellow baboon and her sister was identical to that between her peers and their mothers (Walters, 1981, in preparation). Would this occur were her mother still present? This is an important question for kin selection theory as it applies to interaction within primate groups, but has implications for kin recognition only if answered in the affirmative.

Recognition mechanisms

Only where relatives regularly come into contact can mechanisms to recognize those relatives evolve (Sherman, 1980). In most mammals kin more distant than half-siblings seldom encounter one another in a context conducive to recognition, and recognition is restricted to this level of relatedness (Sherman, 1980; Gouzoules, 1984; see also Chapters 9 and 10). It is not yet clear whether a variety of classes of kin regularly associate in species ranging solitarily or in family groups; therefore the extent of kin recognition within groups (above and beyond the level of group membership, see above) that occurs or is even conceivable in these social systems is uncertain. In both female-bonded and non-female-bonded groups, however, individuals associate in groups with their grandparents, cousins and other distant kin, so that elaborate kin recognition is at least possible. The question is whether such kin associate regularly enough for recognition to have evolved (Gouzoules, 1984).

There is a widespread consensus that maternal kin recognition in primates is based on a predictable social context, specifically familiarity during development (Breed and Bekoff, 1981; Bateson, 1983; Moore and Ali, 1984; Gouzoules, 1984; Gouzoules and Gouzoules, 1986). There is little direct supporting evidence of this, however. Predictable distribution in space (i.e. spatial proximity) is not sufficient to explain some observed kin biases (see above) and therefore can be ruled out as the primary mechanism. There is no evidence of intrinsic recognition of maternal kin, and attempts to demonstrate recognition of paternal kin by such means have failed (see below). That paternal kin are not intrinsically recognized reinforces the contention that such recognition does not occur in primates, but intrinsic recognition of maternal kin has not yet been examined directly. Cues that could be used in such a mechanism include visual ones (field workers involved in longitudinal studies of primates are well aware of close physical resemblance between parents and offspring) and olfactory ones (many species practice intense olfactory exploration of infants, a behavior with no clear function, see Hrdy, 1976; Altmann, 1980; Walters, in preparation). Potential mechanisms include recognition alleles and phenotype matching based on either self or mother as referent (see Chapter 2, etc.).

By default, then, recognition through social context emerges as the most likely mechanism. Mechanisms based on predictable social context appear to be the rule in mammals (Bekoff, 1977, 1978; Sherman, 1980; Bekoff and Byers, 1981; Gouzoules, 1984; see also Chapter 10), so primates apparently are not exceptional in this regard. Such mechanisms are based on a strong correlation between interaction in a particular context and relatedness (Sherman, 1980; Holmes and Sherman, 1983). In species such as ground squirrels or canids the mechanism is straightforward; association during early development is confined to particular kin classes by the rearing situation so that particular kinds of interaction (e.g., play), bond-formation or familiarization during that time are strongly correlated with kinship. Subsequently treating others differently based on familiarity or social bonding then leads to kin bias in behavior (Bekoff, 1977, 1978; Sherman, 1980; Bekoff and Byers, 1981). The kin recognition mechanism must be more complex in non-human primates, because contact early in development is not restricted to a few close kin of a predictable class. Instead, in many species individuals are in at least occasional contact from birth with a variety of others, including the mother, maternal siblings, other maternal kin and non-relatives.

One way in which kin may be recognized is through a reference individual, specifically the mother (Berman, 1978, 1982; Sherman, 1980; Holmes and

Sherman, 1983; Gouzoules, 1984). Young primates learn to recognize their mother soon after birth (McKenna, 1979b); they might then identify other kin from the way in which others interact with their mother. This mechanism requires, first of all, that primates recognize one another individually (Brown, Sanderson and Michod, 1983). There is abundant evidence that they do (Breed and Bekoff, 1981; Cheney and Seyfarth, 1982; Cheney, 1983; Moore and Ali, 1984), and can even retain recognition after long separation (Erwin et al., 1974). The mechanism also requires that certain interactions or social bonds of the mother are correlated with kinship. There is no single behavior that is directed only at close kin in most species, so the mechanism must involve the relative frequency with which the mother exchanges one or more behaviors with other group members.

One can only speculate as to the precise nature of the mechanism. Perhaps the mother has discrete classes of relationships from which the youngster identifies discrete classes of kin, or perhaps the mother has a continuum of relationships that are translated into a continuum of positive associations involving her infant. Bekoff (1977, 1978; Bekoff and Byers, 1981) has proposed that play is particularly important in kin recognition by social interaction in canids. Play, any other behavior characterized by kin bias (see above), or any combination thereof could be used in recognition by primates. The mechanism might also vary, perhaps with the social system. In family groups and solitary species there are few others that might be confused with maternal kin within the social unit, and recognition may be correspondingly simpler than in multi-female groups.

Alternatively, an individual's own interactions rather than those involving its mother could form the basis of kin recognition. Older individuals could directly recognize their younger maternal siblings by observing births. If the infant's developing social tendencies mirrored how others behaved differentially toward them, sibling bonds could then develop very simply. If kin more distant than siblings are indeed recognized, then a mechanism more complex than this last one must be hypothesized. Infants and their mothers clearly experience differential treatment from other group members, and there are correlations with kinship involved (Berman, 1978, 1982; Altmann, 1980). It is therefore possible that an individual could recognize kin through its own interactions.

What behaviors are involved in kin recognition? Do young non-human primates learn precise relationships, that is, identify some others as sisters and some as aunts, as humans do? Or do they assess some degree of affinity between mother and others, or others and themselves, and mirror that in their own interactions? The most basic questions about the mechanism of maternal kin recognition that form the basis for kin bias in behavior in non-human primates have yet to be addressed.

Implications for kin selection

Kin bias in interaction within groups, and associated kin recognition mechanisms, presumably evolve via kin selection (Sherman 1980). This does not preclude the possiblity that alternative mechanisms such as reciprocal altruism may be equally or even more important in the evolution of primate social behavior (Silk, 1986). The magnitude of the kin bias relative to other effects on the distribution of the behavior is one indication of the importance of kin selection. Still, the primary benefits of grooming, despite a large kin bias in its distribution, may be individualistic and based on reciprocity, focusing reciprocal exhanges on kin that add only a slight increment to inclusive fitness. Regardless, demonstrated kin biases indicate that primates are unusual in the pervasiveness of kin selection effects on interaction within social units, and there are hints that the bias may extend to more distant kin than in other kinds of animals.

Patrilineal kin bias within social groups

Most primates live with paternal as well as maternal kin. In certain social systems patrilineal half-siblings may be more common than matrilineal ones (see above). Data on paternal kin bias are, however, much fewer than data on maternal bias. As a result the patterns of bias, let alone the kin recognition mechanisms involved, are not yet known (Gouzoules, 1984).

Recognition of offspring by males

Most of the data that do exist concern recognition of offspring by males. Differential behavior toward infants by adult males is well documented for a variety of species. Species practicing infanticide and living in one-male groups are an extreme example; adult males kill some infants and behave protectively toward others (Hrdy, 1977, 1979; Gouzoules, 1984; Struhsaker and Leland, 1986). In this case differential behavior is highly correlated with paternity, being related to the male's period of residency in the group. Kin recognition may be based simply on period of residency or on prior sexual interaction with the mother (Gouzoules, 1984).

Differential behavior is less pronounced in species living in multi-male groups. In baboons male association with infants is highly differentiated. Association involves spatial proximity, co-feeding, alliance formation, and in some cases reciprocation by the infant in the form of agonistic buffering (Fig. 3) (Packer, 1980; Altmann, 1980; Stein, 1981, 1984; Strum, 1983; Smuts, 1985; Whitten, 1986). Association is more closely correlated with long-term bonds between males and the mothers of the infants than with mating relationships during the period in which the infant was conceived, however

Fig. 3. Paternal kin bias is less conspicuous among non-human primates than maternal kin bias, and more difficult to study. Male baboons preferentially protect certain infants, and this is correlated with paternity, although not absolutely. The infants may in turn assist the adult males by allowing a prefered male to carry them during an agonistic interaction with another male, such as is occurring here

(Stein, 1984; Smuts, 1985). Special relationships between infants and adult males are more subtle in macaques, but they occur in some species (Grewal, 1980b; Berenstain, Rodman and Smith, 1981). Again, however, the weak differential bonding that exists appears more closely correlated with bonds between males and females than with paternity (Berenstain, Rodman and Smith, 1981). Thus, paternal kin bias occurs in the interaction of adult male and infant baboons and macaques, but correlations between kinship and behavior are not as high as for maternal kin bias. The kin recognition mechanism apparently involved, recognition through bonding with a reference individual, the mother, is imprecise.

Differential behavior toward paternal siblings

The great difficulty in documenting kin bias involving paternal siblings has been the establishing of kinship. Genetic techniques now enable identification of paternal siblings in multi-male groups in which females mate with several males, but only one study has yet made use of these techniques to measure differential behavior toward paternal siblings. Small and Smith (1981), studying rhesus macaques, found a difference in the resistance of mothers to approaches by juveniles that were unrelated to the infant and those that were paternal-siblings of the infant. However, the effect was small and did not extend to full-siblings, nor to another measure of infant–juvenile interactions. Considering the inconsistency of these results, and that confounding factors such as dominance relationships were not considered, there is no solid indication of kin bias in these data (Gouzoules, 1984).

Just as researchers have difficulty determining paternity within multi-male primate groups, the monkeys themselves may not be able to recognize paternal-siblings from social interaction. Intrinsic recognition mechanisms are therefore a more compelling possibility for paternal kin than for maternal kin where precise mechanisms based on social interaction are feasible. Sackett and his colleagues have examined intrinsic recognition of paternal siblings in pig-tailed macaques (*Macaca nemestrina*) in two experiments. In the first, juveniles were simultaneously exposed to two peers unknown to them, one of which was a paternal-sibling, the other an unrelated juvenile (Wu et al., 1980). Thirteen of 16 subjects oriented more toward the related juvenile. This was originally construed as evidence of phenotype matching. However, the test situation was not a natural one in that juveniles in the wild do not have to discriminate among strangers but among familiar group members. In the second experiment (Fredrickson and Sackett, 1984), the youngsters were tested with familiar peers as well as strangers. The subjects oriented more toward familiar animals than unfamiliar ones, regardless of paternal kinship, and did not orient more toward unfamiliar relatives than toward unfamiliar non-relatives.

Fredrickson and Sackett (1984) interpreted these data as indicating the absence of intrinsic recognition (see also Gouzoules, 1984). This is true only if orientation in a choice situation is something that would normally be directed preferentially toward kin. A test involving effects of patrilineal kinship on a behavior such as grooming would be more conclusive.

There is thus no evidence of precise recognition of, or kin bias toward, paternal siblings. Imprecise recognition mechanisms, and corresponding kin biases, are nevertheless possible. In both multi-male and one-male groups peers are often paternal siblings (see above). Even if paternal siblings are treated no differently from non-siblings among peers, treating peers in general differently than non-peers should result in some kin bias (J. Altmann, 1979). In most group-living species immatures interact frequently with peers. This interaction includes maintaining spatial proximity (e.g. peer groups, Dunbar and Dunbar, 1975), play, and frequent grooming (Pereira and Altmann, 1985; Walters, 1986). For example, the juvenile yellow baboons that I studied groomed and formed alliances more frequently with other juveniles of similar age than with any other individuals except maternal kin (J. R. Walters, in preparation). Perhaps differential treatment of peers is based solely on effects of age on the costs and benefits of the behaviors involved, but it may also have evolved through kin selection. Whether the special relationships with peers that have been documented for juveniles persist into adulthood is not yet known (Walters and Seyfarth, 1986). This is one possible explanation for the special relationships that have been observed among adult female macaques and baboons that are not maternal kin (Gouzoules, 1984; Silk, 1986).

Kin bias toward paternal kin, whether expressed as differential treatment of peers generally or of siblings specifically, is an important area for future research. A better characterization of the kinds of paternal kin bias that may exist is of primary importance. This knowledge would indicate which possible kin recognition mechanisms are best to explore. If peers are treated specially as a class, then simple recognition by social context, specifically membership in an age cohort within a group, is most likely. If there is discrimination among peers, a more specific mechanism such as phenotype matching or use of male–mother bonding as a referent is more likely.

Recognition of paternal kin more distant than siblings has not been examined. In general it is unlikely to occur because most such kin rarely reside together, but there are exceptions. Examples are mothers of paternal siblings and paternal uncles in species in which brothers often immigrate into the same group. In both these cases if the closer paternal relative could be recognized, its social bonds with other group members could be used to identify more distant relatives. If recognition of paternal siblings is demonstrated, these will become interesting possibilities.

INBREEDING AVOIDANCE

There is considerable evidence of detrimental effects of inbreeding in captive primates (Crawford and O'Rourke, 1978; Smith, 1982; Ralls and Ballou, 1982) and limited evidence from one wild population (Packer, 1979b). There is no evidence from primates of positive effects of breeding with distant relatives such as has been reported for Japanese quail (*Coturnix coturnix japonica*) (Bateson, 1982, 1983).

It is useful to separate avoidance of close kin, known as incest avoidance, from inbreeding avoidance generally. Theoretical expectations are clearer for incest avoidance than other effects of kinship on mating success; mating with close kin is likely to be detrimental, whereas mating with more distant kin may have either negative or positive effects on fitness (Harcourt, 1978; Packer, 1979b; Wade, 1979; Bateson, 1983; Moore and Ali, 1984). The empirical data from primates, too, indicate selection to recognize and avoid mating with close kin, but leave unclear expectations for distant kin.

Incest avoidance

The dispersal patterns exhibited by non-human primates preclude the frequent occurrence of opposite-sex kin in the same social unit as adults. Dispersal, therefore, may be viewed to some extent as a mechanism to avoid incest, and in a few species lack of opportunity for selection may have prevented the evolution of additional incest avoidance mechanisms (see below). However, there is evidence from a wide variety of species of additional mechanisms operating within social units.

The best documented mechanism is discrimination against close maternal kin as mates. As juveniles, young males may direct much of their sexual behavior at their mother (Missakian, 1973; Hanby and Brown, 1974), but in both apes and cercopithecines if adult males reside in the same social unit as their mother they interact sexually with her at much lower rates than with other females (Imanishi, 1965; Sade, 1968; Itani, 1972; Enomoto, 1974, 1978; Harcourt, Stewart and Fossey, 1976; Packer, 1979b; Pusey, 1980; Takahata, 1982; Pusey and Packer, 1986). Evidence that maternal kin other than the mother are avoided in mating is more limited. There are data from chimpanzees indicating reduced sexual activity between maternal siblings (van Lawick-Goodall, 1968; Pusey, 1980), but only for macaques are there data on sexual interaction between more distant kin. These data indicate sharply reduced sexual activity between close kin and significantly reduced activity in kin as distant as cousins (Sade, 1968; Enomoto, 1978; Takahata, 1982; Gouzoules and Gouzoules, 1986). Rates of sexual interaction between even more distant kin are comparable to those between non-relatives.

Avoidance of incest is not absolute among non-human primates. On rare occasions female gorillas remain in a group in which their son inherits breeding status, and breed with him (Stewart and Harcourt, 1986). In gibbons, if a young male's father dies he may fail to disperse from the family group, form a pair-bond with his mother and father her offspring (Chivers and Raemaekers, 1980; Tilson, 1981; Leighton, 1986). It is not known how long these mother–son pairs persist (Pusey and Packer, 1986). Mating between close kin occurs, albeit rarely, even in those species exhibiting pronounced incest avoidance (Takahata, 1982; Stewart and Harcourt, 1986).

Data on paternal kin are again harder to acquire. They suggest that incest avoidance may at least occur between fathers and daughters. In baboons, females appear to avoid as sexual partners males that associated with them as infants (Packer, 1979b; Scott, 1984; J. R. Walters, in preparation). As discussed above, male association with infants is correlated with paternity, although imperfectly. Female chimpanzees also avoid their previous male associates as mating partners when they reach adulthood, although it is not entirely clear that these associates include their probable fathers (Pusey, 1980). However, frequent mating between fathers and daughters in a captive rhesus population (Smith, 1982) suggests that avoidance of paternal kin, and hence perhaps their recognition, may be limited to certain species (Gouzoules, 1984). Perhaps the presence of recognition mechanisms involving paternal kin of opposite sex is related to the frequency with which such individuals live together in the same group.

Smith's (1982) study of rhesus macaques also revealed frequent mating between paternal siblings. There are no other data on sexual interaction between paternal siblings or more distant paternal kin. Thus avoidance of siblings or more distant paternal kin has not been documented (but see below).

The existence of incest avoidance within social units implies kin recognition. For much the same reasons as discussed for recognition related to kin biases in altruistic behavior, and in a similar absence of data, recognition mechanisms based on social context have emerged as the likely bases for incest avoidance. In all cases where incest avoidance has been shown, the individuals involved had some kind of prior affiliative bond. This suggests that positive association prior to maturity leads to reduced sexual activity in adulthood. There are striking data from humans indicating that intimate association during childhood has strong negative effects on later sexual attraction (Shepher, 1971; Young, 1978; van den Berghe, 1983), and it has frequently been suggested that such a negative imprinting process is a general phenomenon in primates, rather than an exclusively human one (Sade, 1968; Demarest 1977; Harcourt, Stewart and Fossey, 1976; Harcourt, 1978; Bateson, 1983). Evidence of this in humans comes from studies of sexual attraction between non-kin reared as kin. No such evidence exists for non-human primates, and no attempt has been made to test for alternative mechanisms.

If a mechanism based on social interaction exists, many questions about its precise nature need to be addressed. The effect of early experience might be either continuous or discrete. Are individuals with the closest prior associations the least attracted to one another sexually, or are all sufficiently familiar individuals avoided equally? Also, it is not clear whether sex is an additional context in which behavior is based on the recognition mechanisms that operate in producing differential deployment of altruistic behavior toward kin. There are indications in the sexual interactions of non-kin within groups that a mechanism specific to sexual behavior might exist (see below).

There are theoretical reasons for expecting stronger selection for avoiding incest in females than in males (Bengsston, 1978; Harcourt, 1978; Packer, 1979b) and there are data from non-human primates indicating that females are primarily responsible for lack of mating between close kin. In baboons and macaques reduction of sexual responsiveness to close maternal kin is more pronounced in females, although it may also occur in males (Enomoto, 1974, 1978; Packer, 1979b). The sexes appear to contribute equally to incest avoidance in chimpanzees, however (Pusey, 1980).

Dispersal and inbreeding

The various dispersal systems exhibited by non-human primates all have the effect of reducing opportunity for breeding between close kin. Dispersal may therefore be viewed as a mechanism to avoid inbreeding. Kin recognition in this case is based on predictable social context, specifically group membership and its correlation with relatedness.

The extent to which inbreeding avoidance is involved in the evolution of dispersal is, however, a controversial issue (Harcourt, 1978; Moore and Ali, 1984; Pusey and Packer, 1986). In primates dispersal reduces close inbreeding, but does not eliminate it, because dispersal distances are short and individuals often disperse to groups containing their same-sex kin (see above). Also, there are several selective forces other than avoiding inbreeding that might affect dispersal, chiefly reproductive competition and resource acquisition (Packer, 1979b; Wade, 1979; Moore and Ali, 1984; Pusey and Packer, 1986). These other factors are clearly involved in dispersal in some species. For example, if one sex disperses, regular dispersal by individuals of the opposite sex cannot be explained by avoidance of inbreeding (Marsh, 1979). This dispersal pattern characterizes several species, for example those living in family groups. Also, in many species dispersal is forced by aggression, presumably related to competition for mates or resources. This is especially characteristic of species living in one-male, age-graded-male and family groups (Kleiman, 1979; Moore and Ali, 1984; Walters, 1986; Pusey and Packer, 1986). Finally, additional factors unrelated to inbreeding such as dominance rank (rhesus macaques, Moore and Ali, 1984) or bonds with same-sex group members (gorillas, Harcourt and Stewart, 1981) affect dispersal in certain species.

Apes and cercopithecines

There is considerable evidence that effects of kinship on sexual interaction play some role in dispersal in apes and cercopithecines that live in multi-male groups. In this regard dispersal from the natal group must be distinguished from subsequent secondary transfers (Pusey and Packer, 1986). Secondary transfers appear to be related to sexual competition; transfers of male baboons, for example, are related to the numbers of estrous females in the relevant groups, and to the male's dominance status and expected reproductive success (Packer, 1979b; Moore and Ali, 1984; Pusey and Packer, 1986). In contrast, initial dispersal appears to be little affected by reproductive competition and greatly affected by the presence of kin within the natal group. Male baboons often attain very high dominance rank, which is correlated with success in competition for females, prior to dispersal (Hausfater, 1975; Packer, 1979a), and their dispersal is unrelated to the availability of estrous females (Packer, 1979b). Similar patterns have been observed in a few other species (e.g. Cheney, 1983). Male Japanese macaques (*Macaca fuscata*) have even been observed to disperse in the absence of other adult males in the natal group (Sugiyama, 1976). On the other hand, male dispersal was greatly reduced in a group of yellow baboons formed by the fusion of two others (Altmann, 1980); presumably an atypically large number of unrelated adult females were available as mates for these males. Similarly, when group fission occurs, males will often remain as adults in the subgroup that does not contain their maternal kin, rather than disperse (Itani, 1972; Demarest, 1977).

There is corresponding evidence that females discriminate against male members of their natal group in their sexual behavior. Discrimination against close maternal kin and fathers has already been discussed. In addition, a generalized sexual attraction toward unfamiliar individuals has been reported for a number of species (Pusey, 1980; Cheney, 1983; Colvin, 1983; Pusey and Packer, 1986, and a generalized discrimination against males of the natal group has been reported in a few (Packer, 1979b). It has been proposed that females may discriminate against all males familiar to them as immatures, rather than just close kin, in their adult sexual behavior (Harcourt, Stewart and Fossey, 1976; Harcourt, 1978; Packer, 1979b; Pusey, 1980; Scott, 1984; Pusey and Packer, 1986). Such discrimination could be viewed either as an imprecise mechanism to avoid incest directed at close paternal kin, especially siblings, or as a more general mechanism to avoid even low levels of inbreeding. The former seems more likely; considering the levels of inbreeding that exist in primate populations, mating with an unrelated group member is unlikely to result in significantly more inbreeding than mating with an immigrant male from an adjoining group.

Evidence of general discrimination against males of the natal group is not yet compelling, because it is based primarily on behavioral data that are open to other interpretations, rather than on mating data (Moore and Ali, 1984). For example, presenting by females may have a greeting as well as sexual function. That females present more to unfamiliar males than to familiar ones may therefore reflect a difference in greeting interaction rather than sexual attraction.

If females do discriminate against males of their natal group, then male dispersal may be viewed as a means of increasing mating opportunities, and inbreeding avoidance by females may be said to drive dispersal (Harcourt, 1978; Henzi and Lucas, 1980; Pusey and Packer, 1986). Inbreeding avoidance by males may also play a role if males prefer unfamiliar females as mates. Incest avoidance contributes to this effect, but limits mate choice significantly only in small groups. An additional more general discrimination by females against males of their natal group would be necessary to drive dispersal in most species.

Inbreeding avoidance may also explain female movement in apes and cercopithecines. Secondary transfers in those species in which females normally remain in their natal groups have been linked to incest avoidance (Grewal, 1980b; Moore, 1984), and female transfer in species such as chimpanzees, in which males remain in their natal groups, may be due to inbreeding avoidance in a manner that parallels male movement in other species (Moore and Ali, 1984).

The recognition mechanisms

If there is a kinship effect on sexual interaction in addition to the incest avoidance discussed previously, it is presumably based on kin recognition by group membership. General discrimination against members of the natal group is usually thought of as an extension of the same negative effect of familiarity during immaturity that results in discrimination against close maternal kin (e.g. Harcourt, 1978). The mechanism may be distinct from that related to incest avoidance however, in so far as it has a smaller effect on behavior. A continuous mechanism could account for both levels of discrimination, but it is equally plausible that two distinct mechanisms are involved. The relation of such mechanisms to those that are the bases of kin biases in altruistic behavior is obscure. Differential behavior toward individuals that become familiar during immaturity as opposed to group members generally has not been demonstrated for altruistic interaction, a fact that suggests at least two independent recognition mechanisms.

Other species

It is difficult to determine whether incest avoidance or a general discrimination against members of the natal group in sexual interactions occurs in species in which dispersal is forced by aggression, because individuals seldom fail to disperse. In some such species there may not be sufficient opportunity for selection to reduce inbreeding beyond that reduction attributable to dispersal. Gibbons may represent such a case. Not only does incest occur when the same-sex parent is not present to prevent it by driving its offspring from the group (see above), but sexual interaction with the opposite sex parent is generally common prior to dispersal (Tilson, 1981), suggesting that the usual inhibitory effects of close kinship on sexual interaction are absent.

It will be interesting to discover if active inbreeding avoidance occurs in other species such as howlers and langurs in which dispersal is forced by aggression (Pusey and Packer, 1986). In some species that live in one-male groups, when an adult male's tenure lasts until his male offspring are mature, he often allows them to remain in the group. That these offspring usually disperse anyway raises the possibility that inbreeding avoidance may occur

(Pusey and Packer, 1986), but whether discrimination against kin specifically or members of the natal group generally occurs on these rare occasions is unknown. Therefore it is not yet clear whether inbreeding avoidance within social units, and accompanying kin recognition, is widespread among primates or restricted to certain taxa or social systems.

CONCLUSIONS AND PROSPECTUS

Research on kin recognition in non-human primates is in its infancy. Close relatives are clearly recognized, but the limits of recognition are not yet certain. Careful research on altruistic and sexual interaction of distant kin is especially needed to determine if they are treated differently from non-kin. It will also be important to collect longitudinal data from species other than apes and cercopithecines to determine if emerging syntheses accurately reflect general patterns among primates or are biased towards these well studied taxa.

As for the recognition mechanisms themselves, there are widely held views as to their identity and function, but these views lack empirical support. Even the most basic experiments remain to be done. Separating kinship and social context by altering rearing conditions as Sackett's group has done can be used to determine if recognition is indeed based on social interaction rather than some intrinsic mechanism. For example, sexual attraction among kin reared apart and non-kin reared together might be compared to that among kin reared together. Elaborate experiments will be required to investigate some proposed recognition mechanisms, such as those involving a reference individual. To test whether siblings are recognized through the mother, one must cause not only the test animal but also its mother to interact with a non-relative in a manner normally typical of kin.

The kin recognition mechanisms that are the basis of particular effects of kinship on behavior must be better specified before research can proceed effectively. Are individuals of varying relatedness recognized as distinct classes, or are there recognition thresholds or continua? Are there independent mechanisms related to altruistic and sexual interaction, or does one recognition mechanism form the basis for both kinds of effects? Specific mechanisms need to be articulated for all observed kinship effects on behavior, especially those operating within groups.

Research on kin recognition in non-human primates has reached an exciting stage in which the critical questions have only recently come into focus. As is typical of such developing areas, the literature is rife with speculation, and the data are too few to refute any of it. The challenge now is to begin the careful and tedious work necessary to provide the answers to these questions.

ACKNOWLEDGMENTS

I thank Dr A. Massey, Dr R. Powell, and Dr M. Walek for reviewing a previous draft of the manuscript, and the editors for suggesting helpful changes. I am grateful to the Kenya Ministry of Tourism and Wildlife, Amboseli National Park Warden Joseph Kioko, Dr S. Altmann, Dr J. Altmann, and Dr G. Hausfater for making my research on yellow baboons possible, and for their kind assistance during the work. The research was funded by the National Institute of Mental Health through training (MH 15181) and research (MH 19617) grants awarded to Dr S. Altmann and Dr J. Altmann. I also thank the Agricultural Research Service of North Carolina State University for support during preparation of the manuscript. This is paper no. 10022 of the journal series of the North Carolina Agricultural Research Service, Raleigh, NC 27695-7601.

LITERATURE CITED

Alexander, R. D. 1974. The evolution of social behavior. *Annual Review of Ecology and Systematics*, **5**, 325-383.

Altmann, J. 1979. Age cohorts as paternal sibships. *Behavioral Ecology and Sociobiology*, **6**, 161-164.

Altmann, J. 1980. *Baboon mothers and infants*. Harvard University Press, Cambridge, Massachusetts.

Altmann, S. A. 1979. Altruistic behaviour: the fallacy of kin deployment. *Animal Behaviour*, **27**, 958-962.

Altmann, S. A. and J. Altmann. 1979. Demographic constraints on behavior and social organization, in Bernstein, I. S. and E. O. Smith, eds., *Primate ecology and human origins*, Garland STPM Press, New York, pp. 47-64.

Altmann, S. A. and J. R. Walters. 1978. Book review: *Kin selection in the Japanese monkey* by J. A. Kurland. *Man*, **13**, 324-325.

Bateson, P. 1982. Preference for cousins in Japanese quail. *Nature*, **295**, 236-237.

Bateson, P. 1983. Optimal outbreeding, in Bateson, P. ed., *Mate choice*, Cambridge University Press, Cambridge, England, pp. 257-277.

Bearder, S. 1986. Lorises, bushbabies and tarsiers: diverse societies in solitary foragers, in Smuts, B., D. L. Cheney, R. M. Seyfarth, R. T. Wrangham and T. T. Struhsaker, eds., *Primate societies*, University of Chicago Press, Chicago, Illinois, in press.

Bekoff, M. 1977. Mammalian dispersal and the ontogeny of individual behavioral phenotypes. *American Naturalist*, **111**, 715-732.

Bekoff, M. 1978. Social play: structure, function, and the evolution of a cooperative social behavior, in Burghardt, G. M. and M. Bekoff, eds., *The development of behavior*. Garland STPM Press, New York, pp. 367-383.

Bekoff, M. 1981. Mammalian sibling interactions. Genes, facilitative environments, and the coefficient of familiarity, in Gubernick, D. and P. H. Klopfer, eds., *Parental care in mammals*. Plenum, New York, pp. 307-346.

Bekoff, M. and J. A. Byers. 1981. A critical reanalysis of the ontogeny and phylogeny of mammalian social and locomotor play: an ethological hornet's nest, in Immelman, K., G. W. Barlow, L. Petrinovich and M. Main, eds. *Behavioral development*. Cambridge University Press, Cambridge, England, pp. 296-337.

Bengtsson, B. O. 1978. Avoiding inbreeding: At what cost? *Journal of Theoretical Biology*, **73**, 439–444.
Berenstain, L., P. S. Rodman and D. G. Smith. 1981. Social relations between fathers and offspring in a captive group of rhesus monkeys (*Macaca mulatta*). *Animal Behaviour*, **29**, 1057–1063.
Berman, C. M. 1978. *Social relationships among free-ranging infant rhesus monkeys*. Ph.D. thesis, Cambridge University, Cambridge, England.
Berman, C. M. 1982. The ontogeny of social relationships with group companions among free-ranging infant rhesus monkeys I. social networks and differentiation. *Animal Behaviour*, **30**, 149–162.
Bertram, B. C. R. 1975. The social system of lions. *Scientific American*, **232**(5), 54–65.
Box, H. O. 1977. Quantitative data on the carrying of young captive monkeys (*Callithrix jacchus*) by other members of their family groups. *Primates*, **18**, 475–484.
Breden, F. and M. J. Wade. 1981. Inbreeding and evolution by kin selection. *Ethology and Sociobiology*, **2**, 3–16.
Breed, M. D. and M. Bekoff. 1981. Individual recognition and social relationships. *Journal of Theoretical Biology*, **88**, 589–593.
Brown, J. S., M. J. Sanderson and R. E. Michod. 1983. Evolution of social behavior by reciprocation. *Journal of Theoretical Biology*, **99**, 319–339.
Brown, K. and D. Mack. 1978. Food sharing among captive *Leontopithecus rosalia*. *Folia Primatologica*, **29**, 268–290.
Busse, C. D. 1978. Do chimpanzees hunt cooperatively? *American Naturalist*, **112**, 771–774.
Bygott, J. D. 1979. Agonistic behavior, dominance and social structure in wild chimpanzees of the Gombe National Park, in Hamburg, D. A. and E. R. McCown, eds., *The great apes*. Benjamin/Cummings Publishing Company, Menlo Park, California, pp. 405–427.
Caine, N. and G. Mitchell. 1979. A review of play in the genus *Macaca:* social correlates. *Primates*, **20**, 535–546.
Charles-Dominique, C. 1977. *Ecology and behaviour of nocturnal primates*. Columbia University Press, New York.
Cheney, D. L. 1978a. Interactions of immature male and female baboons with adult females. *Animal Behaviour*, **26**, 389–402.
Cheney, D. L. 1978b. The play partners of immature baboons. *Animal Behaviour*, **26**, 1038–1050.
Cheney, D. L. 1983. Proximate and ultimate factors related to the distribution of male migration, in Hinde, R. A. ed., *Primate social relationships*. Sinauer Associates, Sunderland, Massachusetts, pp. 241–249.
Cheney, D. L. and R. M. Seyfarth. 1982. Recognition of individuals within and between groups of free-ranging vervet monkeys. *American Zoologist*, **22**, 519–530.
Cheney, D. L. and R. M. Seyfarth. 1983. Non-random dispersal in free-ranging vervet monkeys: social and genetic consequences. *American Naturalist*, **122**, 392–412.
Cheney, D. L. and R. W. Wrangham. 1986. Predation, in Smuts, B., D. L. Cheney, R. M. Seyfarth, R. W. Wrangham and T. T. Struhsaker, eds., *Primate societies*, University of Chicago Press, Chicago, Illinois, in press.
Chivers, D. J. and J. J. Raemaekers. 1980. Long-term changes in behavior, in Chivers, D. J. ed., *Malayan forest primates*. Plenum, New York, pp. 209–260.
Clark, A. 1978. Sex ratio and local resource competition in a prosimian primate. *Science*, **201**, 163–165.
Colvin, J. 1983. Influences of the social situation on male emigration, in Hinde, R. A. ed., *Primate social relationships*. Sinauer Associates, Sunderland, Massachusets, pp. 160–171.

Crawford, M. H. and D. H. O'Rourke. 1978. Inbreeding, lymphoma, genetics and morphology of the *Papio hamadryas* colony of Sukhumi. *Journal of Medical Primatology*, **7**, 355–360.

Crook, J. H. 1970. The socio-ecology of primates, in Crook, J. H., ed. *Social behavior in birds and mammals: essays on the social ethology of animals and man.* Academic Press, New York, pp. 103–166.

Defler, T. R. 1978. Allogrooming in two species of macaque (*Macaca nemestrina* and *Macaca radiata*). *Primates*, **19**, 153–167.

Demarest, W. J. 1977. Incest avoidance among human and non-human primates, in Chevalier-Skolnikoff, S. and F. E. Poirier, eds., *Primate bio-social development.* Garland, New York, pp. 323–342.

Dunbar, R. I. M. 1979. Structure of gelda baboon reproductive units I. Stability of social relationships. *Behaviour*, **69**, 72–87.

Dunbar, R. I. M. 1984. *Reproductive decisions.* Princeton University Press, Princeton, New Jersey.

Dunbar, R. and E. P. Dunbar. 1975. Social dynamics of gelada baboons. *Contributions to Primatology*, Vol. 6, Karger, Basel.

Emlen, S. T. and S. L. Vehrencamp. 1983. Cooperative breeding strategies among birds, in Brush, A. H. and G. A. Clark, Jr., eds., *Perspectives in ornithology*, Cambridge University Press, Cambridge, England, pp. 93–120.

Enomoto, T. 1974. The sexual behavior of Japanese monkeys. *Journal of Human Evolution*, **3**, 351–372.

Enomoto, T. 1978. On social preferences in sexual behavior of Japanese monkeys (*Macaca fuscata*). *Journal of Human Evolution*, **7**, 283–293.

Erwin, J., T. Maple, J. Willott and G. Mitchell. 1974. Persistent peer attachments of rhesus monkeys: responses to reunion after two years of separation. *Psychological Reports*, **34**, 1179–1183.

Fedigan, L. M. 1972. Social and solitary play in a colony of vervet monkeys. *Primates*, **13**, 347–364.

Fossey, D. 1979. Development of the mountain gorilla (*Gorilla gorilla beringei*): the first 36 months, in Hamburg, D. A. and E. R. McCown, eds., *The great apes.* Benjamin/Cummings Publishing Company, Menlo Park, California, pp. 139–184.

Frederickson, W. and G. Sackett. 1984. Kin preferences in primates (*Macaca nemestrina*): relatedness or familiarity? *Journal of Comparative Psychology*, **98**, 29–34.

Goldizen, A. 1986. Tamarins and marmosets: communal care of offspring, in Smuts, B., D. L. Cheney, R. M. Seyfarth, R. W. Wrangham and T. T. Struhsaker, eds., *Primate societies,* University of Chicago Press, Chicago, Illinois, in press.

Goodall, J., A. Bandora, E. Bergman, C. Busse, H. Matama, E. Mpongo, A. Pierce and D. Riss. 1979. Intercommunity interactions in the chimpanzee population of the Gombe National Park, in Hamburg, D. A. and E. R. McCown, eds., *The great apes,* Benjamin/Cummings Publishing Company, Menlo Park, California, pp. 13–53.

Gouzoules, S. 1984. Primate mating systems, kin association, and cooperative behavior: evidence for kin recognition? *Yearbook of Physical Anthropology*, **27**, 99–134.

Gouzoules, S. and H. Gouzoules. 1986. Kinship, in Smuts, B., D. L. Cheney, R. M. Seyfarth, R. W. Wrangham, and T. T. Struhsaker, eds., *Primate societies,* University of Chicago Press, Chicago, Illinois, in press.

Grewal, B. S. 1980a. Changes in relationships of nulliparous and parous females of Japanese monkeys at Arashiyama with some aspects of troop organization. *Primates*, **21**, 330–339.

Grewal, B. S. 1980b. Social relationships between central males and kinship groups of Japanese monkeys at Arashiyama with some aspects of troop organization. *Primates*, **21**, 161–180.

Hamilton, W. D. 1964. The genetical evolution of social behavior. I. II. *Journal of Theoretical Biology*, **7**, 1–52.
Hamilton, W. J., III, R. E. Buskirk and W. H. Buskirk. 1976. Defense of space and resources by chacma (*Papio ursinus*) baboon troops in an African desert and swamp. *Ecology*, **57**, 1264–1272.
Hamilton, W. J., III, C. Busse and K. S. Smith. 1982. Adoption of infant orphan chacma baboons. *Animal Behaviour*, **30**, 29–34.
Hanby, J. and C. E. Brown. 1974. The development of sociosexual behaviors in Japanese macaques *Macaca fuscata*. *Behaviour*, **49**, 152–196.
Harcourt, A. H. 1978. Strategies of emigration and transfer by primates, with particular reference to gorillas. *Zeitschrift für Tierpsychologie*, **48**, 401–420.
Harcourt, A. H. 1979. Social relationships among adult female mountain gorillas. *Animal Behaviour*, **27**, 251–264.
Harcourt, A. H. and K. J. Stewart. 1981. Gorilla male relationships: can differences during immaturity lead to contrasting reproductive tactics in adulthood? *Animal Behaviour*, **29**, 206–210.
Harcourt, A. H., K. S. Stewart and D. Fossey. 1976. Male emigration and female transfer in wild mountain gorilla. *Nature*, **263**, 226–227.
Hasegawa, T. and M. Hiraiwa. 1980. Social interactions of orphans observed in a free-ranging troop of Japanese monkeys. *Folia Primatologica*, **33**, 129–158.
Hausfater, G. 1975. Dominance and reproduction in baboons (*Papio cynocephalus*). *Contributions to Primatology*, Vol. 7, Karger, Basel.
Henzi, S. P. and J. W. Lucas. 1980. Observations on the inter-troop movement of adult vervet monkeys (*Cercopithecus aethiops*). *Folia Primatologica*, **33**, 220–235.
Hoage, R. J. 1977. Parental care in *Leontopithecus rosalia rosalia:* sex and age differences in carrying behavior and the role of prior experience, in Kleiman, D. G. ed., *The biology and conservation of the Callitrichidae*. Smithsonian Institution Press, Washington, pp. 293–305.
Holmes, W. G. and P. W. Sherman. 1983. Kin recognition in animals. *American Scientist*, **71**, 46–55.
Hrdy, S. B. 1976. Care and exploitation of non-human primate infants by conspecifics other than the mother. *Advances in the Study of Behaviour*, **6**, 101–158.
Hrdy, S. B. 1977. *The langurs of Abu*. Harvard University Press, Cambridge, Massachusetts.
Hrdy, S. B. 1979. Infanticide among animals: a review, classification and examination of the implications for the reproductive strategies of females. *Ethology and Sociobiology*, **1**, 13–40.
Imanishi, K. 1965. Identification: a process of socialization in the subhuman society of *Macaca fuscata*, in Altmann, S. A. ed., *Japanese monkeys: a collection of translations*. S. A. Altmann, Atlanta, Georgia, pp. 30–51.
Ingram, J. C. 1977. Interactions between parents and infants, and the development of independence in the common marmoset (*Callithrix jacchus*). *Animal Behaviour*, **25**, 811–827.
Itani, J. 1972. A preliminary essay on the relationship between social organization and incest avoidance in non-human primates, in Poirier, F. E. ed., *Primate socialization*. Random House, New York, pp. 165–171.
Kaplan, J. R. 1978. Fight interference and altruism in rhesus monkeys. *American Journal of Physical Anthropology*, **49**, 241–250.
Kawai, M. 1965. On the system of social ranks in a natural troop of Japanese monkeys (II)—ranking order as observed among the monkeys—on and near the test box, in Altmann, S. A., ed., *Japanese monkeys: a collection of translations*, S. A. Altmann, Atlanta, Georgia, pp. 87–104.

Kleiman, D. G. 1979. Parent-offspring conflict and sibling competition in a monogamous primate. *American Naturalist*, **114**, 753–759.
Kummer, H. 1978. On the value of social relationships to non-human primates: a heuristic scheme. *Social Science Information*, **17**, 687–705.
Kurland, J. A. 1977. Kin selection in the Japanese monkey. *Contributions to Primatology. Vol. 12,* Karger, Basel.
Leighton, D. 1986. Gibbons: territoriality and monogamy, in Smuts, B., D. L. Cheney, R. M. Seyfarth, R. W. Wrangham and T. T. Struhsaker, eds., *Primate societies,* University of Chicago Press, Chicago, Illinois, in press.
Marler, P. 1976. On animal aggression: the roles of strangeness and familiarity. *American Psychologist*, **31**, 239–246.
Marsden, H. M. and S. H. Vessey. 1968. Adoption of an infant green monkey within a social group. *Communications in Behavioral Biology, (A)*, **2**, 275–279.
Marsh, C. W. 1979. Female transference and mate choice among Tana River red colobus. *Nature*, **218**, 568–569.
Massey, A. 1977. Agonistic aids and kinship in a group of pigtail macaques. *Behavioral Ecology and Socibiology*, **2**, 31–40.
McKenna, J. J. 1979a. The evolution of allomothering behavior among colobine monkeys: function and opportunism in evolution. *American Anthropologist*, **81**, 818–840.
McKenna, J. J. 1979b. Aspects of infant socialization, attachment, and maternal caregiving patterns among primates: a cross-disciplinary review. *Yearbook of Physical Anthropology*, **22**, 250–286.
Meikle, D. B. and S. H. Vessey. 1981. Nepotism among rhesus monkey brothers. *Nature*, **294**, 160–161.
Melnick, D. J. and K. K. Kidd. 1983. The genetic consequences of social group fission in a wild population of rhesus monkeys (*Macaca mulatta*). *Behavioral Ecology and Sociobiology*, **12**, 229–236.
Michod, R. E. 1982. The theory of kin selection. *Annual Review of Ecology and Systematics*, **13**, 23–55.
Miller. M. H., A. Kling and D. Dicks. 1973. Familial interactions of male rhesus monkeys in a semi-free-ranging troop. *American Journal of Physical Anthropology*, **38**, 605–612.
Missakian, E. A. 1973. Genealogical mating activity in free-ranging groups of rhesus monkeys (*Macaca mulatta*) on Cayo Santiago. *Behaviour*, **45**, 225–241.
Mitani, J. C. and P. S. Rodman. 1979. Territoriality: the relation of ranging patterns and home range size to defendability, with an analysis of territoriality among primate species. *Behavioral Ecology and Sociobiology*, **5**, 244–251.
Moore, J. 1984. Female transfer in primates. *International Journal of Primatology*, **5**, 537–589.
Moore, J. and R. Ali. 1984. Are inbreeding and dispersal related? *Animal Behaviour*, **32**, 94–112.
Morse, D. H. 1980. *Behavioral mechanisms in ecology.* Harvard University Press, Cambridge, Massachusetts.
Myers, J. P. 1983. Space, time and the pattern of individual associations in a group-living species: sanderlings have no friends. *Behavioral Ecology and Sociobiology*, **12**, 129–134.
Nishida, T. 1979. The social structure of chimpanzees of the Mahali Mountains, in Hamburg, D. A. and E. R. McCown, eds., *The great apes,* Benjamin/Cummings Publishing Company, Menlo Park, California, pp. 73–121.
Packer, C. 1977. Reciprocal altruism in olive baboons. *Nature*, **265**, 441–443.
Packer, C. 1979a. Male dominance and reproductive activity in *Papio anubis. Animal Behaviour*, **27**, 37–45.

Packer, C. 1979b. Inter-troop transfer and inbreeding avoidance in *Papio anubis*. *Animal Behaviour*, **27**, 1–36.
Packer, C. 1980. Male care and exploitation of infants in *Papio anubis*. *Animal Behaviour*, **28**, 512–520.
Packer, C. and A. E. Pusey. 1982. Cooperation and competition within coalitions of male lions: kin selection or game theory? *Nature*, **296**, 740–742.
Pereira, M. E. and J. Altmann. 1985. Development of social behavior in free-living non-human primates, in Watts, E. S. ed., *Non-human primate models for human growth and development*, Alan R. Liss, New York, in press.
Pusey, A. E. 1980. Inbreeding avoidance in chimpanzees. *Animal Behaviour*, **28**, 543–552.
Pusey, A. E. 1983. Mother–offspring relationships in chimpanzees after weaning. *Animal Behaviour*, **31**, 363–377.
Pusey, A. E. and C. Packer. 1986. Dispersal and philopatry, in Smuts, B., D. L. Cheney, R. M. Seyfarth, R. W. Wrangham, and T. T. Struhsaker, eds., *Primate societies*, University of Chicago Press, Chicago, Illinois, in press.
Ralls, K. and J. Ballou. 1982. Effects of inbreeding on infant mortality in captive primates. *International Journal of Primatology*, **3**, 491–505.
Robinson, J. G. 1979. Vocal regulation of use of space by groups of titi monkeys *Callicebus moloch*. *Behavioral Ecology and Sociobiology*, **5**, 1–15.
Rowell, T. E. and D. K. Olson. 1983. Alternative mechanisms of social organization in monkeys. *Behaviour*, **86**, 31–54.
Sade, D. S. 1968. Inhibition of son–mother mating among free-ranging rhesus monkeys. *Science and Psychoanalysis*, **12**, 18–38.
Sade, D. S. 1972. Sociometrics of *Macaca mulatta* I. Linkages and cliques in grooming matrices. *Folia Primatologica*, **18**, 196–223.
Scott, L. M. 1984. Reproductive behavior of adolescent female baboons (*Papio anubis*) in Kenya, in Small, M. F. ed., *Female primates: studies by women primatologists*. Alan R. Liss, New York, pp. 77–100.
Seyfarth, R. M. 1976. Social relationships among adult female baboons. *Animal Behaviour*, **24**, 917–938.
Seyfarth, R. M. 1978, Social relationships among adult male and female baboons. II. Behaviour throughout the female reproductive cycle. *Behaviour*, **64**, 227–247.
Seyfarth, R. M. 1986. Vocal communication and its relation to language, in Smuts, B., D. L. Cheney, R. M. Seyfarth, R. W. Wrangham and T. T. Struhsaker, eds., *Primate societies*, University of Chicago Press, Chicago, Illinois, in press.
Seyfarth, R. M., D. L. Cheney and P. Marler. 1980. Vervet monkey alarm calls: semantic communication in a free-ranging primate. *Animal Behaviour*, **28**, 1070–1094.
Shepher, J. 1971. Mate selection among second generation kibbutz adolescents and adults: incest avoidance and negative imprinting. *Archives of Sexual Behaviour*, **1**, 293–307.
Sherman, P. W. 1980. The limits of ground squirrel nepotism, in Barlow, G. W. and J. Silverberg, eds., *Sociobiology: beyond nature/nurture?* Westview Press, Boulder, Colorado, pp. 505–544.
Sigg, H., A. Stolba, J.-J. Abegglen and V. Dasser. 1982. Life history of hamadryas baboons: physical development, infant mortality, reproductive parameters and family relationships. *Primates*, **23**, 473–487.
Silk, J. B. 1978. Patterns of food sharing among mother and infant chimpanzees at Gombe National Park, Tanzania. *Folia Primatologica*, **29**, 129–141.
Silk, J. B. 1979. Feeding, foraging, and food sharing behavior in immature chimpanzees. *Folia Primatologica*, **31**, 123–142.

Silk, J. B. 1982. Altruism among female *Macaca radiata:* explanations and analysis of patterns of grooming and coalition formation. *Behaviour,* **79,** 162–188.
Silk, J. B. 1986. The evolution and adaptive consequences of altruism, competition, and aggression in primate groups, in Smuts, B., D. L. Cheney, R. M. Seyfarth, R. W. Wrangham and T. T. Struhsaker, eds., *Primate societies,* University of Chicago Press, Chicago, Illinois, in press.
Silk, J. B., A. Samuels and P. S. Rodman. 1981. The influence of kinship, rank, and sex on affiliation and aggression between adult female and immature bonnet macaques (*Macaca radiata*). *Behaviour,* **78,** 111–137.
Small, M. F. and D. G. Smith. 1981. Interactions with infants by full-siblings, paternal half-siblings, and nonrelatives in a captive group of rhesus macaques (*Macaca mulatta*). *American Journal of Primatology,* **1,** 91–94.
Smith, D. G. 1982. Inbreeding in three captive groups of rhesus monkeys. *American Journal of Physical Anthropology,* **58,** 447–451.
Smuts, B. 1985. *Sex and friendship in baboons.* Aldine, Hawthorne, New York.
Starin, E. D. 1978. Food transfer by wild titi monkeys. *Folia Primatologica,* **30,** 145–151.
Stein, D. 1981. *The nature and function of social interactions between infant and adult male yellow baboons* (Papio cynocephalus). Ph.D. thesis, University of Chicago.
Stein, D. 1984. Ontogeny of infant-adult male relations during the first year of life for yellow baboons, in Taub, D. M. ed., *Primate paternalism,* Holt, Rinehart & Winston, New York, pp. 213–243.
Stewart, K. S. and A. H. Harcourt. 1986. Gorillas: variation in female relationships, in Smuts, B., D. L. Cheney, R. M. Seyfarth, R. W. Wrangham and T. T. Struhsaker, eds., *Primate societies,* University of Chicago Press, Chicago, Illinois, in press.
Struhsaker, T. T. 1975. *The red colobus monkey.* University of Chicago Press, Chicago.
Struhsaker, T. T. and L. Leland. 1986. Colobines: male replacement and infanticide, in Smuts, B., D. L. Cheney, R. M. Seyfarth, R. W. Wrangham and T. T. Struhsaker, eds., *Primate societies,* University of Chicago Press, Chicago, Illinois, in press.
Strum, S. 1983. Use of females by male olive baboons (*Papio anubis*). *American Journal of Primatology,* **5,** 93–109.
Sugiyama, Y. 1976. Life history of male Japanese monkeys. *Advances in the Study of Behavior,* **7,** 255–284.
Takahata, Y. 1982. The socio-sexual behavior of Japanese monkeys. *Zeitschrift für Tierpsychologie,* **59,** 89–108.
Tilson, R. L. 1981. Family formation strategies of Kloss's gibbon. *Folia Primatologica,* **35,** 259–287.
van den Berghe, P. L. 1983. Human inbreeding avoidance: culture in nature. *Behavioral and Brain Sciences,* **61,** 91–123.
Van Lawick-Goodall, J. 1968. The behaviour of free-living chimpanzees in the Gombe Stream Reserve. *Animal Behaviour Monographs,* **1,** 165–311.
Wade, M. J. 1980. An experimental study of kin selection. *Evolution,* **34,** 844–855.
Wade, T. C. 1979. Inbreeding, kin selection, and primate social evolution. *Primates,* **20,** 355–370.
Walters, J. R. 1980. Intervention and the development of dominance relationships in female baboons. *Folia Primatologica,* **34,** 61–89.
Walters, J. R. 1981. Inferring kinship from behaviour: maternity determinations in yellow baboons. *Animal Behaviour,* **29,** 126–136.

Walters, J. R. 1986. The transition to adulthood, in Smuts, B., D. L. Cheney, R. M. Seyfarth, R. W. Wrangham and T. T. Struhsaker, eds., *Primate societies*, University of Chicago Press, Chicago, Illinois, in press.

Walters, J. R. and R. M. Seyfarth. 1986. Conflict and cooperation, in Smuts, B., D. L. Cheney, R. M. Seyfarth, R. W. Wrangham and T. T. Struhsaker, eds., *Primate societies*, University of Chicago Press, Chicago, Illinois, in press.

Weigel, R. M. 1981. The distribution of altruism among kin: a mathematical model. *American Naturalist*, **118**, 191–201.

Whitten, P. L. 1986. Infants and adult males, in Smuts, B., D. L. Cheney, R. M. Seyfarth, R. W. Wrangham and T. T. Struhsaker, eds., *Primate societies*, University of Chicago Press, Chicago, Illinois, in press.

Wrangham, R. W. 1980. An ecological model of female-bonded primate groups. *Behaviour*, **75**, 262–300.

Wrangham, R. W. 1982. Mutualism, kinship, and social evolution, in King's College Sociobiology Group, eds., *Current problems in sociobiology*. Cambridge University Press, Cambridge, England, pp. 269–289.

Wu, H. M., W. G. Holmes, S. R. Medina and G. P. Sackett. 1980. Kin preference in infant *Macaca nemestrina*. *Nature*, **285**, 225–227.

Yamada, M. 1963. A study of blood-relationship in the natural society of the Japanese macaque. *Primates*, **4**, 43–65.

Young, J. Z. 1978. *Programs of the Brain*. Oxford University Press, Oxford, England.

Kin Recognition in Animals
Edited by D. J. C. Fletcher and C. D. Michener
© 1987 John Wiley & Sons Ltd

CHAPTER 12

Kin Recognition in Humans

P. A. WELLS
Department of Psychology, University of London Goldsmiths' College, Lewisham Way, London SE14 6NW, England.

> all good kumrads you can tell
> by their altruistic smell
>
> e.e. cummings
> selected poems 1923–1958
> Faber & Faber, London 1969

INTRODUCTION

The study of kin recognition in humans presents an immediate problem. It is that humans possess the concept of kinship. This fact has had two unfortunate, and paradoxically almost contradictory, effects: it has led some people to believe that kin recognition cannot occur in species other than humans (because other species do not possess the concept of kinship); and it has led other people to believe that kin recognition does not occur in humans (because the concept of kinship renders it redundant).

Other chapters in this volume will have removed the first of these misapprehensions. By now, the reader is undoubtedly convinced that kin recognition has been demonstrated in a large number of non-human species. Admittedly, it would have seemed very improbable (in advance of the findings) that tadpoles prefer to associate with siblings rather than with non-siblings (Waldman and Adler, 1979; Blaustein and O'Hara, 1981) or that sweat bees can behave differentially towards conspecifics on the basis of relatedness (Greenberg, 1979; Chapter 7). Few would have guessed that the necessary discriminatory abilities existed in the species concerned, or that the degree of kinship could be so finely judged in the presumed absence of the concept of kin as defined in human terms.

The purpose of the present chapter, however, is to deal with the second issue: the possibility that humans are capable of kin recognition. Kinship as a human concept will be discussed only briefly, on the grounds that it is sufficiently dealt with in the anthropological literature. I would not wish to argue that it is irrelevant, or that it cannot explain many important aspects of nepotistic behavior; but rather that it probably represents just one of the many ways by which certain forms of kin recognition may be mediated.

The first thing to be said about human kin recognition is that for the most part the necessary experiments have simply not been done. By analogy with the literature on non-human species, we would require comparisons of kin recognition abilities in siblings or non-siblings reared together or apart; analyses of social preferences in offspring related only on the paternal side or even embryo transplants carried out to control for pre-natal maternal effects. Some of these procedures are clearly out of the question as far as humans are concerned. Even when they are not (as in the occasional instances of twins or siblings reared apart), the lack of experimental control may cause difficulties: for example, the individuals concerned often possess quite detailed information about each other. Does this mean, then, that kin recognition experiments in humans are in practice impossible?

It may be that the standard experimental techniques cannot be employed, or that direct answers cannot be given to some of the relevant questions. However, there are some issues that can be considered, even though the evidence is indirect.

One relevant approach is to show that humans possess the perceptual abilities which would enable them to discriminate between kin and non-kin. Another is to show that humans possess characteristics of the kind which would enable others to make the relevant discriminations: in other words, that they can be readily differentiated. A third is to demonstrate that there is some relationship between the phenotypic characteristics and the underlying genotype; that is, that the perceived characteristics are in fact facilitating the detection of relatedness. A fourth is to consider whether aspects of human social interactions reflect differing degrees of kinship. These would all be necessary, though not sufficient, conditions for demonstrating differential recognition. Finally (and perhaps more controversially) we may speculate on some of the possible evolutionary functions of kin recognition in humans. Each of these issues will be considered in turn.

DISCRIMINATORY ABILITIES

Even the most cursory glance at standard textbooks on human perception will reveal that humans are capable of making highly sophisticated perceptual judgments based on large quantities of complex information. This is most clearly apparent in the case of auditory and of visual perception, if only because of the amount of experimental evidence available. The important question for present purposes, though, is whether these abilities are indeed used in performing discriminations between individuals of differing degrees of relatedness. In considering the evidence, the reader should bear in mind that very few of the experiments cited below have been designed specifically to investigate kin recognition, and as a result it is not possible to answer some of the questions which one would like to ask in this context.

Let us first consider some examples from the field of visual perception. Curiously, there is a shortage of evidence concerning social recognition based on visual cues. This is unexpected both in view of the amount of evidence concerning visual perception in general and also because humans are generally considered to be a species highly dependent on visual information. It has been shown, however, that even very young infants are capable of performing certain complex social discriminations. For instance, Carpenter has conducted experiments in which both real and representational faces were shown to young infants (Carpenter et al., 1970; Carpenter, 1974). She found that from the age of about two weeks babies spent a significantly longer time in visual fixation on the mother's face than on a stranger's face or that of an inanimate dummy. Thus, the babies could not only discriminate between the various classes of stimuli, but apparently preferred to look at their own mothers.

In a comparable study, Bushnell (1982) presented older infants with slides of the mother's face over a series of trials and, using a habituation procedure, found that infants as young as four to seven weeks of age were able to discriminate between photographs of the mother's face and that of a female stranger. The faces used as stimuli in this experiment were standardized in various ways (for example, in one of the experimental conditions the adult females were photographed wearing a bathing cap in order to control for features such as the contour of the hairline). This enables us to speculate a little about the nature of the cues that the babies might have been using in making the discrimination. Surprisingly, and contrary to Bushnell's previous expectation, they did not seem to be attending in particular to the eyes, but rather to some aspect of the outer contours of the face. It does not follow, however, that this would necessarily be the most salient cue in discriminating between real rather than photographed faces.

None of these experiments would permit the conclusion that infants can discriminate between mother and stranger in the absence of previous exposure to the mother. It is entirely possible that the reported effects occur as a function of familiarity with maternal characteristics. Even so, it is of some interest that infants can perform such complex social discriminations so early in life. It is also possible that if kin recognition operates at all in humans, then familiarity is the very means by which it is achieved.

A similar argument applies to an experiment which suggests that mothers can identify their own infants soon after birth on the basis of the rather limited visual cues provided by photographs (Porter, Cernoch and Balogh, 1984). However, since the authors also found that adult subjects could match photographs of mothers and newborn babies at better than chance level without having seen either of them before, it can reasonably be assumed that there are some physical resemblances between related individuals which enable these judgments to be made.

We now turn to some examples from the area of auditory perception. One group of studies concerns the ability of mothers to identify the cries of their own infants. For instance, Formby (1967) reported that twelve out of 23 women were able to select the tape-recorded cry of their own infant from among five cry samples within 48 h after delivery (no statistical analyses were presented). In a similar study, Murry, Hollien and Müller (1975), using three different types of cry and a total of 96 cry samples, found that mothers could recognize the cries of their own infants with few instances of confusion.

On the whole, then, it seems that some mothers can identify the cries of their own infants even though the amount of previous contact with their offspring may have been limited. But as Murry, Hollien and Müller (1975) point out, we do not know on what features of the cry these perceptual judgments are based, nor do we know what part might be played by familiarity with a particular infant's vocalizations.

Can babies, too, make auditory distinctions between people? An interesting answer to this question is provided in an experiment by Mills and Melhuish (1974). They used an operant procedure whereby young infants were able to elicit the presentation of human voices by sucking on a teat connected to a pressure transducer. By sucking at a given pressure, each baby could hear the mother's voice or a female stranger's voice equated for loudness, but in neither case was the speaker visible. Mills and Melhuish were able to show not only that infants as young as three weeks of age could learn this task, but that they would emit more sucking responses in order to hear the sound of the mother's voice than to hear the voice of a female stranger. A similar technique has been employed in a later experiment by DeCasper and Fifer (1980), who found comparable abilities to be present in babies only 24 h after birth. Again, the possible role of previous familiarity with the mother's voice can neither be established nor disproved in this type of experimental design; neither can the cues upon which the infants were basing their apparent preference be established. There is evidence from other studies, however, that young babies respond more to patterned sounds (such as a bell or speech) than to unpatterned sounds such as a pure tone (Levarie and Rudolph, 1978; Lichtig and Wells, 1980). The extent to which an auditory stimulus varies over time might therefore provide one possible basis for discrimination. Another possible candidate is the fundamental frequency of the voice, although the evidence for this is not conclusive (Miller, Younger and Morse, 1982).

It is, perhaps, surprising that there are several studies relevant to the role of olfaction in human kin recognition, given the common (although probably erroneous) assumption that olfactory cues have little or no part to play in human social interactions. In one such experiment (MacFarlane, 1975), newborn babies were presented with breast pads from their own mother and from an unfamiliar mother. The amount of time spent by each baby with the head turned towards the breast pad was used as a measure of preference. Babies as young as six days of age were found to prefer their own mother's breast pad. As in the case of the auditory and visual experiments, more information on the nature of the relevant cues is needed, particularly as Russell (1976) has speculated that identification of the mother may be mediated by the presence of odors placed on her by the infant during earlier contacts; in other words, the infant may be responding to his or her own olfactory cues.

Cernoch and Porter (1986) have shown that breast-feeding infants can also distinguish their mothers' axillary odors from those of either non-parturient females or lactating females unfamiliar to the infant. They were, however, unable to distinguish between paternal axillary odors and those of an unfamiliar male. The authors argue that the process of breast-feeding may provide access to maternal odors through skin contact; this exposure to the maternal olfactory signature is less likely to occur in the course of bottle-feeding (and, indeed, bottle-feeding babies did not appear to recognize maternal odors). Similarly, the infant is less likely to have been exposed to paternal olfactory cues through prolonged skin contact.

Mothers are also able to identify their own infants on the basis of olfactory cues alone. Russell, Mendelson and Peeke (1982) tested a group of mothers who were blindfolded and asked to distinguish by smell between their own infant and two unfamiliar babies. They were able to make the discrimination as early as six hours post partum, after a single exposure to the baby lasting only half an hour. Fathers, however, were not able to perform the same discrimination even though they too had been given the opportunity to interact with their own infants; this seems to suggest that familiarity alone is not a sufficient condition for recognition. However, it cannot necessarily be argued that mothers had access to phenotypic cues denied to the fathers (namely, their own odors), since Hold and Schleidt (1977) have shown that husbands can recognize their wives' odors. If mothers identify their infants by some process of phenotype matching, then fathers ought in principle to be able to perform the same task, since they too have been exposed to the maternal phenotype.

It has indeed been reported that people are capable of discriminating between their own odors and those of other people (Hold and Schleidt, 1977; Schleidt, 1980). In one sense, this fundamental distinction between 'self' and 'other' can be regarded as the basic paradigm of kin recognition.

Some of the strongest evidence for human kin recognition by olfaction is provided in experiments by Porter and his associates. For example, mothers can distinguish by smell between the garments of their own infants and those of unfamiliar infants within a few days after birth. This ability appears to be present even though the period of contact between mother and baby may have been very limited, as in the case of babies delivered by Caesarian section (Porter, Cernoch and McLaughlin, 1983).

These findings leave open the possibility that recognition of related individuals is achieved by some comparison between the characteristics of 'self' and 'other'. Experiments by Porter and Moore (1981), however, show that this is not necessarily the whole story. In one study, they showed that individual children could be distinguished by their mothers and siblings from unfamiliar agemates through olfactory cues. This could indeed be accounted for in terms of phenotype matching. In a second experiment, however, parents correctly distinguished between the odors of otherwise identical shirts worn by two of their own children. This latter finding is a little more difficult to account for (unless, of course, we assume that one child has greater similarity to the maternal phenotype than does the other), and it also raises some interesting questions concerning the nature of the differences between the children in each family. The siblings were presumably comparable in terms of dietary and other environmental factors (such as household odors), yet they were still discriminable. Porter, Cernoch and Balogh (1985) have also reported that, although the odors of mothers and infants can be matched by adult subjects, the odors of husbands and wives cannot. This is in spite of the fact that environmental cues such as diet are probably quite similar for married couples living in the same household. As Porter and Moore (1981) suggest, it is possible that individual differences in olfactory cues stem from underlying genetic factors. This issue will be discussed more fully below.

INDIVIDUAL DIFFERENCES

Given that some kinds of discrimination between individuals are possible, it is useful to consider in more detail the nature of the distinguishing characteristics. For instance, we have seen from the evidence reviewed above that mothers can distinguish between their offspring on the basis of olfactory cues. What kind of olfactory information are they employing; is it the kind of information which one would expect to differentiate between relatives and non-relatives, or is it for example, attributable to factors such as dietary preferences? If the latter, is there systematic variation which correlates with the degree of relatedness between individuals?

An early study by Kalmus (1955) provides an interesting clue. In a series of small-scale experiments with dogs trained either to follow human tracks or to retrieve objects scented by humans, he found that the animals could distinguish reliably between the sweat odors of different individuals, including the members of a family. However, the dogs were apparently unable to distinguish between identical twins in these tests when the odors were presented in succession (although simultaneous presentation brought about some improvement in performance). Thus it appears that the more closely related individuals were more similar in terms of their olfactory cues. One cannot argue from this, of course, that the human nose would necessarily arrive at the same decision; all we may conclude is that when dogs are required to make the judgment they do so in accordance with the underlying genetic relationship.

There is, in fact, some evidence to suggest that the differences between individuals which are detectable by humans operate in the same manner. Wallace (1977) reported that the ability of women to discriminate between two unfamiliar females on the basis of hand odor was improved when the individuals were not genetically related. Again, identical twins were difficult to distinguish; unrelated individuals, even when on the same diet, were more easily discriminable.

There is certainly enough potential variation in olfactory cues to provide a unique biochemical 'signature' for each individual: given the number of fatty acids that the skin can make, and the number of possible combinations of factors which affect their composition, '... it is extremely unlikely that any two individuals will make exactly the same substances in exactly the same proportions' (Nicolaides, 1974). A similar situation probably obtains in the case of auditory signals. Beecher (1982) has argued that, where individuals associate in large groups, there may be considerable selection pressures to develop mechanisms that can generate a large number of distinctive signatures. His analysis of the calls of bank swallows supports the argument that distinctive signatures are indeed found under natural conditions. The same phenomenon of distinctiveness is also present in human speech, as the literature on voice recognition makes clear; what is not so clear is its functional significance or its role (if any) in human kin recognition.

PHENOTYPES AND RELATEDNESS

Humans possess a sufficient number of complex characteristics to render each individual discriminable, even when the population size is very large. They also possess the perceptual abilities which would be needed to enable them to make the discrimination. However, the mere existence of these does not necessarily imply that phenotypic characteristics are used in practice as an index of relatedness, or that kin are in practice distinguished from non-kin by these rather than by other means.

Some index of the actual degree of relatedness between individuals can be obtained by the use of blood group markers. This technique has been employed by Pakstis *et al.* (1972) and Loehlin, Willerman and Vandenberg (1974) in investigating similarities between siblings, including dizygotic (DZ) co-twins. The interesting question for present purposes, of course, is whether the degree of genetic similarity thus measured correlates with overt phenotypic characteristics.

The question is answered, at least in part, in a study by Carter-Saltzman and Scarr-Salapatek (1975). Citing previous empirical findings, they argue that blood group markers provide a useful index of the degree of genetic similarity between DZ twins or, indeed, between pairs of siblings. Their own experimental hypothesis was that DZ co-twins who have fewer chromosomal differences (as indexed by blood group loci) would also have more similar physical and behavioral characteristics.

The behavioral characteristics which they measured on 145 pairs of DZ twins included five cognitive tests and two tests of personality. Seven physical measures were also obtained by standard anthropometric techniques, together with skin reflectance and an assessment of skeletal age. Similarity between the co-twins was calculated by using twelve blood-group markers, by ratings of photographs of the twins performed by independent judges, and by a short series of questions given to the twins on, for example, the extent to which they were generally mistaken for each other. The findings may be summarized as follows; there was no evidence to suggest that any of the behavioral characteristics were consistently related to blood group differences. The morphological measures, however, were all positively correlated with the blood group measures; three of them were found to be statistically significant. Ratings of physical appearance, and the twins' perceived differences, also correlated significantly with the blood group differences. Interestingly, those DZ twins who mistakenly believed that they were monozygotic (MZ) had fewer blood group differences than did those whose own diagnosis of their zygosity was correct.

Some objections may be raised to the use of rough-and-ready ratings of similarity in the estimate of zygosity. However, there is evidence from other twin studies that judgments of this kind are comparable with more formal assessments of zygosity in their accuracy in distinguishing between MZ and DZ twins. One of the more convincing studies is that by Dumont-Driscoll and Rose (1984). They found that ratings of perceived similarity and behavioral resemblance of like-sex DZ co-twins were related to maximum likelihood estimates of their genetic similarity (as measured by extensive genotyping of the twins and their parents, thus permitting estimates of genes identical by descent).

There are other ways in which phenotypic cues can give some indication of underlying genetic characteristics. One of the most striking examples concerns the major histocompatibility complex (MHC); it has been shown that mating preferences in mice are affected by the presence or absence of certain MHC haplotypes, with the relevant cues apparently transmitted via urinary odors (Yamazaki et al., 1976; Boyse, Beauchamp and Yamazaki, 1983; Chapters 9, 10). One possible advantage of such mating preferences is the maintenance of heterozygosity of genes in the vicinity of, say, the *H-2* region (Yamazaki et al., 1976). As far as humans are concerned, the evidence is less clear-cut, but it has been suggested that women who share large numbers of histocompatibility antigens with their husbands may be more susceptible to high rates of spontaneous abortion (Beer et al., 1981; Taylor and Faulk, 1981). As Hamilton points out (Chapter 13), this represents a complete reversal of what is normally understood by histocompatibility. Nonetheless, there would presumably be some advantage in being able to detect in advance the presence of such antigens. One suggestion is that, in species in which internal fertilization occurs, there may be mechanisms by which compatibility with self is tested by the female immune response before any sperms even reach the egg (Hartung, 1981). Human mate choice mechanisms of the kind reported in laboratory mice would provide an even more efficient way of screening out unprofitable matings.

It is often assumed that behavioral gene detection is inherently improbable in the absence of linkage or pleiotropy (see, for example, Dawkins, 1982). However, the possible importance of these phenomena may have been underrated. If linkage and pleiotropy of behavioral genes are in fact widespread, then gene detection could perhaps occur on an inferential basis. Consider the example of the MHC: in the mouse experiments, it is not argued that mating choices are based on direct inspection of histocompatibility antigens, but the outcome of mate choice based on urinary odors is that in effect certain haplotype preferences are manifested. Interestingly, it has now been shown that there are large numbers of other biological features affected by *H-2* haplotypes; examples include body size, thymus size, the size of T and B cell populations, and '... various aspects of the physiology of steroid hormones whose derivatives are notable odorants' (Boyse, Beauchamp and Yamazaki, 1983). In other words, there appear to be large numbers of potential cues available to the would-be-detector.

There is also some suggestion that, in humans, variation in intellectual abilities and personality may be associated with certain genetic markers. A small-scale study by Gentry, Polzine and Wakefield (1985) has indicated, for example, that scores on a verbal reasoning test are positively correlated with right eye dominance, ability to curl one's tongue, light eye color, inability to taste PTC and with vision correction. There are several reasons for arguing that undue weight should not be attached to these findings, but once again the possibility is raised that behavioral and other phenotypic cues could be used as an index of genotype.

Markers giving information about genetic similarities and differences can also be acquired through learning. Features of bird song dialects may be associated with genetic differences between populations, although geographic constraints also play some part (see Baker and Cunningham, 1985, for a review of this area). Comparable findings have also been reported with respect to humans. For example, Irwin (1985) computed a coefficient of dialect differentiation using linguistic data from six Eskimo tribes living on the west coast of Hudson's Bay. He found that dialect differences between tribes corresponded with genetic differences between tribes, as calculated from genealogies. Thus, acquired knowledge of acquired phenotypic characteristics could in principle be used to give some estimate of relatedness. Irwin also reports that cultural features such as style and ornamentation of dress are used by the Eskimo as indicators of family and extended family membership (a process known to anthropologists as 'badging'); again, these were found to have considerable power to predict the associated biological relationships, even given the small sample employed in this study.

What do these findings tell us? They suggest that appearances do count in the assessment of degree of relatedness, but that not all phenotypic characteristics provide equally useful information. (It will be recalled that, in the study by Carter-Saltzman and Scarr-Salapatek, 1975, measures of cognitive performance and personality, for example, had little predictive value with respect to blood group measures of relatedness, whereas measures of physical appearance were much more informative). With the benefit of hindsight, it may seem obvious that physical characteristics provide a better estimate of the underlying genotype. What is perhaps not so obvious, at least *a priori*, is that they can apparently be used in practice as a rough estimate of relatedness. It may also be worth observing in passing that structural features are presumably less subject to variation for environmental reasons (and, perhaps, less easy to fake) than are relatively volatile features such as personality characteristics. 'Badging' is the most easily modifiable of all; and it provides the basis for the fable of the wolf in sheep's clothing.

SOCIAL INTERACTIONS BASED ON RELATEDNESS

In the previous section, it was argued that it might be possible to estimate the degree of relatedness between individuals by employing phenotypic cues. However, like the other feats of discrimination that were reviewed, this ability is of little more than academic interest (for present purposes) unless it can be shown that interactions between individuals reflect the genetic relationships between them. Indeed, it is perhaps more important to demonstrate this than to consider the perceptual mechanisms that might be involved; effectively, individuals only have to behave as if they were able to discriminate between kin and non-kin. There is certainly no need to postulate any conscious process of discrimination. It may not even be necessary to assume that perceptual discriminations are made at all, as we shall see when discussing some of the mechanisms underlying kin recognition.

The concept of kinship amongst humans is so fundamental as to be virtually universal. This does not necessarily depend on genetically based kin recognition abilities; even if these abilities should be lacking, most cultures have well-documented genealogical systems which ensure that information concerning degrees of kinship is available to the community. It may appear obvious that human social interactions will reflect the degree of relatedness between the individuals concerned (the phenomenon is usually described as nepotism). Yet in practice it is surprisingly difficult to find data which are both relevant and quantifiable.

Some particularly interesting exceptions to this generalization are provided by the work of anthropologists who are not unsympathetic towards insights from other disciplines. Studies such as those by Chagnon (1981) and Chagnon and Bugos (1979) have demonstrated that genealogical relationships have an important part to play in determining social interactions among the Yanomamö Indians. In the first of these, Chagnon has shown that '... no matter how individuals classify people into kinship categories they favor more closely-related genealogical relatives over less-closely related relatives or non-relatives by remaining with them when a village fissions'. In the second, Chagnon and Bugos reported that the help given to other individuals during ax fights depended to a large extent upon relatedness between helper and recipient (although alliances and other affinities also played a part).

The point is even more clearly illustrated in a study by Hames (1979). He collected data on the Ye'kwana Indians of Venezuela with the specific intention of testing the hypothesis that social interactions are a function of coefficients of relatedness. Hames defines the concept of interaction as 'two or more individuals engaged in a coordinated activity'. The behaviors thus included obviously cover a very wide range; the most commonly observed in practice were the sharing of meals, conversation, play and various economic activities such as hunting (in the case of males) and gardening (in the case of females).

Coefficients of relatedness were calculated by obtaining accounts of genealogies from individual members of the village, with as much crosschecking as was necessary to achieve consistency. This resulted in an 99×88 matrix giving the relatedness of each member of the village to each other individual.

The findings were very much in accordance with the hypothesis that relatedness predicts the extent of social interaction. The correlation between the two was highly significant and, coincidentally, surprisingly close to the value reported by Greenberg (1979) in the study on sweat bees mentioned previously (see Chapter 7). Thus, in this society at least, interaction increases with the degree of relatedness.

The sceptical reader will immediately raise a fairly obvious objection to this conclusion, and one that Hames himself has considered. It is that interaction is a function of residential propinquity: those who are closely related tend to live in the same households and would tend to have more frequent interactions for this reason alone. His answer to this is two-fold. In the first place, it could equally well be argued that people who live in close proximity do so because they are related: residential propinquity and social interaction are both functions of relatedness. In the second place, the data give some information on this issue, in that there were many instances of individuals of differing degrees of relatedness living in the same household. Even when the effects of proximity are thus taken into account, there is still a systematic relationship between interaction and relatedness.

Yet another relevant point emerges from this interesting study. In many cultures, kinship terminology does not coincide in all respects with actual degrees of relatedness. This fact has led authors such as Sahlins (1976) to argue that kin terminology and cultural practices associated with such terminology are better predictors of behavior than is 'true' or genetic relatedness. According to Hames, the Ye'kwana have a system of kin terminology which fails to distinguish between degrees of relatedness in several rather striking respects: '... most individuals related at the 0.5, 0.375, 0.25 and 0.125 levels are called "brother" (*udui*) or "sister" (*yaya*). Therefore, it is evident that if we used Ye'kwana kinship terminology we could not predict interaction within these categories'. Using knowledge of 'true' kinship, though, it is evident that we can.

It is always possible that linguistic categories are understood by the users to give only a partial account of true relationships. In Western societies, for example, it is not unusual for the title of 'aunt' to be accorded to females who have no biological relationship with the individual in question; but everybody understands the difference between them and 'true' aunts. Even so, there seems to be little doubt that human social interactions are indeed affected by relatedness.

There is, however, no direct evidence here which would enable us to give an explanation in terms of function, or to look more closely at the nature of the various behaviors involved in these interactions. In order to investigate differences in the quality of behavior involving people of varying degrees of relatedness, let us consider a very different study by Segal (1984). She investigated the cooperative and competitive behaviors of 47 pairs of twins who were aged between six and eleven and comparable in terms of IQ. The twin pairs were given the task of assembling a simple puzzle within a three-minute time limit, and were asked to 'complete the puzzle together'. Their behavior during completion of the joint project was videotaped and categorized by a scorer who was unaware of the purpose of the study. Segal found that 94% of the MZ twin pairs completed the puzzle successfully, whereas only 46% of the DZ pairs did so. In a subsequent task, the participants had the opportunity of showing reciprocal helping behavior (although this was not essential to the task's completion). MZ twins displayed greater cooperation and willingness to work for the benefit of the other partner than did the DZ twins. Again, it is not unreasonable to conclude that interactions between individuals are to some extent a function of the genetic relationships between them; but this time, we may also conclude that greater altruism is shown between those whose degree of relatedness is greater.

RECOGNITION MECHANISMS

Assuming, for the sake of argument, that kin recognition in humans is at least a strong possibility, we now turn to consideration of the mechanisms by which it may be achieved. These have been discussed in general terms by various authors, including Alexander (1979), Dawkins (1982), Blaustein (1983) and Lacy and Sherman (1983). There are two main classes of possible mechanisms involved. In the first case, the definition of relatedness depends on probabilistic cues such as location or prior association. In the second, it depends upon the direct identification of individuals or of categories of kin (such as siblings). Lacy and Sherman (1983) argue that direct identification may be necessary for the occurrence of kin altruism or inbreeding avoidance in circumstances when location no longer serves as a reliable cue (for example, when related individuals disperse or when individuals from different kin groups are intermingled).

On balance, it seems more likely that humans identify kin on the basis of previous familiarity (achieved either through proximity, in the sense of having been reared together, or by cultural means such as the learning of genealogies). However, some of the evidence cited above suggests that the phenomenon known variously as phenotype comparison (Alexander, 1979), phenotype matching (Holmes and Sherman, 1982) or the 'armpit effect' (Dawkins, 1982) might also have some part to play. Lacy and Sherman (1983) argue that phenotype matching is not commonly observed because (according to their model) it requires large numbers of traits. However, if it does occur, greater accuracy can be achieved where there are large numbers of continuously varying traits together with polygenic inheritance. Both these conditions are likely to be fulfilled in the case of humans.

A further possible mechanism for kin recognition—namely, the existence of recognition alleles or 'green-beard' genes (Hamilton, 1964; Dawkins, 1976, 1982; Chapters 2, 3 and 4)—seems intuitively improbable. Again, however, some of the evidence outlined previously suggests that gene detection by behavioral cues is not impossible. The best strategy in the absence of direct evidence is to leave the question open.

FUNCTION AND EVOLUTIONARY SIGNIFICANCE

On the basis of evidence drawn from non-human species, I should predict a number of possible functions for kin recognition in humans. Perhaps the most obvious of these is the avoidance of inbreeding (Bateson, 1983; Partridge, 1983). Indeed, the mechanism which probably subserves the development of kin recognition in humans—namely, prior exposure for some minimal period early in life—seems to be extraordinarily effective in suppressing subsequent tendencies to mate with familiar individuals. Shepher (1983) surveys a considerable body of evidence relating to kibbutz children reared together, and states that there were 'no marriages between people who had been continuously reared together for their first six years'. Similar difficulties arise in 'sim-pua' marriages among the Hokkien-speaking Chinese inhabitants of North Taiwan. (These are marriages in which the future bride is adopted by the bridegroom's family, usually before she has reached the age of three, and often when she is less than one year old.) It is reported that parental coercion is often needed to overcome the couple's reluctance to occupy the same room, and that the incidence of adultery is higher, and fertility rates lower, than in 'ordinary' marriages (Wolf and Huang, 1980).

The immunological studies cited previously also lend support to the motion that recognition systems (whatever the mechanism) are designed in part to prevent mating between those who may be closely related. Beer *et al.* (1981), for example, conclude that '... genetic homozygosity with regard to MHC antigens between female and male is associated with reproductive inefficiency'. However, the relationship is not an absolute one, as the authors make clear.

Bateson's theory of mate choice, based principally upon his work with Japanese quail, suggests that kin recognition functions to enable individuals to select a mate of some intermediate degree of relatedness (Bateson, 1983). There are some difficulties with this proposal (not the least of these is that the quail apparently showed social preferences for first cousins in a relatively inbred laboratory population; whether such preferences are either generalizable or adaptive is open to question). Nevertheless, the theory gives reasons for supposing that genetic similarities and differences are attended to in the choice of a mate. Unfortunately, there is little information concerning human mate choice in this respect. First cousin marriages are actively encouraged in approximately half the societies for which data are available, and discouraged in the remainder (Alexander, 1981). It could be argued that in communities where the degree of outbreeding is so great that matings are effectively random with respect to genotype, there will be decreasing pressure to detect differing degrees of relatedness. Alternatively, potential mates should seek a partner who resembles them in certain respects in order to avoid the disruption of co-adapted gene complexes (Partridge, 1983). Certainly, assortative mating may be based on a variety of physical and psychological characteristics (see, for example, the review by Thiessen and Gregg, 1980). Russell, Rushton and Wells (1984) have argued that this represents a process comparable to that found in some aspects of kin recognition, namely, the detection of genetic similarity between self and other.

It is conceivable that the ability to identify degree of relatedness may be favoured under conditions where paternity is uncertain (Alexander, 1974). Direct evidence on this point appears to be lacking. In common with studies on non-human species, however, it appears there are many different situations in which human offspring whose paternity is in doubt, or who are known to be genetically dissimilar to a parent, may be at risk. For instance, a disproportionate number of battered babies are victims of assaults by step-parents (Lightcap, Kurland and Burgess, 1982). More resources may be invested in a sister's than in a wife's children, or, in extreme cases, adultery may be regarded as grounds for infanticide, especially if the resulting child's physical appearance indicates a 'non-tribal sire' (Daly and Wilson, 1981). It is perhaps not surprising, then, that interested parties (such as the mother and other family members) take great pains to point out phenotypic similarities between newborn offspring and the putative father (Daly and Wilson, 1982). This would presumably serve the dual function of drawing the father's attention to such similarities as might exist, if his own recognition abilities were deficient, and of persuading him of the existence of similarities which in some cases might not in fact be present.

The evidence outlined above gives some indication of potential, if not actual, kin recognition abilities in humans. Yet if such abilities exist, they pose some additional problems. For example, consider the argument that we are uniquely identifiable and distinguishable from each other. Why should this be so? We cannot argue that this individuality has necessarily developed in order to make kin recognition possible (and in fact a moment's thought is sufficient to reveal that uniqueness probably makes kin recognition more difficult, not easier). For instance, we differ from each other in the details of our fingerprints. It happens that police forces have been able to turn this idiosyncrasy to advantage in the detection of criminals; but to argue that this is the evolutionary function of individuality of fingerprints would be to strain the belief of even the most teleologically minded. Similarly, although it has been reported that unique DNA 'fingerprints' can be discerned from blood samples, and that the degree of polymorphism is so great that even closely related individuals can be thus distinguished (Jeffreys, Wilson and Thein, 1985), it is extremely unlikely that these distinctive features have a major role to play in human kin recognition. The characteristics which make kin recognition possible may well be incidental byproducts of some other functional system; indeed, there are currently other, and much more powerful, theories to account for biological uniqueness (see, for example, Hamilton, 1982).

Kin recognition also has its darker side. Although no one would argue that one of the functions of kin recognition is to make infanticide possible, the ability to discriminate between kin and non-kin, and to behave preferentially towards kin, inevitably results in less favorable treatment towards non-kin. Much of this is already implicit in the literature on non-human species, but few have cared to spell out the implications for humans. One controversial exception is Durham (1976), who has considered some of the evolutionary aspects of human aggresive behavior. He speculates that the ability to recognize—and therefore avoid harming—kin and friends may be the crucial factor permitting the evolution of deadly conflict.

The significance of kin recognition cannot be assessed without some understanding of the circumstances under which it evolved. As far as humans are concerned, we are, as is so often the case, confined to speculating about our past evolutionary history. Our best guess is that early humans might have lived in relatively small groups made up of about a dozen to a hundred or so individuals, in which biological relatives would have been fairly closely associated and the average coefficient of relatedness would have been relatively high (say approximately 0.3 to 0.4). Multiply determined relationships would probably also have been common. Actions favoring members of the immediate community would, on the whole, be of benefit to related individuals (although the degree of relatedness would clearly be of some importance). The same cannot be said of many present-day cultures. What are the possible implications of this?

One implication is that kin selection may play a less important role in human affairs than was once the case (see Chapter 13). Another is that indiscriminate altruism, which would on the whole have acted in favor of relatives in the past, may now be less appropriate; perhaps we should even expect increasing pressures towards the development of some system of gene recognition. A further possibility is that, as far as humans are concerned, the balance may have shifted towards reciprocal altruism. This seems especially likely given our excellent capacity for recall of the details of past social interactions.

It is clearly a difficult task to devise appropriate kin recognition experiments where humans are concerned. One ingenious but almost certainly unworkable suggestion is that surnames could be employed as an index of distant relatedness in humans, presumably on the grounds that those who shared a given surname would have a higher probability of being related than those who did not (Cunningham, 1981). If we accept this argument, then in principle kin recognition experiments could be carried out on related individuals who had not previously encountered each other. A major difficulty, of course, is that surnames do not necessarily indicate individuals who share genes by common ancestry. This is most clearly illustrated in the case of surnames derived from occupations (such as Smith, which also happens to be one of the most common of English surnames). A second difficulty is that distant relationships are unlikely to provide convincing evidence concerning kin recognition. A third is that, although some modified form of the technique could perhaps be employed, the attempt would presumably be doomed to failure if kin recognition in humans does in fact depend on proximal mechanisms.

Another possibility would be to consider the concept of relatedness as a kind of perceptual dimension. In this sense, it would be possible to test for generalization in the same sort of way as one would test for generalization of a particular hue or brightness. Let us suppose, for example, that the acoustic characteristics of infants' cries reflect the degree of relatedness between them (although in practice this is rather unlikely). Adults familiar with the cry of a particular infant could then be asked to judge the relatedness of, say, that infant's MZ or DZ twin, full- or half-sibling, and first or second cousin on the basis of the cry sounds. (Experts in experimental design will, of course, immediately raise the difficulty of controlling adequately for age effects; but since the experiment is a hypothetical one, this objection may safely be ignored.) If relatedness does indeed serve as a dimension, and is discernible through behavioral cues, we would expect the rank ordering of the cries to be neatly correlated with the coefficients of relatedness to the 'criterion' infant.

Similar experiments could be performed using olfactory cues (Dawkins, 1982) or visual information. However, in order to deal with the problem that humans are in most cases already familiar with their own relatives, all these experiments require that relatedness should be assessed by an individual who is not a member of the kinship group in question, or who may even be of a different species. The ability to perceive degrees of relatedness is thus divorced to some extent from its possible functional aspects.

CONCLUSIONS

As we have seen, the evidence for kin recognition in humans is for the most part inferential. Nevertheless, it can be argued that the necessary preconditions are fulfilled. Humans both possess and use the necessary discriminatory abilities; and they also possess a sufficient range of variability in phenotypic cues upon which this discrimination can be exercised. These phenotypic cues can in principle be used as an index of relatedness, and relatedness is indeed one of the major determinants of human social interactions.

These conditions may be necessary for the operation of kin recognition, but they are not sufficient. In fact, it is unlikely that kin recognition occurs in humans in the sense in which it has been demonstrated in some other species:

that is, the ability to identify relatives in the absence of proximal mechanisms such as prior exposure to nest or litter mates. It is difficult to envisage an aspect of our evolutionary past which would have provided the impetus for a genetically based kin recognition system. Accidental separations and reunions of long lost (but instantly recognized) kin are, on the whole, the stuff of fairy tale and myth rather than of evolutionary reality. Let us, however, make the not unreasonable assumption that the theory of kin selection is applicable to the behavior of humans as well as to the behavior of other species. How, then, might we expect to see kin recognition implemented? One answer is that this selective pressure might well have favored the brain mechanisms underlying the cultural evolution of the concept of kinship.

In common with most other species, each individual human is unique and biologically distinct from the others. (MZ or identical twins, who come closest to violating this generalization, are as a result the object of particular interest, fascination and occasionally superstition.) We differ in terms of morphology, in the visual signals which we transmit to each other, in the acoustical signals which we provide, and in our biochemical structure. What is more, we possess from a very early age the abilities which allow us to discern these differences; and in practice, we pay great attention to them. We actively seek out people whom we perceive as being similar to ourselves, and interact preferentially with them when we find them. In many cases, these people are our biological relatives. In others, as a result of processes such as exposure learning, unrelated people are, in effect, treated as 'honorary kin'. To argue from this that kin recognition does not operate in humans would, in one sense, be correct. On the other hand, it could equally well be argued that kin play so important a role in human social behavior that we tend to 'invent' them even when they do not exist.

ACKNOWLEDGMENTS

Much of the work for this chapter was carried out during a period of sabbatical leave at the University of Oxford. I am grateful to the Department of Zoology, and in particular to the Animal Behaviour Research Group, for the hospitality and facilities provided. I am also grateful to the following for helpful comments and discussion: R. Dawkins, T. C. Guilford, W. D. Hamilton, C. J. Nicol, R. J. H. Russell and A. L. J. Wells.

LITERATURE CITED

Alexander, R. D. 1974. The evolution of social behavior. *Annual Review of Ecology and Systematics,* **5**, 325–383.

Alexander, R. D. 1979. *Darwinism and human affairs.* University of Washington Press, Seattle, Washington.

Alexander, R. D. 1981. Evolution, culture and human behavior: some general considerations, in Alexander, R. D. and D. W. Tinkle, eds., *Natural selection and social behavior.* Chiron Press, New York, pp. 509–520.
Baker, M. C. and M. A. Cunningham. 1985. The biology of bird song dialects. *Behavioral and Brain Sciences* (in press).
Bateson, P. 1983. Optimal outbreeding, in Bateson, P., ed., *Mate choice,* Cambridge University Press, Cambridge, England, pp. 257–277.
Beecher, M. D. 1982. Signature systems and kin recognition. *American Zoologist,* **22,** 477–490.
Beer, A. E., M. S. Quebbeman, J. W. T. Ayers and R. F. Haines, 1981. Major histocompatibility complex antigens, maternal and paternal immune responses, and chronic habitual abortions in humans. *American Journal of Obstetrics and Gynecology,* **141,** 987–999.
Blaustein, A. R. 1983. Kin recognition mechanisms: phenotypic matching or recognition alleles? *American Naturalist,* **121,** 749–754.
Blaustein, A. R. and R. K. O'Hara. 1981. Genetic control for sibling recognition? *Nature,* **290,** 246–248.
Boyse, E. A., G. K. Beauchamp and K. Yamazaki. 1983. The sensory perception of genotypic polymorphism of the major histocompatibility complex and other genes: some physiological and phylogenetic implications. *Human Immunology,* **6,** 177–183.
Bushnell, I. R. W. 1982. Discrimination of faces by young infants. *Journal of Experimental Child Psychology,* **33,** 298–308.
Carpenter, G. C. 1974. Visual regard of moving and stationary faces in early infancy. *Merrill-Palmer Quarterly,* **20,** 181–194.
Carpenter, G. C., J. J. Tecce, G. Stechler and S. Friedman. 1970. Differential visual behavior to human and humanoid faces in early infancy. *Merrill-Palmer Quarterly,* **16,** 91–108.
Carter-Saltzman, L. and S. Scarr-Salapatek. 1975. Blood group, behavioral and morphological differences among dizygotic twins. *Social Biology,* **22,** 372–374.
Cernoch, J. M. and R. H. Porter. 1986. Recognition of maternal axillary odors by infants. *Child Development,* in press.
Chagnon, N. A. 1981. Terminological kinship, genealogical relatedness and village fissioning among the Yanomamö Indians, in Alexander, R. D. and D. W. Tinkle, eds., *Natural selection and social behavior.* Chiron Press, New York, pp. 490–508.
Chagnon, N. A. and P. E. Bugos. 1979. Kin selection and conflict: an analysis of a Yanomamö ax fight, in Chagnon, N. A. and W. Irons, eds., *Evolutionary biology and social behavior: an anthropological perspective.* Duxbury Press, North Scituate, Massachusetts, pp. 213–238.
Cunningham, M. R. 1981. Sociobiology as a supplementary paradigm for social psychological research, in Wheeler, L., ed., *Review of personality and social psychology, Vol. 2.* Sage Publications, Beverley Hills, California, pp. 69–106.
Daly, M. and M. I. Wilson. 1981. Child maltreatment from a sociobiological perspective. *New Directions for Child Development,* **11,** 93–113.
Daly, M. and M. I. Wilson. 1982. Whom are newborn babies said to resemble? *Ethology and Sociobiology,* **3,** 69–78.
Dawkins, R. 1976. *The Selfish Gene.* Oxford University Press, Oxford, England.
Dawkins, R. 1982. *The extended phenotype.* W. H. Freeman, San Francisco, and Oxford University Press, Oxford, England.
DeCasper, A. J. and W. P. Fifer, 1980. Of human bonding: newborns prefer their mothers' voices. *Science,* **208,** 1174–1176.

Dumont-Driscoll, M. and R. J. Rose. 1984. *Testing the twin model: is perceived similarity due to genetic similarity?* Paper presented at the Thirteenth Annual Meeting of the Behavior Genetics Association, London, England.

Durham, W. H. 1976. Resource competition and human aggression, part I: a review of primitive war. *Quarterly Review of Biology,* **51,** 385–415.

Formby, D. 1967. Maternal recognition of infant's cry. *Developmental Medicine and Child Neurology,* **9,** 293–298.

Gentry, T. A., K. M. Polzine and J. A. Wakefield, Jr. 1985. Human genetic markers associated with variation in intellectual abilities and personality. *Personality and Individual Differences,* **6,** 111–113.

Greenberg, L. 1979. Genetic component of bee odor in kin recognition. *Science,* **206,** 1095–1097.

Hames, R. B. 1979. Relatedness and interaction among the Ye'kwana: a preliminary analysis, in Chagnon, N. A. and W. Irons, eds., *Evolutionary biology and social behavior: an anthropological perspective.* Duxbury Press, North Scituate, Massachusetts, pp. 239–249.

Hamilton, W. D. 1964. The genetical evolution of social behavior. I, II. *Journal of Theoretical Biology,* **7,** 1–52.

Hamilton, W. D. 1982. Pathogens as causes of genetic diversity in their host populations, in Anderson, R. M. and R. M. May, eds., *Population biology of infectious diseases.* Springer-Verlag, Berlin, pp. 269–296.

Hartung, J. 1981. Genome parliaments and sex with the Red Queen, in Alexander, R. D. and D. W. Tinkle, eds., *Natural selection and social behavior.* Chiron Press, New York, pp. 392–402.

Hold, B. and M. Schleidt. 1977. The importance of human odour in non-verbal communication. *Zeitschrift für Tierpsychologie,* **43,** 225–238.

Holmes, W. G. and P. W. Sherman. 1982. The ontogeny of kin recognition in two species of ground squirrels. *American Zoologist,* **22,** 491–517

Irwin, C. J. 1985. *A study in the evolution of ethnocentrism.* Paper presented at the Fifth Meeting of the European Sociobiological Society, Oxford, England.

Jeffreys, A. J., V. Wilson and S. L. Thein. 1985. Hypervariable 'minisatellite' regions in human DNA. *Nature,* **314,** 67–73.

Kalmus, H. 1955. The discrimination by the nose of the dog of individual human odours and in particular of the odours of twins. *British Journal of Animal Behaviour,* **3,** 25–31.

Lacy, R. C. and P. W. Sherman. 1983. Kin recognition by phenotype matching. *American Naturalist,* **121,** 489–512.

Levarie, S. and N. Rudolph. 1978. Can newborn infants distinguish between tone and noise? *Perceptual and Motor Skills,* **47,** 1123–1126.

Lichtig, I. and P. A. Wells. 1980. Behavioural assessment of neonatal responses to auditory stimuli. *British Journal of Audiology,* **14,** 61–68.

Lightcap, J. L., J. A. Kurland and R. L. Burgess. 1982. Child abuse: a test of some predictions from evolutionary theory. *Ethology and Sociobiology,* **3,** 61–67.

Loehlin, J. C., L. Willerman and S. G. Vandenberg. 1974. Blood group and behavioral differences between twins: a failure to replicate. *Social Biology,* **21,** 205–206.

MacFarlane, A. 1975. Olfaction in the development of social preferences in the human neonate, in CIBA Foundation Symposium No. 33: *Parent–infant interaction.* Associated Scientific Publishers, Amsterdam.

Miller, C. L., B. A. Younger and P. A. Morse, 1982. The categorization of male and female voices in infancy. *Infant Behavior and Development,* **5,** 143–159.

Mills, M. and E. Melhuish. 1974. Recognition of mother's voice in early infancy. *Nature*, **252**, 123–124.
Murry, T., H. Hollien and E. Müller. 1975. Perceptual responses to infant crying: maternal recognition and sex judgements. *Journal of Child Language*, **2**, 199–204.
Nicolaides, N. 1974. Skin lipids: their biochemical uniqueness. *Science*, **186**, 19–27.
Pakstis, A., S. Scarr-Salapatek, R. C. Elston and R. Siervogel. 1972. Genetic contributions to morphological and behavioral similarities among sibs and dizygotic twins: linkages and allelic differences. *Social Biology*, **19**, 185–192.
Partridge, L. 1983. Non-random mating and offspring fitness, in Bateson, P., ed., *Mate choice*. Cambridge University Press, Cambridge, England, pp. 227–255.
Porter, R. H., J. M. Cernoch and R. D. Balogh. 1984. Recognition of neonates by facial-visual characteristics. *Pediatrics*, **74**, 501–504.
Porter, R. H., J. M. Cernoch and R. D. Balogh. 1985. Odor signatures and kin recognition. *Physiology and Behavior*, **34**, 445–448.
Porter, R. H., J. M. Cernoch and F. J. McLaughlin. 1983. Maternal recognition of neonates through olfactory cues. *Physiology and Behavior*, **30**, 151–154.
Porter, R. H., and J. D. Moore. 1981. Human kin recognition by olfactory cues. *Physiology and Behavior*, **27**, 493–495.
Russell, M. J. 1976. Human olfactory communication. *Nature*, **260**, 520–522.
Russell, M. J., T. Mendelson and H. V. S. Peeke. 1982. Mothers' identification of their infant's odors. *Ethology and Sociobiology*, **4**, 29–31.
Russell, R. J. H., J. P. Rushton and P. A. Wells. 1984. Sociobiology, personality and genetic similarity detection, in Royce, J. R. and L. P. Mos, eds., *Annals of theoretical psychology*, Vol. 2., Plenum Press, New York and London, pp. 59–63.
Sahlins, M. D. 1976. *The use and abuse of biology: an anthropological critique of sociobiology*. Tavistock Publications, London.
Schleidt, M. 1980. Personal odor and nonverbal communication. *Ethology and Sociobiology*, **1**, 225–231.
Segal, N. L. 1984. Cooperation, competition, and altruism within twin sets: a reappraisal. *Ethology and Sociobiology*, **5**, 163–177.
Shepher, J. 1983. *Incest: a biosocial view*. Academic Press, New York.
Taylor, C. and W. P. Faulk. 1981. Prevention of recurrent abortion with leucocyte transfusions. *The Lancet*, **ii**, 68–70.
Thiessen, D. D. and B. Gregg. 1980. Human assortative mating and genetic equilibrium: an evolutionary perspective. *Ethology and Sociobiology*, **1**, 111–140.
Waldman, B. and K. Adler. 1979. Toad tadpoles associate preferentially with siblings. *Nature*, **282**, 611–613.
Wallace, P. 1977. Individual discrimination of humans by odor. *Physiology and Behavior*, **19**, 577–579.
Wolf, A. P., and C. Huang. 1980. *Marriage and adoption in China 1845–1945*. Stanford University Press, Stanford, California.
Yamazaki, K., E. A. Boyse, V. Mike, H. T. Thaler, B. J. Mathieson, J. Abbott, J. Boyse, Z. A. Zayas and L. Thomas. 1976. Control of mating preferences in mice by genes in the major histocompatibility complex. *Journal of Experimental Medicine*, **144**, 1324–1335.

Kin Recognition in Animals
Edited by D. J. C. Fletcher and C. D. Michener
© 1987 John Wiley & Sons Ltd

CHAPTER 13

Discriminating Nepotism: Expectable, Common, Overlooked

W. D. HAMILTON
Department of Zoology, Oxford University, South Parks Road, Oxford OX1 3PS, England

The ideas and discoveries presented in this book help us to see biology as a whole. They link across disciplines—for example, psychology to immunology—and they also link across the gap that we imagine to separate the human from the rest. Animals are even more like us than we thought. It turns out that they care about kinship as much or more than we do. The present chapter firstly reviews the probable reasons for this caring and the limits to which its evolution can go. After that the chapter gives what I consider probable reasons for the long neglect of nepotism as a worthwhile subject of enquiry. Why did we not see this unity before? Thoughts about the evolution of human attitudes have to be speculative; but beyond even speculation, the chapter ends with some personal opinions about how the new matters of identity and relatedness and recognition, once joined to ecology, enable us to see ourselves reflected in several new ways and so change our world view. These closing opinions are expressed, of course, entirely independently of other authors.

New genes contributing to altruism advance in frequency provided that the altruistic acts they cause create, on average, benefits sufficiently greater than losses, and provided the benefits go to close enough relatives. 'Sufficiently greater' and 'close enough' are quantified in a now well known approximate criterion:

$$br - a > 0$$

Here a is loss to altruist, b gain to beneficiary, and r relatedness. In what follows the relation above will be referred to as the kin-selection criterion or merely as the criterion. It is equivalent to a statement that the inclusive fitness effect of the action must be positive (Hamilton, 1964a). What is happening is that through the enhanced reproduction of others due to altruism, more copies of the causative gene are created than are expected to be lost to the altruist. Because of common ancestry, a relative has an increased chance of carrying the gene or genes. Their presence is never so certain as in the altruist itself but the criterion says that the gain ratio (b/a) of the act may more than compensate.

Various limitations and potential inaccuracies in the criterion have been shown. However, accepting slow Darwinian evolution in the adaptations concerned—evolution mediated by genes of small effect not obligately expressed in their bearers—the criterion works well at least to a point where crudities of current ability to measure fitness and not the failings of theory, must be the overriding limitation in any test (Michod, 1982; Boyd and Richerson, 1980; Cheverud, 1985).

Following a neo-Darwinian view, then, complex traits of altruism can be built up by accumulation of genes that satisfy the criterion. Each adds to a set already there and this set—to which, in effect, the new gene is applying for admission—is a part of the 'environment' over which gene effects must be averaged in determining whether selective forces will allow it to rise in frequency. Interaction with some of the genes already there may be epistatic but this is taken into account in evaluating average selective effects (Fisher, 1930). Thus in the context of inclusive fitness (as with many other criticisms directed at neo-Darwinism) there seems little cause to worry about limitations of one-locus theory. The problems raised by complex epistasis in polymorphic situations are of course interesting, but their importance is probably only in proportion to the occurrence of non-neutral polymorphisms generally.

The set of traits involved in the adaptive suicide of the worker honey bee when it stings a vertebrate attacker of its nest can be taken as example of the claims just made. Some traits are structural (e.g. barbs on the sting, features promoting abscission of a sting apparatus, Herman, 1971) and some are behavioral (e.g. ferocity, Michener, 1975). The queen whose safety the acts of suicide serve is carrying the same gene unexpressed[1], and it is through her breeding success that the genes are maintained. In the case of the genes modifying behavior there is good evidence of difference at a racial level. However, this is easily understandable from the differences in predation to which bees are subject in different continents (Winston, Taylor and Otis, 1983); there is no reason to expect polymorphism in behavior within any one area—or at least, no more reason to expect it than there is for any non-social character. Regarding variability in general, the theory differs only slightly from the classical neo-Darwinism which it extends. Within its sphere of application, compared to the classical theory, it implies that less variability will be due to gene differences and relatively more to environmental conditioning. It has to assume that genetic differences are present from time to time, during the spread of new mutations through the population. However, these transiences will usually go to completion. Most of the time the various morphs, acts, etc., will be observed seeming carefully to aid only relatives that are more likely to carry the same causative genes, whereas in fact all members of the population carry them. Thus contrary to a common misconception the theory does not imply that differences in social behavior have to be genetic differences.

All this, of course, applies not only to altruism. A very similar argument works for the opposite of altruism and implies restraint on the evolution of extremes of selfishness and spite. Genes will not go ahead if they take fitness away from relatives for too little self profit (Hamilton 1964a,b, 1970).

It should by now be clear that, if kin-selection models are a fair image of reality, closeness of relatedness is a crucial factor for determining adaptive courses of action in social situations. Thus an ideally adapted social organism would know the relatedness of every individual around it and would make its behavior finely conditional on the relatedness perceived. However, it does not follow from this that ability to discriminate degrees of relatedness automatically implies that kin selection is the model relevant to its origin. In fact, since even earlier than Darwin, it had been realized that most organisms tend to avoid closely inbred matings. The reasons must have to do with the function of sexuality and this is not yet quite resolved (see e.g. Bell, 1982; Shields, 1982; Hamilton, 1982); but whatever the function is, here must be another set of reasons for discriminating. Some animals certainly do use discrimination for purposes of mate selection. Japanese quail for example evidently use an early imprinting of their chick companions towards obtaining, much later, preferred degrees of consanguinity in their mates (Bateson, 1983). In general, however, finding of ability to discriminate, even if the uses made of it are still obscure, is prima facie suggestion of the importance that kin selection is likely to have in nature. Even if started by other pressures, it is likely that discrimination, once present, will diversify the sociality of a species in both group-beneficial and group-harmful ways. Inevitably, as ability to discriminate refines, new avenues for social behavior open, and the manifestations can be both positive (e.g. eusociality) and negative (e.g. infanticide).

Writing on the kin-selection criterion 20 years ago I pointed out that where substantial increments to inclusive fitness could occur through discrimination, the ability would be expected to evolve. Four pages were given to the topic together with a summarized principle[2] which seemed worth stating as a corollary to the kin selection principle. Examples that I had been able to find at that time, however, were no more than faintly suggestive that a widespread phenomenon might exist in animals other than man. I knew of course of the keen interest which *Homo sapiens* always has in relatives. It did not seem appropriate to discuss this in the paper because culture was so obviously a confusing factor; the human facts, however, had been very important for germinating the idea that a quantitative criterion for social transactions generally might be found.

Apart from man, my best encouragement at that time for the evolution of discrimination lay in examples of birds learning to recognize their own young at exactly the rearing stage where they were beginning to mix with young of other parents (Hamilton, 1964b; Beecher, 1982). Beyond this I only had various scraps of evidence at more remote levels, suggesting a racial or locally nepotistic kind of discrimination in various animals. In retrospect the interest of these scraps is that even at that time, they had been long known, and, by furthering an idea that animal and human cases might run parallel, they could have suggested discrimination within the range of close relatives. One may wonder why, even while a theory was lacking, such evidence was not sought. I shall return to this point later.

Since 1964 evidence for discrimination has expanded greatly, particularly during the past few years. Cases have been found where not only do we not know how animals discriminate, but we can hardly even guess why they do so—i.e. we don't see that use is made of the ability either in mating or in any other actions that plausibly affect inclusive fitness. Many of the new cases are discussed in this book. Waldman's discovery of discrimination in tadpoles may be picked out here as an early surprise. Not only the occurrence but the remarkable fineness of the discrimination is well established in this case: tadpoles distinguish even their half-siblings from their full-siblings. The list of anurans known to discriminate is growing (Chapter 9), yet the function of such abilities in the wild remains quite unknown.

Altruistic or selfish acts are only possible when a suitable social object is available. In this sense behaviors are conditional from the start. Conditionality upon assessment of relatedness of a potential recipient is in fact just one of several ways that the behavior may be refined. An animal might well change its behavior according to the age or state of health of the interactant concerned, for example. In what circumstances, in general, would discrimination be expected to be most developed?

Firstly, it is obviously expected to be highly developed where potential gains to inclusive fitness are great and where along with this goes a high cost when acts are misapplied (see also Chapter 4). This implicates the species that are already fairly social. Adaptive cooperation brings about, ultimately, acts of successful reproduction that would not otherwise have occurred; as we have seen, it is important to an individual participating in such extra reproduction that it should involve relatives, not just any animal in the population.

The second and third factors bearing on high discrimination interact with the first one; however, to see them we have to move round the benevolent group statue of cooperation and look at the daggers and scowls, or perhaps merely blank expressions, that are on the other side.

Wherever extensive preparatory behavior towards reproduction occurs, selfish acts, in the form of usurping the preparations made by other individuals, can be very rewarding. Examples here are egg dumping which parasitizes the brooding efforts of another bird of the same species (Yom-Tov, 1980) and the usurping of nests in the nest-building Hymenoptera (Bohart, 1970). Other examples are legion: think of any object on which an animal expends much preparatory effort; then study the natural history either in the literature if this exists, or in the field. Very likely you will find that this object is sometimes usurped. And on the principles outlined, if an animal is going to parasitize, it should, other things being equal, parasitize the most distant relatives, so that discrimination of degree is expected.

The third and final consideration I will mention is that positive powers are most expected where inactive context-driven (and therefore cheap) discriminations are least likely to be effective. For example an aphid that has just made a dispersal flight might be expected to change to a more selfish and damaging exploitation of the host plant. After its dispersal any companions found feeding on the same plant are unlikely to be its siblings. Therefore it need not 'consider' any more the effects that its actions may have on the welfare of neighbors. It should become more purely selfish. If there is a change in behavior, this aphid is performing a *de facto* discrimination based on context. But the same crude method could not be used where, say, even before any migration, there might be many mixed clones and where it might be advantageous to distinguish sister from non-sister aphids and react to them accordingly. To effect discrimination here (supposing it is needed: one sees no compelling reasons why it should be, but then one sees none for tadpoles either) requires something like a chemical comparison of the target aphids with the self or with a certain set of highly probable siblings, or with the mother before the offspring lost contact with her: all this, with the necessary receptors and neural connections, requires a much more detailed specification in the genome than the reaction triggered by context. In aphids such examples are hypothetical as yet, but in light of what is now known for toads and bees, it would not be very surprising if real examples were to be found.

A common situation in which all the conditions may be fulfilled is that of broods fathered by more than one sire. In cases of mixed insemination it will be impossible to know which are full-siblings in the brood and which are half-siblings unless there are fairly refined powers of discrimination. The more obvious courses of social learning, well discussed by Alexander (1979)—knowing who the mother is, the family context etc.—will not serve here. Yet the difference between the relatedness coefficients (0.5 and 0.25) is considerable, and brood mates are often in situations where both high altruism and easy self-profitable selfishness might be advantageous. Thoughts on these lines lessen the surprise that a ground squirrel that seems to have a high level of promiscuity (Hanken and Sherman, 1981) also proves to have a ready ability to discriminate relatives in various ways, at least up to the level of telling full- from half-siblings (Holmes and Sherman, 1982). With these squirrels it is actually known that they give alarm calls that benefit others and are dangerous to the giver, and that the calls occur in social contexts where it is especially likely that kin will be benefitted. Thus the calls are altruistic, and, in a general way, the expectations of kin selection are confirmed. Honey bees, on the other hand, have long seemed a contrary case to the ground squirrels with the regular mixture of full- and half-siblings rather ignored (Hamilton, 1972; Page and Metcalf, 1982), despite the existence of well developed kin discrimination in other bees (Buckle and Greenberg, 1981); however, at least some evidence of kin discrimination has now been found in the honeybee also (Breed, 1983; Getz and Smith, 1983; Chapter 8). In general, a good test of the thesis of this chapter will be to compare discrimination abilities between close species that are monogamous and those that are promiscuous, given that the broods can do something social in the first place. It should be pointed out that although the discrimination has to be a positive adaptation if it exists at all, it need be nothing mysterious and its arrival in a series of small steps is easy to imagine. As a minimum one might suggest an array of odoriferous chemicals whose production and quantitative mixture depends on a set of varying genes. These genes may be polymorphic for quite other reasons (see Chapter 4) but in the end, via the odoriferous products (possibly wastes) which they contribute, they effect a chemical signature. Next—and this could happen quite separately—there needs to be an ability to habituate to familiar odors: with respect to the chemicals we have suggested, much of the time it would be a matter of habituating to the particular blend of odors that characterizes self. Finally, behavior needs to be made conditional in a graded manner on the degree of familiarity of the odoriferous environment in which an animal finds itself at the moment when it is in a position to do something social. The last step is the only one where kin selection has to be involved.

It is easier with most animals to imagine all this happening for odors than for visual stimuli—for example, one reflects that a vertebrate cannot, without a mirror, habituate to its own face and this is the place where humans most expect to find landmarks for individual and kin recognition. The fact that even humans, normally thought of as relying on vision much more than on smell, can fairly easily recognize both their mates and their kin by smells imparted to clothing (see Chapter 12) suggests that an old mammalian method is still available to us—and available in spite of our about 90% monogamy and in spite also of some probable reversal of evolutionary 'policy' on nepotistic kin discrimination that occurred in recent human evolution, as will be discussed later.

In the above suggestion of how to get discrimination out of habituation, the reader may have noticed that if it works at all, the proposed mechanism has no reason to stop at discriminating full-siblings from half-siblings. Among full-siblings, for example, there are those who happen to have a more similar shake out of chromosomes from the parents[3] and those who have less. Rather surprisingly there is already evidence that human discrimination (here undoubtedly using many media besides smell) can discriminate among full-siblings (Pakstis et al., 1972). So long as the genes involved in all parts of the discrimination are well spread among the chromosomes, all genes except for the 'non-discriminator' allele that is being replaced, benefit from any action that fulfills the kin-selection criterion on basis of the estimated r. We need not expect that other elements in the genome will evolve to suppress the effect. The system of measuring similarity is quite different from one that reacts to a particular trait—i.e. a 'green-beard trait' (Dawkins, 1976; Ridley and Grafen, 1981) as will shortly be seen.

The system outlined above was based on a minimal kind of 'social learning' (Alexander, 1979). Prior to application in behavior, the learning, which in this elemental case was self-habituation, can be done by the individual in total isolation. Could there be any mechanism of kin recognition that does not need learning of any kind? In fact, there do seem to be effects describable as kin recognition which are like this, but at the same time there are also reasons to believe that they cannot evolve into nepotistic patterns of any complexity. Absolute innate discrimination of genotypes, differentiating them as to whether they do or do not overlap with self at one particular locus, is already well known in all those plants whose outcrossing is regulated by the one-locus multiple-allele self-incompatibility system (Lewis, 1979; de Nettancourt, 1977). The reaction happens between the diploid stigmatic tissue of the whole plant and the haploid gametophyte arising from the pollen grain that has alighted on it. Here it is a case of seeming failure to cooperate when the two tissues are 'related', resulting in actual death of the gametophyte. The failure happens whenever the gene in the pollen grain is the same as either one of the genes at the incompatibility locus in the receiving plant. In essence it is as if the pollen grain has a chemical 'green beard' by which it is recognized, and if the host finds it to be at all similar to its own, the germinating pollen grain is rejected. But is it necessarily the reversal of usual kin cooperation that it seems? Maybe it is the pollen grain that auto-destructs, or at least grows non-competitively when it senses, by the 'green beard' of its host, that it has landed on a close relative. In this case the death is altruistic—an evolutionarily 'taught' reaction based on the fact that if a closely related pollen grain does race ahead and claim an ovule, the resulting seed is almost doomed through its homozygosity (or its lack of differing genome) anyway. Thus the genes of the pollen grains propagate better by giving to their probable twins in the ovum that they might have entered, the chance of joining the more suitable gene partners that are brought in by a different pollen grain. At the back of this speculative but not unreasonable interpretation are the still mysterious advantages of sexuality and outcrossing. These advantages, of course, also underly the very existence of kin selection as we know it, since without them there would be no mendelizing genetics. But that line of thought leads off towards longer digressions that will not be pursued here.

The above example will probably be considered a cheat of a 'green beard' anyway. But given that this phenomenon exists—in cherry trees for example—it is possible to imagine the following. When cherry tree roots collide underground they could measure their genetic overlap by the same type of mechanism. Then they could either cooperate by physically uniting their root systems[4] or else they could keep separate and compete[5]. There are other possibilities also of course—for example one could fasten on the other parasitically—but for the present discussion we assume that if they unite, each is able to defend itself against harm.

Suppose that a root connection is made and that subsequently one tree is dying for want of water, to which the other has limited access. Suppose the other supplies the dying tree with water at a benefit-for-cost ratio far more generous than the limit set by the kin selection criterion based on pedigree relatedness. The helping tree as a result sets less seed and the helped tree sets more seed—but the 'more' is not enough to fulfill the criterion except at the recognition locus and possibly at loci very nearby. Suppose this happens again and again in instances of such paired trees. If now, elsewhere in the genome, a mutant gene arises that blocks the outward sap flow at this benefit-for-cost ratio, then that gene is advanced. This is easily seen by the kin-selection principle[6]. Thus as regards its actions as a potential donor, the genome evolves to suppress a 'green-beard' transfer. Effects of this kind would therefore be transitory and be unable to evolve into regular adaptations. If such aid is given regularly through root grafts, it is likely to be established on some basis either of typical kin selection or else of reciprocation.

Potential plant examples like those suggested above may seem remote, even if valid, to animal-oriented readers of this book. The following examples probably won't seem much better, even though they are animal and even though the animal in the first case is—well—woman. There is good evidence that a minor part of human sterility may result from too great a similarity of the partners for genes in the major histocompatibility complex (Beer, Gagnon and Quebbeman, 1981). It appears to result from failure of the fertilized ovum to implant properly. The blocks evidently are at a different stage, but have some sort of parallel to the genetic incompatibility systems of plants. Whether we admit anything 'green-beardy' about this or not, the parallel itself is the more impressive when we learn that a system almost certainly homologous to the major histocompatibility complex reaches back to our own sessile ancestors (or slightly more accurately, our remote 'aunts'), the tunicates (Scofield et al., 1982), and is probably used by them to prevent coalescence of adjacent individuals. On lines of the parallel, referring again to the human zygote and to the possibility that 'altruism' is involved in a decision not to attach and grow, we note here, firstly, the complete reversal of the normal effect of histocompatibility, and secondly, the seemingly completely maladaptive outcome of the failure to implant which even arguments about altruism will not easily explain. Surely both the potential mother and the fertilized ovum should agree that it is better to produce some offspring, even if more than ideally homozygous for HLA, than to produce no offspring at all. Perhaps it is only the unusually maladaptive extremes of such early fetal loss that come to notice.[7] In any case there is at least an interesting hint of a rather innate recognition ability, and this case leads on to notice of yet another related phenomenon on the borders of immunology: the house mice which show mating preferences based on unlikeness of certain genes in the histocompatibility complex (Boyse, Beauchamp and Yamazaki, 1983). This effect is very surprising, perhaps the more so when it is known that it is mediated through odor coming from urine. We cannot do more than note here what appears to provide the basis for a minimally learned discrimination system of the kind already outlined, and that it seems to have a potential to become 'green-beardy' in so far as the effect is concentrated in the small *H-2* region of the chromosome. But once again, we do not expect anything describable as an innate kin recognition adaptation, used for regulating social behavior other than mating, for the reasons already given in the hypothetical case of the trees.

One of the novel outcomes of trying to rationalize social behavior by arguments of inclusive fitness is the realization that not all elements of the genome are expected to 'want' the same thing, and that they may be making contributions to strategies that are actually contradictory within the one genome (Hamilton, 1967). The unlinked gene that suppresses the effects of a 'green-beardy' one in the previous paragraph is an instance of this. Both these genes could be on autosomes. More regular instances of intragenomic conflict can be expected between the sex chromosomes and autosomes. Since sex chromosomes are usually a single pair, and only very occasionally a major part of the genome, and since the rest of the genome goes symmetrically with sex and submits to the very fair and uniform process of meiosis, the effects of conflict of strategy between autosomes and the X chromosome (the Y, being usually inert, does not come into this argument much) are not expected to be very apparent. The manifestations, if ever found, would be discriminatory and so deserve to be mentioned here. Only one example will be given. The behavior of female birds during brood care might be expected to be slightly biased towards assisting their male offspring (Hamilton, 1972). This is because they pass on to sons a very substantial chromosome, the X (if we can extrapolate from those cases where its size and genetic richness are known), whereas they pass on to daughters only a small inert Y. Their male mates might be expected to attempt to compensate for any bias by preferentially feeding the female young. There do seem to be some hints of such biases (Emlen, 1984), but as we would expect from the out-weighting of sex chromosomes by autosomes, they are slight effects found only in some species. It is too early to say whether the inclusive fitness rationale is important.

Perhaps the most interesting thing to come out of the realization of possible conflict within the genome is a philosphical one. We see that we are not even in principle the consistent wholes that some schools of philosophy would have us be. Perhaps this is some comfort when we face agonizing decisions, when we cannot 'make sense' of the decisions we do make, when the bitterness of a civil war seems to be breaking out in our inmost heart.

With evidence now at hand, most of it rapidly accumulated over the past ten years and summarized in this book, it begins to seem that some ability to recognize kin and to react accordingly will be found in any social animal if looked for carefully enough. This potentially implicates almost every animal species, since almost every species is social at least to the extent of mating. If we admit the lower-level phenomena concerned with tissue graft rejection, recognition may encompass much of the metazoa over again. Sponges, sea anemones and corals know their kin (Neigel and Avise, 1983). They graft, avoid and fight accordingly; slime molds (Mycetozoa) fight or do not fight lethal chemical battles according to how they differ in genes, and these genes reflect kinship (Carlile, 1973). Such universality raises a puzzle. At the higher end we have known that our species is inclined to nepotism for as long as we have known anything; we are steeped in its existence and have known of it since, as infants, we began to have any ideas at all. Then we have had it all again from the coded lore of every human culture. Even this is not all that should sensitize us to the issue of relatedness. The idea that animals throw back pale or distorted images of ourselves and we of them is also ancient—far, far older than Darwin. Why, then, have we not long expected to find that animals have ways of discriminating kin? Why were the first recent papers about actual cases reviewed with almost open hostility and suspicion? The reason must have to do with the fact that at least in civilized cultures, nepotism has become an embarrassment.

We admit certain aspects of nepotism, such as parental devotion to offspring, because it is obvious that even high civilizations cannot do without them. I believe that even the most extreme communists would react with horror, like everyone else, to an idea that people's babies should be routinely switched around in the natality wards of hospitals, as a matter of state policy, even though, if put into effect, this policy would immediately bring to reality some of their most ardent dreams. For example, apart from justice in the randomization of opportunity, the policy would, of course, in the space of one or two generations, settle the question of nature and nurture in human ability once and for all. Experiments rather on these lines although less extreme have actually been tried. One of the most daring and recent has been that of the Israeli kibbutzim; but I know of no such experiments that have not had to retreat from the original ambitious aim. Adoption does succeed. It succeeds extremely, surprisingly, nobly—as an outstanding credit to the above-animal nature of man—when it is considered in the light of some of the issues and facts raised in this book. Yet even this doesn't succeed wholly. Virtually no one chooses adoption except as a last resort (Daly and Wilson, 1983).

The same spirit that approves of adoption in principle opposes nepotism rather definitely once outside the circle of closest relationships. When we hear that a son has inherited his father's business we think that this, at the least, is only natural. What else to expect: perhaps actually it is a 'good thing', an expression of strong positive spirit in the nuclear family which we approve. If it is a nephew that inherited the business we are little less certain. If it is a son of a cousin we may mutter something about 'blatant nepotism, unjustified class privilege.' It is extremely hard, perhaps impossible, to know how much of this attitude is natural in us and how much is a product of our teaching. It is difficult to know how much of it is sincere, even in oneself. For sure large parts of such attitudes are taught. One has only to think of how whole cultures have been swayed towards or away from racism within living memory to see that. For sure also, what we express as ideal is not always what we do, and what we do usually errs on the side of being more nepotistic. Such hypocrisy over nepotism is understandable on evolutionary principles. A world where everyone else has been persuaded to be altruistic is a good one to live in from the point of view of pursuing our own selfish ends (Alexander, 1979). This hypocrisy is the more convincing if we don't admit to it even in our thoughts —if only on our deathbeds, so to speak, we change our wills back to favor the carriers of our own genes. However, when all this has been said, a case remains, I believe, that a major reversal of natural selection acting upon the more extreme forms of nepotism took place in recent human evolution. Since the apes we may have become intrinsically more ambivalent about nepotism.

I shall try briefly to explain why I think this has happened. Possibly gearwheels will have been heard grating in this chapter—actually a little back from here—and if the chapter seemed sketchy already it is now going to seem worse. In excuse I must plead that I am trying to explain something that is sure to be controversial in a short space... that is, to explain the puzzle of why people don't want to know about animal nepotism and why nepotism in general has become an embarrassment.

The new pressures which I believe came to affect the issue arose out of the diversity of human activity as it developed, contrasted to the activity of apes. One, the major new pressure, comes from human division of labor. Another comes out of man's increasing intimacy with plants and animals, arising ultimately out of his need to control them in securing his supply of food.

If it makes sense at all to talk about the efficiency of a community, i.e. if the community is integrated enough to have 'functions' that can be measured as more or less efficient, then the ideal road to division of labor is probably that taken by the cells of the embryo or by the worker honey bees of a hive. Here every individual carries the code both for all the varied preliminary paths of ontogeny, and for the detailed acts of living. All these are to combine together for the well-being of the superorganism. Sometimes, especially in the social insect examples, the animal does actually perform several of the activities in its lifetime. In these cases division of labor is all a matter of going to the right part of a general program which every individual carries. It works well so long as storage of very long stretches of unused script is not too much of a problem and the individuals are closely enough related that antagonized interests do not cause the system to break down. In the case of the cells the latter condition is clearly met; in the case of the honey bees it is met, to judge from their behavior, by a rather narrow margin—met with the help perhaps of severe threat to the non-cooperative colony from the outside. In humans the relatedness in a self-supporting group of any size just isn't close enough and division of labor has come about from the very start in a different way.

Here there may come a 'just a minute' from the reader. The human system is indeed different, you may say, but this has nothing to do with relatedness. Rather it is because humans are not, and never were, anywhere near to having the kinds of innate program of development that cells and honey bees have. Instead the information for diverse human activities is stored in culture from which appropriate parts are taken up by learning—taken up into the most broadly impressionable nervous system that is anywhere known. And, you may add, your argument was astray earlier because, by your own standard of emphasis on the individual, what says that the net outcome of human division of labor needs to be efficient—what unit is supposed to benefit if it is?

Well, I agree with these criticisms but not entirely. My answer to the last is that human communities and cultures do replicate sufficiently in their own right (Hamilton, 1975) for there to be appreciable evolution towards efficiency, although I am not sure how this could be measured. The earlier criticism is much harder. But although the impressionability of man is exceptional, it is not unlimited, and learning into a 'blank' memory cannot be the whole story. I must be brief here since this is digression: I will claim that it should be obvious to anyone who has sat in a classroom or played games on a school playing field that all people are not equally teachable in all skills. If the obvious experiments based on infant adoption have been done by those who believe otherwise, then their results have not been reported. Therefore, if done, the results were probably negative. Adoption studies that have been reported, on identical twins reared apart, indicate much heritable idiosyncrasy in personality and abilities. Furthermore, *a priori* it is rather difficult to imagine any 'blank slate' arising by evolution. Admittedly there has been a steady progress in the more flexible and learned approach to adaptation and it is slightly easier to imagine a general mentality that could become a disc jockey or an inventor than it is to imagine a physique that could become a high jumper or a weight lifter—but it is not much easier. The scenario for origin of division of labor that I give below has at least a plausible slant and story. I postulate a mechanism basically analogous to the way interspecies mutualism is expected to arise out of the goalless meandering of interacting species in an ecosystem (Wilson, 1980).

The pressure begins when our ancestors, probably still confined to Africa, expand their exploitation outward from the narrower niches of the apes. It is assumed that brains permit them to understand the past and to imagine the future, that they would understand tit for tat (Axelrod and Hamilton, 1981), and that they easily recognize particular alien groups. Let us imagine two tribes, one of the savannah, one of the forest, which sometimes met along the forest edge. For a long time, if they take notice of each other at all, they snarl and fight. Their traditions—we suppose they already have language—admit no intertribal common ancestry; indeed they would feel insulted at the thought of it for each tribe regards the other as barely simian. Then gradually a change comes. When 'we—the people' go to the forest now we take grass seed and meat with us; these we find easy to gather out here on the plains. We leave them around. Then we find that instead of throwing their spears at us, 'they—the ape pigs' give us their spears—and spear shafts, of course, are what we went there for anyway. And when they howl and mouth at us in that crude way of theirs across the glades, well, now it doesn't sound such a bad sort of uproar when you're in the mood for it. They stand up almost like people when they're howling: actually one even likes to watch the way some of them—the female ones—wave themselves about. One could almost... In short we have the impression that they want us to come back. Of course they wouldn't want to come to us bringing their spears and bow staves and fruit. The sun would roast that pitiful pale skin of theirs right off their backs if they came into the savannah—unless we camped where there were a few more trees, then they might come. But if they did, we would want to make sure they can see that those trees are not a forest they can move into; they're our trees. So is it our forest for that matter—it's just that we don't choose to live there for the present. Those creatures are nothing to us except a clever kind of monkey or pig that can be useful when it decides not to be viscious. So long as they realize that, I suppose, there would be no harm in having them actually around our camp sites; they climb trees so much better and could bring us fruit before the parrots ate it... By some long, long 'rationale' such as this (in reality spread over hundreds of thousands, if not millions, of years, and with the trends I have set out realized, in part, in actual changes of genetic disposition, and at the same time confused much by intermarriage), *Homo economicus* can be imagined to come into existence. More and more complexity would be added as the story goes along. Domestication of animals and the invention of agriculture occasion huge strides. The invention of metal working has been another even more recent incentive to diversity. Here in particular it is in itinerant smiths of recent recorded times working in quite alien cultures, half admired, half despised, and in any case so much needed that their position is relatively secure, that one can see a hint of what has been going on.

Definitely on all fronts it has become imperative not to bristle with hostility every time you encounter a stranger. Instead, observe him, find out what cut of a man he might be. Behave to him with politeness, pretending that you like him more than you do—at least while you find out how he might be of use to you. Wash before you go to talk to him so as to conceal your tribal odor and take great care not to let on that you notice his own, foul as it may be. Talk about human brotherhood. In the end don't even just pretend that you like him (he begins to see through that); instead, really like him. It pays.

As civilization gathers pace and more and more diverse activities open up, it becomes of course more and more impossible that a single human type, either in physique or mental disposition, could do all the tasks equally well. And not just tasks but roles. It is not just the physiques and dispositions for farmers, bankers and watchmakers; civilization needs a more subtle mix, with ingredients drawn from a diversity that was present and used even before the *economicus* stage[8]. Civilization needs leaders determined to stand on top and willing to try novelty; it needs sheep content to be lowly and to be told what to do; it needs introverts to be seers and inventors, and extroverts to keep everyone talking to each other. It is a symbiosis of aptitudes (Hamilton, 1971). It is like the mutualism of a lichen made vastly more complex. It is not a matter of bringing all the needed diversity out of a single genome, any more than this is being done to achieve the complexity of an ecosystem[9]. With this in mind and also looking into the great black pit of non-cooperation that opens at our feet out of the so-called 'prisoner's dilemma' once civilized interdependence has risen (Axelrod, 1984), then surely it is not difficult to see why selection for nepotism has gone partly into reverse and why we are confused about it.

This outline of the present situation and the implied possibility that, once we have caught up with the extra complexity and have it properly figured out, it might be adaptive to pursue nepotism again wholeheartedly, may not seem very likeable. Certainly if it is anything like true, it raises worries about the future of civilization that must be addressed sometime if we are sincere about preserving some human attributes that we now believe to be good. However, I do not intend to face the worries here. Such a discussion would have to consider hypocrisy, both any hypocrisy which may be latent in my own view above and also that which is fairly obvious in the alternatives that are usually preferred. There is indeed here an obvious demagogic value in a talk of generosity rather than of sordid trading; also in a general shouting about brotherhood; also in denial of 'natural' nepotism... However, this is a large subject, and it overlaps many minefields in the human subconscious. I will leave it and turn to the other lighter and more likeable factor that I see as countering nepotism in recent human evolution.

We have grown oddly fond of animals. We have grown oddly fond of plants also. We like them for their own sake; we like to have them around as pets. We like flower gardens. As we isolate them and bring the species close to our gaze, as we look after them almost as if they were our children, we inevitably see their lives with a degree of sympathy that we never gave them before. And then we begin to realize their similarity to ourselves, and beyond this to glimpse the unity of all living nature.

Why we should develop this liking for plants and animals is hard to say. At first sight to connect pets and flowers with the domestication of other species which are being ruthlessly exploited once they are controlled, does not seem very promising. However, it is likely that the truth lies this way. Arbitrary and useless as may seem a Mexican's love pf plants that will cause him to hang a *Sedum morganianum* in an empty tin can from the porch of a building made out of the sides of packing cases and old corrugated iron sheets, it is difficult to imagine that this love is not connected with the immense contribution that meso-Americans have made to the crop-plant flora of the world. Out of the same land also came the first ideas of a botanic garden where every kind of plant would be grown; now of course botanic gardens all over the world store plants and genes in order to protect our food supply from disaster, to say nothing of storing plants whose potential use is as yet unknown. The amerindian contribution to domestic animals is not so great but it exists, especially from the more civilized areas, so that when one finds a tribe in the Amazonian forest with no domestic animals save the dog, and yet with parrots, seriemas, pacas, monkeys, deer scrambling and strolling about the house as pets, one can imagine that the sympathies that make people inclined to this seeming waste of effort have, in other places and times, paid off for them. Meanwhile, if famine strikes one would guess that the pets are sometimes eaten, as the guinea pigs that are kept half as pets by Andean Indians certainly are eaten even without famine.

Here, by the way, we are not looking at the first roads to domestication. Nor is the first road in the Middle East, nor even human. Not far through the forest from huts where Brazilians keep their pets I have watched *Azteca* ants attacking and cutting up a colony of wasps of a species that nests, normally peacefully, inside the nest of ants—and at least in that area, never nests elsewhere. In the case of the attacked colony a recent flood had probably interrupted some of the ants' foraging trails. With little doubt the *Azteca* have been keeping their 'pets' since long before man came to America. Older still, perhaps back in the sunset of the dinosaurs, must be the origin of similar events I have watched in a primitive ant *Hypoponera punctatissima* with its myrmecophilic Collembola. But this is digression[10]. The social insect developments would be on a much more innate and less flexible basis than those of man. My main claim for the present about the pet and plant keeping on the road to domestication is that in the human (if not in the formicine) kind of approach it helps greatly if you are interested in and even love the living things around you.

If you have been saying all your life of the people in the next valley that they are little better than pigs—they eat things that pigs do, gorging with the manners of pigs while also they copulate pig-like in the streets of their villages (one hasn't seen this but everyone knows that it is so), and so only deserve to be speared on sight like pigs—then it may at last trouble you acutely to realize that you love your own pig and are determined that one in particular will not be put to death and eaten. Once this troubling begins, and once there is enough social surplus to make pet keeping and flower growing possible, so that the idea can be constantly present and its effects not too costly, then the process can go rushing ahead in leaps and bounds—in fact perhaps has to rush on until it reaches the point where it says: 'These animals are actually much better than us; they are sort of divine.' Once this stage is reached, then we know how to rationalize and react if it happens that 'they' go on spearing us even when we tried to stop doing it because of their being too pig-like. It can then transpire that actually they are worse than animals, wicked and perverted, fallen from the natural grace of animals—as indeed may be some of our own tribe, come to think of it. Now their unnatural wickedness can be the excuse for our harmful acts. Yet one hopes, when this stage is reached, having once seen them as part of a great living scheme which we might enter gently and mold to our advantage, as we did with our animals, we do not throw spears with quite such abandon as before, instead more defensively. This is all a bit fanciful, but I believe it has at least a grain of truth, and if it has a grain it may help to explain one of the facts that I mentioned early in this chapter; the shock and indignation which has greeted the discovery of nepotistic discrimination in animals. We had become proud of what we can do, but also sorry about it and nostalgic for a supposed golden past. In certain moods we like to idealize the state of nature and we consider ourselves 'fallen.'

There is little reason to expect that *Australopithecus* kept pets. *Homo sapiens neanderthalensis* at least seems to have loved flowers (Solecki, 1971). Just possibly this is connected with the fact that in australopithecine times the hominids in Africa had less morphological diversity than exists in hominids today (i.e. within *sapiens*),[11] whereas after *neanderthalensis*, rather as if the disappearance of that near species came to be regretted, the diversity of human races fans out and, with one or two known blots of extinction, the fan persists down to the present. As for one of the recent blots, I would like to think that the sudden reversal of the policy of the white settlers towards the native Tasmanians, when it was realized—unfortunately too late—that they were nearing extinction, was an expression of a new ethic of conservation and of respect for all forms of life. Started from a small origin along with pet-keeping and horticulture a very long way back, suffering vicissitudes but never extinguished during the explosions of domestication and agriculture, this ethic has been very rapidly expanding in the last hundred years.

It may seem insulting to the Tasmanians to suggest that the emotions that tried to save them were the same as those now trying to save the giant panda. (Maybe as they on Flinders Island, so the panda in our zoos expresses proud contempt for what I write here by refusing to reproduce.) But it is surely better to be saved by patronizing emotions than not to be saved at all. It seems to me that this is a sounder start towards a more peaceful and varied phase of humanity than the impossibly idealistic visions that are commonly advocated. In essence loving everyone as oneself is as unthinkable, and as contrary to the grain of all life, as a program of deliberate baby swapping would be. Yet if we are content with a more realistic objective of a stable 'ecosystem' of man which has integrity and beauty equivalent to that of the natural ecosystem out of which it grows, then the first steps are not far to seek. We must cease to pretend with cries of 'brotherhood' and other nonsense, that there is one ideal way of life, one dominant culture, one right or inevitable program of evolution for our species, whether claiming a way of a noble savage, a Marxist, a Christian, an Englishman or any other. Then, still countering less liberal views, there follows one rule which most different cultures should be willing to enforce, however varied and individualistic their practices may be in other ways; that no race or genotype has right to unlimited numbers, and that every race has a right to be protected against extinction.

NOTES

1. The worst pathologies of the kin-selection criterion arise when genes for social behavior are unconditionally expressed—i.e. expressed by every individual of a given genotype. For example, a dominant gene for suicidal altruism has to exterminate itself in one generation if unconditional, irrespective of recipient relatedness; even if its benefits are always directed to clonal relatives, these themselves have to die as they express the gene. Results with generally the same slant, often concerning multiplicative fitness interactions and genes of potentially large effect, have been discussed by Cavalli-Sforza and Feldman (1978) and others. In situations approaching eusociality, long before caste differences are apparent, there are numerous asymmetries that make conditional expression the natural course (West-Eberhard, 1975), and in my opinion, genes with small effects are likely to be by far the most important. Multiplicative fitness interaction is indeed plausible, but if the effects are small its selection criteria converge with those of additive interaction. Conditionality, although mentioned, was insufficiently emphasized in my previous work (Hamilton, 1964a,b).
2. 'The situations which a species discriminates in its social behaviour tend to evolve and multiply in such a way that the coefficients of relationship involved in each situation become more nearly determinate' (Hamilton, 1964b).
   The focus on the 'behavior of a species' in this statement now seems to me odd. The way of expessing the matter is also indirect and, probably, cowardly—i.e. aiming to divert from its main point and to avoid sounding racist.
3. For any specific relationship, as specified by pedigree connections, the mean relatedness is fixed (barring problems brought in by the sex-linked part of the genome) but variability about the mean depends, roughly speaking, on the number of freely recombining units in the genome—i.e. on the 'recombination index' (see White, 1973).
4. Natural root-grafting between neighboring conspecific trees is common in some species (Graham and Bormann, 1966; Stone and Stone, 1975).

5. The reaction of pollen and stigma is unlikely to carry over to roots in quite the same form because:
   (a) tissues of colliding roots are both diploid whereas pollen is haploid,
   (b) even within the floral parts the pollen reaction is often very localized (de Nettancourt, 1977)—e.g. injecting pollen into the ovary sometimes bypasses the inhibition.
   However, the idea of reactions developed primarily for optimal outbreeding being switched for use in nepotism is generally reasonable and perhaps is the most likely course where genetic markers are being used. There is actually a difficulty about how genetic polymorphism can evolve if nepotistic identification is its main use. This is explained by Crozier in Chapter 4. Strengthening the idea of switching, Crozier refers to the remarkable finding of Smith (1983) on the kin-biased mating reactions of male sweat bees (see Chapter 7). These bees are the same species as studied by Buckle and Greenberg (1981) in their work on discriminative guarding.
6. A claim in Samuelson (1983) to the effect that such an altruism-blocking gene would not supervene completely, but rather would be carried to an intermediate ratio dictated by starting conditions, is an error. Specifically, the statement on p. 8 of Samuelson's paper 'The ultimate proportion of A_2's and a_2's are not predictable, since this depends on the happenstance of initial $[N_{ij}(0)/N(0)]$ ratios,' is not correct. For any gene frequency at the locus of A_1,a_1, the fate of the locus A_2,a_2 must follow by the same reasoning that Samuelson correctly uses for the first locus. The fate is determined by Average(l_{12},l_{22}, > Average(l_{11},l_{21}). Therefore a_2, the suppressor allele, goes to fixation, and, as it does so, expressed altruism in the population vanishes. Linkage between the two loci does not stop this.
7. Against this, heterozygotes might be expected to be much in excess at HLA loci just as they are at incompatibility loci in plants; so far this does not seem to be the case.
8. A possibility that integration of diverse genetical aptitudes may have been part of the group-living of primates in much earlier arboreal times is raised by the polymorphism found for retinal pigments in *Saimiri sciureus*, a New World monkey (Mollon, Bowmaker and Jacobs, 1984).
9. A reverse conjecture, compatible with this, which tries to understand ecosystems and many of their functions by comparison to human economies, is developed by Ghiselin (1974).
10. This footnote partly explains the digression. I want to boast of an animal named for me and even more to express my sense of loss from the death last year of O. W. Richards, who gave me a job at a time when no other biologist that I knew thought that animal nepotism was a 'worthwhile subject of study'.
    The species of the *Azteca* is not known; the wasp is *Stelopolybia hamiltoni* O. W. Richards. The collembolan is some species of *Cyphoderis*, as Richards told me. It was O. W. Richards also who knew, and also spelled for me, *Platyarthrus hoffmannseggii*, which made a white dust as of mealy bugs among the black *Lasius* ants in forked garden soil; who knew *Leptinus testaceus*, when I only described it, beetles blind as the white woodlice but no symphiles of the bumble bees in whose nest they were scrambling, instead parasites of the mice that had been there before. So with many others I learned from the man whose pockets, even whose dinner jacket, never lacked a specimen tube, and whom now I can consult no more.
11. The possibility of genocidal interaction among groups of early hominids has been suggested (Pitt, 1978; Szarski, 1983) and is plausible both theoretically (Vining, 1982) and by comparison with other broad niche social species, including ants (Wilson, 1971) and primates (Eibl-Eibesfeldt, 1979). Some aspects of human behavior seem to require an antecedence of strong kin-group selection, and for this to

work it is almost necessary that groups be mutually hostile as well as genetically closed. Intra-amicable and relatively non-nepotistic groups of Oceana (Sahlins, 1976) are possible examples. Unfortunately, as to distant patterns of group selection in the main stock it is unlikely that the fossil record can reveal much direct evidence.

LITERATURE CITED

Alexander, R. D. 1979. *Darwinism and human affairs*. University of Washington Press, Seattle and London.
Axelrod, R. 1984. *The evolution of cooperation*. Basic Books, New York.
Axelrod, R. and W. D. Hamilton. 1981. The evolution of cooperation. *Science*, **211**, 1390–1396.
Bateson, P. 1983. Optimal outbreeding, in Bateson, P., ed. *Mate choice*. Cambridge University Press, Cambridge, England, pp. 257–277.
Beecher, M. D. 1982. Signature systems and kin recognition. *American Zoologist*, **22**, 477–490.
Bell, G. 1982. *The masterpiece of nature. The evolution and genetics of sexuality*. University of California Press, Berkeley, California.
Beer, A. E., M. Gagnon and J. F. Quebbeman. 1981. Immunologically induced reproductive disorders, in Crosignani, P. G. and B. L. Rubin, eds., *Endocrinology of human infertility: new aspects*. Academic Press, London, pp. 419–439.
Bohart, G. E. 1970. *The evolution of parasitism among bees*. Forty-first Honor Lecture, The Faculty Association, Utah State University, Logan, Utah.
Boyd, R. and P. J. Richerson. 1980. Effects of phenotypic variation on kin selection. *Proceedings of the National Academy of Sciences, USA*, **77**, 7806–7810.
Boyse, E. A., G. K. Beauchamp and K. Yamazaki. 1983. The sensory perception of genotypic polymorphism of the major histocompatibility complex and other genes: some physiological and phylogenetic implications. *Human Immunology*, **6**, 177–183.
Breed, M. D. 1983. Nest mate recognition in honeybees. *Animal Behaviour*, **31**, 86–91.
Buckle, G. R. and L. Greenberg. 1981. Nestmate recognition in sweat bees (*Lasioglossum zephyrum*): Does an individual recognize its own odour or only odours of its nestmates? *Animal Behaviour*, **29**, 802–809.
Buckley, P. A. and F. G. Buckley. 1970. Color variation in the soft parts and down of Royal Tern chicks. *Auk*, **87**, 1–13.
Carlile, M. J. 1973. Cell fusion and somatic incompatibility in myxomycetes. *Berichte der Deutschen Botanischen Gesellschaft Berlin*, **86**, 123–139.
Cavalli-Sforza, L. L. and M. W. Feldman. 1978. Darwinian selection and altruism. *Theoretical Population Biology*, **14**, 268–280.
Cheverud, J. M. 1985. A quantitative genetic model of altruistic selection. *Behavioral Ecology and Sociobiology*, **16**, 239–243.
Daly, M. and M. Wilson. 1983. *Sex, evolution and behavior*. Willard Grant, Boston.
Dawkins, R. 1976. *The selfish gene*. Oxford University Press, Oxford, England.
de Nettancourt, D. 1977. *Incompatibility in angiosperms*. Springer Verlag, Berlin.
Eibl-Eibesfeldt, I. 1979. *The biology of peace and war* (tr. E. Mosbacher). Thames and Hudson, London.
Emlen, S. T. 1984. Cooperative breeding in birds and mammals, in Krebs, J. R. and N. B. Davies, eds., *Behavioural ecology: and evolutionary approach*, Blackwell, Oxford, England, pp. 305–339.
Fisher, R. A. 1930. *The genetical theory of natural selection*. Oxford University Press, Oxford, England.

Getz, W. M. and K. B. Smith. 1983. Genetic kin recognition: honeybees discriminate between full-sisters and half-sisters. *Nature*, **302**, 147–148.

Ghiselin, M. T. 1974. *The economy of nature and the evolution of sex*. University of California Press, Berkeley, California.

Graham, B. F. and J. H. Borman. 1966. Natural root grafts. *Botanical Review*, **32**, 288–292.

Hamilton, W. D. 1964a. The genetical evolution of social behaviour, I. *Journal of Theoretical Biology*, **7**, 1–16.

Hamilton, W. D. 1964b. The genetical evolution of social behaviour, II. *Journal of Theoretical Biology*, **7**, 17–32.

Hamilton, W. D. 1967. Extraordinary sex ratios. *Science*, **156**, 477–488.

Hamilton, W. D. 1970. Selfish and spiteful behaviour in an evolutionary model. *Nature*, **228**, 1218–1220.

Hamilton, W. D. 1971. Selection of selfish and altruistic behavior in some models, in Eisenberg, J. F. and W. S. Dillon, eds., *Man and beast: comparative social behavior*, Smithsonian Press, Washington, DC, pp. 57–91.

Hamilton, W. D. 1972. Altruism and related phenomena, mainly in social insects. *Annual Review of Ecology and Systematics*, **3**, 193–232.

Hamilton, W. D. 1975. Innate social aptitudes of man: an approach from evolutionary genetics, in Fox, R. ed., *Biosocial anthropology*, Malaby Press, London, pp. 133–155.

Hamilton, W. D. 1982. Pathogens as causes of genetic diversity in their host population, in Anderson, R. M. and R. M. May, eds., *Population biology of infectious diseases*, Dahlem Konferenzen, Springer Verlag, Berlin, pp. 269–296.

Hanken, J. and P. W. Sherman. 1981. Multiple paternity in Belding's ground squirrel litters. *Science*, **213**, 351–353.

Herman, H. R. 1971. Sting autotomy, a defensive mechanism in certain social Hymenoptera. *Insectes Sociaux*, **18**, 111–120.

Holmes, W. G. and P. W. Sherman. 1982. The ontogeny of kin recognition in two species of ground squirrels. *American Zoologist*, **22**, 491–517.

Lewis, D. 1979. Genetic versatility of incompatibility in plants. *New Zealand Journal of Botany*, **17**, 637–644.

Lubbock, R. 1980. Clone-specific cellular recognition in a sea anemone. *Proceedings of the National Academy of Sciences, USA*, **77**, 6667–6669.

Michener, C. D. 1975. The Brazilian bee problem. *Annual Review of Entomology*, **20**, 399–416.

Michod, R. 1982. The theory of kin selection. *Annual Review of Ecology and Systematics*, **13**, 23–55.

Mollon, J. D., J. K. Bowmaker and G. H. Jacobs. 1984. Variations in colour vision in a New World Primate can be explained by polymorphism of retinal photopigments. *Proceedings of the Royal Society of London*, (B), **222**, 373–399.

Neigel, J. E. and J. C. Avise. 1983. Histocompatibility bioassays of population structure in marine sponges. Clonal structure in *Verongia longissima* and *Iotrochota birotulata*. *Journal of Heredity*, **74**, 134–140.

Page, R. E. and R. A. Metcalf. 1982. Multiple mating, sperm utilization and social evolution. *American Naturalist*, **119**, 263–281.

Pakstis, A. S., S. Scarr-Salapatek, R. C. Elston and R. Siervogel. 1972. Genetic contributions to morphological and behavioral similarities among sibs and dizygotic twins: linkages and allelic differences. *Social Biology*, **19**, 185–192.

Pitt, R. 1978. Warfare and hominid brain evolution. *Journal of Theoretical Biology*, **72**, 551–575.

Ridley, M. and A. Grafen. 1981. Are green beard genes outlaws? *Animal Behaviour*, **29**, 954–955.

Sahlins, M. D. 1976. *The use and abuse of biology: an anthropological critique of sociobiology*. University of Michigan Press, Ann Arbor, Michigan.

Samuelson, P. A. 1983. Complete genetic models for altruism, kin selection and like-gene selection. *Journal of Social and Biological Structures*, **6**, 3–15.

Scofield, V. L., J. M. Schlumpberger, L. A. West and I. L. Weisman. 1982. Protochordate allorecognition is controlled by a MHC-like gene system. *Nature*, **295**, 499–502.

Shields, W. M. 1982. *Philopatry, inbreeding and the evolution of sex*. State University of New York Press, Albany, New York.

Smith, B. H. 1983. Recognition of female kin by male bees through olfactory signals. *Proceedings of the National Academy of Sciences, USA*, **80**, 4551–4553.

Solecki, R. S. 1971. *Shanidar. The humanity of Neanderthal man*. Allen Lane, London.

Stone, J. E. and E. L. Stone. 1975. The communal root system of red pine: water conduction through root graft. *Forest Service*, **21**, 255–262.

Szarski, H. 1983. Why did the human brain cease to increase 100,000 years ago. *Bulletin de l'Academie Polonaise des Sciences (Sciences biologiques, Cl II)*, **29**, 381–383.

Vining, D. R. 1982. Group selection via genocide. *Mankind Quarterly*, **21**, 27–41.

West-Eberhard, M. J. 1975. The evolution of social behaviour by kin selection. *Quarterly Review of Biology*, **80**, 513–530.

White, M. J. D. 1973. *Animal cytology and evolution*, 3rd Edn. Cambridge University Press, Cambridge, England.

Wilson, D. S. 1980. *The natural selection of populations and communities*. Benjamin/Cummings, Menlo Park, California.

Wilson, E. O. 1971. *The insect societies*. Harvard University Press, Cambridge, Massachusetts.

Winston, M. L., O. R. Taylor and G. W. Otis. 1983. Some differences between temperate European and tropical African and South American Honeybees. *Bee World*, **64**, 12–21.

Yom-Tov, Y. 1980. Intraspecific nest parasitism in birds. *Biological Reviews*, **55**, 93–108.

Author Index

Adler, K., 26, 27, 298, 395
Ågren, G., 323
Ahearn, J. N., 109
Alexander, B. K., 45, 319
Alexander, R. D., 21–23, 33, 55, 57, 58, 68, 288–290, 339, 360, 407–409, 421, 423, 427
Ali, R., 361, 362, 364, 374, 375, 379, 382–384
Allen, J. L., 229
Alloway, T. M., 217, 275
Altmann, J., 361, 363, 368, 369, 372, 374–376, 378, 379, 383
Altmann, S. A., 368, 373
Ambrose, J. T., 249
Anderson, P. K., 315, 320
Anderson, W. W., 100, 103, 104
Andjelković, M., 107
Andrews, E. A., 248
Antony, C., 77, 83–85
Arita, L. H., 109
Armitage, K. B., 288, 330
Averhoff, W. W., 67, 77, 82
Avise, J. C., 61, 426
Axelrod, R., 429, 430
Ayre, D. J., 61

Baker, A. E. M., 320
Baker, M, C., 404
Ballou, J., 379
Balogh, R. D., 398, 400
Barash, D. P., 290
Barlow, G. W., 291
Barnard, C. J., 34, 41, 56, 312, 317–319, 321, 335

Barnett, C., 48, 291
Barrows, E. M., 29, 31, 33, 210, 213, 214, 221, 222, 226
Bartell, R, J., 77, 81–83
Bateson, P. P. G., 8, 19, 29, 75, 217, 287, 306–308, 320, 336, 359, 374, 379, 381, 388, 408, 419
Batra, S. W. T., 212
Bazire-Bénazet, H., 262, 267, 277
Bearder, S., 363
Beauchamp, G. K., 45, 62, 403, 425
Beecher, I. M., 32, 36, 43, 306–309, 334, 342
Beecher, M. D., 32, 33, 36, 43, 55, 65, 126, 306–309, 333, 334, 342, 401, 420
Beer, A. E., 403, 408, 424
Beer, C. G., 32
Beiswenger, R. E., 304, 306
Bekoff, M., 32, 33, 277, 287, 288, 289, 301, 322, 324, 325, 333, 334, 340, 343, 344–346, 362, 365, 374
Bell, G., 419
Bell, W. J., 31, 210, 214, 220–222, 226, 238
Benest, G., 238
Bengtsson, B. O., 381
Bennet-Clark, H. C., 76–80
Bennett, B., 243, 266, 269, 271, 277
Bennett, D., 319
Bentley, D., 58
Berenstain, L., 377
Bergström, G., 275
Berkelhamer, R. C., 254
Berman, C. M., 366, 367, 370–372, 374, 375
Bertram, B. C. R., 306, 362

Birney, E. C., 324
Blackman, R. L., 61
Blaustein, A. R., 31, 34, 40, 41, 44, 57, 287, 288, 292–298, 301, 302, 305, 306, 313, 315, 319, 320, 322–324, 333–335, 337, 338, 340, 343, 395, 407
Blum, M. S., 27, 33, 226, 275
Boch, R., 249
Bohart, G. E., 421
Borgia, G., 33, 55, 57, 58, 68, 290
Borman, J. H., 433
Bornais, K. M., 36, 230
Bowbal, D. A., 97, 98, 102
Bowers, J. M., 45, 319
Bowmaker, J. K., 434
Box, H. O., 364
Boyd, R., 418
Boyd, S. K., 313, 315, 323, 324, 340
Boyse, E. A., 45, 62, 318, 403, 425
Bradshaw, J. W. S., 266, 267
Breden, F., 362
Breed, M. D., 36, 134, 238, 243, 249–251, 269, 333, 343, 374, 375, 422
Brian, M. V., 253, 276
Brodie, E. D., 295
Brothers, D. J., 220, 221, 226
Brown, C. E., 380
Brown, C. J., 103
Brown, J. S., 362, 375
Brown, K., 365
Bruce, H. M., 318
Brückner, D., 27, 31, 249, 251
Bryant, E. H., 82, 99–101
Bryant, P. J., 79, 88, 89, 94
Buckle, G. R., 39, 66, 67, 215, 221, 222, 225, 236, 339, 422, 434
Bugos, P. E., 405
Burgess, R. L., 409
Burk, T., 56
Burnet, B., 76–79, 81, 87
Burtt, E. H., 36
Busack, C. A., 290–292, 334
Bushnell, I. R. W., 397
Buskirk, R. E., 360
Buskirk, W. H., 360
Busse, C. D., 362, 372
Butler, C. G., 23, 36, 46
Butler, L., 250
Byers, J. A., 374, 375
Bygott, J. D., 368, 370

Caine, N., 372
Caire, W., 315, 322
Caldwell, R. L., 33
Calhoun, J. B., 321
Cammaerts, M.-C., 267, 268, 270
Cammaerts-Tricot, M.-C., 267
Carlile, M. J., 426
Carlin, N. F., 67, 68, 256, 259, 260, 271, 272, 276
Carlson. R. G., 227
Carpenter, G. C., 397
Carrick, M. J., 32
Carson, H. L., 89, 94, 108, 113
Carter, C. S., 315, 324
Carter-Saltzman, L., 402, 404
Cavalli-Sforza, L. L., 433
Cernoch, J. M., 398–400
Chagnon, N. A., 405
Chang, H.-C., 77
Charles-Dominique, C., 363
Charnov, E. L., 322, 325
Chauvin, R., 266
Cheney, D. L., 32, 360, 363, 368, 370, 372, 375, 383
Cherix, D., 46
Cherrett, J. M., 260, 266, 270, 271, 277
Cheverud, J. M., 418
Chivers, D. J., 380
Christian, J. J., 341
Clark, A., 363
Clement, J.-L., 248
Clements, F. A., 308
Coenen-Stass, D., 122, 135, 197
Colgan, P., ix, 287, 291, 344
Colvin, J., 370, 383
Connell, J. H., 340
Connolly, K., 76–78, 87
Cooch, F. G., 310
Cook, R. M., 77, 78, 87
Cooke, F., 307, 310, 334
Coppersmith, R., 320
Cornell, J. M., 339
Cornell, T. J., 235
Cowling, D. E., 77, 81
Craig, R., 254
Cramp, S., 307
Crawford, M. H., 379
Creffield, J. W., 248
Creighton, W. S., 265
Crewe, R. M., 252
Crook, J. H., 360

Crossley, S. A., 77, 106
Crowcroft, P., 315, 320
Crozier, R. H., 34, 48, 55, 58, 62, 65, 67–70, 203, 227, 245, 246, 254, 258, 259, 333, 434
Crozier, Y. C., 254
Cunningham, M. A., 404
Cunningham, M. R., 410
Curio, E., 343

Daly, M., 409, 427
Daniels, T. J., 287, 333, 346
Davis, L. S., 315, 329, 330
Dawkins, R., 11, 22, 25, 26, 28, 33–35, 55–57, 248, 288–290, 339, 403, 407, 411, 423
de Nettancourt, D., 423, 434
De Vroey, S. C., 253, 257
DeBenedictis, P. A., 304
DeBruyn, G. J., 266
DeCasper, A. J., 398
Decelles, P., 218
Defler, T. R., 362
Del Rio Pasada, M. G., 275
Delany, M. J., 315, 316
DeLong, K. T., 315
Demarest, W. J., 381, 383
Dew, H. E., 234
Dewsbury, D. A., 30, 312, 313, 322, 323
Diamond, J. M., 340
Dicks, D., 370
Dix, M. W., 48, 55, 58, 62, 65, 67–69, 203, 245, 246, 258, 259, 333
Dodd, D. M. B., 107–109
Doschek, E., 78, 87
Døving, K. B., 292
Dow, M., 77
Downing, H. A., 234
Driessen, G. J. J., 266
D'Udine, B., 313, 319
Dumont-Driscoll, M., 402
Dumpert, K., 273
Dunbar, E. P., 378
Dunbar, R. I. M., 368, 378
Durham, W. H., 410

Eastwood, L., 77, 79
Echols, H. W., 266, 269
Egid, K., 318, 319
Ehrman, L., 43, 77, 82, 92, 100, 102, 103
Eibl-Eibesfeldt, I., 434

Eisenberg, J. F., 315, 322, 341
Elmes, G. W., 255
Emery, C., 274, 276
Emlen, J. T., 32
Emlen, S. T., 362, 425
Enomoto, T., 380, 381
Eoff, M., 105
Erickson, E. H., 27, 249, 251, 252
Errard, C., 271, 272, 277
Erwin, J. T., 375
Espmark, Y., 334
Esswein, U., 164, 202
Evans, H. E., 212
Evans, R. M., 306
Evers, C., 250
Evershed, R. P., 267, 268
Evesham, E. J. M., 270
Ewing, A. W., 76–81, 100

Falls, J. B., 32
Fairbairn, D. J., 315
Farquhar, M., 91
Faulk, W. P., 403
Fedigan, L. M., 372
Feldman, M. W., 433
Felsenstein, J., 106
Fielde, A. M., 257, 262, 277
Fifer, W. P., 398
Finerty, J. P., 322, 325
Finney, G. H., 307, 310, 334
Fisher, R. A., 24, 302, 418
Fletcher, D. J. C., 1, 19, 23, 27, 33, 36, 46, 254, 275, 276
Forbes, G. C., 234
Formby, D., 398
Fossey, D., 368, 380, 381, 383
Fox, L. R., 341
Franks, N. R., 32
Frase, B. A., 288
Frazier, J., 227
Fredrickson, W. T., 336–378
Free, J. B., 235
Freneau, D., 262, 273, 274
Fretwell, S. D., 339
Frumhoff, P., 250

Gagnon, M., 424
Gailey, D. A., 77, 78, 86
Gamboa, G. J., 30, 36, 212, 228–234, 236, 237
Gardner, A., 102

Gavish, L., 313, 315, 323, 324, 340
Gentry, T. A., 403
Gervet, J., 234
Getz, L. L., 315, 323, 324, 340
Getz, W. M., 11, 27, 31, 48, 55, 59, 63–65, 67, 236, 237, 245, 246, 249–251, 313, 323, 333, 422
Ghiselin, M. T., 434
Giddings, L. V., 109
Gilbert, D. G., 102
Gilder, P. M., 45, 313, 319
Gillespie, J. H., 70
Gittleman, J. L., 346
Goetsch, W., 255, 262
Goldizen, A., 364, 365, 372
Goldstein, R. B., 77, 78
Goodall, J., 360, 368
Gottlieb, G., 306
Gould, S. J., 343
Gouzoules, H., 362, 365, 367–371, 374, 380
Gouzoules, S., 362–365, 367–371, 373–376, 378–381
Grafen, A., 288, 290, 346, 423
Graham, B. F., 433
Grau, J. H., 312, 321, 322
Greenberg, L., 26, 29, 31, 39, 43, 66, 67, 134, 215, 218, 221–225, 236, 238, 339, 395, 405, 422, 434
Greenspan, R. J., 77, 78
Gregg, B., 409
Grewell, B. S., 371, 377, 384
Griffin, D. R., 287
Grosberg, R. K., 338
Gross, A. C., 78
Grossfield, J., 78, 86, 319
Gubernick, D. J., 44, 335

Hahn, S., 32, 36, 43, 307–309
Hall, J. C., 77, 78, 85, 86
Hall, L. M., 77, 80, 83, 85
Halliday, R. B., 265, 266
Halpin, Z. T., 32, 319
Hames, R. B., 405, 406
Hamilton, W. D., 1, 7, 19, 22, 24, 33, 34, 36, 55, 62, 243, 288, 290, 306, 346, 360, 403, 407, 410, 417, 419, 420, 422, 425, 428–430, 433
Hamilton W. J., III, 360, 372
Hanby, J., 380
Hangartner, W., 261

Hanken, J., 26, 28, 326, 330, 421
Hansen, R. M., 315, 322
Hanson, S. J., 78, 79, 87, 92
Happold, D. C. D., 315, 316
Harcourt, A. H., 364, 368, 372, 379–384
Hartung, J., 403
Hasegawa, T., 372
Haskins, C. P., 253, 256, 258, 265–267, 269, 272, 274, 276, 277
Haskins, E. F., 253, 256, 258, 265–267, 272, 274, 276, 277
Hasler, A. D., 291
Hassinger, D. D., 304
Hausfater, G., 363, 382
Hawkins, W. A., 220
Hellack, J. J., 313
Hendricks, H., 77
Henzi, S. P., 383
Hepper, P. G., 320
Hering, W., 202
Hermann, H. R., 269, 418
Hess, W. R., 341
Hessler, C. M., 334
Hews, D. K., 302, 306
Hilborn, R., 341
Hill, J. L. 313
Hinde, R. A., 25, 34
Hiraiwa, M., 372
Hoage, R. J., 364
Hodgkin, N. M., 88
Hoese, B., 122, 197
Hoffman, G., 127, 199, 201
Hoffman, J., 313, 323, 340
Hoikala, A., 77, 81
Hold, B., 399
Holdich, D. M., 125, 137
Hölldobler, B., 14, 33, 36, 42, 43, 67, 68, 126, 199, 212, 220, 246, 252, 253, 255, 256, 259, 260, 266–268, 271, 272, 276, 277
Hollien, H., 398
Holmes, W. G., 14, 20, 23, 33, 36, 37, 39, 66, 151, 203, 211, 247, 287–290, 313–315, 325–328, 330, 331, 334, 336–339, 342, 360, 374, 407, 421
Hood, L., 65
Hoogland, J. L., 309, 330, 342
Horvall, R. M., 291
Hotta, Y., 77, 83
Howard, D. F., 47, 254, 255
Howick, L. D., 248

Howse, P. E., 262, 266–268, 271, 277
Hoy, R. R., 58
Hrdy, S. B., 341, 370, 372, 374, 376
Huang, C., 408
Hubbard, M. D., 261

Idoji, H., 77–79
Ikeda, H., 77–80
Imanishi, K., 380
Ingram, J. C., 364
Irwin, C. J., 404
Isingrini, M., 253, 271, 273, 274
Itani, J., 380, 383

Jackson, F., 77, 86
Jacobs, G. H., 434
Jaffe, K., 256, 260, 262, 263, 267, 268, 277
Jaisson, P., 261, 271–274, 277
Jallon, J.-M., 77, 83–85
Jannett, F. J., Jr., 324
Janzen, D. H., 255, 265
Jeanne, R. L., 30, 228–230, 232, 234
Jeffreys, A. J., 409
Jenkins, S. H., 322
Johns, D. W., 330
Jutsum, A. R., 260, 266, 270, 271, 276

Kalmus, H., 249, 400
Kamm, D. R., 221
Kaneshiro, K. Y., 109
Kaplan, J. N., 32
Kaplan, J. R., 369
Kareem, A. M., 34, 41, 312, 317–319, 321, 335
Katz, L. C., 334
Kawai, M., 372
Kence, A., 99–101
Keough, M. J., 341
Keppie, D. M., 341
Kerr, S., 90
Kessler, S., 105
Kidd, K. K., 361, 364
Kimball, K. T., 99–101
Kimsey, R. B., 249
Klahn, J. E., 36, 230, 234
Kleiman, D. G., 382
Kline, J., 318
Kling, A., 370
Knoppien, P., 100, 101
Köhler, F., 46

Krebs, C. J., 315, 322, 340
Krempien, W., 125, 169
Kruckeberg, J. F., 92, 93, 95, 102
Kruse, K. C., 302
Kukuk, P. F., 134, 217, 220, 222, 223
Kummer, H., 372
Kurland, J. A., 27–29, 290, 366, 367, 369–371, 373, 409
Kyriacou, C. P., 77, 78

Labov, J. B., 318
Lack, D., 21
Lacy, R. C., 33, 55, 57, 60, 64, 65, 211, 290, 333, 339, 407
Laidlaw, H. H., 27, 249
Lakovaara, S., 77
Larch, C. M., 30, 229
Lehrman, D. S., 34
Leighton, D., 364, 380
Leland, L., 370, 376
LeMoli, F., 257, 272, 274, 275, 277
Lenington, S., 318, 319
Lenoir, A., 253, 271, 273, 274
Leon, M., 320
Leonard, J., 77, 82, 100, 102
Leroy, Y., 77
Levari, S., 398
Levieux, J., 268
Levine, L., 319
Lewis, D., 423
Lewontin, P. F., 343
Lewontin, R. C., 343
Lichtig, I., 398
Lightcap, J. L., 409
Lincoln, R. J., 121, 132
Linsenmair, C., 33, 121, 129, 158
Linsenmair, K. E., 33, 66, 121, 122, 129, 134, 135, 141–143, 145, 149, 152, 155–158, 164, 178, 181, 189–192, 194–199, 202, 206, 236
Llewellyn, J. B., 322
Loehlin, J. C., 402
Loekle, D. M., 341
Löfqvist, J., 275
Long, C. E., 90, 95
Longhurst, C., 268, 271
Lubbock, J., 255
Lubbock, R., 61, 62
Lucas, J. W., 383
Lumme, J., 77
Lumpkin, S., 36, 307–309, 342

Lumsden, C. J., 266, 267, 277

Mabelis, A. A., 266
MacFarlane, A., 399
Mack, D., 365
Madison, D. M., 341, 345
Makin, J. W., 311, 312, 316, 339
Mane, S. D., 77, 86, 95
Manning, A., 77, 79–81
Marcuse, M., 256, 262, 263, 277
Marikovsky, P. J., 121, 206
Markow, T. A., 78, 79, 87, 90–92, 95, 99
Marler, P., 42, 360
Marsdon, H. M., 372
Marsh, C. W., 382
Maruo, O., 77–79
Massey, A., 28, 369, 370
Matochik, J. A., 311, 312, 316, 339
Matsumoto, H., 78
Maynard Smith, J., 343
McArthur, P. D., 32
McCarley, H., 315
McDonald, J., 77, 79
McGuire, P. R., 103, 323
McKaye, K. R., 291
McKenna, J. J., 363, 375
McLaughlin, F. J., 400
McLean, I. G., 315, 326, 330
McNally, C. M., 310
Meikle, D. B., 370
Melhuish, E., 398
Melnick, D. J., 361, 364
Mendelson, T., 399
Metcalf, R. A., 70, 422
Michener, C. D., 1, 14, 21, 26, 31, 33, 42, 43, 67, 126, 134, 199, 209, 210, 212, 214, 220–222, 226, 235, 243, 246, 248, 252, 418
Michener, G. R., 315, 326, 328–330
Michod, R. E., 360, 361, 375, 418
Miller, C. L., 398
Miller, D. D., 77, 78, 80
Miller, D. E., 32
Miller, M. H., 370
Mills, M., 398
Mintzer, A., 67, 257, 258, 260, 265, 276
Mirsky, P. J., 310
Missakian, E. A., 380
Mitani, J. C., 360, 365
Mitchell, G., 372
Mollon, J. D., 434

Moore, J., 361–364, 368, 374, 375, 379, 382–384
Moore, J. D., 316, 400
Moore, R. A., 249
Morel, L., 258
Morgan, E. D., 267, 268
Mori, A., 273, 277
Morse, D. H., 361
Morse, P. A., 398
Morse, R. A., 249
Morton, L., 67, 83
Müller, E., 398
Murry, T., 398
Myers, C. A., 341
Myers, J. H., 315, 322, 340
Myers, J. P., 361
Myrberg, A. A., 32
Myton, B., 315, 321

Neigel, J. E., 61, 426
Nicolaides, N., 401
Nishida, T., 368
Noonan, K. C., 250, 252
Noonan, K. M., 230
Nordeng, H., 292
Nowbahari, M., 253, 271, 273, 274
Nussbaum, R. A., 295

Oakley, B., 292
O'Donald, P., 93, 94
O'Hara, R. K., 31, 40, 41, 44, 288, 292–298, 301, 304–306, 334, 335, 337, 338, 341, 395
Olson, D. H., 298, 305
Olson, D. K., 368, 371
O'Rourke, D. J., 379
Oster, G. F., 7
Otis, G. W., 418
O'Tousa, J. E., 78, 87

Packer, C., 361–364, 369, 376, 379–385
Page, R. E., 27, 70, 249, 251, 252, 422
Pak, W. L., 78, 87
Pakstis, A. S., 402, 423
Pamelo, P., 254
Pankey, J., 37, 311, 316, 334
Parisian, T. R., 27, 31, 249, 251
Parmigiani, S., 257
Parsons, P. A., 76
Parsetti, M., 272, 277
Partridge, B. L., 334

Partridge, L., 91, 102, 313, 319, 408, 409
Pasteels, J. M., 253, 257
Paterson, H. E., 75
Patty, R. A., 77, 78
Pearson, B., 254, 270
Peeke, H. V. S., 399
Pereira, M. E., 361, 363, 369, 372, 378
Petersen-Braun, M., 264, 270
Petit, C., 76, 92
Petrinovich, L., 32
Pfennig, D. W., 36, 212, 228, 232
Pickens, A. L., 248
Pinsker, W., 78, 87
Pitcher, T. J., 334
Pitt, R., 434
Poindron, P., 32
Polis, G. A., 341
Pollock, G. B., 36
Polzine, K. M., 403
Popper, K. R., 25
Porter, R. H., 37, 38, 288, 311, 312, 316, 334, 335, 339, 398–400
Post, D. C., 30, 228–230, 232, 234
Potel, M. J., 334
Powell, J., 67, 83, 107–109
Pratte, M., 231
Probber, J., 43, 92
Prout, T., 103
Pruzan, A., 77, 82
Pusey, A. E., 361–364, 368, 380–385

Quaid, M., 90
Quebberman, J. F., 424
Quinn, J. F., 338
Quinn, T. P., 290–292, 334

Radesäter, T., 307, 309
Raemaekers, J. J., 380
Ralls, K., 379
Reeve, H. K., 36, 212, 228, 232
Reichson, J. M., 261
Richards, O. W., 434
Richardson, R. H., 67, 77, 82
Richerson, P. J., 418
Richmond, R. C., 77, 86, 95
Riedman, M. L., 342
Ringo, J. M., 77–80, 89, 105, 110
Rheingold, H. L., 289
Ribbands, C. R., 36, 44, 134, 249
Ridley, M., 290, 423
Riggio, R. J., 32

Robertson, H. M., 77–79, 100, 110
Robinson, G., 249, 251
Robinson, J. G., 360
Rockwell, R. F., 307, 310, 319, 334
Rockwood, L. L., 270
Rodman, P. S., 360, 365, 370, 377
Rohwer, S., 339
Romppainen, E., 77
Rongstad, O. J., 328
Rose, R. J., 402
Røskaft, E., 334
Ross, K. G., 23, 27, 235, 254, 276
Ross, N. M., 230–232
Rothstein, S. I., 290, 322, 333, 342, 343
Rowe, F. P., 315, 320
Rowell, T. E., 368
Rudolph, N., 398
Rushton, J. P., 289, 409
Russell, M. J., 399
Russell, R. J. H., 289, 409
Rutter, W. J., 211
Ryan, R. E., 229, 234, 235

Sackett, G. P., 336, 337, 378, 385
Sade, D. S., 366, 380, 381
Sahlins, M. D., 406, 434
Salazar, L. I., 95, 96
Salceda, V. M., 104
Salzen, E. A., 339
Samollow, P. B., 302
Samuels, A., 370
Samuelson, E. A., 434
Sanchez, C., 260, 263, 268, 277
Sanderson, M. J., 362, 375
Saunders, T. S., 260, 266, 270, 271, 277
Sawada, N., 77–79
Scarr-Salapatek, S., 402, 404
Schleidt, M., 399
Schneider, P., 121, 206
Schneirla, T. C., 256, 272, 276, 277
Schoener, T. W., 340
Scholz, T., 291
Schorsch, M., 77, 82
Schulze-Kellman, K., 229
Schwagmeyer, P. L., 315, 328
Schwer, W. A., 93
Scofield, V. L., 62, 338, 424
Scott, L. M., 380, 383
Scovell, E., 32
Sebens, K. P., 61
Seelinger, G., 121, 125, 196

Seelinger, U., 121, 196
Segal, N. L., 460
Seiger, M. B., 310
Selander, R. K., 315, 320
Senior, A., 95
Seyfarth, R. M., 32, 360, 364, 365, 367, 369, 370, 375, 379
Shachak, M., 122, 135, 198
Shapiro, D. Y., 344
Sharp, G. D., 344
Shellman, J. S., 37, 232
Shellman-Reeve, J., 234
Shepher, J., 381, 408
Shepherd, G, M., 211
Sheppard, D. H., 315, 329
Sherman, P. W., 14, 20, 23, 24, 26, 28, 33, 36, 37, 39, 55, 57, 60, 64–66, 151, 203, 211, 247, 288–290, 309, 313–315, 325–328, 330, 333, 334, 336–339, 342, 360, 368, 373–376, 407, 421
Shields, W. M., 287, 306, 326, 328, 419
Shorey, H. H., 76, 81, 82, 83
Sieber, O. J., 32
Siegel, R. W., 77, 78, 86
Sigg, H. A., 368
Silk, J. B., 365, 368, 370, 373, 376, 379
Silverman, J. M., 238
Sim, L., 32
Simon, G. S., 345
Slater, P. J. B., 45, 308, 313, 319
Sloane, C., 77, 81
Small, M. F., 28, 378
Smith, B. H., 31, 70, 209, 215–218, 223, 226, 227, 237, 243
Smith, D. G., 28, 377–379, 381
Smith, K. B., 249, 250, 422
Smith, K. S., 372
Smith, M. V., 252
Smuts, B., 376, 377
Solbrig, O. T., 61
Solecki, R. S., 432
Spencer, H., 269
Spiess, E. B., 75–77, 81, 90, 92–99, 101, 102, 106
Spieth, H. T., 76, 78, 80, 86–89, 105, 110
Stabell, O. B., 292, 334
Starin, E. D., 365
Stebbins, R. C., 302
Steele, E. J., 338
Stein, D., 372, 376, 377

Steinmetz, M., 65
Stewart, K. S., 368, 372, 380–383
Stickel, L. F., 315, 321
Stiller, T. M., 250
Stoddart, D. M., 344
Stoddart, J. A., 61
Stone, B. M., 302
Stone, E. L., 433
Stone, J. E., 433
Storm, R. M., 295, 304
Struhsaker, T. T., 368, 370, 376
Strum, S., 376
Stuart, R. J., 259
Sturtevant, A. H., 81
Sudd, J. H., 253
Sugiyama, Y., 383
Suomalainen, E., 61
Sved, J. A., 106
Sype, W. E., 305
Szarski, H., 434

Taber, S., 27
Takabatake, I., 77–79
Takahata, Y., 380
Takeda, N., 122
Taylor, C., 403
Taylor, O. R., 418
Taylor, R. W., 253, 255, 256, 276, 277
Templeton, A. R., 68, 109
Tepper, C. S., 95
Tepper, V. J., 38, 311, 312, 335
Thein, S. L., 409
Thiessen, D. D., 409
Thomas, G., 77
Thomas, J. A., 324
Thorne, B. L., 248
Tilson, R. L., 380, 383
Tomkins, L., 77–79, 83, 85, 89
Traniello, J. F. A., 264, 269, 276
Treisman, M., 338
Trivers, R. L., 7, 22
Tsacas, L., 77
Tschinkel, W. R., 47, 254, 255
Tuckfield, R. C., 339

van Delden, W., 77
van den Berg, M. J., 77, 83, 85
van den Berghe, P. L., 381
Van der Meer, R. K., 275
van Lawick-Goodall, J., 380
Van Raalte, A. T., 266

Van Wormer, J., 307, 310
Vandenberg, S. G., 402
Varvio-Aho, S.-L., 254
Veal, R., 315, 322
Vehrencamp, S. L., 362
Velthuis, H. H. W., 249, 251
Venard, R., 77, 83, 84
Vessey, S. H., 370, 372
Vestal, B. M., 313, 315
Vinson, S. B., 258, 276
Visscher, P. K., 27, 251
Vom Saal, F. S., 335, 344
von Schilcher, 76–80

Wade, M. J., 361, 362
Wade, T. C., 361, 362, 379, 382
Wakefield, J. A., Jr., 403
Waldman, B., 26, 27, 31, 44, 293–295, 298–300, 304, 335, 395, 420
Wallace, P., 401
Wallis, D.. I., 253, 257, 258, 260, 276
Walters, J. R., 359, 364–370, 372–374, 378–380, 382
Ward, P. S., 254, 276
Wassersug, R. J., 302, 334
Watler, D., 36
Watanabe, T. K., 103
Wcislo, W. T., 218, 227
Weigel, R. M., 373
Weining, D. R., 434
Wells, M. C., 346
Wells, P. A., 289, 395, 398, 409
Wendel, J., 27
West-Eberhard, M. J., 212, 433
Wheeler, W. M., 253, 271
Whelden, R. M., 265
White, D. M., 38, 311, 312, 316, 335
White, M. J. D., 433

Whitten, P. L., 376
Wiens, J. A., 340
Wilbur, H. M., 306
Wilke, C. M., 101, 106
Willerman, L., 402
Wilson, D. S., 428
Wilson, E. O., 7, 21, 36, 134, 199, 244, 253, 254, 260, 261, 264, 267, 272, 274–277, 288, 334, 344, 434
Wilson, M. I., 409, 427
Wilson, S. C., 314, 323, 324
Wilson, V., 409
Winship-Ball, A., 32
Winston, M. L., 418
Winterbottom, S., 261–264, 270
Wojcik, D. P., 275
Wolf, A. P., 408
Wood, D. F., 77, 79, 110
Wrangham, R. W., 360, 361, 363, 364
Wright, A. A., 295, 302
Wright, A. H., 295, 302
Wu, H., 35, 40, 41, 43, 44, 336–338, 378
Wynne-Edwards, V. C., 22
Wyrick, M., 37, 38, 311, 312, 316, 334

Yadava, R. P. S., 252
Yaeger, P., 90, 95
Yair, A., 122, 135
Yamada, M., 368
Yamaguchi, M., 45, 318, 319, 334
Yamazaki, K., 45, 62, 312, 318, 319, 334, 403, 425
Yeaton, R. I., 329
Yom-Tov, Y., 342, 421
Yoshida, S. M., 315, 329
Young, A. M., 269
Young, J. Z., 381
Younger, B. A., 398

Index of Scientific and Common Names

Acacia hindsii, 67
Acomys, 335
   cahirinus, 15, 37, 311–313
Acromyrmex octospinosus, 260, 262, 266, 267, 270, 271
Aenictus, 256
Amblyopone pallipes, 264, 265
Amphibians, 12, 292–306, 334, 345
Anemones, 61, 68, 70, 426
Anser caerulescens, 307, 310
Antelopes, 9
Ants, 8, 10, 55, 210, 243, 244, 252–278, 434
   African weaver, 36, 267
   Acacia, 67, 255, 265
   Australian meat, 265
   European red wood, 265
   fire, 13, 27, 31, 33, 46, 47, 254
   harvester, 268
   leaf-cutting, 260, 266, 269
   Pharoah's, 264
   slave-making, 31, 32, 274, 275
Aphids, 61, 421
Apis mellifera, 4, 13, 17, 23, 27, 248–252
   adansonii, 252
   capensis, 252
   mellifera, 252
Ascidians, 62
Atta
   cephalotes, 260, 262, 267, 270, 277
   columbica, 270
   sexdens, 262, 267
   texana, 266, 269
Australopithecus, 432
Azteca, 272, 431, 434

Baboons, 369, 370, 376, 377, 379–382

Baboons (*cont.*)
   gelada, 368
   yellow, 368, 369, 373, 378, 383
Bees, 10, 67, 209, 210, 213, 220, 235–238, 243, 244, 359, 421, 422
   bumble, 235
   honey, 4, 13, 17, 23, 27, 31, 36, 44, 46, 47, 49, 70, 210, 248–252, 278, 418, 422, 427, 428
   sweat, 3, 13, 17, 31, 33, 39, 212–227, 395, 405, 434
Beetles, 13, 275
Birds, 8, 9, 24, 29, 32, 33, 35, 36, 39, 56, 65, 306–310, 334, 361, 362, 364, 404, 420, 421, 425
   bank swallow, 14, 36, 37, 43, 306–309, 342, 401
   Canada goose, 306, 307, 309, 310
   chicken, 339
   cuckoo, 33, 56
   Japanese quail, 8, 17, 29, 306–308, 339, 379, 408, 419
   lesser snow goose, 307, 310, 339
   sanderlings, 361
   spruce grouse, 9, 341
   zebra finch, 308
Bombus, 213, 223, 238
Botryllus schlosseri, 62, 69
Branta canadensis, 306, 307
Bufo
   americanus, 26, 27, 31, 294, 295, 298–302, 304, 306
   boreas, 292, 294, 295, 298, 300–304

Calidris alba, 361
Callicebus torquatus, 365

Camponotus, 15, 18, 67, 68, 255, 256, 259, 260, 272, 276
  abdominalis, 272
  ligniperda, 273
  pennsylvanicus, 259
  rufipes, 260, 263, 268
  senex, 272
  vagus, 258, 261
Canids, 8, 374, 400, 401
Cataglyphis cursor, 253, 273
Chimpanzees, 18, 364, 365, 368, 370, 380, 381, 384
Cockroaches, 121, 196
Colobus badius, 364
Collembola, 431
Conomyrma bicolor, 254
Coptotermes acinaciformis, 248
Corals, 15, 32, 61, 426
Coturnix coturnix, 8, 17, 29, 306, 307, 379
Crematogaster, 255
Cryptocercus, 121
  punctulatus, 196
Cynomys ludovicianus, 330
Cyphoderis, 434

Dialictus zephyrus, 213
Dolichovespula maculata, 235
Drosophila, 3, 16, 42, 48, 75–119
  affinis, 80
  athabasca, 80
  funebris, 80
  mauritiana, 110
  melanica, 80
  melanogaster, 67, 76, 78–81, 83–88, 90, 93–95, 100–102, 105, 106, 110
  mercatorum, 78, 80
  obscura, 80, 88
  paulistorum, 80
  persimilis, 80, 88, 94, 101, 105, 110
  silvestris, 87, 94
  simulans, 79, 80, 87, 88, 105, 110
  pseudoobscura, 67, 80–83, 87, 88, 100, 103–105, 107, 108, 110
  subobscura, 86, 88
  virilis, 81

Eciton, 256
Erythrocebus patas, 368

Fish, 12, 25, 290–292, 334, 344
  midas, 48
  salmon, 9
  salmon, coho, 290
  salmon, Atlantic, 292
  tuna, 9, 344
Formica, 274
  fusca, 253, 255, 257, 258, 260, 266, 272, 275, 277
  lugubris, 257, 272
  marcida, 266
  podzolica, 266
  polyctena, 261, 265, 272, 273
  rufa, 272
  rufibarbis, 275
  sanguinea, 254, 275
  subaenescens, 266, 271
Frogs, 9, 17, 335, 339
  cascades, 31, 40, 44, 49, 292–298, 335, 337, 339
  red-legged, 292–295, 298, 334
  wood, 294, 295, 299, 300, 334, 341

Galagos, 363
Gerbils, Mongolian, 323
Gibbons, 363, 380, 384
Goats, 44
Gonodactylus festae, 33
Gorilla gorilla, 364
Gorillas, 364, 368, 380, 382
Ground squirrels, 37, 66, 325–331, 336, 338, 339, 374, 421, 422
  Arctic, 39, 314, 315, 325–328
  Belding's, 17, 24, 26, 28, 36, 39, 313, 314, 325–328, 336, 342
  Richardson's, 314, 315, 329, 330
  thirteen-lined, 314, 315, 329–331
Guppies, 9, 341

Halictus zephyrus, 213
Harpagoxenus, 32
  americanus, 31, 32, 275
Hemilepistus reaumuri, 4, 33, 66, 121–208
Homo sapiens, 419
  neanderthalensis, 432
Houseflies, 102
Humans, 13, 17, 21, 23, 28, 65, 395–415, 417, 419, 422, 424, 426, 427–434
  Andean Indians, 431
  Chinese, Hokkien-speaking, 408
  Eskimos, 404

Humans (*cont.*)
  Tasmanians, 432
  Yanamamo Indians, 405
  Ye'kwana Indians, 13, 405, 406
Hymenoptera, 9, 43, 58, 60–62, 70, 209, 210, 237, 270, 421
Hypoponera punctatissima, 431

Iridomyrmex purpureus, 265
Isopods, desert, 33, 44, 121–208

Langurs, 384
  douc, 365
Lasioglossum, 3, 17, 18, 66–68, 211, 228, 237, 238
  zephyrum, 13, 26, 29, 31, 33, 39, 43, 70, 212–227, 232, 233, 236–238
Lasius
  neoniger, 269
  niger, 274, 275, 277
Leptogenys elongata, 271
Leptothorax, 259
  ambiguus, 259
  longispinosus, 259

Macaca
  fuscata, 27, 383
  nemestrina, 28, 336, 378
  mulatta, 28, 372
  radiata, 362
Macaques, 35, 338, 370, 377, 379–381
  bonnet, 362
  Japanese, 27–29, 383
  pigtail, 28, 40, 44, 337, 338, 378
  rhesus, 9, 28, 372, 378, 380–382
Mantis shrimps, 33
Marmots, 330
Megaponera foetans, 268, 271
Meliponinae, 248
Mellivora capensis, 23
Meriones unguiculatus, 323
Messor, 255
Mice, 16, 41, 49, 65, 66, 311–322, 338, 346, 403
  cactus, 313, 314, 321, 322
  deer, 30, 312, 313, 321, 322
  house, 16, 312, 313, 317–320, 335, 338, 339, 425
  spiny, 15, 37–39, 311–314, 316, 319, 335
  white-footed, 312, 313, 321, 322

Microtus
  canicaudus, 313–315, 323, 324
  montanus, 324
  ochrogaster, 313–315, 323, 324
  pennsylvanicus, 323
Mole rats, 8
Monkeys
  howler, 384
  patas, 368
  red colobus, 364
  titi, 365
Monomorium
  ebeninum, 272
  pharaonis, 264, 270
Mus musculus, 16, 41, 312, 313, 317–320
Mycetozoa, 426
Myrmecia, 272, 274, 276
  nigrocincta, 256
  pilosula, 254
  tarsata, 256
Myrmecophodius excavaticollis, 13, 275
Myrmica, 253, 255, 267
  rubra, 253, 254, 257, 261, 264, 268, 270
  ruginodis, 268
  sabuleti, 264, 268
  scabrinodis, 263, 264, 268

Nasutitermes ripperti, 248
Neivamyrmex, 256, 277
Neoponera apicalis, 262
Norway rat, 320, 321
Nothomyrmecia macrops, 253, 255, 276, 277

Odontomachus bauri, 256, 263, 277
Oecophylla longinoda, 36, 267
Oncorhyncus kisutch, 290

Pachycondyla harpax, 271
Pan troglodytes, 364
Panda, 432
Papio cynocephalus, 368
Peromyscus, 313, 314, 323
  eremicus, 30, 322
  leucopus, 312, 313, 321, 322
  maniculatus, 30, 312, 313, 322
Plants, 423, 430, 433, 434
  acacia, 67, 257
  cherry, 424
  dandelion, 61
Poecilia reticulata, 9, 341

Pogonomyrmex, 269
  badius, 261
  barbatus, 268
  maricopa, 268
  rugosus, 268
Polistes, 18, 211–213, 228–238
  carolina, 232
  exclamans, 229
  fuscatus, 30, 36, 228, 230, 232–235, 237
  gallicus, 231
  metricus, 230–232
Porcellio, 121, 196–198, 202, 206
  albinus, 197
Prairie dog, black-tailed, 330
Primates, 8, 12, 13, 27, 35, 42, 44, 46, 49, 359–393, 434
Proceratium croceum, 269
Promyrmecia, 272
Pseudomyrmex
  ferruginea, 15, 67, 257, 258, 265, 272, 276
  venefica, 255, 265
Pygathrix nemaeus, 365

Rana
  aurora, 292, 294, 295, 298, 300, 302, 343
  cascadae, 17, 31, 40, 292–298, 300–302, 304–306, 337
  sylvatica, 294, 295, 299, 301, 304, 306
Ratels, 23
Rattus norvegicus, 320
Reptiles, 344
Reticulitermes, 248
  hesperus, 248
Rhytidoponera
  impressa, 254
  mayri, 254
  metallica, 258, 265, 266
Riparia riparia, 14, 36, 306, 307

Rodents, 9, 339

Saimiri sciurus, 434
Salamanders, 345
Salmo salar, 292
Slime molds, 426
Solenopsis, 13, 31, 275
  geminata, 272
  invicta, 27, 33, 46, 47, 254, 261
  richteri, 275
Spermophilus, 66
  beldingi, 17, 24, 313, 314, 325–328, 330
  parryii, 39, 314, 315, 325–328, 330
  richardsonii, 314, 315, 329
  tridecemlineatus, 314, 315, 328, 330
Sponges, 15, 32, 61, 426
Stelopolybia hamiltoni, 434
Stenamma, 255
  fulvum, 257
Stigmatomma pallipes, 265, 276

Teleogryllus, 16, 58
Termites, 8, 10, 58, 243–245, 248
Theropithecus gelada, 368
Thymus vulgaris, 261
Toads, 9, 334, 421
  American, 26, 27, 31, 44, 294, 295, 298, 299, 335, 341
  western, 292, 294, 295, 298, 341
Trachymesopus stigmus, 272
Tunicates, 424

Vespula maculifrons, 235
Voles, 9, 322–325, 340, 341
  gray-tailed, 313–315, 323
  meadow, 323, 324
  prairie, 313–315, 323, 324

Wasps, 9, 10, 209, 210, 213, 220, 235, 236, 238, 243

Subject Index

(See also Index of Scientific and Common Names)

Abortion, 403
Adoption (*see also* Alloparental care), 8
  in humans, 408, 426, 427
  in non-human primates, 13, 28
Adoptions, experimental
  in ants, 259
  in humans, 428
  in isopods, 141–145, 147–149, 155–157, 174–176
Affiliative behavior (*see also* Affinitive behavior; Aggregation; Preferential association), 301, 304, 317
Affinitive behavior (*see also* Affiliative behavior; Aggregation; Preferential association), 26, 29, 37, 40, 405
Aggregation (*see also* Affiliative behavior; Affinitive behavior; Preferential association)
  of aposematic, distasteful insects, 24, 25, 302
  of birds, 307
  of fish, 9, 12, 25, 292
  of isopods, 122
  of polistine wasps, 229, 230, 232, 235
  of sweat bees, 213
  of tadpoles, 9, 26, 294, 296, 297, 302–306
Aggression (agonism), 13, 21, 26, 48
  in anemones, 61, 62, 68, 70, 426
  in birds, 36, 309
  in cockroaches, 196
  in ground squirrels, 26, 36, 39, 313–315, 326–330, 336
  in humans, 410, 429, 434
  in isopods (*see also* Alienation), 123, 125–134, 138–195, 199–201, 205

Aggression (*cont.*)
  in mantis shrimps, 33
  in mice, 41, 312, 313, 314
  in non-human primates, 28, 360, 363, 365, 370–372, 376, 382, 384
  in social insects, 9, 12, 27, 47, 210, 212, 220–226, 230, 234–238, 244, 248–250, 252–274, 277
  in termites, 248
  in voles, 314, 315, 340
Aggression, inhibition of
  in isopods, 181–188, 190, 192–196, 204, 205
  in sweat bees, 222
Aiding (helping) (*see also* Alliances; Altruism; Cooperative behavior; Nepotism; Reciprocity), 19, 21–23, 56, 288, 289, 340, 419
  in birds, 364
  in ground squirrels, 327, 330
  in humans, 23, 405, 407
  in mice, 312, 316
  in non-human primates, 28, 364
  in tadpoles, 306
  misdirection of, 20, 288, 420
Alarm (warning) behavior, 9, 24, 25
  in birds, 24, 361
  in ground squirrels, 24, 325, 326, 328, 339, 421, 422
  in non-human primates, 360, 361
  in tadpoles, 302, 306, 339
Alienation of family members (isopods)
  and molting, 149–152, 154, 159, 179, 182, 185, 187, 196
  and the Regensberg phenomenon, 136, 163

Alienation of family members (*cont.*)
  by contact with conspecifics, 125, 148, 149, 152, 164, 165, 171, 172, 175, 189, 192, 193, 200
  by extracts, 125, 139, 140
  by fecal material, 126
  by hemolymph, 139, 140
  by members of mixed groups, 171, 172
  by *Porcellio* species, 203
  by rubbing, 145, 162, 163, 183–186
  by separation, 135, 150–152, 154, 157, 179, 180
  by water, 136, 163, 164
  defined, 125
  of exuvia, 185
  of young, 192
Alliance formation, 8
  in humans, 405
  in non-human primates, 13, 362, 363, 365, 369–371, 376, 378
Allogroomimg (*see also* Grooming)
  in mice, 41
  in non-human primates, 362, 364, 365
Alloparental care (*see also* Adoption), 8, 342, 364, 365
Altruism (*see also* Aiding; Nepotism; Reciprocal altruism), 8, 10, 11, 21–25, 34, 56, 57, 61, 68, 69, 75, 288, 305, 306, 330, 359–379, 381, 384, 385, 407, 410, 417–424, 433, 434
Animal thinking, 287
Anthropomorphism, 21
Antiaphrodisiacs, 86
Antigens, 57, 62, 403, 408
Aphrodisiacs, 83, 85
Armpit effect, 57, 288, 407
Assortative mating, 75, 76, 82, 83, 105–110, 112, 290, 307, 310, 339, 409

Behavioral genes, 403, 418
Bird song, 42, 404
Bruce effect, 16, 318
Burrows of isopods
  competition for, 197, 198
  defense of, 123, 125, 128, 130, 132, 198–200
  densities of, 126, 127, 198, 199, 201
  excavation of, 197, 198
  marking of, 129, 202

Burrows of sweat bees
  defense of, 219, 220
  density of, 213
  location of, 213

Cannibalism and avoidance of, 9, 288, 340, 341
  in guppies, 9, 341
  in isopods, 139, 180, 191, 192, 194, 195, 198, 205
  in social insects, 266, 272, 273
Capitalism, 22
Care of young (*see also* Alloparental care; Parental care), 1, 66
Castes (social insects), 10, 209, 210, 220, 235, 238, 243
Causes
  proximate, 8, 19, 20
  ultimate, 8, 19
Cheating, 248
Cleptoparasitism (social insects), 245
Clones, 61, 62, 67, 68, 433
Coadapted gene complexes, 9, 11, 217, 409
Codominance, 64, 69
Colony size (social insects), 209, 210, 235, 243, 264–266
Colony specific odors (social insects), 36, 44, 46, 59, 222, 257, 261
Compatibilities, transitive and intransitive, 61, 62, 67
Competition
  among males for mates, 89–92, 96, 369, 382
  between siblings, 48, 370
  for resources, 197, 246, 278, 288, 340, 341, 370, 382, 406
Competition, intraspecific, 340, 341, 346
  among genetic groups (social Hymenoptera), 243–245, 247, 249–252, 278
Conflict, intragenomic, 425, 426
Contamination with discriminators (isopods)
  of adult conspecifics, 124, 137, 138, 144, 145, 149, 153–156, 158, 172–174, 189, 195
  of feces, 126
  of glass rods, 125, 126, 164, 182, 183
  of substrate, 125, 131, 132
  of young, 193, 196, 205
Conversation as interaction, 13, 405

SUBJECT INDEX 455

Cooperative behavior (*see also* Aiding; Alliances; Altruism; Nepotism; Reciprocity), 9, 11, 36, 48, 66, 68, 71, 420
  in birds, 362
  in ground squirrels, 326
  in humans, 406, 407, 429, 430
  in isopods, 123, 168, 198, 201
  in non-human primates, 360–365, 368
  in social carnivores, 362
  in social insects, 428
  in tadpoles, 302, 305
  in white-footed mice, 322
  versus uncooperative behavior, 25, 26, 29, 31, 41, 47, 48
Courtship, 3, 14, 210
  in *Drosophila*, 30, 76, 79–88, 90–99, 101, 102, 109–112
  in houseflies, 102
  in subsocial arthropods, 121
Cousins
  and mating preference in Japanese quail, 8, 29, 308, 408
  and reduced sexual activity in macaques, 380
  and selective infanticide, 341
  discrimination from siblings, 17
  marriages between, 408
  recognition of in non-human primates, 373
  similarity between in humans, 411
Crossfostering, 37, 38, 49
  of ants, 259, 261, 271–274
  of ground squirrels, 39, 326, 327, 329
  of mice, 37, 38, 46, 311, 317, 322
  of voles, 323
Cue bearer (*see also* Recognized individual), 10, 31, 33, 34, 43, 44, 49, 56, 210, 211, 214, 221, 237, 238
Cues/labels/signals (*see also* Discriminators; Discriminating substances), 13, 14, 16, 30, 31, 58, 60, 66, 76, 90, 93, 98, 99, 112, 210, 333, 334
  anatomical sources in social insects, 13, 227, 234, 252, 261–264, 267, 268, 275, 277
  artificial, 221, 261, 316
  auditory (acoustic), 4, 13, 32, 36, 42, 43, 47, 65, 66, 76–81, 111, 112, 210, 298, 308, 309, 333, 398, 399, 401, 411, 412

Cues/labels/signals (*cont.*)
  behavioral, 4, 14, 66, 210, 407, 411
  brood-specific, 274
  chemical (*see also* Discriminators; Discriminating substances; Pheromones), 12, 13, 16, 32, 36, 42–46, 48, 49, 62, 65, 66, 76, 77, 81–86, 111, 112, 125–127, 134, 169, 202, 204, 210, 214, 215, 221, 222, 233–237, 256–264, 266–269, 271, 275, 278, 289, 291–293, 298, 312, 316, 318–320, 329, 334–336, 344, 345, 374, 399–401, 403, 411, 422, 423, 425
  classification of, 43, 44
  colony-specific (social insects), 261, 274, 277
  complexity of, 41–43, 190, 226, 334, 401
  development of, 12, 20, 36, 333
  dietary-derived, 44, 134, 211, 212, 233, 247, 249, 255, 260, 266, 267, 334, 400
  endogenous, 66, 212, 233, 244, 255, 258, 261, 262
  environmental accidentals, 35, 44, 46, 47
  environmentally-derived (*see also* Cues, extrinsic), 10, 15, 18, 20, 36, 43, 44, 46, 66, 134–136, 199, 210–212, 221, 232–235, 237, 238, 244, 247, 250, 255, 260–262, 276, 278, 400
  extended phenotypes, 35, 44, 49, 66
  extrinsic (*see also* Cues, dietary-derived; Cues, environmentally-derived; Cues, extended phenotypes; Labels, maternal), 10, 20, 36, 43, 44, 46
  fixed versus variable, 334
  genetic (*see also* Cues, endogenous; Cues, intrinsic; Cues, phenotypic), 15, 18, 36, 58, 59, 68–70, 79, 136, 199, 211, 212, 215, 216, 221, 223, 225, 233, 234, 237, 238, 244, 245, 249–252, 255, 261, 276, 278, 289, 333, 334, 400
  inhibitory, 222
  intrinsic (*see also* Cues, endogenous; Cues genetic; Cues, phenotypic), 11, 20, 36, 43, 44, 46, 49

Cues/labels/signals (*cont.*)
  learning of, 11, 14, 15, 17, 33, 35, 46, 47, 57, 66, 67, 70, 86, 153–157, 166, 176–178, 188–196, 203, 204, 211, 212, 214, 217, 223, 225, 232–238, 245–248, 255, 258, 261, 271–274, 278, 304, 309, 336, 339, 425
  maternal, 15, 18, 44, 61, 67, 68, 196, 205, 238, 247, 255–261, 276, 278, 296, 301, 305, 311, 312, 335, 336
  memory of, 12, 46, 47, 151–154, 161, 179, 190, 211, 215, 233, 256–258, 272–274, 277
  modes of selection of, 71
  paternal, 296, 305
  odor (*see* Cues, chemical)
  olfactory (*see* Cues, chemical)
  phenotypic (*see also* Cues, endogenous; Cues, genetic; Cues, intrinsic), 20, 33–35, 40, 46, 49, 56–58, 63, 64, 333, 336–338, 399–404, 409, 411
  polygenic inheritance of, 407
  site-specific, 33
  stability of, 46
  stochastic components of, 211, 212
  tactile, 76, 88, 89, 111
  urinary odors, 16, 45, 49, 62, 312, 403, 425
  visual, 13, 36, 42–45, 48, 66, 76, 78, 86–88, 92, 110–112, 210, 298, 307, 308, 310, 333, 334, 344, 374, 397, 398, 411, 412, 422
  worker-derived (social insects), 15, 18, 59, 60, 68, 233, 255, 257–264, 266–269, 275–278

Decision rules, 10, 11, 63, 245, 246–248
Defense, 66, 75
  in birds, 24, 361
  in cockroaches, 196
  in distasteful insects, 24, 25
  in ground squirrels, 24, 325, 326, 328, 329, 421, 422
  in isopods, 123, 125, 128, 130, 132, 198–200
  in non-human primates, 360–362
  in social insects, 1, 23, 219, 220, 235, 244, 246, 247, 249, 253, 278, 418
  in tadpoles, 302, 306, 339

Dialect differentiation
  in birds, 404
  in humans, 404
Discriminating individual (*see also* Recognizing individual), 10, 31, 33, 34, 40, 43, 44, 46, 127, 247
Discriminating substances
  defined, 43, 212
  in bumble bees, 235
  in isopods, 127
  in polistine wasps, 233
  in sweat bees, 214
Discrimination
  among potential mates, 1, 3, 19, 75–119, 380, 381, 383–385
  and recognition, 61
Discriminators (recognition pheromones), 246
  defined, 10, 43, 212
  in isopods, 124–126, 136–138, 140, 142–149, 151–156, 158, 159, 161–165, 169–175, 178–183, 185–192, 196, 199, 201–205, 236
  in polistine wasps, 233, 234, 236
  in sweat bees, 213, 216–218, 220–223, 225–227, 236
Discriminatory/recognition ability
  of humans, 396–400
  sex differences in, 194, 399
Dispersal/emigration, 49, 335
  in aphids, 421
  in ground squirrels, 24, 26, 313, 325, 326, 328
  in humans, 407
  in isopods, 122, 141, 205
  in mice, 312, 314, 320
  in non-human primates, 361–364, 368, 380, 382–385
  in spruce grouse, 341
  in tadpoles, 304, 305
  in voles, 314, 324, 340, 341
Division of labor
  ethological, 209, 272, 427, 428
  reproductive, 209, 243
Dominance
  gene, 64, 433
  social, 1, 32, 48, 56, 70, 234, 244, 287, 320, 344, 369–371, 378, 382
Dufour's gland (social insects), 227, 263, 264, 267, 268, 275
Dulosis (social insects), 31, 32, 274–277

SUBJECT INDEX 457

Dyadic pairing, experimental, 37, 38, 311, 317, 321–323, 326, 329
Dyads in non-human primates
  kin, 368, 369
  non-kin, 368

Egg-laying workers (social insects), 226, 234, 244, 250, 251, 254, 265
Embryo transplants, 396
Eusociality, 419, 433
  defined, 10, 209
Evolution, 1, 22, 226, 306
  of complex signals, 43
  of cooperative behavior in groups, 361, 362
  of distastefulness in gregarious insects, 24, 25, 302
  of human culture, 417, 428–433
  of humans, 410, 412, 422, 427
  of kin recognition (see Kin recognition)
  of mate recognition systems, 76, 90, 105–112
  of selfishness, 419
  of social parasitism, 275

Familiarity
  coefficient of, 340, 344
  recognition by (see Kin recognition)
Family badges (isopods)
  and stimulation of aggression, 130–133
  chemical analysis of, 164, 169, 189, 202
  defined, 125
  degree of complexity, 169–174, 176–178
  efficiency of, 199
  environmental influence on, 134–136, 147, 199
  evidence for chemical nature, 125, 126, 134
  genetic basis of, 136, 142, 169, 199, 202
  interfamilial variability, 126–135, 142, 171, 199, 201–203
  intrafamilial variability, 127, 140–165, 169, 178–191, 201, 203–205
  learning of, 147, 153–157, 166, 175, 176, 178, 192, 193, 196, 203
  memory of, 151–154, 161, 179
  multicomponent nature of, 140, 202
  number of, 199, 202

Family badges (isopods) (cont.)
  of artificially mixed groups, 165–178
  of young, 192, 206
  origin of, 134–140
  perception of, 123–125, 190, 200
  quantitative differences in, 162–165, 182–188, 202, 205
  secondary adjustment of, 143–145, 147–149, 188, 203
  stability of, 127, 199
  subtraction of components from, 174–176, 178–180, 190, 191, 203, 204, 206
Family groups, 46, 60
  in 13-lined ground squirrels, 328
  in cockroaches, 196
  in deer mice, 322
  in geese, 307, 310, 339
  in house mice, 320
  in isopods, 122, 123, 197–201, 206
  in non-human primates, 363, 368, 372, 375, 382
  in prairie voles, 324
  in salmon, 290, 291
  in social insects, 31, 245, 253
  unilateral acceptance of in isopods, 127, 140, 171, 202
Fecal embankment (isopods), 123, 126–130, 132, 199, 200
Female choice (*Drosophila*), 87, 89–105, 110–112
Fission of social groups
  in baboons, 383
  in honey bees, 31, 244, 249–251, 278
  in humans, 405
Fitness, 11, 22, 68–70, 103, 108, 217, 218, 245, 319, 342, 359, 379, 418, 419, 433
Food sharing
  in humans, 13, 405
  in non-human primates, 362, 365
  in spiny mice, 316
Founder-flush theory, 108, 112
Foundress associations (social insects), 36, 228, 229, 230, 231, 234
Frequency-dependent selection (see also Rare male mating advantage), 68–70
Fusion
  of tissues, 15, 62, 424, 426, 433
  of social groups, 383

Gardening (as interaction), 405
Gene linkage, 142, 403, 434
Genealogical information (sources of), 27, 28, 30, 223, 228, 365, 366, 405
Genetic incompatibility
 in plants, 423, 424, 434
 in tunicates, 424
Genetic markers, 28, 79, 80, 82, 85–87, 90–98, 100–107, 112, 227, 254, 265, 327, 336, 377, 402–404, 409, 434
Genetic models, 55–73, 246
 foreign-label rejection, 64, 67, 69, 236, 246
 genotype recognition, 64, 236, 246
 Gestalt, 10, 11, 12, 15, 59–61, 67, 203, 246, 247, 258, 259
 habituated-label acceptance, 64, 236, 246
 Individualistic, 10–12, 15, 59, 60, 62, 65, 69, 246, 247, 258, 274
 involving metric (continuous) traits, 64, 65
 tests of, 66–68
Genetic potential and environment, 34, 35
Genetic similarity theory, 289
Genocide, 434
Glossary, 10–12
Green-beard effect, 11, 12, 14, 34, 56–58, 248, 290, 407, 423–425
Grooming behavior (non-human primates), 9, 13, 361–363, 365–369, 371, 373, 376, 378
Group selection, 21, 434
Groups (*see also* Family groups; Mixed groups, experimental)
 homo- and heterogeneous, 31, 32, 210
 importance of size of, 61, 66–68, 164, 167–171, 176, 198, 236, 255, 366, 367, 368, 410
 stochastic variation in composition of, 61, 66
Guarding of nests
 in isopods, 122, 125, 128, 129, 133, 198, 200, 205
 in social insects, 26, 31, 210–212, 219–228, 236–238, 249
Gynes (social insects), 210, 229–235, 244, 245, 258, 269

Habituation, 33, 61, 94, 211, 217, 237, 397, 422, 423

Half-siblings, 17, 63, 335, 339, 373, 421
 in American toads, 31, 295, 299, 301, 335
 in ants, 253, 254
 in cascades frog tadpoles, 31, 40, 41, 295, 296, 301, 305, 335, 337
 in ground squirrels, 39, 326–328, 336, 342, 421
 in honey bees, 31, 249, 250, 422
 in house mice, 41, 313, 317, 318, 335
 in humans, 411
 in non-human primates, 40, 336, 376
 in Norway rats, 321
 maternal, 31, 40, 41, 289, 295, 296, 299, 301, 317, 327–329, 335–337, 342, 376
 paternal, 31, 40, 41, 295, 296, 299, 301, 313, 317, 318, 330, 335–337, 339, 376
Helping behavior (*see* Aiding)
2-Heptanone, 262
Hibernation
 in ground squirrels, 329
 in isopods, 122, 141, 158, 159, 200
 in social insects, 229, 230, 232
Highly eusocial
 defined, 243
 in insects, 210, 235, 243–285
Huddling
 in spiny mice, 37, 38, 311, 312
 in white-footed mice, 321
 in voles, 323
Hunting
 in humans, 405
 in non-human primates, 362
Hydrocarbons
 of ants, 13, 275
 of beetles, 275
 of *Drosophila*, 84, 85, 111
 of sweat bees, 227

Immune system (*see also* Major histocompatibility complex), 16, 58, 61, 66, 70, 71, 403, 408, 417, 425
Imprinting, 45, 46, 49, 211, 232, 271–275, 277, 336, 381, 419
Inbred strains, use of
 in *Drosophila*, 79, 82, 83, 94–98, 100–109
 in house mice, 45, 62, 318–320
 in social insects, 216, 223, 252, 258, 276

SUBJECT INDEX

Inbreeding and incest, 287
  in *Drosophila*, 82
  in frogs, 305
  in isopods, 158, 159
  in non-human primates, 361–364, 379–381, 383, 384
  in social insects, 30, 36, 49, 218, 229, 235
Inbreeding and incest avoidance, 7, 8, 19, 56, 61, 66, 68–71, 359, 379, 419
  in black-tailed prairie dogs, 330
  in *Drosophila*, 67
  in humans, 17, 18, 381, 407, 408
  in mice, 16, 322
  in non-human primates, 18, 361, 379–385
  in social insects, 30, 218
  in voles, 324, 325, 340
Inbreeding depression, 9, 11, 340, 379
Inclusive fitness, 2, 9–11, 19, 22, 24, 28, 247, 269, 270, 288, 302, 306, 330, 346, 359, 364, 376, 417–420, 425
Infants, 8
  of humans, 397–399, 426, 428
  of non-human primates, 363, 367, 372, 376–378, 380
Infanticide, 340, 341, 419
  in ground squirrels, 325
  in humans, 409, 410
  in non-human primates, 370, 376
Innate-learned dichotomy (heredity versus environment), 5, 14, 34, 66
Inquilinism (social insects), 274, 275
Intracolonial relatedness (social insects), 27, 247, 250–255, 264, 270, 276
Investment in kin (*see also* Nepotism), 21–25, 28, 33, 245
Isolating mechanisms, 4, 13, 14, 75, 76, 80, 105, 106, 109, 110
Isolation of test subjects, 339
  in chickens, 339
  in isopods, 131, 137, 138, 143–145, 147, 149–155, 160, 178–181, 185, 186, 204
  in Japanese quail, 308
  in mice, 37, 38, 311, 316, 322, 339
  in social insects, 223, 224, 229, 232, 234, 237, 258, 259
  in tadpoles, 40, 292–294, 296, 298, 299, 301

Karyotypes, use of, 103–105
Kibbutzim, 17, 408, 426
Kin altruism (*see also* Nepotism), 22, 23, 407
Kin biased behavior, 20, 25–30, 47, 48, 287, 359–379
Kin colony (social insects), 212, 221–223, 225, 226
Kin recognition
  and brain size, 4, 169–171
  and chemosensory deficiency, 85, 91, 92, 316, 328, 329
  and habitat selection, 9, 14, 306, 340, 341
  and maintenance of gene polymorphism, 68–71, 103, 104, 403, 418, 434
  and metamorphosis, 345
  and optimality theory, 343
  and premating behavior, 30, 383
  and 'reproductive expectancy', 28
  and spatial proximity (*see also* Kin recognition, by location), 13, 18, 365–374, 376, 378, 406, 407
  as a byproduct (*see* Recognition-switching)
  as a class phenomenon, 56
  as a polymorphic trait, 341
  by familiarity, 33, 34, 37–41, 43, 46, 49, 289, 290, 294, 298, 300, 301, 306, 307, 309, 313–316, 318, 320–325, 327, 329–331, 334, 335, 337, 339, 345, 360, 362, 374, 384, 385, 397–399, 407, 412, 421, 423
  by location (spatial distribution), 3, 13, 14, 18, 33, 34, 36, 37, 44, 56, 123, 128, 191, 192, 199, 211, 288–290, 360, 407
  by phenotype matching, 11, 12, 15, 33, 34, 38–40, 43, 46, 56, 57, 59–61, 151, 203, 211, 289–291, 294, 301, 307, 308, 310, 313–316, 318, 321, 327, 330, 331, 334–339, 344, 360, 374, 378, 379, 399, 400, 407
  by recognition alleles (intrinsic or innate mechanisms), 11, 12, 14, 16–18, 34, 35, 40, 49, 56, 57, 59, 60, 66, 68, 69, 245, 247, 248, 289–291, 294, 301, 307, 308, 313–315, 318, 319, 321, 327, 330, 331, 334–339, 360, 374, 378, 385, 405, 407, 410, 412, 423, 425

Kin recognition (cont.)
  defined, 2, 11, 30, 31
  effects of molting on (isopods), 132, 135, 143–151, 154, 157, 166–168, 178–191, 204–206
  environmental influences on, 344–346
  evolution of, 15, 18, 58, 68–71, 206, 301, 302, 343, 344, 376, 408–412, 419, 420, 422, 423, 434
  functions of (see also Mating preference; Defense), 2, 3, 8, 9, 19, 20, 66, 191, 197–201, 205, 206, 220, 230, 235, 238, 245, 246, 249–253, 278, 287, 288, 291, 292, 298, 302–306, 309, 310, 316, 320–323, 325, 339–341, 343, 359, 360, 396, 406, 408–411, 420
  gametic, 338
  genetically based (see Kin recognition, by recognition alleles)
  genetics of, 16, 17, 55–73
  intraspecific variability in, 341–346, 373
  mechanisms of, 13–15, 18, 20, 33, 34, 55–57, 65, 66, 238, 243, 288–290, 333–339, 345, 359, 360, 369, 371–377, 381, 384, 385, 405, 407, 408, 411
  ontogeny of, 20, 35, 37, 38, 40, 66, 271–274, 288, 298, 300, 306, 311, 316, 325, 327, 344–346, 408
  post-natal maternal influences on (see also Cues, maternal), 311–313, 316, 318
  pre-natal influences on, 34, 35, 40, 44, 46, 293, 296, 299, 301, 315, 328, 330, 335, 336, 344, 396
  prevalence of, 2–4, 426
  sex linked, 16, 58, 67, 142, 425
  study of in humans, 10, 396, 410, 411
  without previous experience (see also Kin recognition, by recognition alleles), 34, 35, 40, 41, 43–45, 289, 329, 330, 412, 423
Kin selection, 7–9, 11, 19, 25, 29, 30, 49, 55, 56, 68, 69, 245, 288, 302, 304, 306, 340, 341, 359, 361–364, 371, 373, 376, 379, 412, 419, 422–424
  criterion of, 417–419, 423, 424, 433
Kingrams, 11, 63

Kinship, 2, 10, 13, 287, 337, 346, 366–371, 373–375, 377, 395, 396, 405, 406, 417
Kings (social insects), 243, 245

Learning (see also Cues, learning of; Kin recognition, by familiarity), 14, 33, 34, 56, 60, 65, 211, 289, 404
  of cultural information in humans, 407, 428
  of family badges in isopods, 147, 153–157, 166, 175, 176, 178, 192, 193, 196, 203
  of individual traits in isopods, 188–191, 195, 203–205
  of kin in non-human primates, 44, 49, 359, 360, 375
  of kin in spiny mice, 38, 49
  of recognition in ground squirrels, 330, 338
  of young in birds, 37, 420
Location, recognition by (see Kin recognition)

Macrocyclic lactones, 13, 227
Major histocompatibility complex (MHC)
  in humans (HLA), 16, 65, 403, 408, 424, 425, 434
  in mice (H-2), 16, 45, 46, 49, 62, 65, 318–320, 338, 339, 402, 403, 425
Male-diploid genetic system, 58, 64, 243
Male-haploid genetic system, 58, 64, 243
Males of social insects, 27, 29, 30, 58, 209, 213–219, 227–229, 234, 235, 243, 245, 251, 270, 271
Mandibular glands (social insects), 13, 252, 262–264, 267, 277
Marsupial mancas (isopods), 143, 191, 195
Mate selection (see Mating preference)
Maternal influence on sibling recognition (spiny mice), 311, 312, 316
Mating preference (see also Inbreeding and incest avoidance), 9, 12, 13, 19, 29, 30, 46, 47, 49, 56, 359, 380, 408, 419
  in birds, 8, 29, 307, 308, 310, 408
  in *Drosophila*, 3, 30, 42, 48, 75–119
  in humans, 403, 408

Mating preference (*cont.*)
  in mice, 16, 30, 45, 46, 62, 312, 318–320, 322, 346, 403, 425
  in Mongolian gerbils, 323
  in non-human primates, 336, 361–363, 383, 379–385
  in social insects, 3, 12, 29, 30, 213–220, 228, 229, 235, 237
  in voles, 313
Matrilines
  in ground squirrels, 325
  in non-human primates, 28, 361–376
  in social insects, 246, 247
Methodological reductionism, 2, 3, 5
4-Methyl-3-heptanone, 262
Misidentification of kin, 23, 42, 63, 64, 70, 121, 128, 134, 158, 159, 202, 215, 224, 225, 288, 289, 305, 309, 341–344
Mixed groups, experimental (*see also* Adoptions, experimental; Crossfostering; Transfer experiments)
  in ants, 67, 255, 256, 259, 269–274
  in honey bees, 250, 251
  in humans, 407
  in isopods, 145–150, 152–154, 157, 158, 164–182, 188–190, 195, 203–205
  in sweat bees, 222, 225, 226
  in tadpoles, 292–294, 298–301, 337
  in wasps, 229–232, 235
Monogamy, 422
  in birds, 307
  in cockroaches, 196
  in isopods, 122, 198
  in social insects, 27, 254
Monogyny (social insects), 15, 26, 249, 253, 254, 255, 264–267, 271, 276
Multicolonial ant populations, 264, 265
Multilocus recognition systems (*see also* Genetic models), 15, 59, 60, 62, 65, 70, 71, 142, 276
Multiple paternity (*see* Polyandry)
Mutualism, 428, 430

Neo-Darwinism, 418
Nepotism (*see also* Aiding; Altruism), 8, 10, 11, 21–25, 288, 306, 326, 328, 330, 395, 405, 417–433
Nest entrance marking (social insects), 36, 220, 267

Neutral behavior (non-human primates), 367, 368
Non-kin colony (social insects), 212, 222, 223, 226

Observed individual (*see* Recognized individual)
Orphans, 8, 372, 373
Outbreeding, 75, 229, 287, 338, 423
  optimization of, 2, 8, 9, 11, 19, 75, 217, 287, 292, 288, 306, 308, 320, 322, 339, 434
Outlaw alleles, 57, 290

Pair formation/bonding, 1
  in gibbons, 380
  in isopods, 126, 143, 158, 178, 198, 199
  in snow geese, 307
  in voles, 313, 323, 324
Panselectionism, 2
Parasitism, 23, 33, 56, 198, 271, 274–276, 335, 342, 343, 420, 421, 424
Parental care, 1, 3–5, 8, 10, 19, 66, 197, 198, 313, 324, 328, 425
Parental investment (*see also* Parental care), 23, 33
Parental manipulation, 22, 48
Patrilines
  in non-human primates, 361, 363, 376–379
  in social insects, 27, 31, 246, 247, 250–252
Peer groups (non-human primates), 378
Phenotype matching (*see* Kin recognition)
Pheromones (*see also* Cues, chemical; Discriminators)
  alarm, 9, 263, 277
  and social parasitism, 275
  primer, 13, 16, 212
  recognition, 10–13, 15–17, 32, 36, 43, 47, 59, 61, 62, 66, 67, 70, 111, 124, 210, 212, 216, 246
  releaser, 13, 212, 277
  sex, 81, 83, 95, 214–220
  territory marking, 36, 44, 244, 266–269, 277
  variations in composition (polymorphism), 42, 59, 65, 82, 83, 111, 140–165, 171, 226, 227, 252, 401

Pheromones (*cont.*)
  variations in concentration, 32, 42, 59, 65, 82, 159, 162–165, 182–186, 205–227, 246
Philopatry (*see also* Viscous populations), 14, 230–232, 305
Play
  in humans, 13, 405
  in non-human primates, 365, 372, 375, 378
Play groups, 13, 35, 40
Pleiotropy, 87, 403
Piggy-backing/hitch-hiking (*see also* recognition-switching)
  of kin recognition alleles, 69, 70
  of mate recognition in *Drosophila*, 109, 112
Poison glands (social insects), 263, 264, 267
Polyandry, 335, 421
  in *Drosophila*, 95
  in ground squirrels, 26, 39, 325, 326, 330
  in isopods, 143
  in non-human primates, 377
  in social insects, 27, 31, 60, 70, 243, 245, 248, 249, 253
Polydomy (social insects), 264–266, 275, 276
Polygamy, 335
  in birds, 307
Polygyny (non-human primates), 361, 363
Polygyny (social insects), 15, 60, 243, 245, 248, 252, 254, 258, 264–266, 271, 276
Population fluctuations in voles, 313, 322, 340
Predation, 9, 23–25, 302, 306, 325, 328, 346, 360, 362, 418
Preferential association with familiar kin versus familiar non-kin
  in honey bees, 31, 244, 249, 251
  spiny mice, 15, 311
  tadpoles, 293, 294, 300, 337
Preferential association with familiar non-kin versus unfamiliar kin
  in spiny mice, 38, 311
Preferential association with familiar versus unfamiliar conspecifics
  in ants, 253

Preferential association (*cont.*)
  in Canada geese, 309, 310
  in fish, 291
  in humans, 412
  in non-human primates, 44, 378
  in spiny mice, 15, 37, 38
  in tadpoles, 293, 294, 298–300, 302, 337
Preferential association with kin versus non-kin
  in ants, 253
  in Canada geese, 309, 310
  in fish, 291, 292
  in humans, 412
  in Japanese quail, 308
  in pigtail macaques, 40, 44, 378
  in polistine wasps, 230–232, 235
  in spiny mice, 15, 38, 311, 312
  in tadpoles, 26, 40, 41, 44, 293, 294, 296–302, 304, 337, 395
  in vespine wasps, 235
Preferential association with maternal versus paternal half-siblings
  in tadpoles, 296
Preferential association with siblings versus half-siblings
  in tadpoles, 40, 41, 295, 296, 337
Preferential association with unfamiliar kin versus unfamiliar non-kin
  in fish, 291
  in Japanese quail, 307, 308
  in pigtail macaques, 40, 44, 378
  in spiny mice, 311
  in tadpoles, 40, 41, 294, 296, 299, 301, 337
  in white-footed mice, 321
Preferential feeding in social insects
  of brood, 250, 253
  of queens, 244
  of sisters, 250
Preferential grooming in non-human primates, 365–369
Pregnancy blocking (*see* Bruce effect)
Primitively eusocial
  defined, 210
  in insects, 209–241, 243
Promiscuity (*see also* Polyandry), 421, 422
Prosopagnosia, 17
Pseudokin (social insects), 212, 223, 232, 254

SUBJECT INDEX 463

Pseudolearning, 56

Qa-Tla locus in mice, 62
Queens of social insects, 12, 13, 15, 18, 23, 26, 27, 31, 33, 46, 47, 59, 61, 67, 209, 210, 220, 222, 226, 234, 238, 243–245, 249, 251–261, 264–266, 269–272, 275–278, 418

Racism, 427
Rare male mating advantage (*Drosophila*), 82, 90, 92–105, 112
Reciprocal altruism (*see also* Reciprocity), 22, 56, 376, 410
Reciprocity, 10, 22, 24, 362, 376, 424
Recognition and discrimination, 61
Recognition alleles (*see* Kin recognition)
Recognition/discrimination (*see also* Kin recognition)
  among brood of social insects, 18, 232–234, 238, 244, 245, 249–253, 271–278
  by male social insects, 213, 227, 234, 235
  levels of, 20, 31–33
  of classes within groups, 31, 32, 43, 44, 56, 58, 67, 246, 373, 375, 385, 407
  of degree of relatedness, 1, 4, 11, 13, 17, 19, 20, 31, 41, 43, 46, 60, 75, 210, 211, 215, 216, 223, 224, 244, 249–251, 294–296, 298, 299, 301, 305, 307, 308, 313, 314, 317, 333, 337, 372, 373, 397, 401, 404, 408, 409, 411, 419–421
  of groups, 1, 4, 10, 20, 31, 56, 66, 121, 333, 360, 429
  of individuals, 1, 4, 16, 20, 32, 33, 35, 42, 43, 45, 46, 49, 55, 58, 64, 70, 71, 123, 141, 152, 153, 190, 200, 211, 214, 215, 247, 249, 287, 289, 333, 334, 360, 375, 397, 400, 401, 407, 409, 422
  of kin and non-kin, 1, 2, 7, 8–12, 15, 17, 19, 33, 36, 38, 40, 46, 64, 66, 123, 125, 136, 141, 198, 287, 292–295, 298, 300, 301, 308, 309, 316, 320, 329, 330, 337, 343, 345, 346, 372, 375, 396, 400, 407, 421, 423

Recognition/discrimination (*cont.*)
  of males by workers in social insects, 270, 271
  of mates, 33, 198
  of nestmates in social insects, 3, 36, 39, 42, 58, 59, 67, 210, 213, 215, 216, 218, 220–227, 229–235, 238, 243–278
  of offspring by males, 194, 195, 376, 377, 409
  of offspring by parents, 4, 8, 17, 30, 32, 33, 35–37, 43, 56, 68, 69, 191–196, 198, 203, 205, 206, 287, 288, 291, 308, 309, 342, 344, 397–400, 420
  of parents by offspring, 4, 17, 32, 43, 287, 291, 397–399
  of paternal half-siblings and non-kin, 40, 296, 301, 313, 317, 335–337
  of queens by workers in social insects, 13, 27, 46, 47, 244, 251, 252, 255, 257, 269, 270, 275, 277
  of sexes, 1, 56, 88, 111, 344
  of siblings and half-siblings, 17, 31, 40, 41, 64, 250, 295, 296, 301, 327, 328, 330, 335, 337, 420–422
  of siblings reared together and apart (*see also* Crossfostering; Preferential association), 225, 257–261, 320, 396
  of species, 1, 4, 88, 99, 105, 110–112, 266, 271–276, 287, 344
Recognition-switching (*see also* Piggybacking/hitch-hiking), 70, 71, 409, 410, 422, 434
Recognized individual (*see also* Cue bearer), 10, 210, 221
Recognizing individual (*see also* Discriminating individual), 10, 12, 210, 221, 238
Referents
  defined, 12, 57
  differential weighting of, 61
  diversity of, 66
  group, 12, 60, 67, 204, 211, 245, 247, 251
  parent (*see also* Cues, maternal), 15, 18, 61, 64, 67, 310, 374, 375, 377, 379, 385, 421
  relatives, 33, 34, 60, 63, 64, 67–69, 141, 204, 211, 225, 236, 238, 247, 309, 421

Referents (cont.)
  self, 12, 14, 33, 34, 57, 59, 60, 64–67, 151, 152, 211, 225, 236, 237, 245, 247, 248, 250, 251, 301, 308, 309, 336–339, 345, 360, 374, 399, 400, 403, 409, 421–423
  subset of a group, 12, 60
Relatedness
  and baby battering, 409
  and human social interactions, 404–407
  and kinship terminology, 406
  and surnames, 410, 411
  average in social groups, 361–364
  coefficients (r) of, 8, 24, 26, 27, 31, 58, 64, 216, 223, 224, 226, 253, 254, 270, 340, 344, 405, 406, 410, 411, 421–423, 433.
  electrophoretic analysis of, 28, 227, 254, 265, 327, 336
  in social insect colonies, 253, 254
Resource-adapted populations (*Drosophila*), 107–109, 112
Responding individual (*see* Recognizing individual)

Schooling (*see* Aggregation)
Self-fertilization, prevention of, 338
Selfishness
  behavioral, 22, 419, 420, 421, 427
  constraints upon evolution of, 419
  of genes, 10, 22
Semisocial
  defined, 210
  in bees, 222
Separation experiments (*see also* Cross-fostering; Isolation of test subjects; Transfer experiments), 37
  in isopods, 135, 150–154, 160, 161, 189–191
  in social insects, 255–258, 260, 262, 266, 269
Sensory deficiencies and courtship (*Drosophila*), 85, 87, 90–93, 112
Sexual selection, 89
Sexuality, function of, 419, 423
Sibling species
  of ants, 276
  of *Drosophila*, 80, 88, 105, 110

Signatures
  defined, 124
  in ants, 255
  in bees, 422
  in birds, 14, 36, 401
  in humans, 399, 401
  in isopods, 124–128, 134–136, 140–165, 173, 178, 179, 188–196, 200–206
Single locus recognition systems, 57, 58, 61, 62, 64, 68, 69, 318, 338, 423
Slave-making ants (*see* Dulosis)
Sleeping groups (non-human primates), 363
Social bonding (*see also* Pair formation/bonding), 374, 375, 377, 382
Social systems of non-human primates
  age-graded groups, 363, 382
  family groups, 363, 368, 372, 375, 382
  female-bonded groups, 361–364, 369, 370, 373
  multi-female groups, 375
  multi-male groups, 363, 368, 376–378, 382
  non-female-bonded groups, 363, 364, 369, 370, 373
  one-male groups, 363, 365, 368, 376, 382, 384
  solitary, 363, 368, 375
Sociobiology, 7, 25
Sperm clumping, 143
Sperm precedence, 95
Sterility in humans, 424
Steroid hormones and odorant derivatives, 16, 403
Sternal glands (social insects), 234
Stimulus intensities, 32
Suicidal behavior, 23–25, 418
Supercolonies (ants), 265, 266
Superorganism, 427
Swarming in honey bees, 23, 31, 47, 244, 249–251, 278

Templates
  common feature template, 247, 255
  continuously modifiable templates, 35, 49
  defined, 12
  different in same individual, 46, 61, 247
  innate, 12

Templates (*cont.*)
  learned, 11, 12, 15, 34, 49, 211, 237, 247, 248, 251, 278
  matching cues with, 30, 31, 244, 245, 337
  mean template, 246, 247, 253, 263, 277
  multiple-mean template, 247, 251, 253, 274, 277
  multiple-template, 246, 247, 253
Terminology, 10–12, 20–25
Territorial behavior, 1, 60, 66, 68, 288
  in ground squirrels, 26, 39, 314, 326, 328, 329, 336
  in mice, 41, 312, 320, 322
  in non-human primates, 360, 364, 365
  in social insects, 36, 228, 244, 266–269, 277
Tissue rejection, 15, 423, 426
T-locus in mice, 16, 318, 319, 346
Transfer experiments (*see also* Adoptions, experimental, Cross-fostering; Mixed groups, experimental), 36, 37
  in birds, 14, 36
  in ground squirrels, 36, 327
  in isopods, 125, 127, 128, 130, 151, 154–159, 170, 171, 174, 176, 181, 189–192, 194, 195, 202
  in social insects, 225, 258–263, 265, 266, 273
Transfer/exchange of discriminators
  in isopods (*see also* Alienation; Contamination), 126, 137, 138, 145–147, 153–155, 159, 161, 164, 173, 185, 187, 190, 191, 196, 199, 203, 205
  in sweat bees, 222
Transfer of cues in ants, 262
Transfers (natural) among social groups, 363, 382, 384
Trunk trails of ants, 268, 269
Twins, 396, 401, 406, 423
  dizygotic, 402, 407, 411
  monozygotic, 402, 406, 407, 411, 412, 428

Uncooperative behavior, 39, 430
  versus cooperative behavior, 25, 26, 29, 31, 41, 47, 48
Unicolonial ant populations, 264–266
Uniqueness in humans, 409, 410, 412
Usurpation of nests and resources, 42, 234, 245, 248, 420, 421

Valves gland (social insects), 267
Variables influencing kin recognition
  age, 28, 36–38, 40, 47, 128, 249, 323, 367, 369, 370, 372, 373, 379, 400, 411, 420
  aggression, 26, 27, 48, 370, 371
  body size, 40
  dominance relationships, 48, 370, 378
  health, 28, 47, 420
  ownership of resources, 28
  rearing conditions, 40, 47
  reproductive status, 28, 47
  sex, 28, 37, 40, 47, 323, 367, 369, 370, 372
  spatial proximity, 366–373, 406
Viscous populations (*see also* Philopatry), 36, 218, 254

Wing clipping
  of *Drosophila*, 30, 78, 79, 99–102, 107
  of houseflies, 99
Wing displays (*Drosophila*), 80, 110
  flicking, 88
  vibration, 76, 78–81, 83, 84, 88, 93, 110, 111
Wingless males (*Drosophila*), 79, 80
Workers of social insects, 8, 12, 13, 15, 18, 23, 24, 27, 59, 60, 67, 209, 215, 222, 226, 232, 234, 238, 243, 245, 249–278

Y-maze olfactometer, 16, 45, 84, 318, 320